Current Progress in Animal Breeding

Current Progress in Animal Breeding

Editor: Carlos Hassey

CALLISTO REFERENCE

www.callistoreference.com

Callisto Reference,
118-35 Queens Blvd., Suite 400,
Forest Hills, NY 11375, USA

Visit us on the World Wide Web at:
www.callistoreference.com

ISBN: 978-1-64116-048-3 (Hardback)

Cataloging-in-Publication Data

Current progress in animal breeding / edited by Carlos Hassey.
    p. cm.
Includes bibliographical references and index.
ISBN 978-1-64116-048-3
1. Animal breeding. 2. Breeding. I. Hassey, Carlos.
SF105 .C87 2019
636.082--dc23

Table of Contents

Preface

The science of animal breeding deals with the study of evaluation of the genetic value of livestock. It also explores selective breeding methodologies of animals to produce desirable traits in subsequent generations of the target species. The scientific techniques of animal breeding incorporate the principles of population genetics, quantitative genetics and molecular genomics. In the era of rapid population inflation, implementation of animal breeding programs is essential for meeting rising global demands of food and resources. This book is compiled in such a manner, that it will provide in-depth knowledge about the principles and practices of animal breeding. The researches included in this book discuss the most vital concepts and emerging trends in this field. It includes contributions of experts and scientists which will provide innovative insights into this field. This book is a research guide for experts as well as students.

This book is a comprehensive compilation of works of different researchers from varied parts of the world. It includes valuable experiences of the researchers with the sole objective of providing the readers (learners) with a proper knowledge of the concerned field. This book will be beneficial in evoking inspiration and enhancing the knowledge of the interested readers.

In the end, I would like to extend my heartiest thanks to the authors who worked with great determination on their chapters. I also appreciate the publisher's support in the course of the book. I would also like to deeply acknowledge my family who stood by me as a source of inspiration during the project.

Editor

Disentangling the relative roles of resource acquisition and allocation on animal feed efficiency: insights from a dairy cow model

Laurence Puillet[1*], Denis Réale[2] and Nicolas C. Friggens[1]

Abstract

Background: Feed efficiency of farm animals has greatly improved through genetic selection for production. Today, we are faced with the limits of our ability to predict the effect of selection on feed efficiency, partly because the relative importance of the components of this complex phenotype changes across environments. Thus, we developed a dairy cow model that incorporates the dynamic interplay between life functions and evaluated its behaviour with a global sensitivity analysis on two definitions of feed efficiency. A key model feature is to consider feed efficiency as the result of two processes, acquisition and allocation of resources. Acquisition encapsulates intake and digestion, and allocation encapsulates partitioning rules between physiological functions. The model generates genetically-driven trajectories of energy acquisition and allocation, with four genetic-scaling parameters controlling these processes. Model sensitivity to these parameters was assessed with a complete factorial design.

Results: Acquisition and allocation had contrasting effects on feed efficiency (ratio between energy in milk and energy acquired from the environment). When measured over a lactation period, feed efficiency was increased by increasing allocation to lactation. However, at the lifetime level, efficiency was increased by decreasing allocation to growth and increasing lactation acquisition. While there is a strong linear increase in feed efficiency with more allocation to lactation within a lactation cycle, our results suggest that there is an optimal level of allocation to lactation beyond which increasing allocation to lactation negatively affects lifetime feed efficiency.

Conclusions: We developed a model to predict lactation and lifetime feed efficiency and show that breaking-down feed conversion into acquisition and allocation, and introducing genetically-driven trajectories that control these mechanisms, permitted quantification of their relative roles on feed efficiency. The life stage at which feed efficiency is evaluated appears to be a key aspect for selection. In this model, body reserves are also a key component in the prediction of lifetime feed efficiency since they integrate the feedback of acquisition and allocation on survival and reproduction. This modelling approach provided new insights into the processes that underpin lifetime feed efficiency in dairy cows.

Background

Improving feed efficiency (FE) is a longstanding goal of the livestock sector and is still highly relevant in the current context. Indeed, more efficient animals will produce the same amount of products using less resource and generating less waste in the environment, such as methane or nitrogen. As a result, both pressure on resources (e.g. land use that competes with human food production) and environmental impacts (e.g. greenhouse gas emissions) will decrease. In the past decades, FE of farm animals has increased substantially. For example, Capper et al. [1] reported that, in the USA, the amount of feedstuffs needed to produce one billion kg of milk reached 8.26×10^9 kg in 1944 and only 1.88×10^9 kg in 2007, which corresponds to a 77 % increase in FE. This huge increase in FE was obtained by selecting high-producing genotypes and providing them a high-quality

*Correspondence: laurence.puillet@agroparistech.fr
[1] UMR Modélisation Systémique Appliquée aux Ruminants, INRA, AgroParisTech, Université Paris-Saclay, 75005 Paris, France
Full list of author information is available at the end of the article

environment to maximize the expression of their production potential. A high level of production leads to a dilution of the fixed costs of production (maintenance requirements and non-productive stages of life) and thus an increase in FE. However, there is growing evidence that this means of increasing FE is not sustainable, particularly for dairy cattle females. The first reason is that a high level of production is negatively associated with other dairy female traits, such as fertility and health [2, 3]. Selection for high production has led to undesired responses by indirect selection that result in greater negative energy balance, i.e. greater body reserve mobilization during early lactation that leads to more reproductive or health problems. As a result, the expected dilution effect linked to higher production may be offset by a decline in productive lifespan because of poor health and/or fertility. If one considers the non-productive period (the phase prior to first calving) of the cow's life as an efficiency cost to be diluted by the productive part of the cow's lifespan, then it is clear that reducing the productive lifespan of the cow will decrease lifetime FE. Even if the integration of functional traits into selection indices has, to some extent, limited these negative associations [4, 5], it is far from clear what is the optimal pattern of body reserve usage across the lactation cycle to maximize lifetime FE [3]. A second reason that limits our capacity to sustainably improve FE relates to the role of genotype-by-environment (G × E) interactions on FE and its component traits. The environment in which production occurs will change in the future and breeding objectives will have to account for such changes (for instance, performance under low levels of nutrition or heat stress conditions [6]). In the context of genetic selection for feed efficiency in a future changing environment, we need to know how the environment in which selection is performed shapes the genetic correlations between the component traits of FE. For instance, a strong genetic propensity to accumulate body reserves prior to calving may be negatively correlated with FE in rich environments (where those reserves are less needed), but the converse may be expected in poor or variable environments. These G × E interactions still need to be better experimentally quantified in dairy cows, which until now have been kept in relatively controlled environments, although there is a considerable amount of data for other mammalian species (e.g. rabbits [7], mice [8], and pigs [9]).

Simulations can be a useful tool to explore such contrasting scenarios, provided that the design of the animal "building block" at the heart of the simulation is an appropriate representation of the main biological processes that contribute to, in this case, FE. In animal nutrition, FE is generally considered as the product of digestive efficiency and metabolic efficiency. Digestive efficiency reflects the

animal's ability to acquire nutrients, i.e. intake and digestion, while metabolic efficiency reflects nutrient partitioning and utilization for physiological functions. These two steps in the conversion process can be broadly designated as resource acquisition and allocation. They are both affected by genetic variation and thus contribute to variation in FE [10]. However, in the relatively few nutritional models that include animal genotype, the genotype is invariably included via the concept of production potential, i.e. the maximum amount of a product such as milk that the animal can produce. This is typically used to estimate nutrient requirements and thereby the required diet composition for a given intake level. Given that the total production produced is the product of nutrient intake and nutrient partition, this way of representing the animals' genotype does not allow the study of the genetic variation in acquisition and in allocation, separately. Thus, to improve our ability to predict the effect of selection on different components of FE, we need to develop simulation models that account for genetic and environmental effects at both the level of acquisition and the level of allocation.

Thus, we developed a mathematical description of the interplay between the main life functions of a dairy cow. This systemic model explicitly integrates energy acquisition and allocation as processes that drive the expression of phenotypic traits, and therefore FE. The model accounts for genetic components in both processes and therefore allows the simulation of genotypes that result from different combinations of acquisition and allocation trajectories.

The aim of this paper is to present the basic assumptions, ideas and design of the model, and the evaluation of its behaviour to variation in four key parameters related to acquisition and allocation. Simulations were used to quantify how changes in parameters that drive acquisition and allocation affect the different definitions of efficiency, thus providing proof-of-concept of the importance of breaking-down FE into these components.

Methods
Model description

The model description follows the overview, design concepts, and details (ODD) protocol for describing individual- and agent-based models [11]. The model is currently implemented with Modelmaker version 3.0 (Cherwell Scientific Ltd, 2000).

Overview

In order to design a model that represents the animal building block for predicting G × E interactions on feed efficiency (FE), we chose to break the overall process of resource conversion down into three elementary processes: resource acquisition, allocation and utilization.

Resource acquisition is ultimately defined as the input of energy in the organism, resulting from the intake of dry matter (DM) from the environment and its conversion into metabolizable energy (ME) through digestion. Acquisition depends on resource availability (environmental component) and on genetic capacity to acquire resource (animal component). Resource allocation is defined as the partitioning of ME among the following physiological functions: growth, gestation, lactation, maintenance and reserves. Allocation depends on a genetic component and on changes in physiological states. Finally, resource utilization is defined as the conversion of quantities of energy allocated to physiological functions into phenotypes (body mass components, milk, conception and survival probabilities). With this structure based on a decomposition of the processes that generate phenotypes, the model is flexible enough to represent responses to resource availability through variation of acquisition, variation of allocation or a combination of both.

As proposed by [12], we consider that gene regulations give rise to meta-mechanisms at the animal level, which can be represented by a set of parameters in a dynamic model of life functions. Our model is based on this principle. It was not designed to capture all the physiological mechanisms that underpin life functions. We consider the dairy cattle female as an active biological entity with its own agenda [13], rather than being a passive convertor of resource into products. This view reflects the fact that gene expression changes with age and physiological state, and thereby the relative priorities among life functions change throughout the female lifespan. For example, cows in early lactation partition energy towards the mammary gland and mobilize body reserves, irrespective of the quality of the feed available. As lactation progresses, cows increasingly partition energy away from milk towards body reserves. These changes in priorities reflect temporal differences in gene expression through the life of the animal that are the result of evolution and that have been further shaped by selection. To capture these changes, genetically-driven lifetime trajectories of acquisition and allocation (DM intake and energy partition) are assumed. They provide the dynamics that control the flow of resources to different life functions; the efficiency of utilization of these resources is assumed not to change with time and physiological state. Both resource acquisition and resource allocation trajectories can be modulated via genetic-scaling parameters, which allow the representation of the between-animal innate variability in these processes. These are not breeding values per se, rather they are multipliers on acquisition and allocation trajectories and thus provide the means to represent differences between genotypes in acquisition

and allocation, as proposed by [12]. For acquisition, the genetic-scaling parameters operate on the maximum intake reached at maturity and during lactation. For allocation, the genetic-scaling parameters operate on the rate of transfer of priorities between life functions. In our study, the model represents a single cow with genetic-scaling parameters as independent inputs that reflect its genotype. We do not consider that the model represents the mean of a population but rather, that it provides the elementary animal unit for building virtual populations in an individual-based population model to study the effects of selection. In this context, it will be possible to set different heritability values and different genetic correlations between the parameters of the model to study how genetic constraints will affect the evolution of the cow's FE. On the basis of acquisition and allocation trajectories that are driven by genetic-scaling parameters, and resource availability, the model simulates trajectories of phenotypes (DM intake, quantities of energy, body mass components and milk production) and timings of reproductive events throughout the lifespan of an individual cow. With this representation of the animal, the phenotypic expression of a genotype permitted by the environment can be simulated during different phases over which FE is determined and different sources of variability in FE can be better decomposed.

Design and concepts

The model structure is made up of four sub-models: acquisition, allocation, utilization and physiological status (Fig. 1). Acquisition and allocation sub-models are core modules that integrate the genetic determinants and lifetime dynamic changes. Utilization and physiological status are supporting modules that are based on simple principles and existing approaches. They are not the focus of the modelling effort since our aim is not to study the mechanisms that are associated with energy utilization and reproduction.

The allocation sub-model is the core of the model and drives the partitioning of ME between physiological functions. It accounts for changes in priorities during the lifespan of the animal by generating genetically-driven dynamic changes in coefficients of partition among four life functions: growth, future progeny, current progeny and survival. As proposed by [14], genetically-driven changes refer to any change that occurs in cows kept in a non-constraining environment. Dynamic changes of these compartments are illustrated in Fig. 2.

The compartment *AllocG* represents the female priority for growth. Its level gives the coefficient of partition for growth function, i.e. the proportion of acquired energy that is allocated to growth. The compartment *AllocPf* represents the female priority for its future progeny.

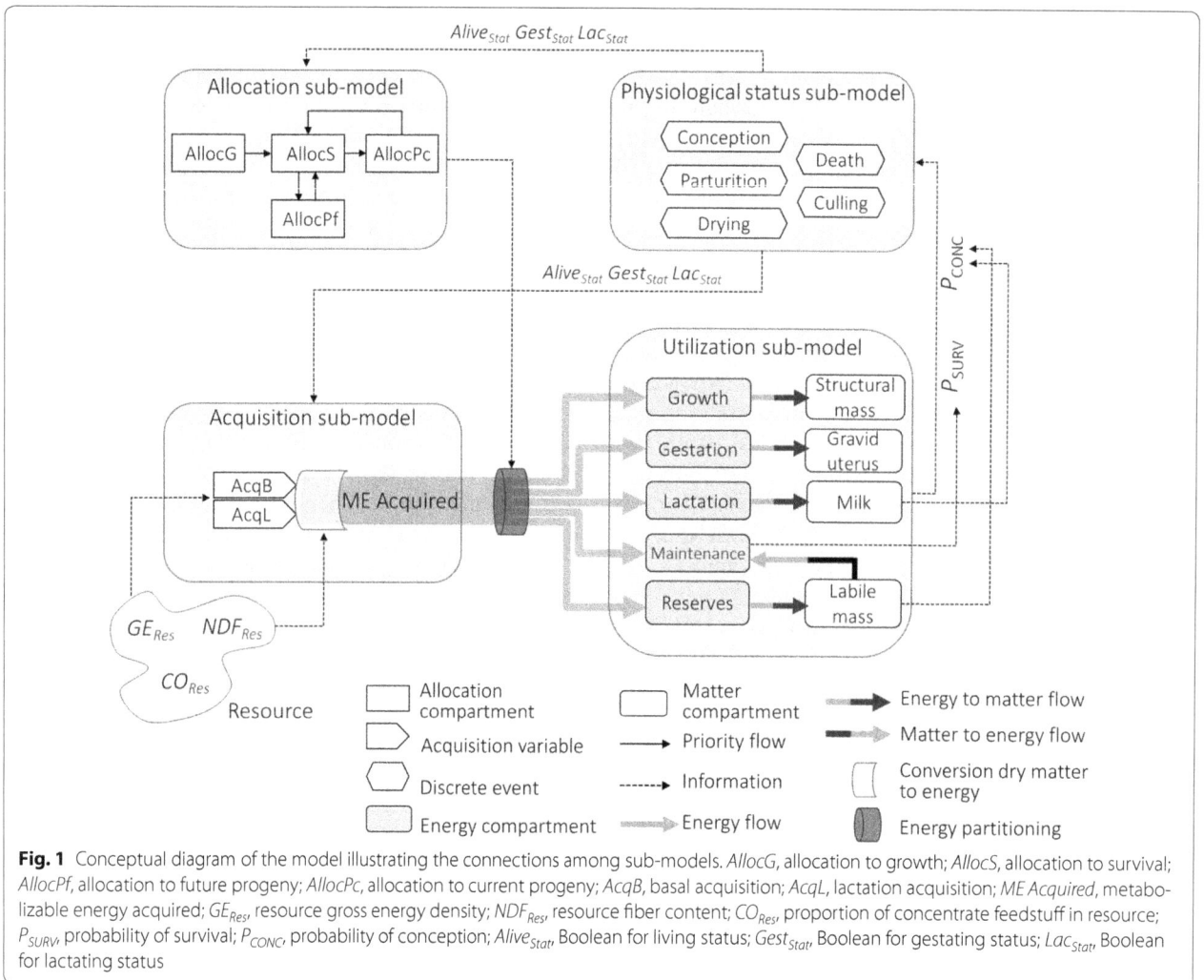

Fig. 1 Conceptual diagram of the model illustrating the connections among sub-models. *AllocG*, allocation to growth; *AllocS*, allocation to survival; *AllocPf*, allocation to future progeny; *AllocPc*, allocation to current progeny; *AcqB*, basal acquisition; *AcqL*, lactation acquisition; *ME Acquired*, metabolizable energy acquired; *GE_Res*, resource gross energy density; *NDF_Res*, resource fiber content; *CO_Res*, proportion of concentrate feedstuff in resource; *P_SURV*, probability of survival; *P_CONC*, probability of conception; *Alive_Stat*, Boolean for living status; *Gest_Stat*, Boolean for gestating status; *Lac_Stat*, Boolean for lactating status

Its level gives the coefficient of partition for the gestation function. The compartment *AllocPc* represents the female priority for its current progeny. Its level gives the coefficient of partition for the lactation function. Finally, the compartment *AllocS* represents the priority for survival. Its level gives the coefficient of partition for somatic functions, defined as body mass maintenance and body reserves. Compartments are linked by flows that represent transfers of priorities among life functions. These transfers lead to changes in coefficients of partition (level of compartments) and to a switch in energy investment when the resulting coefficients are used in the utilization sub-model. A dimensionless quantity of one is moving in the network of compartments to represent transfers of priorities. Therefore, by construction, the sum of the partitioning coefficients is equal to 1. This ensures a neutral balance between energy acquired and energy allocated to functions. During early life, the female's priority for growing progressively switches to survival with age.

The proportion of energy for growth is high after birth but the priority for growth progressively declines as it approaches maturity by transferring priority towards survival functions with an increasing allocation to the benefit of energy for somatic functions. At first conception, the female's priority switches from survival to future progeny. An increasing proportion of energy is invested in gestation function, at the expense of the proportion of energy for somatic functions. At parturition, the female's priority switches from future progeny to current progeny. No more energy is invested for gestation and an increasing proportion of energy is invested for lactation. The female's priority for its current progeny decreases as lactation progresses, to the benefit of the priority for the female's own survival. The proportion of energy invested for lactation decreases while the proportion for somatic functions increases. When drying-off occurs, there is a discrete shift in priority between current progeny and survival: energy is no longer invested in lactation. In

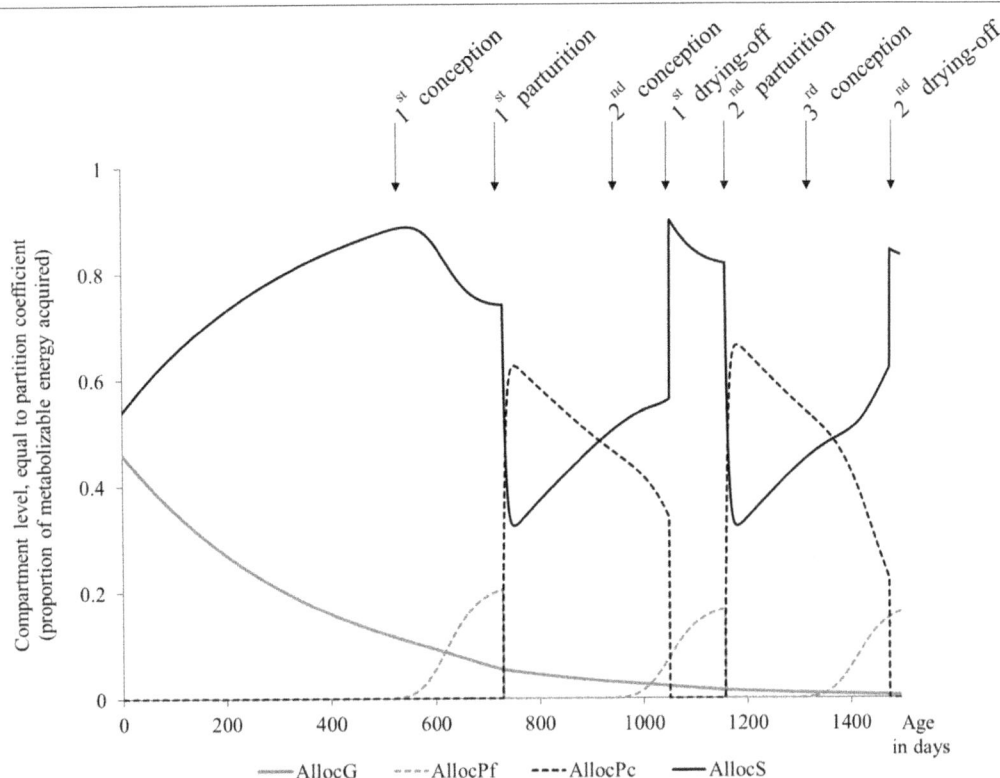

Fig. 2 Dynamic changes in the allocation sub-model over two reproductive cycles of a dairy cow. *AllocG*, allocation to growth; *AllocPf*, allocation to future progeny; *AllocPc*, allocation to current progeny; *AllocS*, allocation to survival

addition to the control of priority flows by changes in the female's physiological status (conception and parturition), the priority flows of the allocation sub-model are driven by genetic-scaling parameters. They control priority flows between *AllocG*, *AllocS*, *AllocPf*, and *AllocPc*. Implementation of different values for these parameters corresponds to different rates of priority transfers among functions and this allows the simulation of genetic differences in the profiles of allocation to growth, to gestation and to lactation. The values of these parameters are independent model inputs, therefore, in our study, the genetic differences for each allocation profile are independent of each other.

The acquisition sub-model is the second core sub-model since it simulates dynamic changes in dry matter intake throughout lifetime. Acquisition is made up of a basal acquisition component, *AcqB* and a lactation acquisition component, *AcqL* as illustrated on Fig. 3. The basal component describes the maturation of the biological structures linked to resource acquisition as the female matures. The lactation component represents the increase in resource acquisition that is induced by the lactating status. As for allocation, in addition to changes in physiological status, dynamic changes in acquisition are driven by the genetic-scaling parameters, $AcqB_{GEN}$

and $AcqL_{GEN}$, allowing the scaling of DM intake curves. Different values can be implemented to simulate genetic differences in the acquisition profiles among individuals.

The dynamic variables that are generated by acquisition and allocation sub-models are combined in the utilization sub-model. This sub-model encodes the conversion of the energy allocated to physiological functions into matter, based on efficiency coefficients and energy contents of this matter. Material variables such as the level of milk production are then used to compute conception probability and survival probability. The physiological status sub-model uses these probabilities to determine the female's status.

The model uses a time step of 1 day. Simulation starts at birth and stops at the female's death or culling. At each time step, all the elements are updated simultaneously (Runge–Kutta 4 numerical integration with a fixed time step for compartments). Processes that occur within a time step represent daily biotransformation, from acquisition of dry matter to phenotypes. Updated phenotypes are then used by the discrete events of the physiological status sub-model and may lead to a change in the female's status (gestating, lactating, alive), that is effective at the next time step. In this study, we consider a constant nutritional environment (resource availability, energy density, proportion

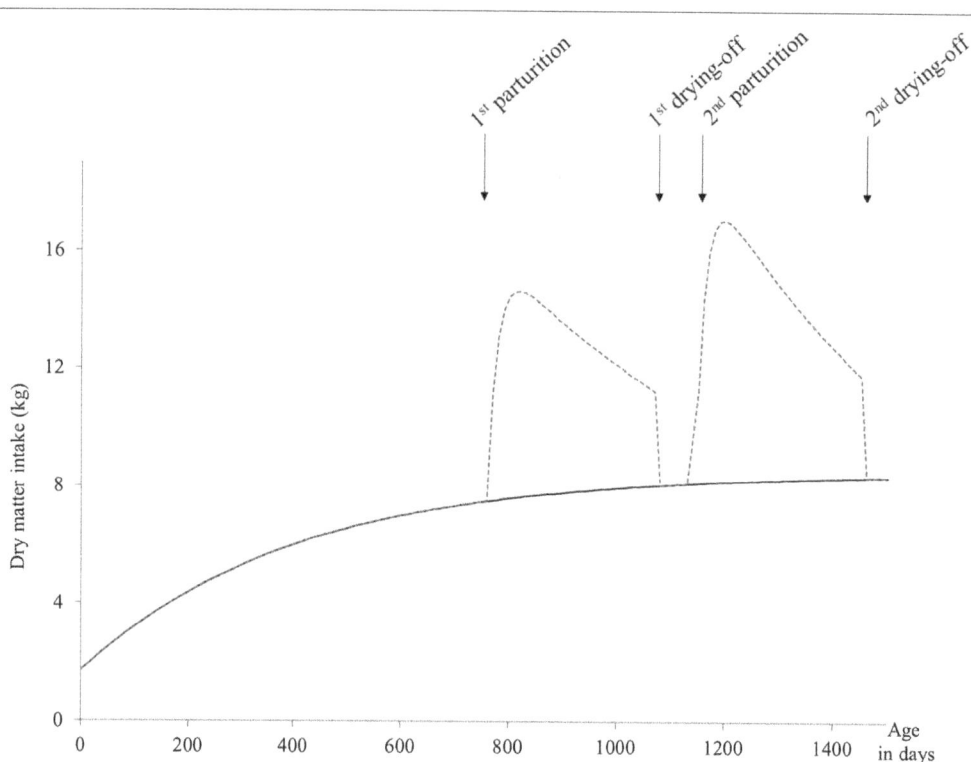

Fig. 3 Dynamic changes in dry matter intake in the acquisition sub-model over two reproductive cycles of a dairy cow. Total dry matter intake is made up of a basal component (*AcqB, solid line*) and a lactation component (*AcqL, dotted line*)

of fibres and concentrate in the diet) but temporal variation in the environment could easily be implemented.

Model details and assumptions

To better understand the dynamic nature of the model, we present the main details of allocation and acquisition and we highlight the linkage, and thus in-built coherence, between life functions. All parameters, compartments, flows and variables are defined in Tables S1 and S2 [see Additional file 1: Table S1 and Additional file 2: Table S2]. Discrete events are described in Table S3 [see Additional file 3: Table S3].

Allocation sub-model The allocation sub-model is made up of four compartments that reflect the priorities for four life functions: growth, future progeny, current progeny and survival. Dynamics of these compartments are based on mass action laws to represent the progressive transfers of priority among functions across various physiological states. The structure of the allocation sub-model is in Fig. 4, where the amounts of priority for the different life functions are given by *AllocG* for growth (priority for growing), *AllocS* for somatic functions (priority for survival), *AllocPf* for gestation (priority to future offspring), and *AllocPc* for lactation (priority to current offspring).

The transfers of priority are given by the flows $f_{prio}G2S$ (priority transfer from growth to survival), $f_{prio}S2Pf$ (priority transfer from survival to future progeny), $f_{prio}S2Pc$ (priority transfer from survival to current progeny) and $f_{prio}Pc2S$ (priority transfer from current progeny to survival). These flows generate the changes in compartment levels and thus the dynamic changes in the allocation of energy to the associated function. They are activated or inactivated depending on the physiological state. They are modulated by the two genetic-scaling parameters. The rate of change in the proportion of energy allocated to growth, *AllocG* is defined by the following differential equation:

$$\frac{dAllocG}{dt} = -f_{prio}G2S. \qquad (1)$$

The flow $f_{prio}G2S$ represents the decrease in allocation to growth as the female ages. It is given by:

$$f_{prio}G2S = AllocG \cdot G2S_{GEN} + 0.01 \cdot AllocPf, \qquad (2)$$

where parameter $G2S_{GEN}$ is the genetic-scaling parameter that defines the rate of priority transfer from growth to survival. An increase in $G2S_{GEN}$ leads to a greater priority transfer and a larger decrease in the *AllocG* level and thus a decrease in the proportion of energy allocated to

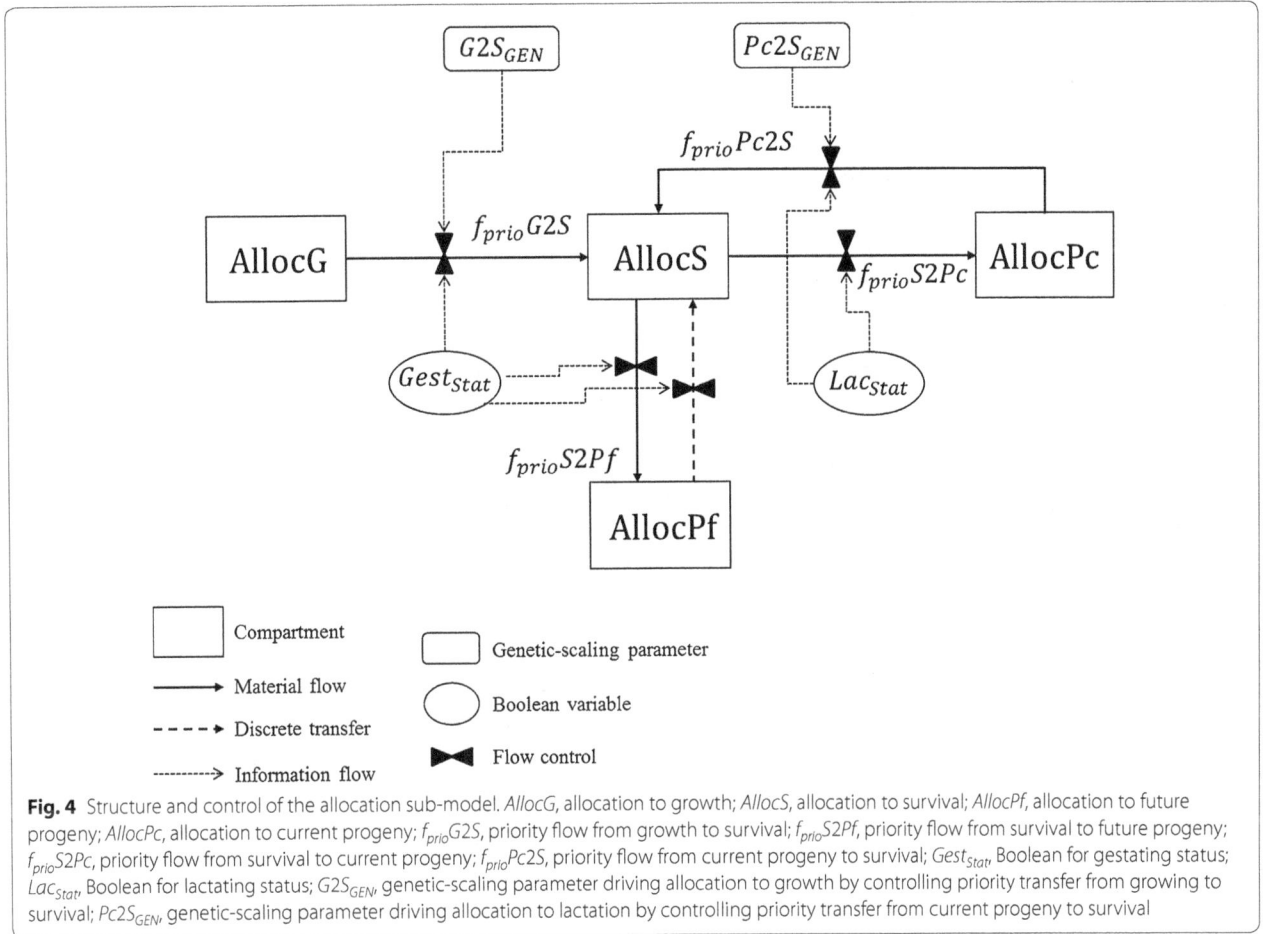

Fig. 4 Structure and control of the allocation sub-model. *AllocG*, allocation to growth; *AllocS*, allocation to survival; *AllocPf*, allocation to future progeny; *AllocPc*, allocation to current progeny; $f_{prio}G2S$, priority flow from growth to survival; $f_{prio}S2Pf$, priority flow from survival to future progeny; $f_{prio}S2Pc$, priority flow from survival to current progeny; $f_{prio}Pc2S$, priority flow from current progeny to survival; *Gest_{Stat}*, Boolean for gestating status; *Lac_{Stat}*, Boolean for lactating status; $G2S_{GEN}$, genetic-scaling parameter driving allocation to growth by controlling priority transfer from growing to survival; $Pc2S_{GEN}$, genetic-scaling parameter driving allocation to lactation by controlling priority transfer from current progeny to survival

growth. The priority transfer from growth to survival is further increased by *AllocPf*. Thus, when gestation starts, decrease in allocation to growth is accelerated to enable a greater priority transfer towards future progeny. It is assumed that gestation slows down growth.

The proportion of energy allocated to gestation is given by *AllocPf*, the rate of change of which is defined by the following differential equation:

$$\frac{dAllocPf}{dt} = +f_{prio}S2Pf. \tag{3}$$

The flow $f_{prio}S2Pf$ represents the increase of allocation to gestation as gestation time increases. It is modelled with a rising sigmoid function that depends on gestation time and on four fixed parameters (k_HPf_0, k_HPf_1, k_HPf_2 and k_HPf_3), as we currently assumed no variability for allocation to gestation [see Additional file 1: Table S1]. The proportion of energy allocated to lactation is given by *AllocPc* that is defined by the following differential equation:

$$\frac{dAllocPc}{dt} = \left(+f_{prio}S2Pc - f_{prio}Pc2S\right). \tag{4}$$

The increase in the proportion of energy allocated to milk production at the beginning of lactation is described by $f_{prio}S2Pc$ as given by:

$$f_{prio}S2Pc = AllocS_2 \cdot S2P_c \cdot Lac\_Stat. \tag{5}$$

The parameter $S2P_c$ drives the priority transfer from survival to current progeny at the beginning of lactation. In the current version, it is fixed since no genetic variance in the priority transfer from survival to current progeny is assumed.

The decrease in energy for milk production as lactation progresses is described by the flow $f_{prio}Pc2S$, given by:

$$\begin{aligned} f_{prio}Pc2S = AllocPc \cdot Lac_{Stat} \\ \cdot \left(Pc2S_{GEN} + AllocPf \cdot Gest_{Stat} \cdot 0.06\right). \end{aligned} \tag{6}$$

The parameter $Pc2S_{GEN}$ is the genetic component for the priority transfer from current progeny to somatic functions. Changing its value affects the dynamics of *AllocPc* and allows the representation of different strategies of lactation allocation. An increase in $Pc2S_{GEN}$

accelerates the rate of priority transfer from lactation to survival $f_{prio}Pc2S$, and thus, decreases the level of allocation to lactation. This priority flow is also affected by the allocation level to future progeny. As gestation progresses, $AllocPf$ increases and accelerates the return of priority from $AllocPc$ to $AllocS$. This effect accounts for the depressive effect of gestation on lactation [15]. Finally, the proportion of energy allocated to somatic functions is given by $AllocS$ defined by the following differential equation:

$$\frac{dAllocS}{dt} = \left(f_{prio}G2S - f_{prio}S2Pf - f\_prioS2Pc + f\_prioPc2S\right). \quad (7)$$

Since there is no loss of the dimensionless quantity of one in the compartment network, at each time step, the sum of the compartment's levels is equal to 1, thus ensuring a neutral balance between energy acquired and energy allocated to functions.

Acquisition sub-model The acquisition sub-model simulates the quantity of DM acquired from the nutritional environment, depending on resource characteristics (gross energy density, fibre content and proportion of concentrate feedstuff), and its conversion into ME through digestion.

The total daily intake of DM, is given by:

$$AcqT = AcqB + AcqL. \quad (8)$$

The variable $AcqB$ is defined by:

$$AcqB = \left(AcqB_{GEN} - 0.8 \cdot AcqB_{GEN} \cdot e^{-k_{AcqB_{MAT}} \cdot t}\right). \quad (9)$$

The parameter $AcqB_{GEN}$ is the genetic-scaling parameter that drives basal acquisition. It represents the asymptote of the curve, which corresponds to the maximum intake of DM at maturity for a non-lactating animal. Although intake is frequently expressed as a percentage of body weight, we explicitly chose not to do this. It would create an a priori correlation between allocation to growth and acquisition, and thus, prevent the study of the relative roles of these components on the phenotypic traits and FE. The variable $AcqL$ is defined by:

$$AcqL = Lac_{Stat} \cdot AcqL_{Max} \cdot AcqL_{Dyn}. \quad (10)$$

The shape of the curve during lactation is given by the $AcqL_{Dyn}$ component of Equation E15 in Table S2 [see Additional file 2: Table S2]. The maximum DM that is reached during lactation is given by $AcqL_{Max}$, which depends on the genetic-scaling parameter $AcqL_{GEN}$. This latter is the maximum DM reached at maturity to account for the maturation of the potential to acquire resource as the female ages [see Additional file 2: Table S2]. The total

intake of DM is converted into ME available for allocation, ME_{Acq}, depending on the energy density of the DM available in the environment, GE_{Res} in Mcal/kg, and the metabolizability of the diet, ME_{PctGE}, representing the energy losses through faeces, urine and enteric methane during digestion. It is affected by the level of dry matter intake and the proportion of concentrate as proposed by [16].

Utilization and physiological status sub-models The detailed description of utilization and physiological status sub-models is in Additional file 4. The energy utilization sub-model combines ME from the acquisition sub-model with partition coefficients from the allocation sub-model and simulates the conversion of the energy, which is allocated to physiological functions (growth, gestation, lactation and somatic functions), into traits. The quantity of energy allocated to growth is converted into structural mass, which corresponds to the non-labile part of the body mass. The quantity of energy allocated to somatic functions is primarily used for maintenance and the remainder used for body reserves. The quantity of energy allocated to body reserves is converted into labile mass. This body compartment can subsequently be used to provide energy through mobilization, contrary to the structural mass. The quantity of energy allocated to gestation is converted into gravid uterus mass. Finally, the quantity of energy allocated to lactation is converted into milk production. Traits resulting from this conversion of energy into kg of matter are further used by the utilization sub-model to compute survival probability and conception probability, which are used in the physiological status sub-model. We assumed that survival probability became null when the female was not able to cover its maintenance requirements during 15 consecutive days or when it was selected for culling. Culling occurred after the second lactation, if conception did not occur 200 days after calving. Based on [17], we assumed that probability of conception is influenced by milk production, body condition score and energy balance.

Model calibration
The aim of this study was to evaluate the behaviour of the model in response to the variation of two parameters related to acquisition ($AcqB_{GEN}$ and $AcqL_{GEN}$) and of two parameters related to allocation ($G2S_{GEN}$ and $Pc2S_{GEN}$). Consequently, all other parameters were set at fixed values during the simulations. The values of parameters related to animal nutrition (diet characteristics and conversion of energy into body mass components and milk production) and reproduction (timing of events and probability of conception) were taken from previously published data based on the analysis of large datasets

[see Additional file 1: Table S1]. The values of parameters related to dynamic and structural aspects of the model were determined during a calibration step. For some aspects such as the rates of transfer of priority, the model's parameters cannot be measured directly from experimental data, but they can be inferred from data that track all the relevant traits throughout the animal's lifespan. Unfortunately, very few studies have reported time series values for the full set of body mass, body reserves, milk production, and gestation mass through both young and adult phases of life. To overcome this limitation, it is possible to piece together consistent lifetime curves from a large number of suitably chosen studies over shorter time periods using a meta-analysis approach. This was done in a previously reported study to calibrate another model of dairy cow performance, called GARUNS [18, 19]. This GARUNS model is not suited to our current study on the dissociation of allocation and acquisition, but in the latter study, its use provided reference trajectories throughout life for the above-mentioned traits that represent the average curves from the literature. Accordingly, we used the GARUNS reference trajectories to calibrate the current model. Detailed aspects of the calibration are in Additional file 5. The comparison of the body mass, milk production and body condition score trajectories that were simulated by the model in the current paper, the GARUNS reference curves and the compilation of data from the literature are presented in Figures S10, S11 and S12 [see Additional file 6: Figures S10, S11 and S12]. The calibration step was done by iterative changes in parameter values until the model's simulated trajectories converged with the GARUNS reference curves, and the data from the literature.

Finally, we evaluated the impact of the stochastic processes associated with the simulation of reproduction events. The use of a probability of conception P_{CONC} to determine if the simulated female becomes pregnant implies the use of a random process (see *CONCEPTION* event in Table S3 [see Additional file 3: Table S3]). As a result, for the same model parameterization, the time at which conception occurs, resulting from the random process, can vary among simulations and lead to slightly different outputs. To account for this stochastic aspect and stabilize the variance of the model's outputs, each simulation had to be replicated 20 times.

Model simulations for sensitivity analysis
Model behaviour was explored by a global sensitivity analysis that aimed at evaluating how variation in model inputs, i.e. the four genetic-scaling parameters, affects FE. Two FE definitions were used, one at the lactation level and one at the end of the animal's life. FE_Lac2 corresponds to the ratio between energy acquired and

energy produced in milk, cumulated over the second lactation. FE_life corresponds to the same ratio, cumulated from birth to death. The four parameters ($G2S_{GEN}$, $Pc2S_{GEN}$, $AcqB_{GEN}$ and $AcqL_{GEN}$) were set at three different levels (L: low; M: medium and H: high) and combined in a complete factorial design [see Additional file 6: Table S4]. This led to 81 simulations, with 20 replications to account for the stochastic processes of reproduction. The discretization of parameters into levels allowed a reasonable computation time while enabling the exploration of the model behaviour in response to different combinations of values for genetic-scaling parameters. For each of the four genetic-scaling parameters, the medium level corresponded to the value that was determined in the calibration step [see Additional file 1: Table S1]. Values for low and high levels of parameters corresponded to equidistant deviations in percentage of the medium level. The percentages of deviations were chosen to simulate trajectories of traits that were consistent with the range of trait values observed in the existing data. The detailed description of the parameters levels is in Additional file 6. By testing all the combinations of parameter values, different individual profiles of acquisition and allocation were simulated and the corresponding lifetime trajectories of traits were used to compute FE. The sensitivity of the model's output to variation in inputs was evaluated with sensitivity indices based on variance decomposition: output variability is decomposed into the main effects of parameters and interactions. Given our factorial simulation design, analysis of variance is a natural method for this variance decomposition [20].

Results
The model simulates credible lifetime trajectories of acquisition and allocation and is sensitive to changes in genetic-scaling parameters, as shown in Additional file 6. Figure 5 shows the boxplots for the two definitions of FE. Outputs related to body mass, energy utilization and reproductive performance were also computed for each simulated cow, at the lactation level and at the end of life. Table 1 summarises the results for the two FE criteria and for the energy acquired and allocated to milk. Table 2 summarises outputs at the lactation level and Table 3 at the lifetime level. The analysis of variance used to compute sensitivity indices for the two definitions of FE is in Additional file 6.

Allocation to growth: sensitivity of phenotypic traits to the variation in the genetic-scaling parameter $G2S_{GEN}$
Lactation level
During second lactation, increasing allocation to growth had a negligible positive effect on FE_lac2 (Fig. 5). It also had a negligible positive effect on cumulative energy allocated to milk production (E_milk_lac2 in 10^3 MJ) and

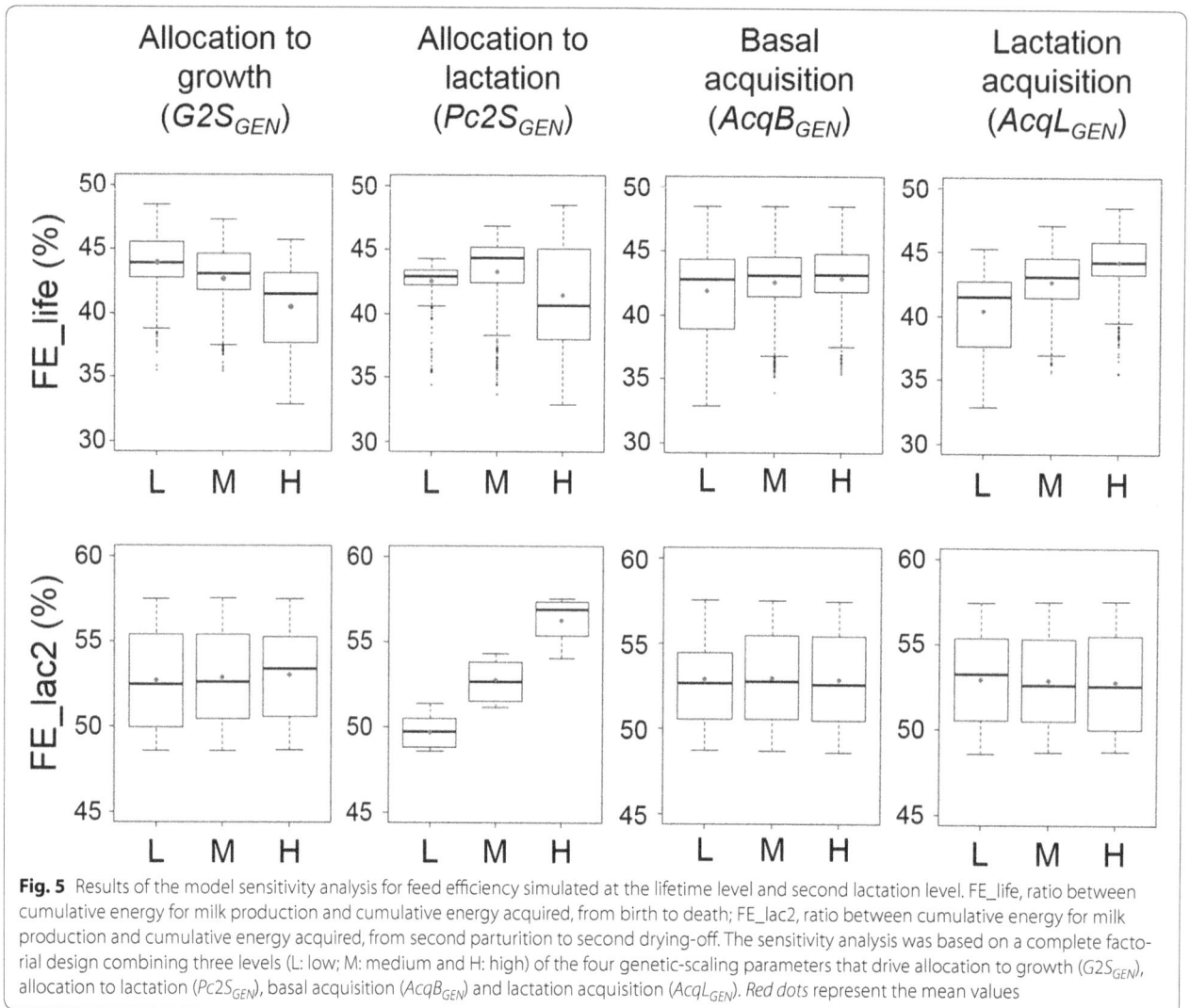

Fig. 5 Results of the model sensitivity analysis for feed efficiency simulated at the lifetime level and second lactation level. FE_life, ratio between cumulative energy for milk production and cumulative energy acquired, from birth to death; FE_lac2, ratio between cumulative energy for milk production and cumulative energy acquired, from second parturition to second drying-off. The sensitivity analysis was based on a complete factorial design combining three levels (L: low; M: medium and H: high) of the four genetic-scaling parameters that drive allocation to growth ($G2S_{GEN}$), allocation to lactation ($Pc2S_{GEN}$), basal acquisition ($AcqB_{GEN}$) and lactation acquisition ($AcqL_{GEN}$). *Red dots* represent the mean values

on cumulative energy acquired during second lactation (E_acq_lac2 in 10^3 MJ) as shown in Table 1. Changes in E_milk_lac2 for different levels of allocation to growth (−0.58 and +0.59 % for L and H compared to M) were larger than changes in energy acquired (−0.15 and +0.33 % for L and H compared to M). As a result, the ratio FE_lac2 slightly increased. The small increase in E_acq_lac2 for increasing allocation to growth was due to an increase in age at second parturition (Table 2). When individuals were older at parturition, they also had a higher level of basal acquisition because it increased with age, until maturity. The slight increase in E_milk_lac2 for increasing allocation to growth was due to an increase in the interval between second and third parturitions (Table 2). Increasing allocation to growth resulted in a decrease in labile mass at the second drying (from 137 to 115 kg, see Table 2), and consequently a delay in the time to next gestation. When gestation occurred later, the period during which this

function coexisted with lactation decreased. In the model, allocation to gestation decreased allocation to lactation (see Equation E2 in Table S2 [see Additional file 2: Table S2]). The shorter the period of coexistence was, the lower was the depressive effect of gestation on lactation, allowing a slight increase in E_milk_lac2. Regarding the energy used for growing during second lactation (E_grow_lac2 in 10^3 MJ), increasing allocation to growth led to very small differences in energy expenditure for growth (Table 2). As most of the energy for growing was spent before second lactation, the different levels of allocation to growth led to only small differences during second lactation.

Lifetime level

Increasing allocation to growth resulted in a decrease in FE_life (Fig. 5) associated with a decrease in energy used for milk production (E_milk_life in 10^3 MJ) and a decrease in energy acquired (E_acq_life in 10^3 MJ). As

Table 1 Measures of efficiency, energy acquired, and milk energy output for model sensitivity analysis

	Growth allocation			Lactation allocation			Basal acquisition			Lactation acquisition		
	L	M	H	L	M	H	L	M	H	L	M	H
FE_lac2												
Value (%)	52.70	52.92	53.04	49.65	52.76	56.24	52.87	52.94	52.84	52.95	52.88	52.82
Deviation from M (%)	−0.41	0.00	0.23	−5.90	0.00	6.61	−0.13	0.00	−0.20	0.13	0.00	−0.12
E_acq_lac2												
Value (10^3 MJ)	54.58	54.66	54.84	54.70	54.67	54.71	51.55	54.68	57.85	52.49	54.68	56.90
Deviation from M (%)	−0.15	0.00	0.33	0.05	0.00	0.08	−5.72	0.00	5.81	−4.00	0.00	4.05
E_milk_lac2												
Value (10^3 MJ)	28.75	28.92	29.09	27.15	28.84	30.77	27.25	28.95	30.56	27.80	28.92	30.05
Deviation from M (%)	−0.58	0.00	0.59	−5.84	0.00	6.71	−5.85	0.00	5.59	−3.87	0.00	3.93
FE_life												
Value (%)	43.97	42.71	40.55	42.53	43.27	41.43	41.86	42.51	42.85	40.38	42.62	44.23
Deviation from M (%)	2.96	0.00	−5.04	−1.72	0.00	−4.27	−1.54	0.00	0.80	−5.26	0.00	3.79
E_acq_life												
Value (10^3 MJ)	636.20	559.59	464.99	708.03	573.46	379.28	481.09	557.99	621.69	452.51	551.67	656.60
Deviation from M (%)	13.69	0.00	−16.91	23.47	0.00	−33.86	−13.78	0.00	11.42	−17.98	0.00	19.02
E_milk_life												
Value (10^3 MJ)	281.99	242.72	194.09	302.90	252.88	163.03	207.05	241.76	269.98	187.40	238.70	292.70
Deviation from M (%)	16.18	0.00	−20.03	19.78	0.00	−35.53	−14.36	0.00	11.67	−21.49	0.00	22.63

Measures are for the periods from birth to death (_life), and from parturition to drying off in second lactation (_lac2) for the three levels (L: low; M: medium and H: high) of the four genetic parameters that drive allocation to growth ($G2S_{GEN}$), allocation to lactation ($Pc2S_{GEN}$), basal acquisition ($AcqB_{GEN}$) and lactation acquisition ($AcqL_{GEN}$)

Values are in absolute terms and as percentages of the M level of the corresponding parameter

Parameters values are in Additional file 1: Table S1

Table 2 Outputs of model sensitivity analysis at the second lactation level

	Growth allocation			Lactation allocation			Basal acquisition			Lactation acquisition		
	L	M	H	L	M	H	L	M	H	L	M	H
Age at second parturition (days)	1138	1163	1225	1144	1164	1218	1214	1167	1145	1202	1172	1153
Interval between parturitions 2–3 (days)	406	418	423	386	404	455	409	419	417	429	413	404
E_grow_lac2 (10^3 MJ)	0.501	0.544	0.544	0.568	0.541	0.480	0.450	0.538	0.600	0.476	0.533	0.579
E_mobilized_lac2 (% total energy use during lac2)	1.54	1.74	2.13	1.61	1.65	2.14	1.83	1.80	1.78	2.21	1.74	1.47
Labile mass at drying 2 (kg)	137	124	115	157	125	94	113	124	139	110	125	141

Outputs correspond to measures of age at second parturition, interval between second and third parturitions, cumulative energy for growth, cumulative energy mobilized for the period between parturition and drying-off of the second lactation and labile mass at the second drying for the three levels (L: low; M: medium and H: high) of the four genetic-scaling parameters that drive allocation to growth ($G2S_{GEN}$), allocation to lactation ($Pc2S_{GEN}$), basal acquisition ($AcqB_{GEN}$) and lactation acquisition ($AcqL_{GEN}$)

shown in Table 1, changes in E_milk_life were greater than changes in E_acq_life, leading to a decreased ratio FE_life. The two components of FE_life both decreased because the increase in allocation to growth led to a shorter lifespan and a shorter productive life with a smaller number of lactations (Table 3). Increasing growth allocation had negative effects on survival and reproduction, which were due to the trade-off between structural mass and labile mass. In the model, the energy allocated to growth fuels the structural mass rather than the labile mass, which is fuelled by energy allocated to somatic functions. Consistently, an increase in allocation to growth led to an increase in structural mass (Table 3) because the energy used for growth increased (E_grow_life in 10^3 MJ). At the same time, an increase in allocation to growth led to a decrease in labile mass, as illustrated in

Table 3 Outputs of model sensitivity analysis at the lifespan level

	Growth allocation			Lactation allocation			Basal acquisition			Lactation acquisition		
	L	M	H	L	M	H	L	M	H	L	M	H
Longevity (years)	11.3	10.2	8.8	12.4	10.4	7.6	9.6	10.2	10.6	8.9	10.1	11.3
Productive longevity (years)	7.7	6.7	5.4	8.6	6.8	4.3	6.1	6.7	7.0	5.5	6.6	7.7
Number of lactations	8.8	7.6	6.1	9.9	7.8	4.9	6.9	7.6	8.0	6.3	7.5	8.7
Structural mass (kg)	425	446	468	446	446	446	406	446	487	446	446	446
Mean labile mass at drying (kg)	138	126	114	160	126	93	113	126	140	109	126	144
E_grow_life (10^3 MJ)	10.24	10.79	11.37	10.83	10.81	10.76	9.80	10.81	11.80	10.69	10.80	10.92
E_balance_life (10^3 MJ)	12.81	11.99	10.88	15.49	11.71	8.47	10.82	11.89	12.97	10.51	11.78	13.38
E_maintenance (% total energy expenditure)	48.30	49.09	50.44	49.58	48.73	49.51	49.70	49.22	48.91	50.64	49.14	48.05
E_mobilized_life (% total energy use)	1.59	1.63	1.70	1.82	1.52	1.57	1.64	1.63	1.63	1.74	1.59	1.58
E_milk_lactation (10^3 MJ)	31.92	31.69	31.15	30.54	31.87	32.36	29.63	31.62	33.51	29.73	31.60	33.43

Outputs correspond to measures of longevity, productive longevity, number of lactations, structural mass, mean labile mass at drying-off and cumulative energy outputs for the period between birth and death for the three levels (L: low; M: medium and H: high) of the four genetic-scaling parameters that drive allocation to growth ($G2S_{GEN}$), allocation to lactation ($Pc2S_{GEN}$), basal acquisition ($AcqB_{GEN}$) and lactation acquisition ($AcqL_{GEN}$). Energy outputs are: cumulative energy for growth, total energy balance (difference between cumulative energy for reconstitution of reserves and cumulative energy mobilized), cumulative energy mobilized and average cumulative energy per lactation. Parameters values are in Additional file 1: Table S1

Table 3 by the mean labile mass at drying. Furthermore, it led to an increase in labile mass mobilization as shown by the proportion of energy mobilized (Table 3). The level and the use of labile mass are involved in survival (ability to cover maintenance requirements) and reproduction (effect of body condition score and energy balance on conception probability). As a result, the ratio between productive lifespan and longevity decreased from 0.68 to 0.61 as allocation to growth increased, which increased the proportion of energy spent for maintenance over the lifetime (E_maintenance in % of total energy expenditure, Table 3), thus reducing the dilution of maintenance costs and decreasing FE.

Allocation to lactation: sensitivity of phenotypic traits to variation in the genetic-scaling parameter $Pc2S_{GEN}$

Lactation level

During second lactation, increasing allocation to lactation strongly increased FE_lac2 (Fig. 5). This effect was due to an increase in E_milk_lac2 and a stagnation of E_acq_lac2 (Table 1). The increase in energy used for milk was made possible by a higher mobilization of body reserves (Table 2).

Lifetime level

Contrary to effects at the lactation level, increasing allocation to lactation was not beneficial at the lifetime level (Fig. 5). When $Pc2S_{GEN}$ increased from L to M, FE_life increased slightly. When it increased from M to H, FE_life decreased. When allocation to lactation increased, both E_acq_life and E_milk_life decreased. On the one

hand, the increase in allocation to lactation from L to M led to a larger decrease in E_acq_life than in E_milk_life, and thus to a slight increase in FE_life. On the other hand, the increase in allocation to lactation from M to H led to a smaller decrease in E_acq_life than in E_milk_life, thus to a decrease in FE_life. Increasing allocation to lactation decreased E_acq_life and E_milk_life (Table 1). These effects were modulated by a decrease in the number of lactations associated to shorter lifespan and productive life (Table 3). This effect was due to a lower labile mass at drying, which impaired survival and reproductive performance. This lower labile mass was not linked to growth, i.e. the structural mass was similar for all levels of allocation to lactation. Within the full factorial design, variations of growth allocation and basal acquisition, which both determine structural mass, were the same across various levels of allocation to lactation. This lower labile mass was due to a decrease in life energy balance (E_balance_life, defined as the cumulative energy for labile repletion minus the cumulative energy for mobilization in 10^3 MJ) when allocation to lactation increased. In spite of the overall decrease in E_acq_life and E_milk_life that is caused by a shorter life, increasing allocation to lactation logically led to an increase in the energy used for milk production per lactation (E_milk_lactation in 10^3 MJ). When increasing allocation to lactation from L to M, the higher productivity per lactation compensated the decrease in lifespan and productive life. The proportion of energy spent on maintenance decreased slightly from L to M (Table 3), which improved the dilution of maintenance costs and increased FE_life. When increasing allocation to lactation from M to H, the higher productivity per lactation did not compensate the reduction in lifespan and productive life (the ratio productive life/longevity dropped from 0.65 to 0.57). The proportion of energy spent on maintenance increased, which resulted in less dilution of maintenance costs and a decrease in FE_life.

Basal acquisition: sensitivity of phenotypic traits to variation in the genetic-scaling parameter $AcqB_{GEN}$

Lactation level

During second lactation, increasing basal acquisition had almost no effect on FE_lac2 (Fig. 5). Increasing basal acquisition increased E_acq_lac2 and E_milk_lac2 in the same proportion (Table 2), thus leading to the same ratio.

Lifetime level

Increasing basal acquisition led to a very small increase in FE_life (Fig. 5). Both E_acq_life and E_milk_life increased substantially when basal acquisition increased. This effect was due to a small increase in labile mass made possible by the increased acquisition (Table 3), which favoured reproduction and therefore productive life. As a result, energy for maintenance represented a slightly smaller part of the energy budget (Table 3). This slight improvement in dilution of maintenance costs explained the small increase in FE_life.

Lactation acquisition: sensitivity of phenotypic traits to variation in the genetic-scaling parameter $AcqL_{GEN}$

Lactation level

During second lactation, increasing lactation acquisition had almost no effect on FE_lac2 (Fig. 5). As for basal acquisition, increasing lactation acquisition increased E_acq_lac2 and E_milk_lac2 in the same proportion (Table 1), thus leading to the same ratio.

Lifetime level

Contrary to the lactation level, increasing lactation acquisition resulted in a substantial increase in FE_life (Table 1). Both E_milk_life and E_acq_life increased when lactation acquisition increased but E_milk_life increased proportionally more than E_acq_life, thus increasing the ratio. The overall increase in E_milk_life and E_acq_life was due to an increase in lifespan and productive life, because of a larger number of lactations. Increasing lactation acquisition led to a higher level of labile mass, which favoured survival and reproduction (the ratio between productive life and longevity increased from 0.62 to 0.68). With a longer lifespan and more lactations, the energy spent on maintenance represented a smaller part of the energy budget (50.64, 49.14 and

48.05 % for L, M and H), which improved the dilution of fixed costs and increased FE_life.

Discussion

Effects of acquisition and allocation on feed efficiency

Our simulation results show that the mechanisms of acquisition and allocation had contrasted effects on FE depending on the time scale over which efficiency was calculated, as shown in Fig. 5. At the lactation level, improvement in FE was achieved by increasing allocation to lactation. This result is consistent with past selection strategies on milk production, leading to a dilution of maintenance costs [21]. At the lifetime level, improvement in FE was achieved by decreasing allocation to growth and increasing lactation acquisition. This improvement was caused by a higher level of body reserves (see Sections Allocation to growth: sensitivity of phenotypic traits to the variation in the genetic-scaling parameter $G2S_{GEN}$ and Lactation acquisition: sensitivity of phenotypic traits to variation in the genetic-scaling parameter $AcqL_{GEN}$), which resulted in improved reproduction and survival and thus favoured the dilution of the non-productive part of the lifespan. In contrast to the effect at the lactation level, increasing allocation to lactation improved lifetime FE up to an optimal level, beyond which the effect on FE became negative.

These findings highlight the crucial role of life stages when considering selection for FE and the metric associated with the phenotype selected for. Based on our simulation results, selecting for FE at the lactation level will result in the selection of females with high allocation to lactation. These females are not those that maximize FE at the lifetime level, i.e. short-term efficiency may come at the cost of reduced sustainability. In contrast, selecting for FE at the lifetime level will result in the selection of females with low allocation to growth and high lactation acquisition. These females are not those that maximize FE at the lactation level. The importance of the life stages in interpreting efficiency measures was previously mentioned by [3].

The lifetime effects of acquisition and allocation mechanisms on FE were mainly due to the central role of body reserves. This component is pivotal in the phenotypic feedback of acquisition and allocation on survival and reproduction. The impact of allocation to growth was due to a direct trade-off between structural mass and body reserves. Increasing allocation to growth decreased body reserves, which had a negative effect on survival and reproduction. The impact of lactation acquisition on FE was due to a positive effect on body reserves. Increasing acquisition, with equivalent maintenance costs, resulted in larger body reserves, and thus had a positive effect on survival and reproduction. This emphasizes the importance of body reserves for animal resilience [22–24]. The feedback of body reserves on survival and reproduction led to differences in lifespan and number of production cycles. Such variations modulate the dilution of the fixed costs, through maintenance expenditure and non-productive stages, and therefore affect FE. The role of body reserves in improving FE is not straightforward, as shown by the lifetime effect of allocation to lactation on FE. Our results suggested an optimal level of allocation to lactation. Increasing allocation to lactation from a low to a medium level had a positive effect on FE in spite of reducing lifespan and lactation number. This result agrees with the recent work of [17] who reported that dairy cows with high genetic merit for milk have a shorter lifespan and lower reproductive performance, but they have a slightly higher lifetime FE than cows with low genetic merit. The higher level of production per lactation cycle compensated for the smaller number of lactation cycles. When allocation to lactation increased from a medium to a high level, this positive effect disappeared and FE decreased. The previous offset of production per lactation cycle was not sufficient to balance the reduction in number of lactation cycles. These results highlight the complexity of FE as a phenotype, which results from a combination of production time and production level. It is clear that improving FE implies the dilution of fixed costs linked to maintenance expenditure and non-productive stages. In the past, the strategy was based on paying back fixed costs with a higher level of production. However, this resulted in negative effects on other time components of FE (lifespan, reproductive success) and this strategy is not adapted to variable or low quality environments [25]. Thus, the challenge is to find the optimal strategy for the dilution of costs depending on the environmental conditions. In addition, more research is needed to quantify the genetic variability in the relation between body reserves and their effect on reproductive performance. This relation can greatly affect FE over the lifetime and therefore should be included in future selection strategies for FE. The model structure is flexible enough to incorporate such future findings. The parameters that determine how body condition score, milk and energy balance influence the probability of conception can be considered as genetic-scaling parameters and set at different values. Thus, as discussed in the following section, simulations will allow the comparison of FE associated with genotypes that reflect various effects of body reserves on reproductive performance.

Modelling approach

We propose a model that demonstrated the relevance of breaking-down FE into acquisition and allocation mechanisms. They play different roles on FE depending on the female's life-stage and this may impact future selection strategies. This alone justifies the conceptual break with the majority of the models for performance prediction that use production potential, i.e. the product of acquisition and allocation (to that output), as the genetic driver. In our model, the genetic-scaling parameters are dissociated into independent parameters that control acquisition and allocation separately. Thus, our model offers opportunities to explore different biological strategies of paying back fixed costs to improve FE, and to investigate which strategy is better adapted to a given environment. The idea here is to identify a point of diminishing returns, that is to say the point after which increasing lifespan or productivity no longer increases FE. Other genetic-scaling parameters could be explored (for instance shape of lactation curve or gestation allocation for prolific species) to further evaluate which combinations of acquisition and allocation maximize FE. Furthermore, other environments could be explored to better use the ability of the model to represent G × E interactions. In this study, we considered a constant environment with fixed resources, both in quantity and quality. The next step could be to carry out a sensitivity analysis of the model to different levels of energy density and evaluate if acquisition and allocation mechanisms have the same impact on feed efficiency. Using the model for various environments will raise the larger issue of available data for validation of the model components that simulate environmental effects. In this study, we validated the consistency of phenotypic trajectories (body mass, body condition score and milk production) simulated by the model with the trajectories corresponding to a compilation of data from the literature, and carried out a global sensitivity analysis, which are two key steps of the model validation [26, 27]. To go further than this partial validation, we will need data corresponding to frequent and long-term measurements and data related to feed intake, body mass and milk production. Even if not yet available, such datasets are likely to be provided by on-going projects on feed efficiency.

The model presented here represents a single cow, defined by its genetic-scaling parameters. By simulating different values of these parameters, the model allows the comparison of phenotypic performances expressed by different genotypes, which reflect different strategies of acquisition and allocation. Clearly, further exploration of the genetic aspects would require a population model. Using the techniques of individual-based modelling, the present animal model could be multiplied to create a virtual population within which each animal has its own genetic-scaling parameters. Such a population model could either specify additive genetic (co)variance structures among the genetic-scaling parameters, or let them (co)evolve naturally. In the present animal model, we broke the phenotypic correlations by decomposing mechanisms and then studied the effects of parameters that drive these mechanisms by setting independent values. In the population context, we need to re-introduce genetic correlations at the level of mechanisms (for instance, the correlation between the value of the growth allocation parameter and the value of the basal acquisition parameter). Using the model as the building block of an individual-based population model will also allow the incorporation of trans-generational aspects. By simulating populations of genotypes under selection, we will be able to evaluate which genetic-scaling parameters combinations are selected depending on environmental conditions and thus to improve the prediction of the effects of selection strategies in different environments.

Conclusions

Feed efficiency is a complex phenotype, which in physiological terms combines the acquisition of resources from the environment with the allocation of resources between physiological functions, including production and non-productive functions. Our results show that breaking-down feed conversion into acquisition and allocation, and introducing genetically-driven trajectories that control these mechanisms, permitted quantification of their relative roles on feed efficiency. Furthermore, our results show that the life stage at which feed efficiency is evaluated appears to be a key aspect for selection. When feed efficiency is evaluated over the second lactation, it is mainly affected by allocation to lactation. When feed efficiency is evaluated in the long-term, i.e. over the whole lifespan, it is mainly affected by allocation to growth and acquisition of resource during lactation. While there is a strong linear increase in feed efficiency with more allocation to lactation within a lactation cycle, our results suggest that there is an optimal level of allocation to lactation beyond which increasing allocation to lactation negatively affects lifetime feed efficiency. Our modelling approach highlights the role of body reserves in the prediction of lifetime feed efficiency since they integrate the feedback of acquisition and allocation on survival and reproduction. It also provides new insights into the processes that underpin lifetime feed efficiency in dairy cows.

Additional files

Additional file 1: Table S1. Parameters of the different sub-models. The table provides the listing of the symbols used for the model's parameters, the description of the parameters, their values and the source used to define these values [16–18, 28].

Additional file 2: Table S2. Elements and equations of the allocation, acquisition and utilization sub-models. The table provides the listing of the elements included in the model (compartments, flows, variables) with their symbols and units. It also provides the equations that define them or indicate the discrete events that change their values [16–19, 29].

Additional file 3: Table S3. Description of the discrete events in the physiological sub-model. Description: The table provides the description of the triggers that activate the discrete events and the actions on model's elements implemented by the events.

Additional file 4. Details of model description. Description: The document provides the description of all model's equations and detailed aspects of the discrete events functioning. Figure S1 shows an example of the flows of energy simulated by the model [15–17, 28–30].

Additional file 5. Details of the calibration of the model. The document provides a description of some detailed aspects of the model's calibration. It indicates the assumptions that were made to introduce biological relationships in the model and how the values of the parameters for these relationships were defined. The relationships relate to the effect of the degree of maturity on the energetic value of structural mass gain and growth function efficiency, the proportion of energy allocated to gestation (Figure S2), the age-dependence of the energetic value for maintenance (Figures S3, S4 and S5), the effect of parity on lactation acquisition (Figures S6, S7 and S8) and the effect of parity on lactation allocation (Figure S9) [18, 19, 31, 32].

Additional file 6. Calibration results and sensitivity analysis details. The document provides the results of the calibration step in the model development. This step was based on a comparison with the GARUNS model and completed by a comparison with a compilation of data from literature related to body mass (Figure S10), milk production (Figure S11) and body condition score (Figure S12). The document also provides additional information on the global sensitivity analysis with the values of the parameters used in the complete factorial design (Table S4) and on how these values of the parameters were determined. Finally, the document provides additional results of the sensitivity analysis with the variance decomposition of the two feed efficiencies depending on the input parameters (Table S5) and series of figures that show the effects of parameters on the dynamic changes of empty body mass, dry matter intake and milk production (Figures S13, S14, S15, S16, S17, S18, S19, S20, S21, S22, S23 and S24) [18, 19, 33–41].

Authors' contributions
LP, DR and NCF participated in the design of the model and virtual experiments. LP implemented the model, ran virtual experiments and analysed the results. All authors read and approved the final manuscript.

Author details
[1] UMR Modélisation Systémique Appliquée aux Ruminants, INRA, AgroParis-Tech, Université Paris-Saclay, 75005 Paris, France. [2] Département des Sciences Biologiques, Université du Québec à Montréal, Montréal, QC H3C 3P8, Canada.

Acknowledgements
The authors are grateful to the PHASE division of the French National Institute for Agricultural Research (INRA) for the financial support of this interdisciplinary research. They thank Dr Olivier Martin and Dr Phuong Ho Ngoc for their technical help with the GARUNS model and Dr Melanie Dammhahn, Dr Muriel Tichit and Dr Daniel Sauvant for their valuable comments on this work. They also would like to thank the two anonymous reviewers and journal's editors for their helpful comments.

Competing interests

The authors declare that they have no competing interests.

References

1. Capper JL, Cady RA, Bauman DE. The environmental impact of dairy production: 1944 compared with 2007. J Anim Sci. 2009;87:2160–7.
2. Rauw WM, Kanis E, Noordhuizen-Stassen EN, Grommers FJ. Undesirable side effects of selection for high production efficiency in farm animals: a review. Livest Prod Sci. 1998;56:15–33.
3. Berry DP, Crowley JJ. Cell biology symposium. Genetics of feed efficiency in dairy and beef cattle. J Anim Sci. 2013;91:1594–613.
4. Miglior F, Muir BL, Van Doormaal BJ. Selection indices in Holstein cattle of various countries. J Dairy Sci. 2005;88:1255–63.
5. Egger-Danner C, Cole JB, Pryce JE, Gengler N, Heringstad B, Bradley A, et al. Invited review: overview of new traits and phenotyping strategies in dairy cattle with a focus on functional traits. Animal. 2015;9:191–207.
6. Hayes BJ, Lewin HA, Goddard ME. The future of livestock breeding: genomic selection for efficiency, reduced emissions intensity, and adaptation. Trends Genet. 2013;29:206–14.
7. Savietto D, Friggens NC, Pascual JJ. Reproductive robustness differs between generalist and specialist maternal rabbit lines: the role of acquisition and allocation of resources. Genet Sel Evol. 2015;47:2.
8. Rauw WM, Luiting P, Beilharz RG, Verstegen MWA, Vangen O. Selection for litter size and its consequences for the allocation of feed resources: a concept and its implications illustrated by mice selection experiments. Livest Prod Sci. 1999;60:329–42.
9. Knap PW, Su G. Genotype by environment interaction for litter size in pigs as quantified by reaction norms analysis. Animal. 2008;2:1742–7.
10. Arthur PF, Archer JA, Johnston DJ, Herd RM, Richardson EC, Parnell PF. Genetic and phenotypic variance and covariance components for feed intake, feed efficiency, and other postweaning traits in Angus cattle. J Anim Sci. 2001;79:2805–11.
11. Grimm V, Berger U, Bastiansen F, Eliassen S, Ginot V, Giske J, et al. A standard protocol for describing individual-based and agent-based models. Ecol Model. 2006;198:115–26.
12. Tardieu F. Virtual plants: modelling as a tool for the genomics of tolerance to water deficit. Trends Plant Sci. 2003;8:9–14.
13. Friggens NC, Newbold JR. Towards a biological basis for predicting nutrient partitioning: the dairy cow as an example. Animal. 2007;1:87–97.
14. Friggens NC, Berg P, Theilgaard P, Korsgaard IR, Ingvartsen KL, Løvendahl P, et al. Breed and parity effects on energy balance profiles through lactation: evidence of genetically driven body energy change. J Dairy Sci. 2007;90:5291–305.
15. Coulon JB, Pérochon L, Lescourret F. Modelling the effect of the stage of pregnancy on dairy cows' milk yield. Anim Sci. 2010;60:401–8.
16. Sauvant D, Giger-Reverdin S. Modélisation des interactions digestives et de la production de méthane chez les ruminants. Inra Prod Anim. 2009;22:375–84.
17. Phuong HN, Blavy P, Martin O, Schmidely P, Friggens NC. Modelling impacts of performance on the probability of reproducing, and thereby on productive lifespan, allow prediction of lifetime efficiency in dairy cows. Animal. 2016;10:106–16.
18. Martin O, Sauvant D. A teleonomic model describing performance (body, milk and intake) during growth and over repeated reproductive cycles throughout the lifespan of dairy cattle. 1. Trajectories of life function priorities and genetic scaling. Animal. 2010;4:2030–47.
19. Martin O, Sauvant D. A teleonomic model describing performance (body, milk and intake) during growth and over repeated reproductive cycles throughout the lifespan of dairy cattle. 2. Voluntary intake and energy partitioning. Animal. 2010;4:2048–56.
20. Ginot V, Gaba S, Beaudouin R, Aries F, Monod H. Combined use of local and ANOVA-based global sensitivity analyses for the investigation of a stochastic dynamic model: application to the case study of an individual-based model of a fish population. Ecol Model. 2006;193:479–91.
21. VandeHaar MJ, St-Pierre N. Major advances in nutrition: relevance to the sustainability of the dairy industry. J Dairy Sci. 2006;89:1280–91.
22. Blanc F, Bocquier F, Agabriel J, D'hour P, Chilliard Y. Adaptive abilities of the females and sustainability of ruminant livestock systems. A review. Anim Res. 2006;55:489–510.
23. Ollion E, Ingrand S, Delaby L, Trommenschlager JM, Colette-Leurent S, Blanc F. Assessing the diversity of trade-offs between life functions in early lactation dairy cows. Livest Sci. 2016;183:98–107.
24. Jenet A, Fernandez-Rivera S, Tegegne A, Wettstein HR, Senn M, Saurer M, et al. Evidence for different nutrient partitioning in Boran (Bos indicus) and Boran × Holstein cows when re-allocated from low to high or from high to low feeding level. J Vet Med A Physiol Pathol Clin Med. 2006;53:383–93.
25. Douhard F, Tichit M, Amer PR, Friggens NC. Synergy between selection for production and longevity and the use of extended lactation: insights from a resource allocation model in a dairy goat herd. J Anim Sci. 2014;92:5251–66.
26. Kleijnen JPC. Verification and validation of simulation models. Eur J Oper Res. 1995;82:145–62.
27. Sargent RG. Verification and validation of simulation models. In: Proceedings of the 2011 winter simulation conference, 11–14 Dec 2011; Phoenix. 2011. p. 183–98.
28. Hou C, Amunugama K. On the complex relationship between energy expenditure and longevity: reconciling the contradictory empirical results with a simple theoretical model. Mech Ageing Dev. 2015;149:50–64.
29. Martin O, Sauvant D. Modeling digestive tract contents in cattle. Renc Rech Rumin. 2003;10:167–70.
30. Kuhn MT, Hutchison JL, Wiggans GR. Characterization of Holstein heifer fertility in the United States. J Dairy Sci. 2006;89:4907–20.
31. Brouwer E. Report of sub-committee on constants and factors. In: Blaxter K, editor. Energy metabolism. London: Academic Press; 1965. p. 441–3.
32. Geay Y. Energy and protein utilization in growing cattle. J Anim Sci. 1984;58:766–78.
33. Hoffman PC. Optimum body size of Holstein replacement heifers. J Anim Sci. 1997;75:836–45.
34. Kertz AF, Reutzel LF, Barton BA, Ely RL. Body weight, body condition score, and wither height of prepartum Holstein cows and birth weight and sex of calves by parity: a database and summary. J Dairy Sci. 1997;80:525–9.
35. Schutz M, Hansen L, Steuernagel G, Kuck A. Variation of milk, fat, protein and somatic-cells for dairy cattle. J Dairy Sci. 1990;73:484–93.
36. Drame E, Hanzen C, Houtain J, Laurent Y, Fall A. Evolution of body condition score after calving in dairy cows. Ann Med Vet. 1999;143:265–70.
37. Berry DP, Veerkamp RF, Dillon P. Phenotypic profiles for body weight, body condition score, energy intake, and energy balance across different parities and concentrate feeding levels. Livest Sci. 2006;104:1–12.
38. Mao IL, Sloniewski K, Madsen P, Jensen J. Changes in body condition score and in its genetic variation during lactation. Livest Prod Sci. 2004;89:55–65.
39. de Haas Y, Pryce JE, Calus MPL, Wall E, Berry DP, Løvendahl P, et al. Genomic prediction of dry matter intake in dairy cattle from an international data set consisting of research herds in Europe, North America, and Australasia. J Dairy Sci. 2015;98:6522–34.
40. Buckley F, O'Sullivan K, Mee JF, Evans RD, Dillon P. Relationships among milk yield, body condition, cow weight, and reproduction in spring-calved Holstein–Friesians. J Dairy Sci. 2003;86:2308–19.
41. Saltelli A, Ratto M, Andres T, Campolongo F, Cariboni J, Gatelli D, et al. Global sensitivity analysis: the primer. Chichester: Wiley; 2008.

2

Exploring the genetic architecture and improving genomic prediction accuracy for mastitis and milk production traits in dairy cattle by mapping variants to hepatic transcriptomic regions responsive to intra-mammary infection

Lingzhao Fang[1,2]* , Goutam Sahana[1], Peipei Ma[1], Guosheng Su[1], Ying Yu[2], Shengli Zhang[2], Mogens Sandø Lund[1] and Peter Sørensen[1]

Abstract

Background: A better understanding of the genetic architecture of complex traits can contribute to improve genomic prediction. We hypothesized that genomic variants associated with mastitis and milk production traits in dairy cattle are enriched in hepatic transcriptomic regions that are responsive to intra-mammary infection (IMI). Genomic markers [e.g. single nucleotide polymorphisms (SNPs)] from those regions, if included, may improve the predictive ability of a genomic model.

Results: We applied a genomic feature best linear unbiased prediction model (GFBLUP) to implement the above strategy by considering the hepatic transcriptomic regions responsive to IMI as genomic features. GFBLUP, an extension of GBLUP, includes a separate genomic effect of SNPs within a genomic feature, and allows differential weighting of the individual marker relationships in the prediction equation. Since GFBLUP is computationally intensive, we investigated whether a SNP set test could be a computationally fast way to preselect predictive genomic features. The SNP set test assesses the association between a genomic feature and a trait based on single-SNP genome-wide association studies. We applied these two approaches to mastitis and milk production traits (milk, fat and protein yield) in Holstein (HOL, n = 5056) and Jersey (JER, n = 1231) cattle. We observed that a majority of genomic features were enriched in genomic variants that were associated with mastitis and milk production traits. Compared to GBLUP, the accuracy of genomic prediction with GFBLUP was marginally improved (3.2 to 3.9%) in within-breed prediction. The highest increase (164.4%) in prediction accuracy was observed in across-breed prediction. The significance of genomic features based on the SNP set test were correlated with changes in prediction accuracy of GFBLUP ($P < 0.05$).

Conclusions: GFBLUP provides a framework for integrating multiple layers of biological knowledge to provide novel insights into the biological basis of complex traits, and to improve the accuracy of genomic prediction. The SNP set test might be used as a first-step to improve GFBLUP models. Approaches like GFBLUP and SNP set test will become increasingly useful, as the functional annotations of genomes keep accumulating for a range of species and traits.

*Correspondence: lingzhao.fang@mbg.au.dk
[1] Department of Molecular Biology and Genetics, Center for Quantitative Genetics and Genomics, Aarhus University, 8830 Tjele, Denmark
Full list of author information is available at the end of the article

Background

In general, genetic variation in complex or quantitative traits is considered to be governed by a large number of loci with small to moderate effects, which are individually undetectable by genome-wide association studies (GWAS) with stringent significance thresholds [1–5]. A better understanding of the genetic architecture that underlies complex traits (e.g. the distribution of causal variants and their effects) could improve the predictive ability of models [4, 6–9]. This would be beneficial for genomic prediction of disease risk in humans and for estimating genetic values in livestock and plant species of agricultural importance [4, 6–9].

The genomic best linear unbiased prediction (GBLUP) assumes that all genomic markers contribute equally to variability of a trait [10] and ignores any prior biological knowledge on genetic architecture of the trait. However, genomic markers that are associated with a complex trait may not be uniformly and randomly distributed over the genome, but rather be clustered in genes that are part of interconnected biological pathways and networks [2, 11, 12]. The genomic regions that are likely to be enriched in variants affecting a trait are defined as genomic features. Based on different biological hypotheses, genomic features can be defined from various sources of biological knowledge, such as genes, gene ontologies, biological pathways, or other types of external evidence. Incorporating this biological information may improve the predictive abilities of models. We extended the GBLUP model to implement this strategy by including a separate random effect for the joint action of single nucleotide polymorphisms (SNPs) within a genomic feature [8], which we call a genomic feature BLUP (GFBLUP) model. As a result, individual SNP relationships can be weighted differently in GFBLUP according to the variance explained by SNPs within and outside the genomic feature [8]. The GFBLUP model has been applied to three complex traits (i.e. chill coma recovery, starvation resistance and startle response) in the unrelated inbred lines of *Drosophila melanogaster* populations [8]. Compared to GBLUP, the prediction accuracy with GFBLUP was substantially improved when incorporating several gene ontology (GO) categories as genomic features [8]. A possible increase in prediction accuracy with GFBLUP would depend on whether the genomic feature is enriched in causal mutations.

The GFBLUP model is computationally intensive for evaluating many genomic features [8]. Therefore, it is important to develop a computationally fast approach. The SNP set test based on GWAS-derived single-SNP test statistics could be one such approach. It would be of interest to investigate the relationship between the significance of a genomic feature based on the SNP set test and the predictive ability of the GFBLUP model.

To date, there are many genes that are yet neither functionally characterized nor mapped to any biological databases [13–16], in particular in livestock populations. For example, in cattle only ~20% of the genes are annotated in Kyoto Encyclopaedia of Genes and Genomes (KEGG) pathways [17]. However, transcriptomics studies have been conducted on small-scale experimental populations to investigate the dynamic state of the transcriptome in particular tissues, revealing thousands of genomic features (e.g. genes and pathways) that are engaged in the biological processes of complex traits [18–20]. Such transcriptomics studies provide tissue-specific genomic features that are likely to be enriched in genomic variants affecting specific traits.

Mastitis, an inflammatory condition of the mammary gland, is often caused by invading pathogens. It is the most costly disease in the dairy industry due to treatment cost, reduction in milk production and milk quality, and in some cases culling of the affected cows [21]. Gram-negative *Escherichia coli* (*E. coli*) is a common mastitis-causing bacteria [22], and the lipopolysaccharides (LPS) released by *E. coli* induce acute inflammatory responses [23]. Genes with expression levels that are significantly affected during the early stage of infection have also been suggested to be involved in overall metabolism [19, 23–26]. Moreover, it is well established that mastitis is unfavorably correlated with milk production traits [25]. Since liver plays key roles in innate immune response and metabolic regulation [27], we hypothesized that hepatic transcriptomic regions that are responsive to intra-mammary infection (IMI) may be enriched in genomic variants that impact mastitis and milk production traits. Using these regions as genomic features might provide more predictive GFBLUP models compared to the GBLUP model. In addition, since gene expression patterns and molecular interaction networks are consistent across breeds [28], we further hypothesized that the use of transcriptomic data obtained on one breed may contribute to improve genomic prediction in other breeds.

In the current study, mastitis and three milk production traits (i.e. milk, fat and protein yield) from Nordic Holstein (HOL, n = 5056) and Jersey (JER, n = 1231) cattle were analyzed using imputed sequence genotype data (~15 million SNPs) and hepatic transcriptome data from an IMI study. Our main objectives were to apply the GFBLUP model and SNP set test: (1) to investigate the genomic variance explained by transcriptomic regions that are responsive to IMI; (2) to improve the accuracy of within-breed and across-breed genomic prediction using GFBLUP compared to GBLUP; and (3) to investigate the relationship between the predictive ability of GFBLUP and the significance of genomic features based on the SNP set test.

Methods

Intra-mammary infection (IMI) study

The IMI experimental design and collection of liver biopsies were reported previously [23, 29]. In brief, eight healthy HOL dairy cows in their first lactation (9 to 12 weeks after calving) were selected for the experiment. The udder quarters of all studied cows were free from mastitis pathogens based on bacteriological examinations. Milk somatic cell count (SCC) for each studied quarter was <100,000. The right front quarter was infected with 200 μg of *E. coli* LPS (0111:B4) (Sigma-Aldrich, Brøndby, Denmark) dissolved in 10 mL of a 0.9% NaCl solution, while the left front quarter was used as a control and challenged with 10 mL of 0.9% NaCl solution only. Clinical signs, data on production traits together with milk and blood parameters associated with LPS infection were recorded throughout the trial and confirmed that mastitis inflammation was induced. Liver biopsies collected 22 h before and 3, 6, 9, 12 and 48 h after LPS infection in three cows were used for RNA extraction. Sampling procedures for liver biopsies were described previously [30]. Finally, 18 RNA-Seq libraries (at each time point with three biological replicates) were sequenced using 100-bp paired-end sequencing in Illumina Hiseq2000 sequencing technology.

Statistical analysis of RNA-Seq data

Statistical approaches used for analysing RNA-Seq data were described previously [31]. Briefly, sequence reads of each sample were aligned to the bovine reference genome assembly (UMD 3.1), using a sensitive and efficient mapping program based on the seed-and-vote algorithm implemented in the Rsubread package in R/Bioconductor [32] _ENREF_65. The number of reads that were mapped to 24,616 Ensemble genes **(ftp://ftp.ensembl.org/pub/release-86/gtf/bos_taurus) was counted using the function Feature-Counts in the Rsubread package with default settings. The average mapping rate across all samples was approximately 68%. Analysis of differential gene expression was conducted using edgeR [33]. A small number of highly expressed genes in a sample can cause an RNA composition effect, i.e. a substantial proportion of the total library size could be consumed by these highly expressed genes, which results in the remaining genes to be under-sampled [33]. Therefore, the most recommended weighted trimmed means of M-values (TMM) were used to normalize the total count data (i.e. the total library size) between each pair of samples, in order to adjust for RNA composition effect [33]. After normalization of the total library size, a negative binomial generalized linear model (GLM) was applied for each gene, because the count data of genes follow non-normal distributions, which commonly exhibit a quadratic mean–variance relationship [33]. The relevant factors in the experimental design were also adjusted by the GLM, and gene differential expression was determined using a likelihood ratio test [33]. In the GLM model, where the number of reads mapped to gene g in sample i is denoted as y_{gi} and the total number of mapped reads is denoted as N_i, it is assumed that $y_{gi} \sim NB(\mu_{gi}, \phi_g)$, where μ_{gi} and ϕ_g are the location and the dispersion parameters of the negative binomial distribution, respectively. To ensure stable inference for each gene, an empirical Bayes method was used to compress gene-wise dispersions towards a common dispersion for all genes [33]. Statistical tests for each analysis were adjusted for multiple-testing using the FDR method as implemented in R (version 3.2.4).

Defining genomic features

The differentially-expressed genes (DEG) (i.e. the hepatic transcriptome regions responsive to IMI) that were obtained from the above RNA-Seq analyses were used to define genomic features. First, 30 genomic features were defined using six false discovery rate (FDR) cut-off values (i.e. $\leq 5 \times 10^{-2}$, 10^{-2}, 10^{-3}, 10^{-6}, 10^{-8}, and 10^{-10}) in each of the five experimental comparisons (i.e. 3 vs. −22 h, 6 vs. −22 h, 9 vs. −22 h, 12 vs. −22 h and 48 vs. −22 h), respectively. In addition, since the biological functions of up-regulated and down-regulated genes can be quite different, each of these 30 genomic features was further divided into four subsets based on four $\log_2$(fold-change)s cut-off values (i.e. ≤ -2, ≤ -1, ≤ 1, and >2). Therefore, another 115 genomic features were built, because five conditions were without DEG. In total, 145 genomic features were defined. The number of DEG in each genomic feature is summarized in Table S1 (see Additional file 1: Table S1).

Phenotypic data

The phenotypes were de-regressed breeding values (DRP) from routine genetic evaluations by the Nordic Cattle Genetic Evaluation (NAV, http://www.nordicebv.info/), and were available for 5056 HOL and 1231 JER cattle. Detailed information of these phenotypes was previously described in [34, 35]. Heritabilities for milk, fat and protein yields and mastitis were equal to 0.39, 0.39, 0.39 and 0.04, respectively in HOL, and very similar in JER [34, 35]. The average reliabilities of the DRP for milk, fat and protein yields and mastitis were equal to 0.95, 0.95, 0.95 and 0.83, respectively in HOL; and 0.92, 0.92, 0.92, and 0.76, respectively in JER.

Genotypic data

Imputation from Illumina BovineSNP50 BeadChip (50 K) to Illumina BovineHD BeadChip (high-density, HD) genotypes for these individuals and further to whole-genome sequence variants was described previously [36, 37]. Briefly, genotypes from the 50 K SNP chip for each individual were first imputed to HD genotypes using a multi-breed reference of 3383 animals (1222 HOL, 1326 Nordic Red, and 835 JER). A total of 648,219 SNPs were obtained after imputation to the HD chip. These imputed HD genotypes were then imputed to the whole-genome sequence level using a multi-breed reference population of 1228 individuals from *Run4* of the 1000 Bull Genomes Project [38] and additional whole-genome sequences from Aarhus University including 368 HOL, 86 Nordic Red, and 88 JER individuals [39]. Genotype imputation was done using *Minimac2* [40]. In total, 22,751,039 biallelic variants (SNPs and Indel) were included in the imputed sequence genotypic data. The accuracy of imputation was above 0.85 for the across-breed imputation of 19,498,365 SNPs. Detailed information about imputation accuracy was previously reported in [37]. For each breed, SNPs with a large deviation from Hardy–Weinberg proportions ($P < 10^{-6}$) or with minor allele frequency (MAF) <0.01 were further excluded. A total of 15,355,382 and 13,403,916 SNPs remained for the HOL and JER datasets, respectively. The SNP locations were based on the UMD3.1 reference genome (http://www.ensembl.org/Bos_taurus/Info/Index). A SNP was considered to be linked with a genomic feature if its chromosome position was within the open reading frame of DEG in the particular genomic feature.

Training and validation populations

For within-breed prediction, each of the datasets (i.e. HOL and JER) was divided into training and validation sets based on birth-year of the animal to access prediction accuracy. The birth-year cut-off was 2006 for HOL and 2004 for JER, and the younger animals were assigned to the validation dataset (Table 1). We chose this validation strategy considering routine animal breeding practice where the young bulls breeding values are predicted using a training population of older animals. For across-breed prediction, the complete HOL population (n = 5056) was used as training data to predict

breeding values for all JER bulls (n = 1231). Both GBLUP and GFBLUP models were fitted to compare prediction accuracies.

Genomic models

For each genomic feature as defined before, SNPs were partitioned into two sets (i.e. within and outside the genomic feature), followed by the GFBLUP model analysis:

$$\mathbf{y} = \mathbf{1}\mu + \mathbf{g_f} + \mathbf{g_{-f}} + \mathbf{e},$$

where $\mathbf{y}$ is the vector of phenotypic observations, $\mathbf{1}$ is a vector of 1s, μ is the overall mean, $\mathbf{g_f}$ is the vector of genomic values captured by the SNPs within a genomic feature, $\mathbf{g_{-f}}$ is the vector of genomic values captured by SNPs outside the genomic feature (i.e. the rest of genome), and $\mathbf{e}$ is the vector of residuals. Assumptions for all random effects are given by:

$$\begin{pmatrix} \mathbf{g_f} \\ \mathbf{g_{-f}} \\ \mathbf{e} \end{pmatrix} \sim N \left[\begin{pmatrix} 0 \\ 0 \\ 0 \end{pmatrix}, \begin{pmatrix} \mathbf{G_f}\sigma_f^2 & 0 & 0 \\ 0 & \mathbf{G_{-f}}\sigma_{-f}^2 & 0 \\ 0 & 0 & \mathbf{D}\sigma_e^2 \end{pmatrix} \right],$$

where $\mathbf{G_f}$ and $\mathbf{G_{-f}}$ are genomic relationship matrices that are built using the SNPs within and outside the genomic feature, respectively, which were calculated using the second method described in [41]. Briefly, let $\mathbf{M}$ be the marker matrix that specifies which alleles the individual inherits, and $\mathbf{P}$ be the matrix that contains the frequencies of the second allele at locus (p_i) expressed as a difference from the 0.5 value and multiplied by 2, that is, the column i of $\mathbf{P}$ is $2(p_i - 0.5)$. Matrix $\mathbf{Z}$ was obtained as $\mathbf{M} - \mathbf{P}$, which allows mean values of the allele effects to be equal to 0. Then, $\mathbf{G} = \mathbf{ZTZ'}$, where $\mathbf{T}$ is a diagonal matrix with $T_{ii} = \frac{1}{m[2p_i(1-p_i)]}$. $\mathbf{D}$ is a diagonal matrix with diagonal elements equal to $\frac{1-r^2}{r^2}$, where r^2 is the reliability of DRP, $\sigma_f^2, \sigma_{-f}^2$ and σ_e^2 are the variance components accounted for by the SNPs within and outside the genomic feature, and by the residuals, respectively.

The standard GBLUP model includes only one random genomic effect:

$$\mathbf{y} = \mathbf{1}\mu + \mathbf{g} + \mathbf{e},$$

with the same notation as above except for $\mathbf{g}$, which is the vector of genomic values captured by all genomic SNPs. The random genomic values and the residuals were

Table 1 Overview of training and validation population sizes for genomic predictions

Breed	Number of training individuals	Number of validation individuals	Total number
Within HOL	4011	1054	5056
Within JER	975	256	1231
Across breeds	5056	1231	6287

assumed to be independently distributed: $\mathbf{g} \sim N\left(\mathbf{0}, \mathbf{G}\sigma_g^2\right)$ and $\mathbf{e} \sim N\left(\mathbf{0}, \mathbf{D}\sigma_e^2\right)$.

Estimation of genomic parameters

The variance components, σ_f^2, σ_{-f}^2, σ_g^2 and σ_e^2, were estimated using an average information restricted maximum-likelihood (AI-REML) procedure [42] implemented in DMU [43]. The proportion of genomic variance explained by a genomic feature in the GFBLUP model: $H_f^2 = \frac{\sigma_f^2}{\sigma_f^2 + \sigma_{-f}^2}$. The proportion of phenotypic variance explained by all SNPs: $h_{GFBLUP}^2 = \frac{\sigma_f^2 + \sigma_{-f}^2}{\sigma_f^2 + \sigma_{-f}^2 + \sigma_e^2}$ for GFBLUP, and $h_{GBLUP}^2 = \frac{\sigma_g^2}{\sigma_g^2 + \sigma_e^2}$ for GBLUP.

Validation of genomic prediction

Genomic breeding values (GEBV) were predicted using both GFBLUP and GBLUP models. In the GFBLUP and GBLUP models, GEBV is $\hat{\mathbf{g}}_{total} = \hat{\mathbf{g}}_f + \hat{\mathbf{g}}_{-f}$ and $\hat{\mathbf{g}}_{total} = \hat{\mathbf{g}}$, respectively. Accuracy of predicted genomic breeding values (r) is calculated as the correlation between GEBV and DRP in the validation population. The bias of the genomic predictions with both GFBLUP and GBLUP was evaluated by the regression of DRP on the GEBV, i.e. bias = $\text{cov}(\text{DRP}, \text{GEBV})/\sigma_{GEBV}^2$.

Single-marker GWAS

Single-marker GWAS analyses for four traits were only conducted in the HOL training population, followed by SNP set test analyses for testing the associations between genomic features and traits. Single-marker GWAS was performed using a two-step variance component-based method, to account for population stratification, as implemented in EMMAX [44]. In the first step, the polygenic and residual variances were estimated using the following model:

$$\mathbf{y} = \mathbf{1}\mu + \mathbf{a} + \mathbf{e},$$

where $\mathbf{y}$ is a vector of phenotypes; $\mathbf{1}$ is a vector of 1s; μ is the overall mean; $\mathbf{a}$ is a vector of breeding values, where $\mathbf{a} \sim N\left(0, \mathbf{G}\sigma_a^2\right)$, and $\mathbf{G}$ is the genome relationship matrix estimated using EMMAX based on HD SNP genotypes, but excluding the SNPs on the chromosome that harbours the SNP the effect of which is being estimated; and $\mathbf{e}$ is the vector of residuals, where $\mathbf{e} \sim N\left(0, \mathbf{I}\sigma_e^2\right)$ and $\mathbf{I}$ is an identity matrix. In the second step, the individual effects of SNPs were obtained using a linear regression model:

$$\mathbf{y} = \mathbf{1}\mu + \mathbf{x}\mathbf{b} + \mathbf{\eta},$$

where $\mathbf{y}$, $\mathbf{1}$ and μ are as defined above; $\mathbf{x}$ is a vector of imputed genotype dosages (ranging from 0 to 2), $\mathbf{b}$ is the vector of allele substitution effects (b), and $\mathbf{\eta}$ is a vector of random residual deviates with (co)variance structure $\mathbf{G}\sigma_a^2 + \mathbf{I}\sigma_e^2$.

SNP set test

Summary statistic for a genomic feature

The summary statistic of a genomic feature was calculated as the sum of the test statistics (i.e. t^2) of all SNPs within DEG (i.e. open reading frame) that belonged to the genomic feature:

$$T_{sum} = \sum_{i=1}^{m_f} t_m^2,$$

where m_f is the number of SNPs located in a genomic feature, and t_m^2 is the square of the t-statistics for each SNP in the genomic feature. The t-statistics was calculated as the estimate of the SNP effect (i.e. b) from single-marker GWAS divided by its standard error. This summary statistic is more powerful compared to count-based summary statistics, particularly in situations where genomic features harbor many SNPs each having a small to moderate effect [9, 45].

Testing for association between a genomic feature and a trait

Under the null hypothesis, all SNPs in a genome feature have the same joint effect as those in the randomly selected genomic features. To ensure a null hypothesis is competitive to the alternative hypothesis, the random genomic features must contain the same number of SNPs as the genomic feature being analysed, and the linkage disequilibrium (LD) structure among SNPs should be retained. An empirical distribution of the summary statistics of a genomic feature was therefore obtained by using the following cyclical permutation procedure as described previously [9, 46]. Briefly, the test statistics of SNPs (i.e. t^2) were first ordered based on the chromosome position of the SNPs. A test statistic was randomly selected from this vector. All test statistics were then shifted to new positions, where the selected SNP became the first one, and the other SNPs shifted to new positions, but retained their original order. This uncouples any associations between SNPs and the genomic feature, while retaining the LD structure among SNPs. A new summary statistic was then calculated according to the original position of the genomic feature. The permutation was repeated 1000 times for each genomic feature, and an empirical P value was then calculated based on one-tailed tests of the proportion of randomly sampled summary statistics that were larger than that observed.

Biological function enrichment analysis

In order to investigate the biological function of a genomic feature, functional enrichment analysis of DEG in the particular genomic feature was conducted using a web-based tool, KOBAS2.0 (http://kobas.cbi.pku.edu.cn/home.do) [47], where a hypergeometric gene set enrichment test, based on a gene ontology (GO) database, was applied. The FDR method [48] was used for adjusting multiple tests.

Results

The results for RNA-Seq analyses at different time-point comparisons (i.e. 3 vs. −22 h, 6 vs. −22 h, 9 vs. −22 h, 12 vs. −22 h and 48 vs. −22 h) are summarized in Table S2 (see Additional file 2: Table S2). The $-\log_{10}(P)$ values of imputed sequence-level SNPs from single-marker GWAS for mastitis and milk production traits on the HOL training population are shown in the Manhattan plots of Figure S1 (see Additional file 3: Figure S1). The GFBLUP and GBLUP models were compared for all four traits in within-breed (i.e. HOL and JER) genomic prediction, followed by across-breed prediction (i.e. HOL as the training population and JER as the validation population). The degree of enrichment (i.e. $-\log_{10}(P$ values)) of genomic features based on the SNP set test in the HOL training population was compared with the changes in prediction accuracy of GFBLUP within- and across-breed predictions, respectively.

GBLUP, GFBLUP and SNP set test analyses for Holstein population
Genomic parameters

As shown in Fig. 1a, 128, 106, 99, and 90 of the 145 genomic features explained larger proportions of the total genomic variance (H_f^2) compared to their

SNP-proportion over the whole genome for mastitis, protein, milk and fat yield, respectively. Detailed information is summarized in Tables S3, S4, S5 and S6 (see Additional file 4: Tables S3, S4, S5 and S6). These results demonstrated that the genomic variance of the traits studied is not uniformly distributed along the genome, but appears to be enriched in a subset of hepatic transcriptomic regions that are responsive to IMI. Therefore, the assumption of the GBLUP approach that a priori all markers contribute equally to trait variability does not hold good.

Prediction accuracy

Prediction accuracy of GBLUP was equal to 0.504 (bias = 0.864) for mastitis, 0.602 (bias = 0.775) for protein yield, 0.635 (bias = 0.862) for milk yield, and 0.607 (bias = 0.808) for fat yield. Compared to the GBLUP model, 27, 44, 17 and 13 of the 145 genomic features resulted in higher prediction accuracies with GFBLUP ($\Delta r \geq 0.01$) for mastitis, protein, milk and fat yield, respectively (see Additional file 4: Tables S3, S4, S5 and S6). Among these, we found 8 (9) up- (down-) regulated genomic features for mastitis, 26 (4) for protein yield, 2 (9) for milk yield, and 4 (9) for fat yield (Fig. 2). These results indicate that down-regulated genes could be more often associated with milk and fat yield than up-regulated genes during IMI. The regression coefficient of DRP on

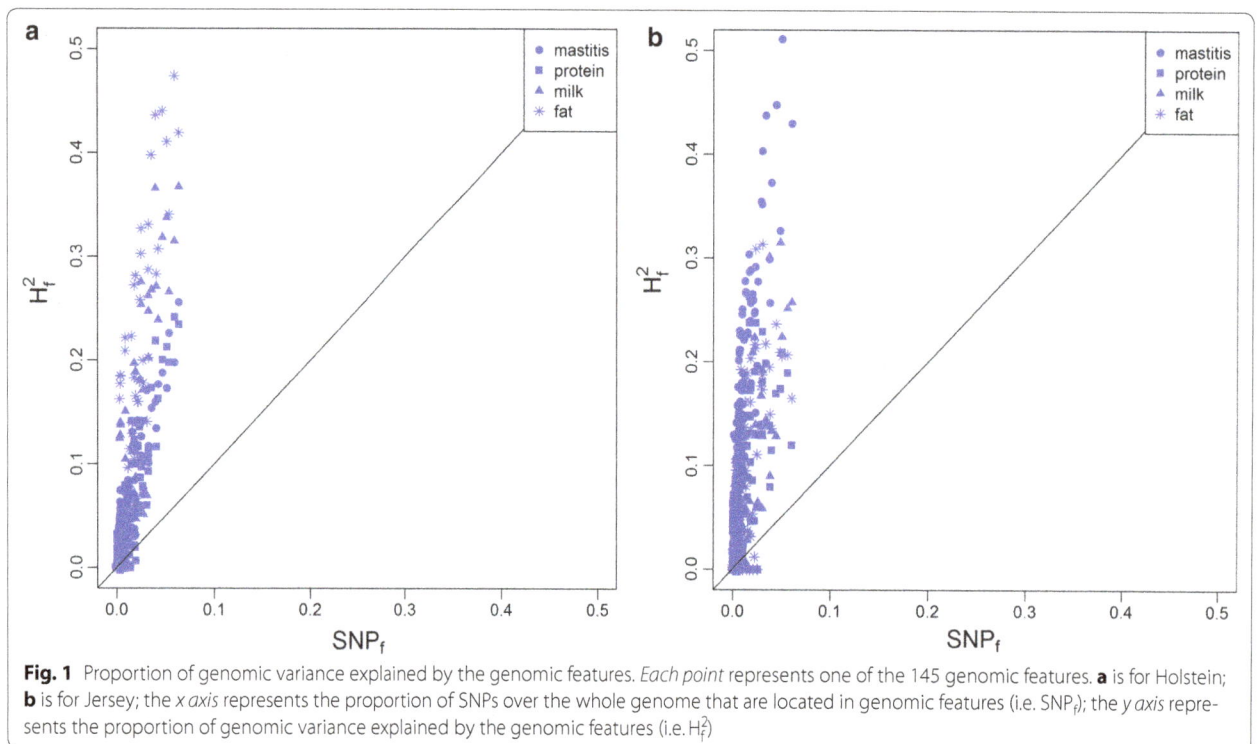

Fig. 1 Proportion of genomic variance explained by the genomic features. *Each point* represents one of the 145 genomic features. **a** is for Holstein; **b** is for Jersey; the *x axis* represents the proportion of SNPs over the whole genome that are located in genomic features (i.e. SNP$_f$); the *y axis* represents the proportion of genomic variance explained by the genomic features (i.e. H$_f^2$)

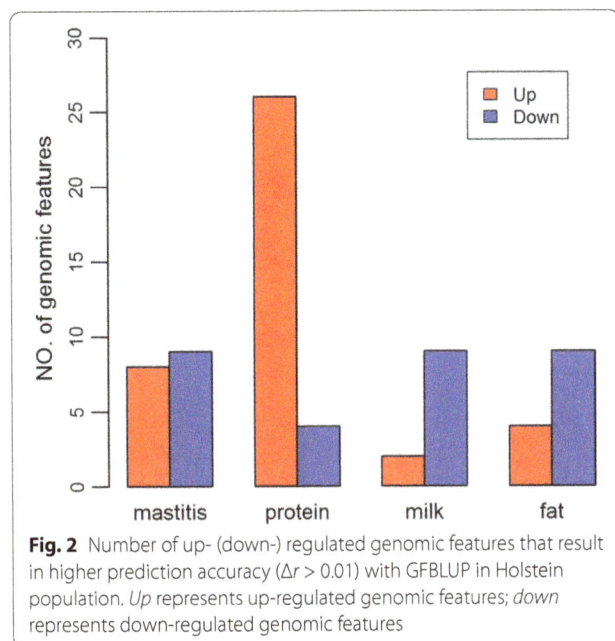

Fig. 2 Number of up- (down-) regulated genomic features that result in higher prediction accuracy ($\Delta r > 0.01$) with GFBLUP in Holstein population. *Up* represents up-regulated genomic features; *down* represents down-regulated genomic features

GBLUP and GFBLUP analyses for the Jersey population
Genomic parameters
As in the analyses for the HOL population (Fig. 1b), we observed that 125, 115, 99, and 83 of the 145 genomic features for the JER population explained a larger proportion of the total genomic variance relative to their SNP-proportion over the whole genome for mastitis, protein yield, milk yield, and fat yield, respectively. Detailed information is in Tables S7, S8, S9 and S10 (see Additional file 6: Tables S7, S8, S9 and S10). It should be noted that all genomic features were defined based on gene expression data that were obtained in HOL cattle. These results imply that a subset of hepatic transcriptomic regions responsive to IMI found for HOL were also enriched in genomic variants for mastitis, protein, milk and fat yield in JER.

Prediction accuracy
Prediction accuracy of the GBLUP model was equal to 0.549 (bias = 0.916) for mastitis, 0.530 (bias = 0.760) for protein yield, 0.597 (bias = 0.796) for milk yield, and 0.433 (bias = 0.669) for fat yield. Compared to the GBLUP model, 21, 14 and 2 genomic features resulted in higher prediction accuracy ($\Delta r \geq 0.01$) with GFBLUP for mastitis, protein, and milk yield, respectively (see Additional file 6: Tables S7, S8, S9 and S10), among which 7, 13 and 0 were in common with those found for HOL, respectively. No genomic features resulted in an increase >0.005 in prediction accuracy for fat yield in JER. The regression coefficient of DRP on GEBV (i.e. bias) for all the GFBLUP analyses ranged from 0.891 to 0.930 for mastitis, from 0.727 to 0.807 for protein yield, from 0.760 to 0.809 for milk yield, and from 0.599 to 0.677 for fat yield. As observed in HOL, the absolute value of (1-bias) was negatively correlated with the change in prediction accuracy for all four traits in JER (see Additional file 7: Figure S3). The top five predictive genomic features for each of the four traits are summarized in Table 3. The average increase in prediction accuracy (Δr) with the best-performing genomic feature across the four traits was 0.020, which corresponds to a 3.9% increase compared to GBLUP. These results indicate that the use of gene expression data obtained from one breed may improve marginally the genomic prediction accuracy in other breeds. It should be noted that, for JER, the increase in prediction accuracy with GFBLUP for milk and fat yield was very small (Table 3).

Comparisons between degree of enrichment from the SNP set test and changes in prediction accuracy of GFBLUP
The changes in prediction accuracy with GFBLUP on the JER validation population were also significantly positively correlated with $-\log_{10}(P)$ based on the SNP set test on the HOL training population for mastitis and protein yield (Fig. 4). Correlations of 0.59 ($P = 3.0 \times 10^{-15}$),

GEBV (bias) for all GFBLUP analyses ranged from 0.862 to 0.873 for mastitis, from 0.772 to 0.783 for protein yield, from 0.857 to 0.866 for milk yield, and from 0.778 to 0.821 for fat yield (see Additional file 4: Tables S3, S4, S5 and S6). The absolute value of (1-bias) tended to be negatively correlated with the change in genomic prediction accuracy with GFBLUP across four traits (see Additional file 5: Figure S2), which indicates that more predictive genomic features lead to less biased predictions. The top five predictive genomic features for each of the four traits are presented in Table 2. The average increase in prediction accuracy with the best-performing genomic feature across the four traits was 0.018, which corresponds to an increase of 3.2% relative to GBLUP.

Comparisons between degrees of enrichment based on the SNP set test and changes in prediction accuracy of GFBLUP
The results of SNP set tests for all 145 genomic features across four traits in the HOL training population are summarized in Tables S3, S4, S5 and S6 (see Additional file 4: Tables S3, S4, S5 and S6). The changes in prediction accuracy of GFBLUP (Δr) were significantly ($P < 0.05$) positively correlated with $-\log_{10}(P)$ of genomic features based on the SNP set test across all four traits (Fig. 3). Correlations of 0.69 ($P < 2.2 \times 10^{-16}$), 0.46 ($P = 4.4 \times 10^{-9}$), 0.46 ($P = 4.4 \times 10^{-9}$) and 0.44 ($P = 3.6 \times 10^{-8}$) were found between changes in accuracy and $-\log_{10}(P$ value) for mastitis, protein yield, milk yield, and fat yield, respectively. These results demonstrated that the SNP set test could be used as a computationally simple way to develop more predictive GFBLUP models.

Table 2 Top five predictive genomic features for mastitis, protein, milk and fat yield in Holstein cattle

Trait	Time (h)[a]	FDR$_{exp}^{b}$	Log$_2$(FC)[c]	P$_{set-test}^{d}$	SNP$_f$ (%)[e]	H$_f^2$ (%)[f]	r_{GFBLUP}^{g}	bias[h]	$\Delta r^{[i]}$
Mastitis	9	5×10^{-2}	NA[j]	0.013	6.36	25.60	0.520	0.872	0.016
	9	5×10^{-2}	>1	0.027	2.32	13.71	0.519	0.872	0.015
	6	5×10^{-2}	NA	0.040	5.92	19.81	0.519	0.873	0.015
	6	10^{-2}	NA	0.043	4.68	18.83	0.518	0.871	0.014
	6	10^{-3}	NA	0.034	3.54	15.39	0.518	0.871	0.014
Protein	48	10^{-6}	>2	0.021	<0.01	1.85	0.622	0.783	0.020
	48	10^{-8}	>2	0.029	<0.01	1.75	0.621	0.782	0.019
	48	10^{-2}	>2	0.023	0.02	3.28	0.621	0.779	0.019
	48	10^{-8}	>1	0.027	<.01	1.71	0.621	0.782	0.019
	48	10^{-10}	>2	0.026	<0.01	1.37	0.620	0.782	0.018
Milk	6	10^{-2}	NA	0.026	4.68	31.90	0.651	0.863	0.016
	6	10^{-3}	NA	0.027	3.54	26.82	0.651	0.865	0.016
	6	10^{-3}	<−1	0.024	1.76	19.74	0.650	0.862	0.015
	6	10^{-6}	<−2	0.022	0.28	12.49	0.649	0.866	0.014
	6	10^{-2}	<−1	0.030	2.49	25.39	0.649	0.859	0.014
Fat	6	10^{-6}	<−2	0.027	0.28	16.28	0.629	0.804	0.022
	6	10^{-3}	<−2	0.028	0.33	17.76	0.626	0.800	0.019
	6	10^{-2}	<−2	0.032	0.36	18.57	0.625	0.798	0.018
	6	5×10^{-2}	<−2	0.032	0.37	18.51	0.625	0.799	0.018
	9	10^{-6}	>1	0.055	0.84	20.94	0.621	0.815	0.014

[a] Time points post intra-mammary infection with *E. coli* LPS

[b] FDR values used to define genomic features from RNA-Seq analysis

[c] Log$_2$(fold-change) values used to define up- (down-) regulated genomic features from RNA-Seq analysis

[d] P values from SNP set test on HOL training population

[e] Proportion of SNPs in genomic features over the whole genome

[f] Proportion of the total genomic variance explained by genomic features

[g] Prediction accuracy with GFBLUP

[h] The regression coefficient of de-regressed proofs (DRP) on predicted genomic breeding values (GEBV)

[i] The change of prediction accuracy with GFBLUP relative to GBLUP

[j] The genomic feature defined without log$_2$(fold-change)

0.52 ($P = 3.1 \times 10^{-11}$), 0.19 ($P = 0.02$) and 0.06 ($P = 0.5$) were found between changes in accuracy and $-\log_{10}(P)$ for mastitis, protein yield, milk yield, and fat yield, respectively.

GBLUP and GFBLUP for across-breed genomic prediction

When the complete HOL population was considered as training population to predict the genomic values of individuals in the JER population, prediction accuracy of GBLUP was very low, i.e. prediction accuracies were equal to −0.058 (bias = −0.343) for mastitis, 0.098 (bias = 0.622) for protein yield, 0.160 (bias = 0.762) for milk yield, and 0.070 (bias = 0.482) for fat yield. Compared to the GBLUP model, 60, 68, 71 and 44 of the 145 genomic features resulted in higher prediction accuracy with GFBLUP ($\Delta r \geq 0.01$) for mastitis, protein, milk and fat yield, respectively (see Additional file 8: Tables S11, S12, S13 and S14). The regression coefficient (i.e. bias)

of DRP on GEBV for all GFBLUP analyses ranged from −0.463 to 0.277 for mastitis, from 0.151 to 1.265 for protein yield, from 0.413 to 0.826 for milk yield, and from 0.002 to 0.577 for fat yield. It should be noted that more predictive genomic features lead to less biased predictions across the four traits (see Additional file 9: Figure S4). In addition, for mastitis, protein and milk yield, the changes in accuracy with GFBLUP in across-breed prediction were significantly correlated with the $-\log_{10}(P)$ of SNP set test in the HOL training population (Fig. 5). The top five predictive genomic features for each of the four traits are summarized in Table 4. The absolute average increase in prediction accuracy (Δr) with the best-performing genomic feature across four traits was 0.111, which corresponds to a 164.4% increase relative to GBLUP. Compared to within-breed prediction, the relative improvement in genomic prediction accuracy seems to be clearer in across-breed prediction.

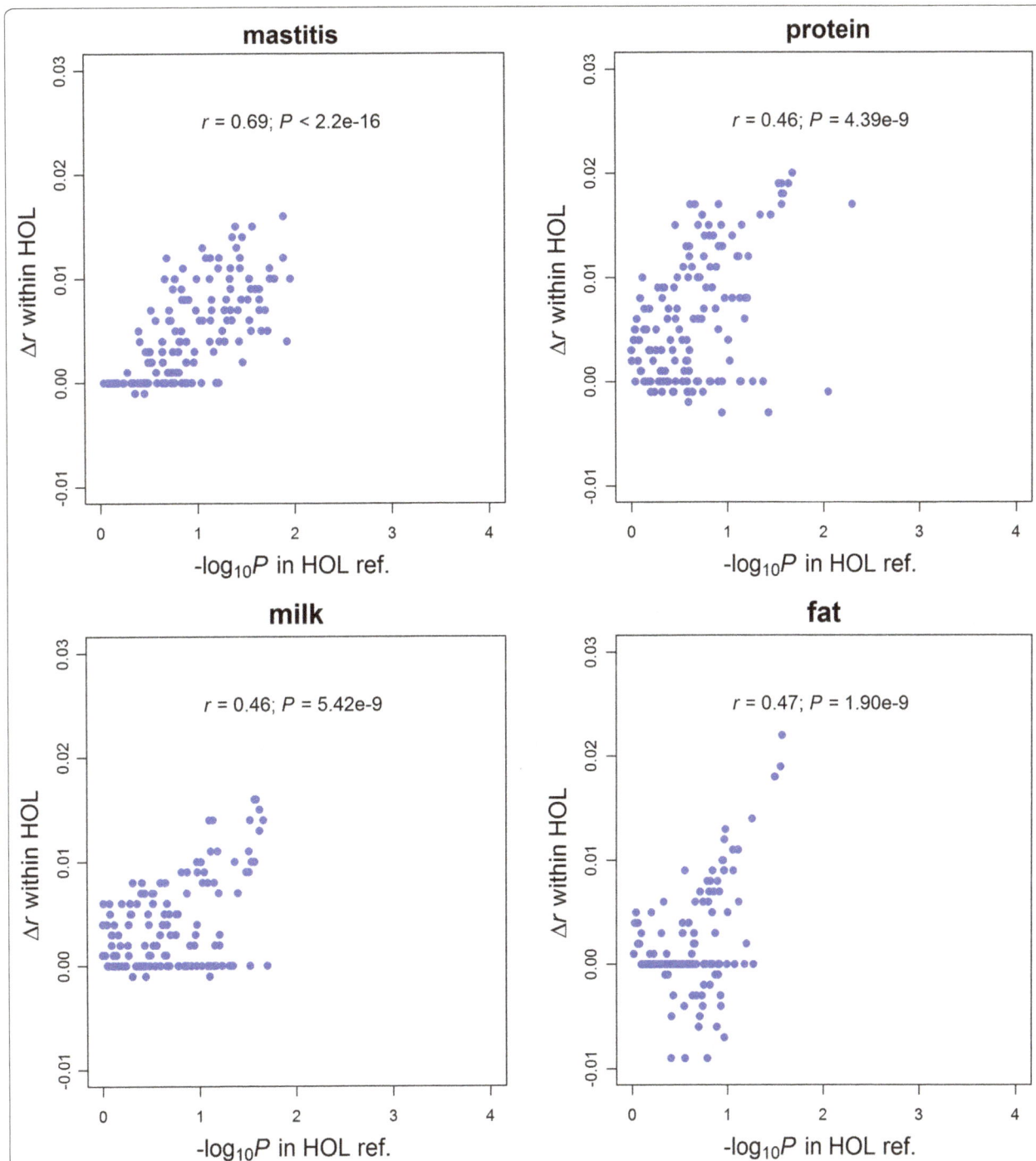

Fig. 3 Comparisons between degree of enrichment from the SNP set test in the Holstein (HOL) training (reference) population and changes in prediction accuracy with GFBLUP in the HOL validation population. *Each point* represents one of the 145 genomic features

Discovery of gene sets associated with protein yield

Genomic features can be ranked based on the predictive ability of GFBLUP. Therefore, our GFBLUP can also be used to map gene sets that are associated with complex traits. For instance, a highly up-regulated genomic feature

with 34 DEG (FDR < 10^{-6}; $\log_2$(fold-change) > 2) that were detected in the 48 vs. −22 h comparison resulted in an increase of 0.204, 0.020 and 0.041 in prediction accuracy for protein yield among across-breed, and within HOL and JER predictions, respectively (see Additional file 10: Table

Table 3 Top five predictive genomic features for mastitis, protein, milk and fat yield in Jersey cattle

Trait	Time (h)[a]	FDR_{exp}^b	$Log_2(FC)^c$	$SNP_f (\%)^d$	$H_f^2 (\%)^e$	r_{GFBLUP}^f	bias[g]	Δr^h
Mastitis	9	10^{-10}	>1	0.46	15.79	0.567	0.927	0.018
	12	10^{-2}	NA[i]	3.98	37.31	0.566	0.930	0.017
	9	10^{-10}	NA	1.31	26.64	0.564	0.921	0.015
	12	10^{-10}	<−1	0.71	16.15	0.564	0.925	0.015
	6	10^{-3}	<−1	1.67	28.69	0.563	0.923	0.014
Protein	48	10^{-2}	>2	0.02	6.42	0.576	0.807	0.046
	48	10^{-6}	>2	<0.01	4.59	0.571	0.797	0.041
	48	10^{-10}	>2	<0.01	4.11	0.569	0.787	0.039
	48	10^{-8}	>2	<0.01	4.28	0.569	0.796	0.039
	48	5×10^{-2}	>2	0.03	6.74	0.568	0.804	0.038
Milk	48	0.01	>2	0.02	2.19	0.608	0.805	0.011
	9	10^{-2}	<−1	3.02	12.85	0.607	0.801	0.010
	12	10^{-8}	<−1	0.88	10.39	0.606	0.809	0.009
	48	5×10^{-2}	>2	0.03	1.38	0.605	0.805	0.008
	9	10^{-3}	<−1	2.31	13.94	0.604	0.800	0.007
Fat	48	5×10^{-2}	>1	0.30	4.04×10^{-7}	0.438	0.672	0.005
	6	5×10^{-2}	>1	2.57	2.00×10^{-7}	0.437	0.672	0.004
	48	5×10^{-2}	NA	0.35	2.24×10^{-6}	0.437	0.672	0.004
	9	10^{-6}	>2	0.32	5.93×10^{-7}	0.437	0.672	0.004
	9	10^{-8}	>2	0.28	5.68×10^{-7}	0.437	0.672	0.004

[a] Time points post intra-mammary infection with *E. coli* LPS

[b] FDR values used to define genomic features from RNA-Seq analysis

[c] Log_2(fold-change) values used to define up- (down-) regulated genomic features from RNA-Seq analysis

[d] Proportion of SNPs in genomic features over the whole genome

[e] Proportion of the total genomic variance explained by genomic features

[f] Prediction accuracy with GFBLUP

[g] The regression coefficient of de-regressed proofs (DRP) on predicted genomic breeding values (GEBV)

[h] The change of prediction accuracy with GFBLUP relative to GBLUP

[i] The genomic feature defined without log_2(fold-change)

S15). These 34 DEG, which include <0.01% of the total number of SNPs, explained 1.84 and 4.59% of the genomic variance for protein yield in HOL and JER, respectively. In addition, they explained 0.44 and 0.50% of the genomic variance for mastitis in HOL and JER, respectively, but did not improve genomic predictions for mastitis. Detailed information of GFBLUP analyses for these 34 DEG across three prediction scenarios is in Table 5. The *P* values based on the SNP set test were 0.021 and 0.18 for protein yield and mastitis, respectively, on the HOL training population. The functional enrichment analysis of these 34 DEG revealed that they were significantly (FDR < 0.05) enriched in innate immune response and negative regulation of endopeptidase activity and protein metabolism (Fig. 6).

Discussion

In the current study, we demonstrated that a subset of the hepatic transcriptomic regions responsive to IMI was enriched in genomic variants associated with mastitis and milk production traits. When using these regions as genomic features, the genomic prediction accuracy with GFBLUP was improved marginally compared to GBLUP. In theory, both the GFBLUP model and SNP set test can easily be extended to incorporate other types of biological information as genomic features, such as sequence annotation, biological pathways and eQTL.

Dissection of the genetic architecture and improvement of prediction accuracy for mastitis and milk production traits in dairy cattle

It has been suggested that milk production and disease resistance traits are controlled by several hundred up to several thousand loci in cattle, most of which have a very small effect [4, 49, 50]. Multiple studies, using different strategies, have been conducted to investigate the genetic architecture that underlies such complex phenotypes, and to improve genomic prediction accuracy within and across breeds [6, 17, 49, 51, 52].

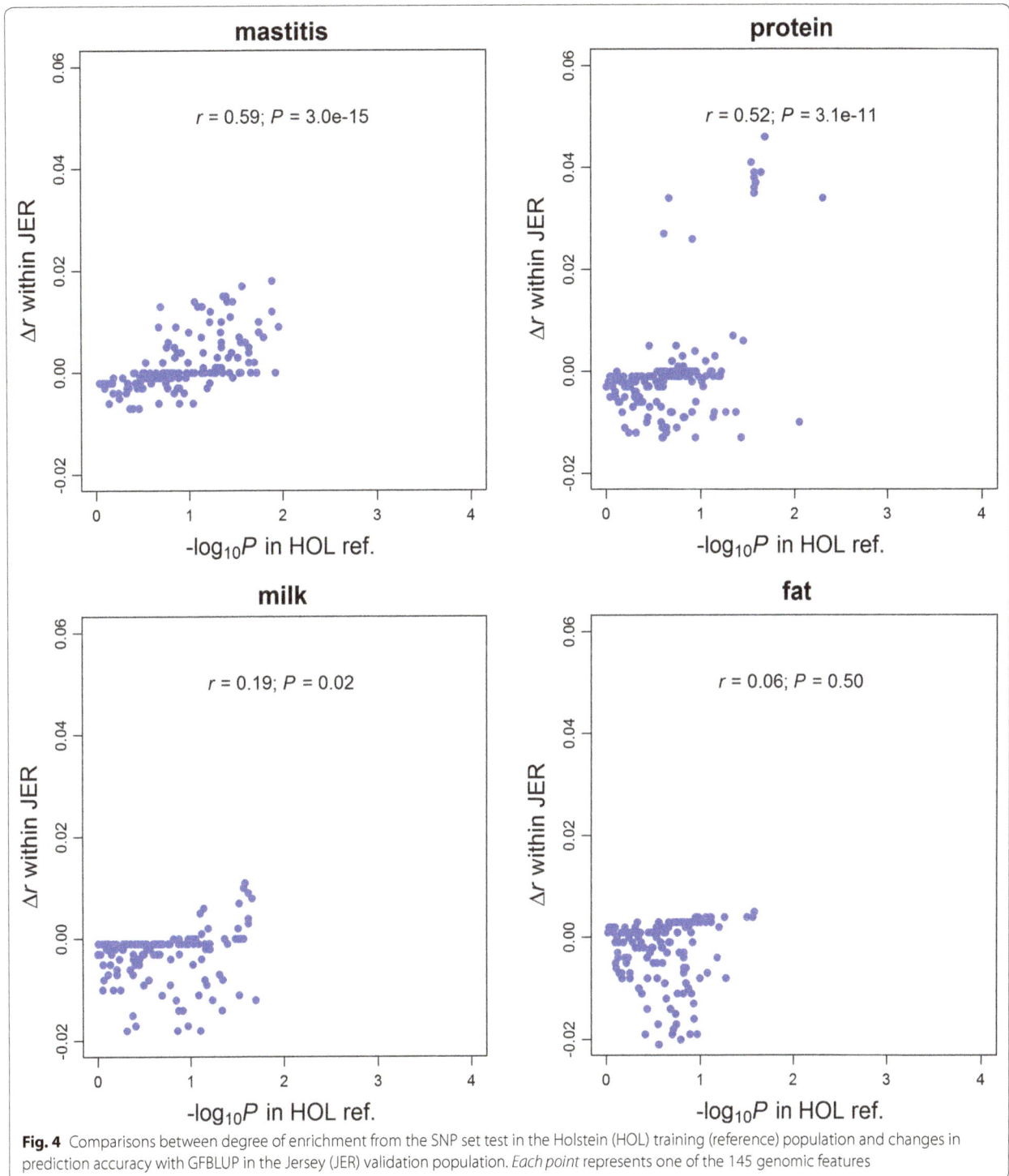

Fig. 4 Comparisons between degree of enrichment from the SNP set test in the Holstein (HOL) training (reference) population and changes in prediction accuracy with GFBLUP in the Jersey (JER) validation population. *Each point* represents one of the 145 genomic features

Genetic architecture and biological interpretation

The approaches that partition genomic variance based on adjacent genomic regions (e.g. 50-SNP genomic segments) or single chromosomes may not provide enough biological insights into the genetic architecture of a trait

[6, 51, 53]. Our results provide evidence that results from gene expression experiments can give additional information about the biological and genetic basis of complex traits. In the current study, we used RNA-Seq data from an IMI experiment as an example to study the

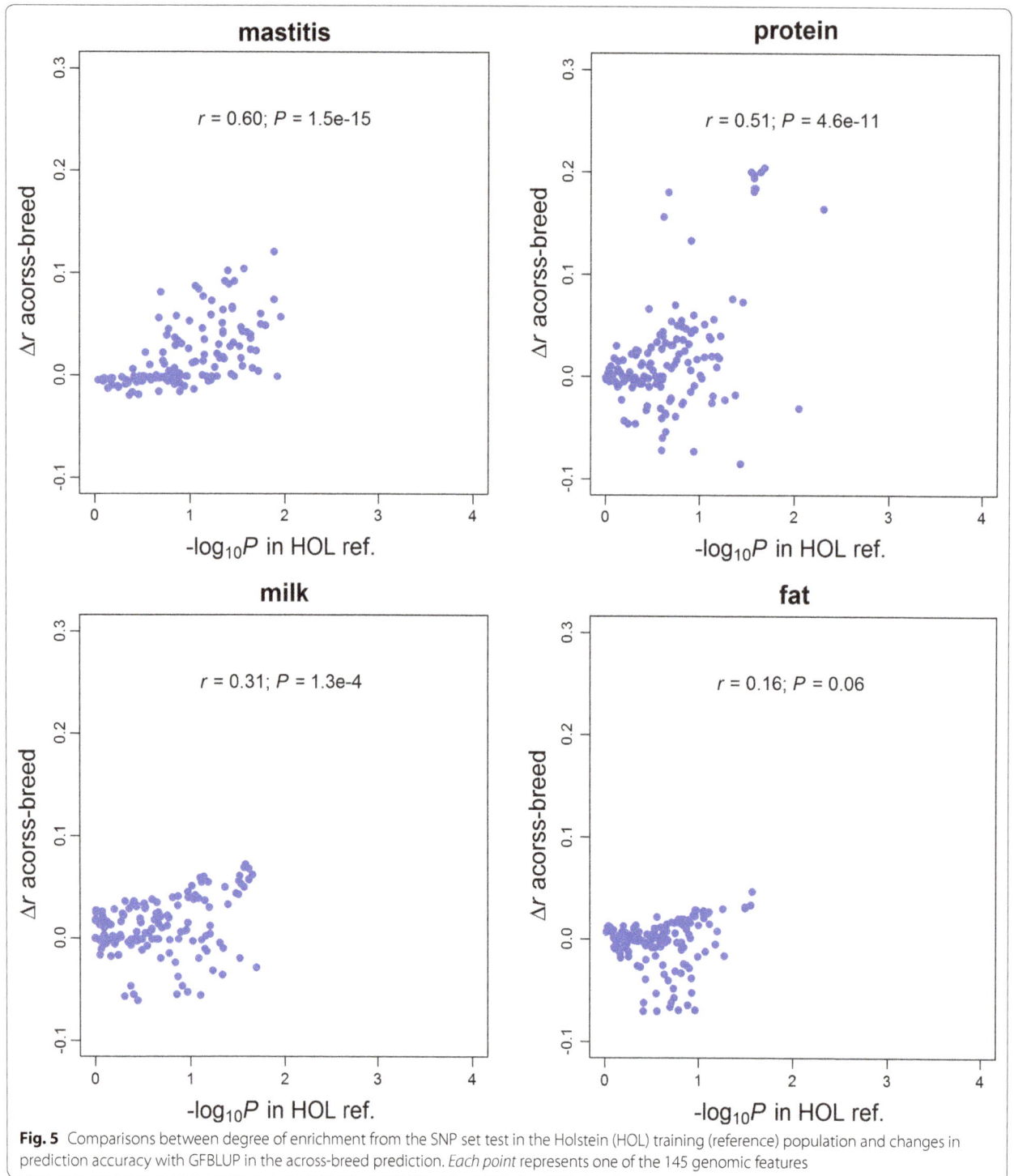

Fig. 5 Comparisons between degree of enrichment from the SNP set test in the Holstein (HOL) training (reference) population and changes in prediction accuracy with GFBLUP in the across-breed prediction. *Each point* represents one of the 145 genomic features

genetic and biological basis of mastitis and milk production traits. We found that a subset of hepatic transcriptomic regions responsive to IMI is enriched in genomic variants associated with these traits. We also found that

down-regulated genes are more often associated with milk and fat yield, which together with the fact that the liver is a crucial organ for host immune responses and metabolism, including lipogenesis, gluconeogenesis, and

Table 4 Top five predictive genomic features for mastitis, protein, milk and fat yield in across-breed prediction

Trait	Time (h)[a]	FDR_{exp}^b	$Log_2(FC)^c$	SNP_f (%)[d]	H_f^2 (%)[e]	r_{GFBLUP}^f	bias[g]	Δr^h
Mastitis	6	10^{-3}	<-1	1.94	9.98	0.063	0.277	0.121
	6	5×10^{-2}	<-1	3.53	14.03	0.046	0.178	0.104
	6	10^{-2}	<-1	2.72	12.68	0.044	0.171	0.102
	9	5×10^{-2}	NA[i]	6.99	25.98	0.034	0.115	0.092
	12	5×10^{-2}	>1	2.34	12.84	0.034	0.112	0.092
Protein	48	10^{-6}	>2	0.01	2.24	0.302	1.250	0.204
	48	10^{-8}	NA	0.01	2.04	0.298	1.264	0.200
	48	10^{-8}	>2	<0.01	2.09	0.295	1.265	0.197
	48	10^{-3}	>2	0.01	2.66	0.292	1.245	0.194
	48	10^{-10}	NA	<0.01	1.60	0.282	1.172	0.184
Milk	9	10^{-3}	<-1	2.69	24.65	0.232	0.798	0.072
	9	10^{-6}	NA	2.60	14.41	0.229	0.805	0.069
	9	10^{-6}	<-1	1.67	8.20	0.228	0.808	0.068
	48	10^{-6}	>2	0.01	0.25	0.222	0.826	0.062
	12	10^{-8}	<-1	1.02	3.95	0.221	0.802	0.061
Fat	6	10^{-3}	>1	1.98	19.66	0.117	0.577	0.047
	9	10^{-6}	NA	2.61	24.48	0.104	0.477	0.034
	6	10^{-6}	<-1	0.95	20.29	0.102	0.446	0.032
	3	5×10^{-2}	>2	0.11	0.85	0.101	0.567	0.031
	3	10^{-2}	>2	0.11	0.72	0.100	0.560	0.030

[a] Time points post intra-mammary infection with *E. coli* LPS

[b] FDR values used to define genomic features from RNA-Seq analysis

[c] Log_2(fold-change) values used to define up- (down-) regulated genomic features from RNA-Seq analysis

[d] Proportion of SNPs in genomic features over the whole genome

[e] Proportion of the total genomic variance explained by genomic features

[f] Prediction accuracy with GFBLUP

[g] The regression coefficient of de-regressed proofs (DRP) on predicted genomic breeding values (GEBV)

[h] The change of prediction accuracy with GFBLUP relative to GBLUP

[i] The genomic feature defined without log_2(fold-change)

cholesterol metabolism [54, 55], implies that the immune responses in the liver during mastitis impair milk production. This is in agreement with a recent study that demonstrated that immune relevant pathways (e.g. leukocyte trans endothelial migration and chemokine signalling pathways) are strongly associated with milk and fat yield in HOL [17].

Within-breed prediction

In populations with a high degree of linkage disequilibrium (LD), such as highly selected dairy cattle breeds, the genomic relationship based on genome-wide markers provides accurate information about the genomic variation of the traits [56], although it does not use any prior biological information. In addition, the LD structure makes it more difficult to partition genomic variance based on genomic features. Therefore, the increase in

prediction accuracy with GFBLUP is small compared to GBLUP, i.e. we observed average increases of 0.018 and 0.022 across four traits within HOL and JER, respectively. This is consistent with a recent study [52] that applied a Bayesian genomic feature model (i.e. BayesRC) to milk production traits. Incorporating 790 candidate genes associated with milk production traits as a genomic feature, they found that the increases in within-breed prediction accuracy with BayesRC were quite small (<0.01) compared to BayesR, which ignores any prior biological information [52].

Across-breed prediction

Across-breed genomic prediction accuracies for milk production traits were close to zero, when HOL was used as training population to predict genomic values for JER using the GBLUP approach. This is in agreement with

Table 5 GFBLUP analyses of 34 genes detected in the comparison 48 h vs. −22 h (FDR < 10⁻⁶; log₂(fold-change) > 2) for mastitis, protein, milk and fat yield

Scenario	Trait	H_f^2 (%)[a]	r^b_{GFBLUP}	bias[c]	Δr[d]
Within HOL	Mastitis	0.44	0.505	0.865	0.001
	Protein	1.84	0.622	0.783	0.020
	Milk	0.32	0.643	0.863	0.008
	Fat	0.15	0.607	0.809	0.000
Within JER	Mastitis	0.50	0.550	0.918	0.001
	Protein	4.59	0.571	0.797	0.041
	Milk	0.00	0.596	0.789	−0.001
	Fat	0.00	0.434	0.671	0.001
Across-breed	Mastitis	0.46	−0.063	−0.373	−0.005
	Protein	2.24	0.302	1.250	0.204
	Milk	0.25	0.222	0.826	0.062
	Fat	0.09	0.079	0.491	0.009

[a] Proportion of total genomic variance explained by the genomic feature

[b] Prediction accuracy with GFBLUP

[c] Regression of coefficient of de-regressed proofs (DRP) on predicted genomic breeding values (GEBV)

[d] Change in prediction accuracy with GFBLUP relative to GBLUP

observations in [50, 56]. When validation and training populations are distantly related (i.e. the LD structure becomes weak), genomic feature modelling approaches such as GFBLUP and BayesRC are expected to perform better than models that ignore prior biological information such as GBLUP and BayesR, provided that the genomic feature is enriched in the genomic variants of the traits across breeds [8, 52]. Therefore, shifting the focus from the complete set of genomic markers to those that are more likely to have functional effects might contribute to improve across-breed genomic predictions [7], as observed in our study. However, breed differences in the segregation of quantitative trait loci (QTL), minor allele frequencies and breed-specific SNP effects could add to the complexity in across-breed prediction.

GFBLUP and alternatives
Factors that influence the performance of GFBLUP
The assumption made in the GBLUP model (i.e. the genomic variance is evenly distributed along the whole genome) does not match the real genetic architecture that underlies the traits. It puts equal weights to the elements in the genomic relationship, whereas the GFBLUP allows putting different weights to the individual genomic relationships in the prediction equation according to the estimated genomic parameters [8]. Prediction accuracy of GFBLUP is influenced both by the genomic variance

explained by the genomic features and by the number of non-causal SNPs in the feature [8, 9]. The GFBLUP model performs better as the genomic feature contains more causal variants (i.e. explaining more genomic variance) and less non-causal markers [8, 9]. However, if the estimated genomic parameters deviate from the true values, it will lead to reduced prediction accuracy, as shown in the current study (Figs. 3, 4, 5), because too much weight is put on the "wrong" genomic relationships in the prediction equations. Our GFBLUP has two components for genomic effects (i.e. f and −f), but in theory it is possible to include multiple genomic feature effects [57, 58], which might improve genomic predictions more compared to the current GFBLUP. However, when the correlations among multiple genomic relationship matrices are high, the variance components are not reliably estimated and thus there is no improvement in prediction accuracy [8, 57]. Therefore, further work is needed to investigate the performance of the GFBLUP model with multiple genomic features, in particular in livestock populations with large LD structures.

Bayesian mixture model and Bayesian GF mixture model
Bayesian mixture models, such as BayesR [50], which ignore prior genomic feature information, are considered to be relevant alternative methods. Both GFBLUP and Bayesian mixture models allow assigning markers to different distributions. GFBLUP assigns a marker set (i.e. genomic feature) to a certain distribution [i.e. $f \sim N(0, \mathbf{G}_f \sigma_f^2)$ or $-f \sim N(0, \mathbf{G}_{-f} \sigma_{-f}^2)$] using prior biological knowledge, whereas Bayesian mixture models attempt to assign markers to predefined distributions based on the data themselves. Previous studies demonstrated that an externally informed genomic feature is necessary for a successful partitioning of genomic variance, while the data themselves may not necessarily suggest which marker should have the greatest weight [8, 50]. The external biological information can also be incorporated into Bayesian mixture models, such as BayesRC [52]. All genomic feature models including GFBLUP and BayesRC are computationally intensive, and they do not necessarily perform better than standard models (i.e. GBLUP and BayesR) when genomic features are less enriched in causal variants [8, 59].

SNP set test
The SNP set test based on single-marker test statistics derived from GWAS is a computationally fast way to evaluate a large number of genomic features [60]. The results of the SNP set test could be used to develop more predictive GFBLUP and similar models. The

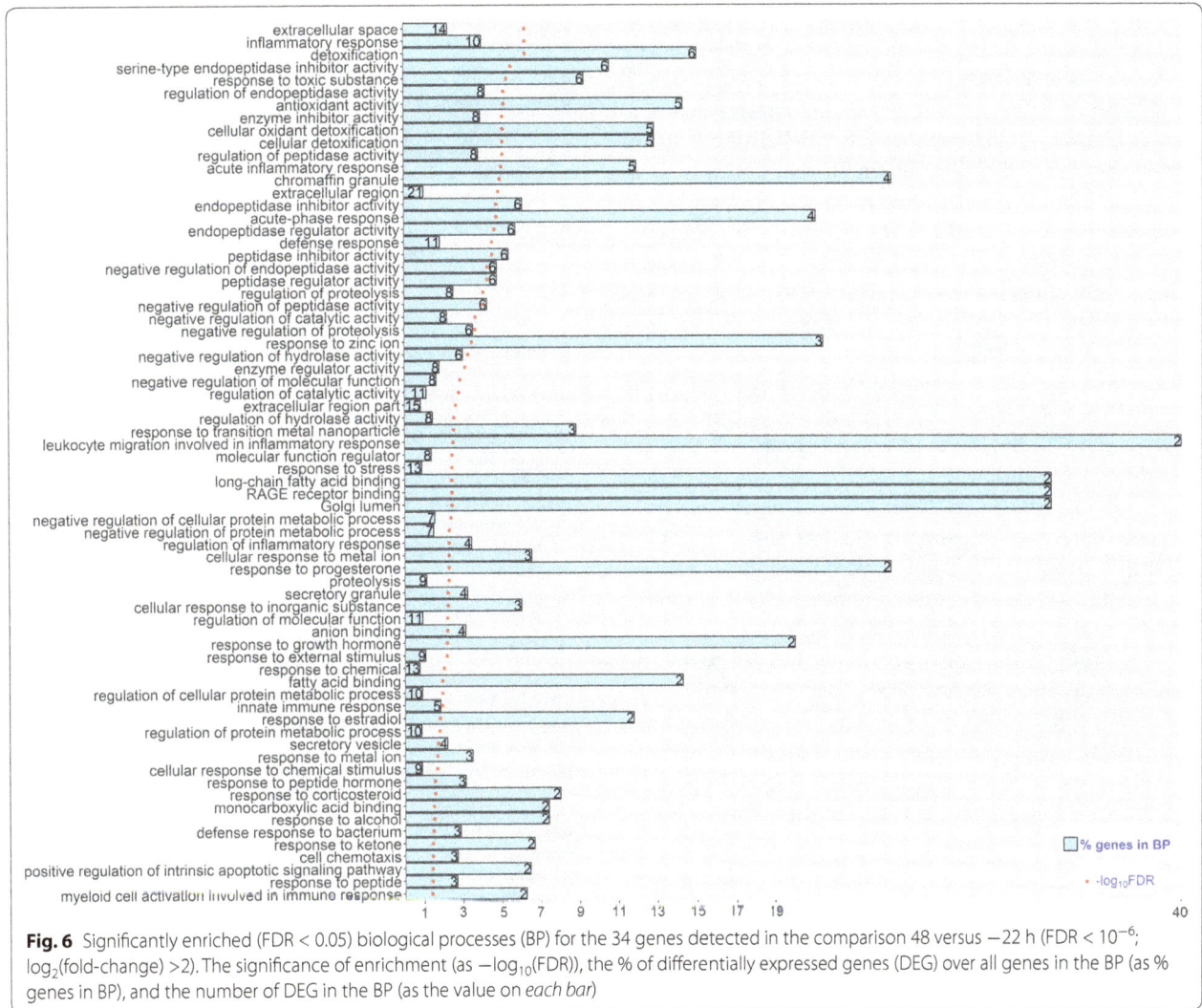

Fig. 6 Significantly enriched (FDR < 0.05) biological processes (BP) for the 34 genes detected in the comparison 48 versus −22 h (FDR < 10^{-6}; $\log_2$(fold-change) >2). The significance of enrichment (as −$\log_{10}$(FDR)), the % of differentially expressed genes (DEG) over all genes in the BP (as % genes in BP), and the number of DEG in the BP (as the value on *each bar*)

current SNP set test method assesses the association between a genomic feature and a trait based on the sum of t^2 of SNPs within the genomic feature. Another commonly used approach for the SNP set test is based on counting associations exceeding a pre-defined significance threshold within the genomic feature [61–63]. One important limitation of this count-based approach is the dichotomization of association signals into significant and non-significant sets, based on a pre-specified significance level, which ignores information regarding the strength of association. Since the genomic variance of mastitis and milk production traits is typically governed by very many markers, each with a small effect [4, 49, 50], the current SNP set test is more likely to match the genetic basis of complex

phenotypes, and is more powerful than the count-based approach [9, 45, 46].

Appropriate genomic features facilitate improved biological interpretation

In order to test different biological hypotheses, many genomic features can be constructed using different sources of prior information, such as prior QTL regions, chromosomes, sequence, biological pathways, and other types of external evidence. The gain in biological knowledge of complex traits relies highly on the genomic feature classification strategies. Since associated genomic markers are not evenly, or necessarily physically, clustered along the genome [2, 51], partitioning genomic variance based on adjacent genomic regions (e.g. haplotypes and chromosomes) is not an ideal way to facilitate the interpretation of biological mechanisms underlying the traits. Biological interpretation may be better served by the use of pathways and gene ontologies as genomic features; however, the quantity and quality of the genes that are functionally annotated in current pathway databases are limited [15], particularly for livestock and plant genomes. Here, we used information from gene expression data to define genomic features, providing novel insights into the genetic and biological basis of mastitis and milk production traits and improving genomic prediction accuracy with GFBLUP.

Since mastitis can be caused by various pathogens, the current RNA-Seq data that originate only from *E. coli* mastitis may be limited to detect all the genes that are functionally relevant with mastitis. Thus, more RNA-Seq data from infections with other types of pathogens could help the detection of genomic features that are associated with mastitis and milk production. In addition, since gene expression patterns depend highly on time, cell types, and tissues, some trait-associated genes might not show differential expression in certain cell types and tissues at a certain physiological stage. Therefore, incorporating more molecular biological information from more tissues (e.g. mammary gland, blood and adipose tissue) and more physiological stages could be important to define the appropriate genomic features that are highly enriched in causal variants.

Conclusions

Compared to GBLUP, GFBLUP models increased the accuracy of genomic prediction for mastitis and milk production traits in dairy cattle by incorporating biological information from gene expression data, and thus provide novel biological insights into the genetic basis of such complex traits. Compared to within-breed prediction, the increase in prediction accuracy seems to be more apparent in across-breed prediction. In addition, the SNP set test can be used as a computationally fast way to develop more predictive GFBLUP or similar models. The current genomic feature modelling approaches provide a general framework for incorporating biological knowledge from independent functional genomics studies to study the genetic architecture and to improve genomic prediction for complex traits. Approaches such as GFBLUP and SNP set test will be increasingly useful as the biological knowledge of functional genomic regions keep accumulating for a range of traits and species.

Additional files

Additional file 1: Table S1. The genomic features defined by RNA-Seq analysis. The data provided represent the number of genes in each of the 145 genomic features defined by using six different FDR cut-off values (i.e. $\leq 5 \times 10^{-2}$, 10^{-2}, 10^{-3}, 10^{-6}, 10^{-8}, and 10^{-10}) and four $\log_2$(fold-change)s (≤ -2, ≤ -1, ≥ 1, and ≥ 2).

Additional file 2: Table S2. Gene differential expression analysis of RNA-Seq data. The data provided represent the results of gene differential expression analysis of RNA-Seq data in five different comparisons, i.e. 3 versus -22 h, 6 versus -22 h, 9 versus -22 h, 12 versus -22 h and 48 versus -22 h.

Additional file 3: Figure S1. Manhattan plots of single-marker genome-wide association analyses (GWAS) with imputed sequence SNPs. The figure provided represents the P values of all imputed sequence SNPs from GWAS for mastitis, protein, milk and fat yield in the HOL training population. Each point represents one SNP.

Additional file 4: Tables S3, S4, S5 and S6. GFBLUP and SNP set test analyses in the Holstein (HOL) population. The data provided represent the detailed results of GFBLUP and SNP set test analyses for mastitis (Table S3), protein (Table S4), milk (Table S5) and fat (Table S6) yield in the HOL population.

Additional file 5: Figure S2. Relationship between bias of genomic predictions and changes in prediction accuracy with GFBLUP for four traits in the Holstein population. Each point represents one of the 145 genomic features. The y axis is the absolute values of (1-bias (b)) for GFBLUP, and the x axis is the changes in prediction accuracy with GFBLUP relative to GBLUP.

Additional file 6: Tables S7, S8, S9 and S10. GFBLUP analyses in the Jersey (JER) population. The data provided represent the detailed results of GFBLUP analyses for mastitis (Table S7), protein (Table S8), milk (Table S9) and fat (Table S10) yield in the JER population.

Additional file 7: Figure S3. Relationship between bias of genomic predictions and changes in prediction accuracy for four traits in the Jersey (JER) population. Each point represents one of the 145 genomic features. The y axis is the absolute values of (1-bias (b)) for GFBLUP, and the x axis is the changes in prediction accuracy with GFBLUP relative to GBLUP.

Additional file 8: Tables S11, S12, S13 and S14. Results of GFBLUP analyses across breeds. The data provided represent the detailed results of GFBLUP analyses for mastitis (Table S11), protein (Table S12), milk (Table S13) and fat (Table S14) yield across breeds.

Additional file 9: Figure S4. Relationship between bias of genomic predictions and changes in prediction accuracy for four traits in across-breed prediction. Each point represents one of the 145 genomic features. The y axis is the absolute values of (1-bias (b)) for GFBLUP, and the x axis is the changes in prediction accuracy with GFBLUP relative to GBLUP.

Additional file 10: Table S15. The 34 up-regulated genes associated with protein yield. The data represent the detailed information of the 34 highly up-regulated genes detected in the liver at 48h post intra-mammary infection (IMI).

Authors' contributions

LF conceived the study, designed, performed and evaluated the experiments, analyzed the data and drafted the manuscript. PM contributed to the study design and analyzed the data. GSa, GSu, YY, SZ and ML contributed to the study design and drafted the manuscript. PS conceived and designed the study, and drafted the manuscript. All authors read and approved the final manuscript.

Author details

[1] Department of Molecular Biology and Genetics, Center for Quantitative Genetics and Genomics, Aarhus University, 8830 Tjele, Denmark. [2] Key Laboratory of Animal Genetics, Breeding and Reproduction, Ministry of Agriculture and National Engineering Laboratory for Animal Breeding, College of Animal Science and Technology, China Agricultural University, Beijing 100193, China.

Competing interests

The authors declare that they have no competing interests.

Funding

This study was funded by the Danish Strategic Research Council (GenSAP: Centre for Genomic Selection in Animals and Plants, Contract 12-132452).

References

1. Wang WY, Barratt BJ, Clayton DG, Todd JA. Genome-wide association studies: theoretical and practical concerns. Nat Rev Genet. 2005;6:109–18.
2. Lango Allen H, Estrada K, Lettre G, Berndt SI, Weedon MN, Rivadeneira F, et al. Hundreds of variants clustered in genomic loci and biological pathways affect human height. Nature. 2010;467:832–8.
3. Yang J, Manolio TA, Pasquale LR, Boerwinkle E, Caporaso N, Cunningham JM, et al. Genome partitioning of genetic variation for complex traits using common SNPs. Nat Genet. 2011;43:519–25.
4. Kemper KE, Goddard ME. Understanding and predicting complex traits: knowledge from cattle. Hum Mol Genet. 2012;21:R45–51.
5. Visscher PM, Brown MA, McCarthy MI, Yang J. Five years of GWAS discovery. Am J Hum Genet. 2012;90:7–24.
6. Hayes BJ, Pryce J, Chamberlain AJ, Bowman PJ, Goddard ME. Genetic architecture of complex traits and accuracy of genomic prediction: coat colour, milk-fat percentage, and type in Holstein cattle as contrasting model traits. PLoS Genet. 2010;6:e1001139.
7. Snelling WM, Cushman RA, Keele JW, Maltecca C, Thomas MG, Fortes MR, et al. Breeding and genetics symposium: networks and pathways to guide genomic selection. J Anim Sci. 2013;91:537–52.
8. Edwards SM, Sørensen IF, Sarup P, Mackay TF, Sørensen P. Genomic prediction for quantitative traits is improved by mapping variants to gene ontology categories in Drosophila melanogaster. Genetics. 2016;203:1871–83.
9. Sarup P, Jensen J, Ostersen T, Henryon M, Sørensen P. Increased prediction accuracy using a genomic feature model including prior information on quantitative trait locus regions in purebred Danish Duroc pigs. BMC Genet. 2016;17:11.
10. Meuwissen THE, Hayes BJ, Goddard ME. Prediction of total genetic value using genome-wide dense marker maps. Genetics. 2001;157:1819–29.
11. Maurano MT, Humbert R, Rynes E, Thurman RE, Haugen E, Wang H, et al. Systematic localization of common disease-associated variation in regulatory DNA. Science. 2012;337:1190–5.
12. O'Roak BJ, Vives L, Girirajan S, Karakoc E, Krumm N, Coe BP, et al. Sporadic autism exomes reveal a highly interconnected protein network of de novo mutations. Nature. 2012;485:246–50.
13. Wang K, Li M, Hakonarson H. Analysing biological pathways in genome-wide association studies. Nat Rev Genet. 2010;11:843–54.
14. Khatri P, Sirota M, Butte AJ. Ten years of pathway analysis: current approaches and outstanding challenges. PLoS Comput Biol. 2012;8:e1002375.
15. Ramanan VK, Shen L, Moore JH, Saykin AJ. Pathway analysis of genomic data: concepts, methods, and prospects for future development. Trends Genet. 2012;28:323–32.
16. Sedeño-Cortés AE, Pavlidis P. Pitfalls in the application of gene-set analysis to genetics studies. Trends Genet. 2014;30:513–4.
17. Edwards SM, Thomsen B, Madsen P, Sorensen P. Partitioning of genomic variance reveals biological pathways associated with udder health and milk production traits in dairy cattle. Genet Sel Evol. 2015;47:60.
18. Mitterhuemer S, Petzl W, Krebs S, Mehne D, Klanner A, Wolf E, et al. Escherichia coli infection induces distinct local and systemic transcriptome responses in the mammary gland. BMC Genomics. 2010;11:138.
19. Buitenhuis B, Rontved CM, Edwards SM, Ingvartsen KL, Sorensen P. In depth analysis of genes and pathways of the mammary gland involved in the pathogenesis of bovine Escherichia coli-mastitis. BMC Genomics. 2011;12:130.
20. Costa V, Aprile M, Esposito R, Ciccodicola A. RNA-Seq and human complex diseases: recent accomplishments and future perspectives. Eur J Hum Genet. 2013;21:134–42.
21. Aitken SL, Corl CM, Sordillo LM. Immunopathology of mastitis: insights into disease recognition and resolution. J Mammary Gland Biol Neoplasia. 2011;16:291–304.
22. Wellnitz O, Bruckmaier RM. The innate immune response of the bovine mammary gland to bacterial infection. Vet J. 2012;192:148–52.
23. Jiang L, Sørensen P, Røntved C, Vels L, Ingvartsen KL. Gene expression profiling of liver from dairy cows treated intra-mammary with lipopolysaccharide. BMC Genomics. 2008;9:443.
24. Pisoni G, Moroni P, Genini S, Stella A, Boettcher PJ, Cremonesi P, et al. Differentially expressed genes associated with Staphylococcus aureus mastitis in dairy goats. Vet Immunol Immunopathol. 2010;135:208–17.
25. Loor JJ, Moyes KM, Bionaz M. Functional adaptations of the transcriptome to mastitis-causing pathogens: the mammary gland and beyond. J Mammary Gland Biol Neoplasia. 2011;16:305–22.
26. Jiang L, Sorensen P, Thomsen B, Edwards SM, Skarman A, Rontved CM, et al. Gene prioritization for livestock diseases by data integration. Physiol Genomics. 2012;44:305–17.
27. Hotamisligil GS. Inflammation and metabolic disorders. Nature. 2006;444:860–7.
28. Huang W, Richards S, Carbone MA, Zhu D, Anholt RR, Ayroles JF, et al. Epistasis dominates the genetic architecture of Drosophila quantitative traits. Proc Nat Acad Sci USA. 2012;109:15553–9.
29. Vels L, Rontved CM, Bjerring M, Ingvartsen KL. Cytokine and acute phase protein gene expression in repeated liver biopsies of dairy cows with a lipopolysaccharide-induced mastitis. J Dairy Sci. 2009;92:922–34.
30. Andersen JB, Mashek DG, Larsen T, Nielsen MO, Ingvartsen KL. Effects of hyperinsulinaemia under euglycaemic condition on liver fat metabolism in dairy cows in early and mid-lactation. J Vet Med Physiol Pathol Clin Med. 2002;49:65–71.
31. Moyes KM, Sørensen P, Bionaz M. The impact of intramammary Escherichia coli challenge on liver and mammary transcriptome and cross-talk in dairy cows during early lactation using RNAseq. PLoS One. 2016;11.e0157400.
32. Liao Y, Smyth GK, Shi W. The Subread aligner: fast, accurate and scalable read mapping by seed-and-vote. Nucleic Acids Res. 2013;41:e108.
33. Robinson MD, McCarthy DJ, Smyth GK. edgeR: a Bioconductor package for differential expression analysis of digital gene expression data. Bioinformatics. 2010;26:139–40.
34. Gao H, Christensen OF, Madsen P, Nielsen US, Zhang Y, Lund MS, et al. Comparison on genomic predictions using three GBLUP methods and two single-step blending methods in the Nordic Holstein population. Genet Sel Evol. 2012;44:8.
35. Thomasen JR, Guldbrandtsen B, Su G, Brøndum RF, Lund MS. Reliabilities of genomic estimated breeding values in Danish Jersey. Animal. 2012;6:789–96.
36. Brondum RF, Guldbrandtsen B, Sahana G, Lund MS, Su G. Strategies for imputation to whole genome sequence using a single or multi-breed reference population in cattle. BMC Genomics. 2014;15:728.
37. Wu X, Guldbrandtsen B, Lund MS, Sahana G. Association analysis for feet and legs disorders with whole-genome sequence variants in 3 dairy cattle breeds. J Dairy Sci. 2016;99:7221–31.
38. Daetwyler HD, Capitan A, Pausch H, Stothard P, Van Binsbergen R, Brøndum RF, et al. Whole-genome sequencing of 234 bulls facilitates mapping of monogenic and complex traits in cattle. Nat Genet. 2014;46:858–65.
39. Höglund JK, Sahana G, Brøndum RF, Guldbrandtsen B, Buitenhuis B, Lund MS. Fine mapping QTL for female fertility on BTA04 and BTA13 in dairy cattle using HD SNP and sequence data. BMC Genomics. 2014;15:790.
40. Fuchsberger C, Abecasis GR, Hinds DA. minimac2: faster genotype imputation. Bioinformatics. 2015;31:782–4.
41. VanRaden PM. Efficient methods to compute genomic predictions. J Dairy Sci. 2008;91:4414–23.

42. Johnson DL, Thompson R. Restricted maximum likelihood estimation of variance components for univariate animal models using sparse matrix techniques and average information. J Dairy Sci. 1995;78:449–56.

43. Madsen P, Jensen J, Labouriau R, Christensen OF, Sahana G. DMU-A package for analyzing multivariate mixed models in quantitative genetics and genomics. In: Proceedings of the 10th world congress of genetics applied to livestock production: 18–22 August 2014; Vancouver; 2014. https://asas.org/docs/default-source/wcgalp-posters/699_paper_9580_manuscript_758_0.pdf?sfvrsn=2.

44. Kang HM, Sul JH, Service SK, Zaitlen NA, Kong SY, Freimer NB, et al. Variance component model to account for sample structure in genome-wide association studies. Nat Genet. 2010;42:348–54.

45. Newton MA, Quintana FA, Den Boon JA, Sengupta S, Ahlquist P. Random-set methods identify distinct aspects of the enrichment signal in gene-set analysis. Ann Appl Stat. 2007;1:85–106.

46. Rohde PD, Demontis D, Cuyabano BCD, Børglum AD, Sørensen P, The GEMS Group. Covariance association test (CVAT) identify genetic markers associated with schizophrenia in functionally associated biological processes. Genetics. 2016;203:1901–13.

47. Xie C, Mao X, Huang J, Ding Y, Wu J, Dong S, et al. KOBAS 2.0: a web server for annotation and identification of enriched pathways and diseases. Nucleic Acids Res. 2011;39:W316–22.

48. Benjamini Y, Hochberg Y. Controlling the false discovery rate: a practical and powerful approach to multiple testing. J R Statist Soc Series B Stat Methodol. 1995;57:289–300.

49. Pimentel Eda G, Erbe M, König S, Simianer H. Genome partitioning of genetic variation for milk production and composition traits in Holstein cattle. Front Genet. 2011;2:19.

50. Erbe M, Hayes B, Matukumalli LK, Goswami S, Bowman PJ, Reich CM, et al. Improving accuracy of genomic predictions within and between dairy cattle breeds with imputed high-density single nucleotide polymorphism panels. J Dairy Sci. 2012;95:4114–29.

51. Jensen J, Su G, Madsen P. Partitioning additive genetic variance into genomic and remaining polygenic components for complex traits in dairy cattle. BMC Genet. 2012;13:44.

52. MacLeod IM, Bowman PJ, Vander Jagt CJ, Haile-Mariam M, Kemper KE, Chamberlain AJ, et al. Exploiting biological priors and sequence variants enhances QTL discovery and genomic prediction of complex traits. BMC Genomics. 2016;17:144.

53. Tiezzi F, Parker-Gaddis KL, Cole JB, Clay JS, Maltecca C. A genome-wide association study for clinical mastitis in first parity US Holstein cows using single-step approach and genomic matrix re-weighting procedure. PLoS One. 2015;10:e0114919.

54. Gao B, Jeong WI, Tian Z. Liver: an organ with predominant innate immunity. Hepatology. 2008;47:729–36.

55. Bechmann LP, Hannivoort RA, Gerken G, Hotamisligil GS, Trauner M, Canbay A. The interaction of hepatic lipid and glucose metabolism in liver diseases. J Hepatol. 2012;56:952–64.

56. Hayes BJ, Bowman PJ, Chamberlain AC, Verbyla K, Goddard ME. Accuracy of genomic breeding values in multi-breed dairy cattle populations. Genet Sel Evol. 2009;41:51.

57. Gusev A, Lee SH, Trynka G, Finucane H, Vilhjálmsson BJ, Xu H, et al. Partitioning heritability of regulatory and cell-type-specific variants across 11 common diseases. Am J Hum Genet. 2014;95:535–52.

58. Speed D, Balding DJ. MultiBLUP: improved SNP-based prediction for complex traits. Genome Res. 2014;24:1550–7.

59. Ober U, Ayroles JF, Stone EA, Richards S, Zhu D, Gibbs RA, et al. Using whole-genome sequence data to predict quantitative trait phenotypes in Drosophila melanogaster. PLoS Genet. 2012;8:e1002685.

60. Fridley BL, Biernacka JM. Gene set analysis of SNP data: benefits, challenges, and future directions. Eur J Hum Genet. 2011;19:837–43.

61. Holmans P, Green EK, Pahwa JS, Ferreira MA, Purcell SM, Sklar P, et al. Gene ontology analysis of GWA study data sets provides insights into the biology of bipolar disorder. Am J Hum Genet. 2009;85:13–24.

62. Medina I, Montaner D, Bonifaci N, Pujana MA, Carbonell J, Tarraga J, et al. Gene set-based analysis of polymorphisms: finding pathways or biological processes associated to traits in genome-wide association studies. Nucleic Acids Res. 2009;37:W340–4.

63. O'Dushlaine C, Kenny E, Heron EA, Segurado R, Gill M, Morris DW, et al. The SNP ratio test: pathway analysis of genome-wide association datasets. Bioinformatics. 2009;25:2762–3.

3

Selection of performance-tested young bulls and indirect responses in commercial beef cattle herds on pasture and in feedlots

Fernanda S. S. Raidan[1,2], Dalinne C. C. Santos[1], Mariana M. Moraes[1], Andresa E. M. Araújo[1], Henrique T. Ventura[3], José A. G. Bergmann[1], Eduardo M. Turra[1] and Fabio L. B. Toral[1]* ⓘ

Abstract

Background: Central testing is used to select young bulls which are likely to contribute to increased net income of the commercial beef cattle herd. We present genetic parameters for growth and reproductive traits on performance-tested young bulls and commercial animals that are raised on pasture and in feedlots.

Methods: Records on young bulls and heifers in performance tests or commercial herds were used. Genetic parameters for growth and reproductive traits were estimated. Correlated responses for commercial animals when selection was applied on performance-tested young bulls were computed.

Results: The 90% highest posterior density (HPD90) intervals for heritabilities of final weight (FW), average daily gain (ADG) and scrotal circumference (SC) ranged from 0.41 to 0.49, 0.23 to 0.30 and 0.47 to 0.57, respectively, for perfor-mance-tested young bulls on pasture, from 0.45 to 0.60, 0.20 to 0.32 and 0.56 to 0.70, respectively, for performance-tested young bulls in feedlots, from 0.29 to 0.33, 0.14 to 0.18 and 0.35 to 0.45, respectively, for commercial animals on pasture, and from 0.24 to 0.44, 0.13 to 0.24 and 0.35 to 0.57 respectively, for commercial animals in feedlots. The HPD90 intervals for genetic correlations of FW, ADG and SC in performance-tested young bulls on pasture (feedlots) with FW, ADG and SC in commercial animals on pasture (feedlots) ranged from 0.86 to 0.96 (0.83 to 0.94), 0.78 to 0.90 (0.40 to 0.79) and from 0.92 to 0.97 (0.50 to 0.83), respectively. Age at first calving was genetically related to ADG (HPD90 interval = −0.48 to −0.06) and SC (HPD90 interval = −0.41 to −0.05) for performance-tested young bulls on pasture, however it was not related to ADG (HPD90 interval = −0.29 to 0.10) and SC (HPD90 interval = −0.35 to 0.13) for performance-tested young bulls in feedlots.

Conclusions: Heritabilities for growth and SC are higher for performance-tested young bulls than for commercial animals. Evaluating and selecting for increased growth and SC on performance-tested young bulls is efficient to improve growth, SC and age at first calving in commercial animals. Evaluating and selecting performance-tested young bulls is more efficient for young bulls on pasture than in feedlots.

Background

Central testing of beef cattle is used quite widely world-wide since the 1950s, especially in the United States and Canada [1], Europe [2] and Brazil [3]. The aim of central testing is to identify young bulls as parents of the next generation which are likely to contribute to increased net income of commercial herds. Young bulls need to be raised under uniform housing, feeding, management and data recording to more accurately estimate the genetic merit of each animal. Growth, carcass, feed efficiency and scrotal circumference are measured during the test or at the end of the test [4–8]. Performance tests can be conducted on pasture or in feedlots. Feeding costs (per day and total cost) for testing young bulls are lower on

*Correspondence: flbtoral@ufmg.br
[1] Departamento de Zootecnia, Escola de Veterinária, Universidade Federal de Minas Gerais, Belo Horizonte, MG 31270-901, Brazil
Full list of author information is available at the end of the article

pasture than in feedlots. However, pasture performance tests take longer than feedlot performance tests [6, 8–12].

After individual testing, outstanding young bulls can be used for breeding, either with or without progeny test, or sold to cow-calf producers. Therefore, the impact of selection for improved economic traits in performance-tested young bulls on growth and reproductive traits of young bulls and heifers in commercial herds is of particular importance. The genetic correlations (±standard error) of average daily gain and mid-test body weight of performance-tested young bulls in feedlots with post-weaning weight (12 to 36 months of age) of commercial animals on pasture present moderate magnitude (0.33 ± 0.15 and 0.56 ± 0.14, respectively) [4]. However, genetic correlations of growth in performance-tested young bulls in feedlots with age at first calving in commercial herds are weak (0.21 ± 0.15 and −0.18 ± 0.13, respectively) [5]. Furthermore, genetic correlations between growth and reproductive traits in performance test and commercial herds both on pasture or in feedlot are unknown. Availability of such data would be useful to evaluate the efficiency of selection in performance tests for the improvement of economic traits in commercial herds and to determine the best environment to carry out performance tests of young bulls. Thus, our aim was to estimate genetic parameters for growth and reproductive traits in performance-tested young bulls and commercial young bulls and heifers on pasture and in feedlots. In addition, we analyzed the impact of selecting performance-tested young bulls for growth and scrotal circumference on growth and reproductive traits in young bulls and heifers in commercial herds, both on pasture and in feedlots.

Methods

Data

Approval by the ethics committee was not necessary for this study because the data were obtained from an existing database. We used records from official performance tests on growth traits and scrotal circumference (SC) of Nellore young bulls on pasture and in feedlots and records from a joint official performance recording scheme on growth and reproductive traits (SC and age at first calving, AFC) of young bulls and heifers. Performance records and pedigree information were provided by Associação Brasileira de Criadores de Zebu (ABCZ).

The performance of 33,013 animals was evaluated in 751 performance tests that were carried out from 2003 to 2012 in the North (Acre, Rondônia, Pará, and Tocantins), Northeast (Bahia and Maranhão), Central West (Goiás, Mato Grosso and Mato Grosso do Sul), Southeast (Espírito Santo, Minas Gerais and São Paulo) and South (Paraná and Rio Grande do Sul) regions of Brazil. Our study included 24,910 animals from 538 tests that were

conducted on pasture and 8103 animals from 213 tests that were conducted in feedlots. Pasture tests lasted 294 days (70 days for adaptation and 224 days for testing) and feedlot tests lasted 168 days (56 days for adaptation and 112 days for testing). Animals were weighed at the beginning and end of the adaptation period and at the end of the testing period. The assessed traits included final weight (FW), average daily gain (ADG) and SC. ADG was calculated as the difference between body weight at the end of the testing period (WEndT) and body weight at the end of the adaptation period (WEndA), divided by the difference between age at the end of the testing period and age at the end of the adaptation period (AEndA). FW was calculated using the following equations FW = WEndA + [ADG × (550 − AEndA)] and FW = WEndA + [ADG × (426 − AEndA)] for performance-tested young bulls on pasture and in feedlots, respectively. The values 550 and 426 are the official standard final ages (in days) according to ABCZ. Individual records for each trait that exceeded the intervals given by the means of the performance tests plus or minus 3.5 standard deviations were excluded, and growth and SC records of animals from performance test on pasture and in feedlots that included less than 20 and 8 animals, respectively, were also excluded.

Performance records of commercial young bulls and heifers were from the official performance recording scheme of ABCZ for commercial purebred herds in Central West (Goiás, Mato Grosso and Mato Grosso do Sul) and Southeast (Minas Gerais and São Paulo) regions of Brazil. These records were collected from 2005 to 2010. The animals were weighed at weaning (from 145 to 265 days of age, mean age of 205 days) and at yearling (from 490 to 610 days of age, mean age of 550 days). The assessed traits included FW and ADG of young bulls and heifers, SC of young bulls, both on pasture and in feedlots, and AFC of heifers on pasture. ADG was calculated as the difference between body weight at yearling (YW) and body weight at weaning (WW), divided by the difference between age at yearling and age at weaning (AW). FW was calculated as follows: FW = WW + [ADG + (550 − AW)]. Individual records for each trait that exceeded the intervals given by the means of contemporary groups plus or minus 4 standard deviations were excluded, and animals from contemporary groups that included less than 10 animals were also excluded. Contemporary groups included animals from the same herd, year and month of birth, sex, and feeding regimen at weaning and yearling (on pasture with or without mineral supplementation or in feedlots). The levels of energy and/or protein supplementation were not available in the dataset, and the feeding regimen at yearling of animals that were fed with any type of energy and/or protein supplementation was considered as a feedlot. A total of 84,565 animals (from 4148 contemporary groups on pasture) and 4468 animals (from 266 contemporary groups in feedlots) were used in this study. Records on AFC were from heifers with growth records (FW and ADG) in the dataset, which originated from 540 contemporary groups on pasture. Heifers with AFC records represented 17.7% of the heifers with growth records. Summary statistics of these data are in Table 1 and the distributions of animals and sires across geographical regions are in Table 2.

The numerator relationship matrix considered pedigree data on 122,046 animals with records and ancestors of recorded animals, which resulted in 377,217 animals. The mean, minimum and maximum numbers of known generations for animals with at least one available record were 6.4, 1.5 and 8.9, respectively. The environmental connectedness through the use of common sires is shown in Fig. 1.

Statistical analyses

Samples of the posterior distributions of the genetic parameters were obtained using a Bayesian approach and Gibbs sampler in multiple-trait analyses. The following general statistical model was used:

$$y_{hijk} = u_h + CG_{hj} + b_{h_{(j)}}\left(A_k - \overline{A_j}\right) + a_{hi} + e_{hijk},$$

where y_{hijk} is the observation for trait h on animal i in performance test (or contemporary group) j with final age k; u_h is the general constant present in each observation for trait h; CG_{hj} is the effect of performance test (or contemporary group) j for trait h; $b_{h_{(j)}}$ is the linear regression coefficient of final age for trait h, nested in the performance test (or contemporary group) j; A_k is the age k; $\overline{A_j}$ is the mean of the final ages of the animals from the contemporary group j; a_{hi} is the breeding value of animal i for trait h; and e_{hijk} is the residual effect for each observation. The effect of age was not included for AFC.

In matrix notation, the following general model was used in multiple-trait analyses:

$$
\begin{bmatrix} \underset{\sim}{\mathbf{y}_1} \\ \underset{\sim}{\mathbf{y}_2} \\ \vdots \\ \underset{\sim}{\mathbf{y}_8} \end{bmatrix}
=
\begin{bmatrix} \mathbf{X}_1 & \boldsymbol{\Phi} & \cdots & \boldsymbol{\Phi} \\ \boldsymbol{\Phi} & \mathbf{X}_2 & \cdots & \boldsymbol{\Phi} \\ \vdots & \vdots & \ddots & \vdots \\ \boldsymbol{\Phi} & \boldsymbol{\Phi} & \cdots & \mathbf{X}_8 \end{bmatrix}
\begin{bmatrix} \underset{\sim}{\boldsymbol{\beta}_1} \\ \underset{\sim}{\boldsymbol{\beta}_2} \\ \vdots \\ \underset{\sim}{\boldsymbol{\beta}_8} \end{bmatrix}
$$

$$
+
\begin{bmatrix} \mathbf{Z}_1 & \boldsymbol{\Phi} & \cdots & \boldsymbol{\Phi} \\ \boldsymbol{\Phi} & \mathbf{Z}_2 & \cdots & \boldsymbol{\Phi} \\ \vdots & \vdots & \ddots & \vdots \\ \boldsymbol{\Phi} & \boldsymbol{\Phi} & \cdots & \mathbf{Z}_8 \end{bmatrix}
\begin{bmatrix} \underset{\sim}{\mathbf{a}_1} \\ \underset{\sim}{\mathbf{a}_2} \\ \vdots \\ \underset{\sim}{\mathbf{a}_8} \end{bmatrix}
+
\begin{bmatrix} \underset{\sim}{\mathbf{e}_1} \\ \underset{\sim}{\mathbf{e}_2} \\ \vdots \\ \underset{\sim}{\mathbf{e}_8} \end{bmatrix},
$$

where $\mathbf{y}_h$ is the vector of records for trait h, $\mathbf{X}_h$ is the incidence matrix of fixed effects; $\boldsymbol{\beta}_h$ is the vector of fixed effects, $\mathbf{Z}_h$ is the incidence matrix of random effects; $\mathbf{a}_h$ is

Table 1 Summary statistics for growth and reproductive traits in performance-tested and commercial young bulls and heifers on pasture and in feedlots

Trait	N	Mean	SD	CV (%)
Performance test on pasture				
Final age (days)[a]	24,910	553.05	24.39	4.41
Final age (days)[b]	14,888	552.72	25.24	4.57
FW (kg)	24,910	350.35	53.09	15.15
ADG (kg/day)	24,910	0.54	0.16	29.63
SC (cm)	14,888	26.61	3.38	12.70
Commercial on pasture				
Final age (days)[a]	84,565	549.46	24.30	4.42
Final age (days)[b]	14,663	548.35	24.39	4.45
FW (kg)	84,565	312.54	58.05	18.57
ADG (kg/day)	84,565	0.36	0.14	38.89
SC (cm)	14,663	25.91	3.67	14.16
AFC (days)	8060	1164.83	180.52	15.50
Performance test in feedlots				
Final age (days)[a]	8103	423.59	26.41	6.23
Final age (days)[b]	4676	420.73	28.01	6.66
FW (kg)	8103	371.65	57.13	15.37
ADG (kg/day)	8103	0.83	0.27	32.53
SC (cm)	4676	25.41	3.31	13.03
Commercial in feedlots				
Final age (days)[a]	4468	549.62	24.17	4.40
Final age (days)[b]	1365	548.59	24.16	4.40
FW (kg)	4468	389.41	71.41	18.34
ADG (kg/day)	4468	0.54	0.18	33.33
SC (cm)	1365	28.46	3.95	13.88

N number of records, *SD* standard deviation, *CV* coefficient of variation (in %), *FW* final weight, *ADG* average daily gain, *SC* scrotal circumference, *AFC* age at first calving

[a] Only for animals with FW and ADG data

[b] Only for animals with SC data

the vector of breeding values for trait h and $\mathbf{e}_h$ is the vector of residuals for trait h. $\boldsymbol{\Phi}$ is the symbol for an empty matrix. The indexes h are as follows: FW, ADG and SC in performance-tested animals on pasture or in feedlots were defined as trait 1, FW, ADG, SC, AFC in commercial animals on pasture were defined as traits 2, 3, 4 and 5, respectively, and FW, ADG and SC in commercial animals in feedlots were defined as traits 6, 7 and 8, respectively. Thereby, six multiple-trait analyses were carried out.

Flat prior distributions were assumed for fixed effects $\left(\begin{bmatrix} \underset{\sim}{\boldsymbol{\beta}_1} & \underset{\sim}{\boldsymbol{\beta}_2} & \cdots & \underset{\sim}{\boldsymbol{\beta}_8} \end{bmatrix}^t\right)$, and normal distributions were assumed for random effects $\left(\begin{bmatrix} \underset{\sim}{\mathbf{a}_1} & \underset{\sim}{\mathbf{a}_2} & \cdots & \underset{\sim}{\mathbf{a}_8} \end{bmatrix}^t \middle| \mathbf{G}\right)$ and $\left(\begin{bmatrix} \underset{\sim}{\mathbf{e}_1} & \underset{\sim}{\mathbf{e}_2} & \cdots & \underset{\sim}{\mathbf{e}_8} \end{bmatrix}^t \middle| \mathbf{R}\right)$, whereas inverted Wishart distributions were assumed for (co)variance

Table 2 Distribution of animals and sires across geographical regions

Trait	Animals					Sires					
	NO	NE	CW	SE	SO	NO	NE	CW	SE	SO	Total
Performance tests on pasture											
Growth	4874	1317	7816	9769	1134	672	288	903	901	120	2047
SC	3243	1094	4581	5413	557	480	236	571	579	72	1347
Commercial on pasture											
Growth	–	–	46,878	37,687	–	–	–	2136	1423	–	3021
SC	–	–	8090	6573	–	–	–	958	578	–	1313
AFC	–	–	4456	753	–	–	–	3604	510	–	1053
Performance tests in feedlots											
Growth	69	–	4307	3051	676	20	–	463	303	80	688
SC	69	–	3281	1288	38	20	–	369	170	10	469
Commercial in feedlots											
Growth	–	–	2458	2010	–	–	–	325	308	–	527
SC	–	–	760	605	–	–	–	146	133	–	227

NO north, *NE* northeast, *CW* central west, *SE* southeast, *SO* south, *Growth* includes final weight and average daily gain, *SC* scrotal circumference, *AFC* age at first calving

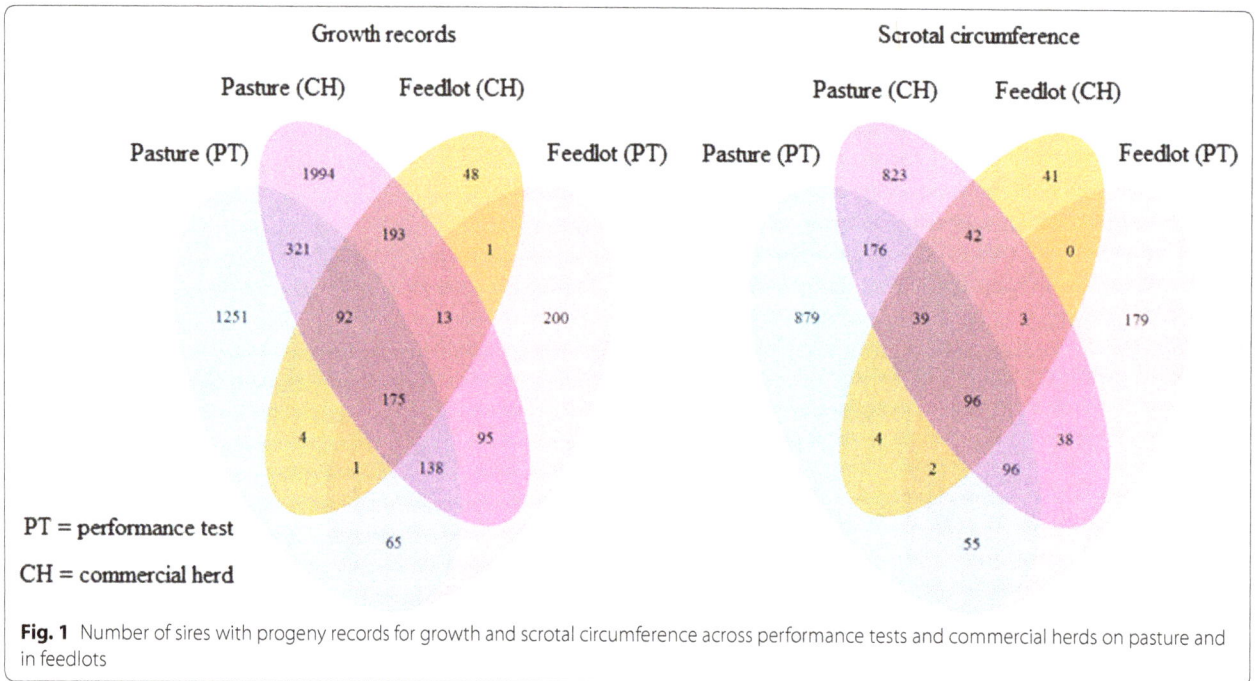

Fig. 1 Number of sires with progeny records for growth and scrotal circumference across performance tests and commercial herds on pasture and in feedlots

matrices $(\mathbf{G_0}|\mathbf{v_a}, \mathbf{S_a})$ and $(\mathbf{R}|\mathbf{v_e}, \mathbf{S_e})$, where $\mathbf{G} = \mathbf{G_0} \otimes \mathbf{A}$ represents the genetic (co)variance matrix and

$$\mathbf{G_0} = \begin{bmatrix} \sigma_{a_1}^2 & \sigma_{a_1a_2} & \cdots & \sigma_{a_1a_8} \\ \sigma_{a_1a_2} & \sigma_{a_2}^2 & \cdots & \sigma_{a_2a_8} \\ \vdots & \vdots & \ddots & \vdots \\ \sigma_{a_1a_8} & \sigma_{a_2a_8} & \cdots & \sigma_{a_8}^2 \end{bmatrix}$$ represents the matrix of

the genetic (co)variances between traits 1 to 8; σ_{ah}^2 represents the additive genetic variance for trait h; $\sigma_{a_h a_{h'}}$ represents the additive genetic covariance between traits h and h'; $\mathbf{R} = \mathbf{R_0} \otimes \mathbf{A}$ represents the residual variance matrix;

$$\mathbf{R_0} = \begin{bmatrix} \sigma_{e_1}^2 & 0 & 0 & 0 & 0 & 0 & 0 & 0 \\ 0 & \sigma_{e_2}^2 & \sigma_{e_2e_3} & \sigma_{e_2e_4} & \sigma_{e_2e_5} & 0 & 0 & 0 \\ 0 & \sigma_{e_2e_3} & \sigma_{e_3}^2 & \sigma_{e_3e_4} & \sigma_{e_3e_5} & 0 & 0 & 0 \\ 0 & \sigma_{e_2e_4} & \sigma_{e_3e_4} & \sigma_{e_4}^2 & 0 & 0 & 0 & 0 \\ 0 & \sigma_{e_2e_5} & \sigma_{e_3e_5} & 0 & \sigma_{e_5}^2 & 0 & 0 & 0 \\ 0 & 0 & 0 & 0 & 0 & \sigma_{e_6}^2 & \sigma_{e_6e_7} & \sigma_{e_6e_8} \\ 0 & 0 & 0 & 0 & 0 & \sigma_{e_6e_7} & \sigma_{e_7}^2 & \sigma_{e_7e_8} \\ 0 & 0 & 0 & 0 & 0 & \sigma_{e_6e_8} & \sigma_{e_7e_8} & \sigma_{e_8}^2 \end{bmatrix}$$

represents the matrix of residual variance of traits 1 to 8; $\sigma^2_{e_h}$ is the residual variance for trait h; $\sigma_{e_h e_{h'}}$ is the residual covariance between traits h and h'; $\mathbf{v_a}$ and $\mathbf{v_e}$ (degrees of freedom of the inverted Wishart distributions) and $\mathbf{S_a}$ and $\mathbf{S_e}$ (8×8 matrices of (co)variance components obtained from preliminary analyses) are the hyper-parameters of the inverted Wishart distributions of genetic and residual (co)variances; and the other terms are the same as those described above. The complete conditional posterior distributions are in Sorensen and Gianola [13].

Gibbs chains of 410,000 iterations were generated for each parameter, with a burn-in period of 10,000 iterations and a sampling interval of 200 iterations in the GIBBS1F90 program [14]. Gibbs chain size, burn-in period and sampling interval were those determined in previous analyses. The genetic and residual variances for FW, ADG, SC and AFC of commercial animals on pasture and FW, ADG and SC of commercial animals in feedlots that are shown in this paper were the means of 12,000 samples obtained in six multiple-trait analyses. Convergence diagnostics were performed by following Geweke's [15] and Heidelberger and Welch's [16] techniques and a visual analysis of the trace plots was performed by using the Bayesian Output Analysis [17] program in R software 3.2.3 [18].

Samples of posterior distributions for efficiency of correlated response (ECR), considering the same intensity of selection for traits in performance-tested and commercial animals, were obtained by the following equation available in Falconer and Mackay [19]:

$$\mathrm{ECR}_{hh'} = \frac{\Delta \mathbf{G}_{hh'}}{\Delta \mathbf{G}_h} = r_{a_{hh'}} \frac{h_{h'}}{h_h},$$

where $\Delta \mathbf{G}_{hh'}$ is the expected genetic gain per generation for trait h in commercial animals when selection was applied for trait h' in performance-tested animals; $\Delta \mathbf{G}_h$ is the expected genetic gain per generation for trait h in commercial animals; h' is the trait under selection in performance-tested animals; h is the indirectly selected trait in commercial animals; $r_{a_{hh'}}$ is the genetic correlation between traits h and h'; and $h_{h'}$ and h_h are the square roots of the heritabilities for traits h' and h, respectively.

In addition to the analyses previously described, two multiple-trait analyses were performed in which FW or ADG of performance-tested animals on pasture were defined as trait 1, FW and ADG of male commercial animals on pasture were defined as traits 2 and 3, respectively, and FW, ADG and AFC of female commercial animals on pasture were defined as traits 4, 5 and 6, respectively. These analyses were performed to estimate the genetic correlations for the same trait between young bulls and heifers. Furthermore, we carried out another two analyses for the same traits measured on

performance-tested and commercial animals in feedlots. A single-trait analysis for AFC was run to compare the results from single and multiple-trait analyses for this trait.

Results

Genetic variation of growth and reproductive traits

Posterior means and the 90% highest posterior density (HPD90) intervals of the variances and heritabilities for growth and reproductive traits in performance-tested and commercial young bulls and heifers are in Table 3. The posterior means of the additive genetic variances for FW and ADG were higher for performance-tested young bulls than for commercial animals on pasture or in feedlots (Table 3).

The posterior means of the additive genetic variance for SC were higher for performance-tested young bulls on pasture than for commercial animals on pasture. However, the additive genetic variances for SC were similar between performance-tested young bulls in feedlots and animals in commercial herds in feedlots (Table 3), because of overlapping HDP90 intervals. In addition, residual variances for FW and SC were smaller for performance-tested young bulls than for commercial animals, and the posterior mean of the residual variance for ADG was higher for performance-tested animals in feedlots than for commercial animals in feedlots (Table 3). Estimated heritabilities were higher for traits in performance-tested young bulls than in commercial animals (Table 3).

The posterior means of the additive genetic and residual variances for FW and ADG were higher for males than for females in commercial herds on pasture and estimated residual variances for FW and ADG were higher for males than for females in commercial herds in feedlots (Table 3). Estimated heritabilities for FW and ADG were similar between males and females in commercial herds on pasture, those for FW were higher for females than for males in commercial herds in feedlots, but with overlapping HDP90 intervals and those for ADG were similar between males and females in commercial herds in feedlots (Table 3).

The additive genetic variance and heritability for AFC were lower in the single-trait than in the multiple-trait analyses (Table 3).

Genetic correlations between male and female traits

Posterior means (and the lower and upper limits of the HDP90 intervals between brackets) of the genetic correlations between male and female FW and ADG in commercial herds on pasture were equal to 0.96 (0.94; 0.98) and 0.75 (0.58; 0.88), respectively. Genetic correlations between male and female FW and ADG in commercial

Table 3 Variance components for growth and reproductive traits in performance-tested and commercial young bulls and heifers on pasture and in feedlots

Trait	σ_a^2	σ_e^2	h^2
Performance test on pasture			
FW	421.03 (380.00; 461.80)	514.38 (487.00; 547.60)	0.45 (0.41; 0.49)
ADG	0.019 (0.016; 0.022)	0.053 (0.051; 0.055)	0.26 (0.23; 0.30)
SC	3.34 (2.94; 3.69)	3.05 (2.79; 3.33)	0.52 (0.47; 0.57)
Commercial on pasture			
FW	322.26 (295.70; 345.30)	721.84 (702.80; 739.90)	0.31 (0.29; 0.33)
M_FW	321.08 (281.90; 358.30)	887.12 (857.30; 916.10)	0.27 (0.24; 0.29)
F_FW	264.14 (238.10; 286.90)	604.12 (585.20; 623.30)	0.30 (0.27; 0.33)
ADG	0.010 (0.009; 0.011)	0.051 (0.050; 0.055)	0.16 (0.14; 0.18)
M_ADG	0.012 (0.011; 0.014)	0.058 (0.057; 0.060)	0.18 (0.15; 0.20)
F_ADG	0.009 (0.008; 0.010)	0.044 (0.042; 0.045)	0.17 (0.15; 0.20)
SC	2.58 (2.20; 2.91)	3.86 (3.59; 4.13)	0.40 (0.35; 0.45)
AFC	3.65 (1.93; 4.36)	15.50 (14.69; 16.91)	0.18 (0.10; 0.22)
AFC[a]	1.68 (1.20; 2.16)	16.96 (16.33; 17.57)	0.09 (0.06; 0.11)
Performance test in feedlots			
FW	756.70 (626.30; 895.80)	689.82 (590.40; 780.30)	0.52 (0.45; 0.60)
ADG	0.064 (0.048; 0.082)	0.181 (0.168; 0.195)	0.26 (0.20; 0.32)
SC	4.27 (3.64; 4.88)	2.49 (2.07; 2.97)	0.63 (0.56; 0.70)
Commercial in feedlots			
FW	426.53 (308.00; 586.90)	860.56 (749.80; 976.40)	0.33 (0.24; 0.44)
M_FW	355.59 (298.10; 432.20)	984.17 (915.50; 1060.00)	0.27 (0.22; 0.31)
F_FW	473.95 (319.40; 645.20)	687.18 (549.70; 803.20)	0.41 (0.28; 0.53)
ADG	0.015 (0.010; 0.019)	0.064 (0.060; 0.070)	0.19 (0.13; 0.24)
M_ADG	0.013 (0.008; 0.018)	0.069 (0.065; 0.075)	0.16 (0.09; 0.22)
F_ADG	0.013 (0.007; 0.018)	0.060 (0.054; 0.066)	0.17 (0.09; 0.23)
SC	3.62 (2.65; 4.63)	4.16 (3.39; 4.99)	0.46 (0.35; 0.57)

Lower and upper limits of the highest posterior density intervals with 90% of the samples are listed between brackets

Posterior means of σ_a^2 additive genetic variance, σ_e^2 residual variance, h^2 heritability, *FW* final weight, *M_FW* male FW, *F_FW* female FW, *ADG* average daily gain, *M_ADG* male ADG, *F_ADG* female ADG, *SC* scrotal circumference, *AFC* age at first calving

[a] Results from single trait analysis. Variances for AFC were multiplied by 10^{-3}

herds in feedlots were equal to 0.96 (0.93; 0.99) and 0.74 (0.63; 0.85), respectively.

Genetic correlations between performance test and commercial herds traits

Posterior means of the genetic correlations of FW, ADG and SC between performance-tested and commercial animals were positive (Table 4), which indicates that selection for either of these traits in performance-tested young bulls will result in improved growth and SC in commercial animals.

The posterior mean of the genetic correlation between FW in performance-tested young bulls on pasture and FW in commercial animals on pasture was higher than the genetic correlation between FW in performance-tested young bulls on pasture and FW in commercial animals in feedlots (Table 4). The same results were observed for ADG and SC (Table 4). These differences were not observed for genetic correlations of FW, ADG and SC between performance-tested young bulls in feedlots and commercial animals on pasture or in feedlots (Table 4).

Genetic correlations of ADG and SC in performance-tested young bulls on pasture with AFC in heifers on pasture were negative (Table 4). However, genetic correlations of FW in performance-tested young bulls on pasture and of FW, ADG and SC in performance-tested young bulls in feedlots with AFC were almost zero (Table 4). Thus, selection for ADG and SC in performance-tested young bulls on pasture will result in decreased AFC in commercial heifers but selection for FW in performance-tested young bulls on pasture or growth and SC in performance-tested young bulls in feedlots will have no effect on AFC in commercial heifers on pasture.

Efficiency of correlated responses

Table 5 presents the efficiencies of correlated responses for FW, ADG, SC and AFC in commercial animals when FW, ADG and SC were selected in performance-tested young bulls.

Correlated responses for (1) FW, ADG or SC in commercial animals on pasture when FW, ADG or SC were selected in performance-tested young bulls on pasture were similar or higher than the direct responses for FW, ADG or SC in commercial animals on pasture, respectively; (2) FW in commercial animals (on pasture or in feedlots) when FW was selected in performance-tested young bulls in feedlots were similar or higher than the direct responses for FW in commercial animals (on pasture or in feedlots); (3) SC in commercial animals on pasture when SC was selected in performance-tested young bulls in feedlots were similar to the direct response for SC in commercial animals on pasture; (4) ADG in commercial animals in feedlots when ADG was selected in performance-tested young bulls on pasture or in feedlots were similar; and (5) SC in commercial animals in feedlots when SC was selected in performance-tested young bulls on pasture or in feedlots were also similar.

Table 4 Genetic correlation between growth and reproductive traits in performance-tested young bulls on pasture and feedlot with growth and reproductive traits in commercial young bulls and heifers on pasture and in feedlots

Traits	Performance-tested young bulls on pasture			Performance-tested young bulls in feedlots		
	FW	ADG	SC	FW	ADG	SC
FW[a]	0.91 (0.86; 0.96)	0.63 (0.54; 0.78)	0.37 (0.27; 0.46)	0.87 (0.82; 0.91)	0.60 (0.47; 0.71)	0.53 (0.44; 0.63)
ADG[a]	0.69 (0.62; 0.76)	0.84 (0.78; 0.90)	0.27 (0.18; 0.37)	0.40 (0.30; 0.51)	0.39 (0.27; 0.52)	0.24 (0.11; 0.36)
SC[a]	0.32 (0.22; 0.40)	0.27 (0.16; 0.37)	0.94 (0.92; 0.97)	0.28 (0.16; 0.40)	0.17 (0.00; 0.33)	0.80 (0.73; 0.88)
AFC[a]	−0.19 (−0.38; 0.09)	−0.26 (−0.48; −0.06)	−0.23 (−0.41; −0.05)	0.02 (−0.17; 0.18)	−0.06 (−0.29; 0.10)	−0.11 (−0.35; 0.13)
FW[b]	0.66 (0.54; 0.78)	0.33 (0.17; 0.54)	0.25 (0.10; 0.38)	0.88 (0.83; 0.94)	0.65 (0.52; 0.77)	0.33 (0.18; 0.47)
ADG[b]	0.54 (0.38; 0.71)	0.39 (0.23; 0.56)	0.23 (0.03; 0.42)	0.72 (0.60; 0.85)	0.58 (0.40; 0.79)	0.26 (0.12; 0.40)
SC[b]	0.12 (−0.10; 0.34)	0.12 (−0.10; 0.28)	0.73 (0.63; 0.83)	0.49 (0.38; 0.61)	0.56 (0.45; 0.70)	0.67 (0.50; 0.83)

Lower and upper limits of the highest posterior density intervals with 90% of the samples are listed between brackets

FW final weight, *ADG* average daily gain, *SC* scrotal circumference, *AFC* age at first calving

[a] FW, ADG, SC and AFC in commercial young bulls and heifers on pasture

[b] FW, ADG and SC in commercial young bulls and heifers in feedlots

Table 5 Efficiency of correlated responses for growth and reproductive traits in commercial young bulls and heifers on pasture and in feedlots when the selection is applied for increased growth and reproductive traits in performance-tested young bulls on pasture and in feedlots

Traits	Performance-tested young bulls on pasture			Performance-tested young bulls in feedlots		
	FW	ADG	SC	FW	ADG	SC
FW[a]	1.10 (1.03; 1.19)	0.58 (0.48; 0.68)	0.48 (0.35; 0.60)	1.12 (1.03; 1.22)	0.55 (0.43; 0.67)	0.74 (0.60; 0.90)
ADG[a]	1.16 (1.00; 1.13)	1.08 (0.94; 1.19)	0.49 (0.32; 0.67)	0.71 (0.54; 0.89)	0.50 (0.34; 0.68)	0.46 (0.24; 0.72)
SC[a]	0.34 (0.24; 0.43)	0.22 (0.13; 0.31)	1.08 (1.01; 1.16)	0.32 (0.17; 0.44)	0.14 (0.02; 0.27)	1.00 (0.90; 1.13)
AFC[a]	−0.33 (−0.68; 0.44)	−0.33 (−0.63; −0.03)	−0.44 (−0.85; −0.05)	0.04 (−0.33; 0.31)	−0.07 (−0.36; 0.14)	−0.20 (−0.73; 0.20)
FW[b]	0.78 (0.47; 0.99)	0.30 (0.09; 0.52)	0.32 (0.16; 0.49)	1.11 (0.98; 1.25)	0.59 (0.46; 0.75)	0.46 (0.24; 0.65)
ADG[b]	0.84 (0.56; 1.19)	0.47 (0.26; 0.71)	0.44 (0.06; 0.81)	1.25 (1.01; 1.53)	0.70 (0.38; 0.95)	0.50 (0.20; 0.84)
SC[b]	0.12 (−0.09; 0.32)	0.09 (−0.06; 0.22)	0.78 (0.64; 0.96)	0.50 (0.37; 0.65)	0.41 (0.29; 0.53)	0.76 (0.50; 0.99)

Lower and upper limits of the highest posterior density intervals with 90% of the samples are listed between brackets

FW final weight, *ADG* average daily gain, *SC* scrotal circumference, *AFC* age at first calving

[a] FW, ADG, SC and AFC in commercial young bulls and heifers on pasture

[b] FW, ADG and SC in commercial young bulls and heifers in feedlots

Discussion

Genetic variation of growth and reproductive traits

The heritabilities, genetic correlations and response to selection for growth and SC in performance-tested young bulls on pasture and in feedlots were presented and discussed previously [8]. The results about these genetic parameters in commercial animals on pasture and in feedlots are quite similar to those presented in Raidan et al. [8]. In most cases, response to selection will be greater for animals in feedlots than on pasture (if selection intensities are the same) because feeding conditions are better and variances are larger for animals in feedlots than on pasture [8, 20, 21].

Genetic variances and heritabilities for growth and SC were higher for performance-tested young bulls than for commercial animals (Table 3). This higher genetic variance for performance-tested animals (except for FW and SC for animals in feedlots because there is difference in the final age between performance-tested and commercial animals) might be a consequence of overall conditions being better and of phenotypic means being higher in performance tests. In general, the environmental conditions (nutrition, sanitary management, etc.) are better for performance-tested young bulls than for commercial animals and they could be responsible for differences in the mean of each trait and in the expression of genetic differences [20, 21]. In addition, temporary random effects in performance testing are lower than in commercial herds because the changes in management conditions are less frequent, and the process of data recording is stricter in performance tests than in commercial conditions [12]. Moreover, the number of young bulls in each performance test was larger than the number of animals in each contemporary group of the commercial herds, which contributes to reduce the error associated with the estimation of systematic effects that are included in the statistical models. The residual variance for ADG is larger for performance-tested young bulls in feedlots than for commercial animals in feedlots because the mean ADG is more than 50% greater in performance tests in feedlots than in performance tests on pasture or in commercial herds on pasture and in feedlots.

AFC records probably originated from a selected group of heifers because the females that have a low weaning weight could have been culled at weaning and some heifers with a low body weight at yearling did not get pregnant during the first breeding season. Thus, the lowest posterior means of genetic variances and heritabilities were obtained from the single-trait analyses. However, the multiple-trait analyses were effective in reducing the bias from selection, as previously stated by Schaeffer [22]. In addition, the posterior mean of the heritability for AFC of commercial animals on pasture obtained from the multiple-trait analysis was similar to the mean heritability of 0.17 obtained from three different samples of Nellore heifers [23–25].

Genetic correlations between male and female traits

Posterior means of heritabilities for growth traits were similar between males and females and genetic correlations between male and female growth traits were high (>0.74). These results agree with those of Garrick et al. [26], Rodríguez-Almeida et al. [27] and Van Vleck and Cundiff [28]. A large fraction of the additive genes for growth traits has the same effect with regard to controlling variation in each sex [26], and there is no evidence of genotype X sex interaction in commercial herds.

Genetic correlations between performance test and commercial herd traits

The genetic correlation between the same trait in different environments has been one of the parameters used to indicate the existence of genotype X environment interaction. Falconer [29] suggested that a genetic correlation between the same trait in different environments lower than 1 is an evidence of genotype X environment interaction. In addition, James [30] and Mulder et al. [31] showed that it is important to have environment-specific breeding programs of progeny testing when the genetic correlations between the same trait in different environments are smaller than the thresholds of 0.70 and 0.61, respectively.

The genetic correlations between the same traits measured in performance-tested animals or in commercial herds were lower than 1, however the upper limits of the HDP90 intervals were higher than 0.79 (Table 4). Therefore, there is no practical effect of genotype X

environment interaction for growth and SC of performance-tested and commercial beef cattle. In addition, heritabilities for traits of performance-tested young bulls were higher than heritabilities for the same traits in commercial animals (Table 3). Moreover, a combination of strong genetic correlation between direct and indirect selected traits and higher heritabilities for indirect traits suggest that indirect selection in performance tests is as efficient as direct selection in commercial herds.

Selection for increased ADG and SC in performance-tested young bulls on pasture will result in reduced AFC in commercial females on pasture. In the literature, estimates of genetic correlations between ADG and AFC range from −0.38 to −0.32 [23, 32] and between SC (at 12 or 18 months of age) and AFC from −0.42 to −0.22 [32, 33]. These results indicate that genes related to ADG and SC could also be related to AFC. In fact, at least one single nucleotide polymorphism (SNP) within the region between 78.85 and 79.85 Mb on chromosome 10, and one SNP within the region between 23.4 and 33.85 Mb on chromosome 14 have been reported to affect both SC and AFC in Nellore cattle [34, 35].

The posterior mean of the genetic correlation between FW in performance-tested young bulls on pasture and AFC was negative (Table 4), but the HDP90 interval included zero, which means that this genetic correlation is not different of zero. The genetic correlation between growth traits in performance-tested young bulls and commercial young bulls and heifers on pasture was sufficiently high to consider these traits in different environments as the same trait. The results in Table 4 suggest that AFC is more strongly correlated with ADG than with FW. The relationships between growth rate, age and live weight at puberty are very complex and it is virtually impossible to separate the effects of growth rate per se from those of live weight and/or age [36]. However, the genetic correlations of ADG and maturation rate with AFC (−0.32 and −0.83, respectively) are stronger than the genetic correlations of FW and weight at maturity with AFC (−0.26 and 0.52, respectively) [23, 37]. In addition, the selection for high growth rate results in a younger and heavier population at puberty [38]. A high growth rate before puberty would involve a considerably higher rate of accumulation of adipose tissue than a low growth rate [36], and this change in body composition can be an effective trigger for puberty [38]. However, the control of reproduction involves a wide variety of interacting mechanisms and it is unlikely that there is only one mechanism involved in the onset of puberty. In addition, the evidence for a relationship between body composition and puberty is not sufficient. A genetic correlation of −0.29 for fat trim from one-half carcass with age at puberty was reported in *Bos taurus* crossbred animals [39], a genetic correlation (±standard error) of 0.13 ± 0.09 for intramuscular fat percentage with heifer pregnancy was reported in the Red Angus breed [40], and an estimated genetic correlation of 0.11 of backfat thickness at 18 months with age at first calving was reported in Nellore breed with a HDP95 interval that ranged from −0.10 to 0.28 [41]. Moreover, based on these results [39–41], differences between breeds might be involved in accumulation of adipose tissue and onset of puberty.

Selection for increased FW, ADG and SC in performance-tested young bulls in feedlots will not change AFC (Tables 4, 5). The estimated genetic correlations of mid-test body weight and ADG in performance-tested young bulls in feedlots with AFC were equal to −0.18 ± 0.13 and 0.21 ± 0.15, respectively [5]. The large standard errors associated with these genetic correlations made it difficult to reach definitive conclusions on the implication of selection for increased growth in performance-tested young bulls in feedlots on AFC. However, the results of the selection experiment presented by Mercadante et al. [42] confirmed that a genetic correlation of almost 0 was found between FW in performance-tested young bulls in feedlots (378 days of age) and days to calving of the first mating, an indicative trait of AFC [43], in Nellore cattle. Similar results were observed for the Angus breed in Australia [44, 45]. Mercadante et al. [42] estimated significant genetic trends of 1.78 ± 0.20 and 2.39 ± 0.20 kg/year for FW and non-significant genetic trends of 0.03 ± 0.16 and 0.19 ± 0.17 days/year for days to calving of the first mating in two lines that were selected for increased FW, respectively. Later, Monteiro et al. [46] showed that selection for increased FW had no effect either on the development of the ovaries and the endometrium or the onset of puberty at 24 months of age in heifers. The selection for increased growth in performance-tested young bulls in feedlots will not change AFC in commercial heifers.

As stated above, AFC is more strongly correlated with ADG than with FW, but only a moderate genetic correlation between ADG of performance-tested young bulls in feedlots and ADG of commercial young bulls and heifers on pasture was observed (0.39, Table 4). Consequently, ADG of performance-tested young bulls in feedlots is not an efficient selection criterion for indirect improvement of ADG and AFC in commercial heifers on pasture.

Genetic correlations of ADG and FW between performance-tested young bulls on pasture (0.74) and in feedlots (0.67) are high [47], but the selection for one or the other had different consequences in commercial herds. Heritability of FW was higher than that of ADG (Table 3) and changes in FW or ADG were obtained in commercial animals when selection is for FW or ADG in performance-tested young bulls (Tables 4, 5), but selection for increased ADG will result in reduced AFC whereas

selection for increased FW will not. FW is more correlated to body weight at the beginning of performance tests than ADG [6, 48], and currently there is no limit for differences in body weight at the beginning of performance tests. Consequently, FW is more affected by body weight at the beginning of the test and herd-of-origin effects than ADG. FW might be more correlated to adult body weight than ADG and increased adult body weight will result in increased energy requirements for the maintenance of cows [49]. These results suggest that ADG is better than FW as a post-weaning selection criterion.

Correlated responses and implications for breeding

Performance testing can be used as a tool to evaluate and select bulls for commercial herds. Furthermore, the results obtained in our study and those obtained by Falconer [50] and Mascioli [51] show that pasture, compared to feedlot, is the best environment for the evaluation and selection of Nellore young bulls. Selection will be more efficient in an environment that allows the maximum expression of the genetic differences [8, 20, 21]. However, Falconer and Latyszewski [52] showed that the improvement obtained by selecting for growth traits on a high plane of nutrition did not carry over when the animals were transferred to a low plane of nutrition, but the improvement made on the low plane of nutrition was retained when the animals were transferred to a high plane of nutrition. Falconer [50] obtained direct and correlated responses for growth traits in mice on two planes of nutrition. The animals selected on a low plane of nutrition were heavier, had less fat and more protein, and females were better dams than those selected on the high plane of nutrition when the two groups were raised on the high plane of nutrition. Thus, selection should be made under the conditions that are the least favorable for the expression of the trait. This author observed the following differences in carcass composition: mice for which growth had been increased by selection on the low plane of nutrition were leaner than those for which growth had been increased by selection on the high plane of nutrition. These results indicate that increases in growth traits of mice on a high or low plane of nutrition were reached by using different physiological pathways [50].

Mascioli [51] conducted individual performance and progeny tests of Canchim young bulls on pasture and in feedlots. These bulls were ranked as superior, intermediate and inferior according to their FW in individual performance tests on pasture or in feedlots (approximately 400 days old). After individual performance tests, the bulls were submitted to progeny tests and their progenies were raised on pasture and in feedlots (progeny test). There is no effect of feedlot performance tests bull's rank (superior, intermediate and inferior) on weaning weight and post-weaning growth of their progeny. However, the bull's progeny that were ranked as superior on pasture performance test were heavier than other classes (intermediate and inferior) for birth weight, weaning weight and weight at 12 months. Mascioli [51] concluded that the selection of Canchim young bulls in favorable environments (feedlots) did not produce the same response to selection in restricted environments (pasture). Similarly, the results presented in Table 5 support the hypothesis that selection for ADG and SC of performance-tested animals on pasture is better than selection for ADG and SC of performance-tested animals in feedlots to improve the means of growth and reproductive traits in commercial animals on pasture or in feedlots.

Conclusions

Heritabilities for growth and scrotal circumference are higher in performance-tested young bulls than in commercial young bulls and heifers, whereas the correlations between the same traits expressed in the different environments are high, implying that indirect selection based on performance test is efficient. Evaluation and selection for increased growth and scrotal circumference on performance-tested young bulls are efficient to improve growth, scrotal circumference and age at first calving in commercial animals. Average daily gain is a better post-weaning selection criterion than final weight in performance tests. Evaluating and selecting performance-tested young bulls is more efficient for animals on pasture than in feedlots.

Authors' contributions
FSSR, HTV, JAGB, EMT, and FLBT discussed the aims and the design of the study. HTV contributed to the data retrieval. FSSR, DCCS, MMM, and AEMA performed data analysis. FSSR and FLBT drafted the manuscript and all authors contributed to this manuscript in its final version. All authors read and approved the final manuscript.

Author details
[1] Departamento de Zootecnia, Escola de Veterinária, Universidade Federal de Minas Gerais, Belo Horizonte, MG 31270-901, Brazil. [2] Present Address: School of Chemistry and Molecular Biosciences, The University of Queensland, 4072 Brisbane, QLD, Australia. [3] Associação Brasileira dos Criadores de Zebu, Uberaba, MG 38022-330, Brazil.

Acknowledgements
The authors thank Associação Brasileira dos Criadores de Zebu (ABCZ) for allowing access to their data; Breno Fragomeni (University of Georgia) for contributions; Ignacy Misztal (University of Georgia) for providing the software GIBBS1F90; the editors and two anonymous reviewers of GSE for their suggestions.

Competing interests
The authors declare that they have no competing interests.

Funding
Coordenação de Aperfeiçoamento de Pessoal de Nível Superior (CAPES) provided fellowships to FSSR and DCCS. Conselho Nacional de Desenvolvimento Científico e Tecnológico (CNPq) provided fellowships to MMM, AEMA, and FLBT. CNPq and Fundação de Amparo à Pesquisa do Estado de Minas Gerais (FAPEMIG) funded the equipments and other facilities used in this study. Pró-Reitoria de Pesquisa da UFMG funded the translation of this paper.

References
1. Cain MF, Wilson LL. Factors influencing individual bull performance in central test stations. J Anim Sci. 1983;57:1059–66.
2. Simm G. Genetic improvement of cattle and sheep. Tonbridge: Farming Press; 2000.
3. Tundisi AGA, Lima FP, Kalil EB, Villares JB, Corrêa A, Vidal MEP. Novas interpretações sobre a eficiência das provas de ganho de peso e a viabilidade da produção econômica de novilhos zebus próximo dos 24 meses de idade. Bol Ind Anim. 1965;23:67–81.
4. Crowley JJ, Evans RD, Mc Hugh N, Pabiou T, Kenny DA, McGee M, et al. Genetic associations between feed efficiency measured in a performance test station and performance of growing cattle in commercial beef herds. J Anim Sci. 2011;89:3382–93.
5. Crowley JJ, Evans RD, Mc Hugh N, Kenny DA, McGee M, Crews DH Jr, et al. Genetic relationships between feed efficiency in growing males and beef cow performance. J Anim Sci. 2011;89:3372–81.
6. de Rezendes Neves HH, Polin dos Reis F, Motta Paterno F, Rocha Guarini A, Carvalheiro R, da Silva LR, et al. Herd-of-origin effect on the post-weaning performance of centrally tested Nellore beef cattle. Trop Anim Health Prod. 2014;46:1235–41.
7. Grion AL, Mercadante MEZ, Cyrillo JNSG, Bonilha SFM, Magnani E, Branco RH. Selection for feed efficiency traits and correlated genetic responses in feed intake and weight gain of Nellore cattle. J Anim Sci. 2014;92:955–65.
8. Raidan FS, Passafaro TL, Fragomeni BO, Josahkian LA, Pereira IG, Toral FLB. Genotype × environment interaction in individual performance and progeny tests in beef cattle. J Anim Sci. 2015;93:920–33.
9. Schenkel FS, Miller SP, Jamrozik J, Wilton JW. Two-step and random regression analyses of weight gain of station-tested beef bulls. J Anim Sci. 2002;80:1497–507.
10. Riley DG, Coleman SW, Chase CC Jr, Olson TA, Hammond AC. Genetic parameters for body weight, hip height, and the ratio of weight to hip height from random regression analyses of Brahman feedlot cattle. J Anim Sci. 2007;85:42–52.
11. Baldi F, Albuquerque LG, Cyrillo JNSG, Branco RH, Oliveira BC Jr, Mercadante MEZ. Genetic parameter estimates for live weight and daily live weight gain obtained for Nellore bulls in a test station using different models. Livest Sci. 2012;144:148–56.
12. Fragomeni BO, Scalez DCB, Toral FLB, Bergmann JAG, Pereira IG, Costa PST. Genetic parameters and alternatives for evaluation and ranking of Nellore young bulls in pasture performance tests. Rev Bras Zootec. 2013;42:559–64.
13. Sorensen D, Gianola D. Likelihood, bayesian, and MCMC methods in quantitative genetics. Statistics for biology and health. 2nd ed. New York: Springer; 2002.
14. Misztal I, Tsuruta S, Lourenço D, Aguilar I, Legara A, Vitezica Z. Manual for BLUPF90 family of programs. Athens: University of Georgia; 2014.
15. Geweke J. Bayesian statistics 4. In: Bernardo JM, Berger JO, Dawid AP, Smith AFM, editors. Evaluating the accuracy of sampling-based approaches to the calculation of posterior moments (with discussion). Oxford: University Press; 1992. p. 164–93.
16. Heidelberger P, Welch PD. Simulation run length control in the presence of an initial transient. Oper Res. 1983;31:1109–44.
17. Smith BJ. Bayesian output analysis program (BOA) version 1.1 user's manual. 2005. http://www.public-health.uiowa.edu/boa/BOA.pdf. Accessed 29 July 2013.

18. R Development Core Team. R: a language and environment for statistical computing. R Foundation for Statistical Computing. Vienna: the R Foundation for Statistical Computing; 2015.

19. Falconer DS, Mackay TFC. Introduction to quantitative genetics. 4th ed. Essex: Longman; 1996.

20. Hammond J. Animal breeding in relation to nutrition and environmental conditions. Biol Rev. 1947;22:195–213.

21. Kearney JF, Schutz MM, Boettcher PJ, Weigel KA. Genotype × environment interaction for grazing versus confinement. I. Production traits. J Dairy Sci. 2004;87:501–9.

22. Schaeffer LR. Sire and cow evaluation under multiple trait models. J Dairy Sci. 1984;67:1567–80.

23. Boligon AA, Albuquerque LG, Mercadante MEZ, Lôbo RB. Study of relations among age at first calving, average weight gains and weights from weaning to maturity in Nellore cattle. Rev Bras Zootec. 2010;39:746–51.

24. Regatieri IC, Boligon AA, Baldi F, Albuquerque LG. Genetic correlations between mature cow weight and productive and reproductive traits in Nellore cattle. Genet Mol Res. 2012;11:2979–86.

25. Eler JP, Bignardi AB, Ferraz JBS, Santana ML Jr. Genetic relationships among traits related to reproduction and growth of Nelore females. Theriogenology. 2014;82:708–14.

26. Garrick DJ, Pollak EJ, Quaas RL, Van Vleck LD. Variance heterogeneity in direct and maternal weight traits by sex and percent purebred for Simmental-sired calves. J Anim Sci. 1989;67:2515–28.

27. Rodríguez-Almeida FA, Van Vleck LD, Cundiff LV, Kachman SD. Heterogeneity of variance by sire breed, sex, and dam breed in 200- and 365-day weights of beef cattle from a top cross experiment. J Anim Sci. 1995;73:2579–88.

28. Van Vleck LD, Cundiff LV. Sex effects on breed of sire differences for birth, weaning, and yearling weights. J Anim Sci. 1998;76:1528–34.

29. Falconer DS. The problem of environment and selection. Am Nat. 1952;86:293–8.

30. James JW. Selection in two environments. Heredity. 1961;16:145–52.

31. Mulder HA, Veerkamp RF, Ducro BJ, Van Arendonk JA, Bijma P. Optimization of dairy cattle breeding programs for different environments with genotype by environment interaction. J Dairy Sci. 2006;89:1740–52.

32. Castro-Pereira VM, Alencar MM, Barbosa PF. Estimativas de parâmetros genéticos e de ganhos direto e indireto à seleção para características reprodutivas e de crescimento em um rebanho da raça Canchim. Rev Bras Zootec. 2007;36:1029–36.

33. Terakado APN, Boligon AA, Baldi F, Silva JA, Albuquerque LG. Genetic associations between scrotal circumference and female reproductive traits in Nellore cattle. J Anim Sci. 2015;93:2706–13.

34. Utsunomiya YT, Carmo AS, Neves HHR, Carvalheiro R, Matos MC, Zavarez LB, et al. Genome-wide mapping of loci explaining variance in scrotal circumference in Nellore cattle. PLoS One. 2014;9:e88561.

35. Costa RB, Camargo GMF, Diaz IDPS, Irano N, Dias MM, Carvalheiro R, et al. Genome-wide association study of reproductive traits in Nellore heifers using Bayesian inference. Genet Sel Evol. 2015;47:67.

36. Lawrence TLJ, Fowler VR. Growth of farm animals. 2nd ed. Wallingford: CABI Publishing; 2002.

37. Gaviolli VRN, Buzanskas ME, Cruz VAR, Savegnago RP, Munari DP, Freitas AR, et al. Genetic associations between weight at maturity and maturation rate with ages and weights at first and second calving in Canchim beef cattle. J Appl Genet. 2012;53:331–5.

38. Foxcroft GR. Growth and breeding performance in animals and birds. In: Lawrence TLJ, editor. Growth in animals. London: Butterworth & Co; 1980. p. 229–47.

39. MacNeil MD, Cundiff LV, Dinkel CA, Koch RM. Genetic correlations among sex-limited traits in beef cattle. J Anim Sci. 1984;58:1171–80.

40. McAllister CM, Speidel SE, Crews DH Jr, Enns RM. Genetic parameters for intramuscular fat percentage, marbling score, scrotal circumference, and heifer pregnancy in Red Angus cattle. J Anim Sci. 2011;89:2068–72.

41. Yokoo MJ, Lôbo RB, Magnabosco CL, Rosa GJM, Forni S, Sainz RD, et al. Genetic correlation of traits measured by ultrasound at yearling and 18 months of age in Nellore beef cattle. Livest Sci. 2015;180:34–40.

42. Mercadante MEZ, Packer IU, Razook AG, Cyrillo JNSG, Figueiredo LA. Direct and correlated responses to selection for yearling weight on reproductive performance of Nelore cows. J Anim Sci. 2003;81:376–84.

43. Forni S, Albuquerque LG. Estimates of genetic correlations between days to calving and reproductive and weight traits in Nelore cattle. J Anim Sci. 2005;83:1511–5.

44. Meyer K, Hammond K, Mackinnon MJ, Parnell PF. Estimates of covariances between reproduction and growth in Australian beef cattle. J Anim Sci. 1991;69:3533–43.

45. Johnston DJ, Bunter KL. Days to calving in Angus cattle: genetic and environmental effects, and covariances with other traits. Livest Prod Sci. 1996;45:13–22.

46. Monteiro FM, Mercadante MEZ, Barros CM, Satrapa RA, Silva JAV, Oliveira LZ, et al. Reproductive tract development and puberty in two lines of Nellore heifers selected for postweaning weight. Theriogenology. 2013;80:10–7.

47. Raidan FSS, Tineo JSA, Moraes MM, Escarce TC, Araujo AEM, Gomes MMC, et al. Associations between growth, scrotal circumference and visual scores of beef cattle in performance tests on pasture or in feedlots. Rev Bras Zootec. In press.

48. Tineo JSA, Raidan FSS, Santos DCC, Toral FLB. Influência da idade e do peso no início do teste na análise genética de características de crescimento, reprodução e escores visuais de tourinhos Nelore em provas de ganho em peso a pasto. Arch Zootec. 2016;65:29–34.

49. National Research Council Nutrient requirement of beef cattle. 7 revised ed. Washington: The National Academic Press; 2000. doi:10.17226/9791.

50. Falconer DS. Selection of mice for growth on high and low planes of nutrition. Genet Res. 1960;1:91–113.

51. Mascioli AS. Interação genótipo x ambiente sobre o desempenho de animais Canchim e cruzados Canchim × Nelore. Jaboticabal: Faculdade de Ciências Agrárias e Veterinárias. Universidade Estadual Paulista; 2000.

52. Falconer DS, Latyszewski M. The environment in relation to selection for size in mice. J Genet. 1952;51:67–80.

4

Genomic evaluation by including dominance effects and inbreeding depression for purebred and crossbred performance with an application in pigs

Tao Xiang[1,2]* ⓘ, Ole Fredslund Christensen[1], Zulma Gladis Vitezica[3] and Andres Legarra[2]

Abstract

Background: Improved performance of crossbred animals is partly due to heterosis. One of the major genetic bases of heterosis is dominance, but it is seldom used in pedigree-based genetic evaluation of livestock. Recently, a trivariate genomic best linear unbiased prediction (GBLUP) model including dominance was developed, which can distinguish purebreds from crossbred animals explicitly. The objectives of this study were: (1) methodological, to show that inclusion of marker-based inbreeding accounts for directional dominance and inbreeding depression in purebred and crossbred animals, to revisit variance components of additive and dominance genetic effects using this model, and to develop marker-based estimators of genetic correlations between purebred and crossbred animals and of correlations of allele substitution effects between breeds; (2) to evaluate the impact of accounting for dominance effects and inbreeding depression on predictive ability for total number of piglets born (TNB) in a pig dataset composed of two purebred populations and their crossbreds. We also developed an equivalent model that makes the estimation of variance components tractable.

Results: For TNB in Danish Landrace and Yorkshire populations and their reciprocal crosses, the estimated proportions of dominance genetic variance to additive genetic variance ranged from 5 to 11%. Genetic correlations between breeding values for purebred and crossbred performances for TNB ranged from 0.79 to 0.95 for Landrace and from 0.43 to 0.54 for Yorkshire across models. The estimated correlation of allele substitution effects between Landrace and Yorkshire was low for purebred performances, but high for crossbred performances. Predictive ability for crossbred animals was similar with or without dominance. The inbreeding depression effect increased predictive ability and the estimated inbreeding depression parameter was more negative for Landrace than for Yorkshire animals and was in between for crossbred animals.

Conclusions: Methodological developments led to closed-form estimators of inbreeding depression, variance components and correlations that can be easily interpreted in a quantitative genetics context. Our results confirm that genetic correlations of breeding values between purebred and crossbred performances within breed are positive and moderate. Inclusion of dominance in the GBLUP model does not improve predictive ability for crossbred animals, whereas inclusion of inbreeding depression does.

*Correspondence: Tao.Xiang@mbg.au.dk
[1] Department of Molecular Biology and Genetics, Center for Quantitative Genetics and Genomics, Aarhus University, 8830 Tjele, Denmark
Full list of author information is available at the end of the article

Background

Crossbreeding is primarily and intensively applied in meat production systems [1], especially for swine and poultry. Crossbreeding capitalizes on heterosis effects and complementarity between breeds, and results in an increased performance of crossbred animals compared to purebred animals [1]. In terminal crossbreeding systems, selection on purebred animals to maximize their crossbred performance is the ultimate goal [2, 3]. Due to the existence of genotype-by-environment interaction effects and non-additive genetic effects in combination with different allele frequencies in different breeds [3, 4], the genetic correlation of breeding values between purebred and crossbred performances (r_{PC}) is usually lower than 1 [1, 5], and therefore, purebred performance under nucleus conditions may not be an optimal predictor for crossbred performance in commercial animals [4, 6].

One of the major genetic bases of heterosis is dominance [7, 8]. At the level of gene action, dominance is due to interactions between alleles at the same locus [9]. In pedigree-based genetic evaluation, dominance is rarely included because large-scale datasets that comprise a high proportion of full sibs are required to obtain accurate estimates and because the computational complexity is high [10]. With the recent availability of single nucleotide polymorphism (SNP) information and the development of genomic selection, estimation of the dominance effects of SNPs has become more feasible [11, 12].

Genomic evaluation has been successfully used in purebred [13, 14] and crossbred populations [15–17]. However, these studies generally ignore the dominance effects. A number of studies have been carried out on genomic evaluation including dominance effects using either simulated [18] or real purebred data [9, 12].

Recently, several studies [19, 20] have tried to extend genomic evaluation including dominance effects from purebred performance to crossbred performance. However, they either used genomic information on purebred animals only [19] or applied a genomic model that assumed that all animals belong to a single population, and thus the variance components were estimated based only on the genotyped crossbred animals [20]. Nevertheless, combining purebred and crossbred information is essential to implement genetic evaluation for crossbred performance [1, 19]. Furthermore, because of genotype-by-environment interaction effects and different patterns of linkage disequilibrium (LD) between SNPs and quantitative trait loci (QTL), the effects of SNPs may be breed-specific [21]. To overcome these issues, a trivariate genomic best linear unbiased predictor (GBLUP) model that explicitly distinguishes between purebred and crossbred data and includes dominance was recently developed by Vitezica et al. [22]. This model allowed the estimation of different, yet correlated, additive and dominance marker effects in crossbred and purebred individuals. However, the empirical predictive ability of the trivariate GBLUP model has not been evaluated yet.

Thus, the current study had the following objectives: (1) to show how genomic inbreeding can be meaningfully included in GBLUP, even for crossbred animals; (2) to estimate the variance components of additive and dominance genetic effects by using data on total number of piglets born (TNB) in two Danish purebred and one crossbred pig populations using the trivariate GBLUP model; (3) to show how to derive, from variance component estimates, estimated genetic correlations of breeding values between purebred and crossbred performances in each pure breed, and also correlations of allele substitution effects between the two pure breeds; and (4) to evaluate the impact of dominance effects from genomic information on genomic evaluation by comparing accuracies of estimated genomic values in different cross-validation scenarios.

Methods

Animals and genotypes

We begin this section with a short presentation of the data used in the study, with the aim of defining the notation for the methodological developments that follow. For this study, all datasets were provided by the Danish Pig Research Centre. Data from three Danish pig populations were analyzed simultaneously: Landrace (L), Yorkshire (Y) and their reciprocal crosses (LY). Only data on TNB data for the first parity of sows in the three populations were used. In total, there were 2126, 2218 and 5143 genotyped sows with own records on TNB for L, Y and LY, respectively. Instead of using original records, corrected phenotypic values of TNB were used as dependent variables for the trivariate GBLUP model, because the pre-correction for non-genetic effects, such as herd-year-season, month at farrowing, and service sire was more accurately achieved on a larger dataset (293,339 L, 180,112 Y, and 10,974 LY). Among the crossbred animals, 7407 LY had a Landrace sire and a Yorkshire dam, while 3567 LY had a Yorkshire dam and a Landrace sire; L and Y populations were from nucleus farms and LY from a commercial farm. The litters of purebred sows were both purebred and crossbred litters. The relationship between LY-L and LY-Y are comparable since, in both cases, parents of the F1 animals are in the purebred datasets; further details about the model used for the pre-correction are in [17]. All the purebred sows had first farrowing dates between 2003 and 2013, while the crossbred sows first farrowed between 2010 and 2013. Only five of these purebred L and Y sows were dams of the LY.

The pedigrees for both purebred and crossbred sows were available and all crossbred animals were traced back to their purebred ancestors until 1994 by the DMU Trace program [23], as was done for the larger dataset used for pre-correction. Consequently, 8227 L, 9851 Y and 5143 LY individuals were in the pedigree. The dataset of pre-corrected TNB records for genotyped individuals is termed "full genomic dataset" throughout the whole paper, and it should not be confused with the larger dataset used to do the pre-correction.

For the "full genomic dataset", purebred sows were genotyped with the Illumina PorcineSNP60 Genotyping BeadChip [24], while the crossbred sows were genotyped with a 8.5 K GGP-Porcine Low Density Illumina Bead SNP chip [25]. SNP quality controls (such as: call rate for individuals $\geq$80%; call rate for SNPs $\geq$90%; minor allele frequencies $\geq$0.01; etc.) were applied on the same dataset in a previous study [26], which provides more details. Then, for the crossbred individuals, imputation from low density to moderate density was done by using a joint reference panel of the two pure breeds [26] using the software Beagle version 3.3.2 [27] (imputation accuracies $\geq$95% in terms of correlation coefficients and $\geq$99% in terms of correct rates between imputed and true genotypes). Finally, 41,009 SNPs were available for all the recorded purebred and crossbred sows.

Considering genomic inbreeding and heterosis

Inbreeding can be defined as the proportion of homozygous SNPs across all loci for each animal, as suggested by several authors (e.g., [28]). If there is directional dominance causing inbreeding depression [29], then inbreeding should be considered in the genetic evaluation models [30]. Otherwise, using pedigree or marker data, estimates of genetic parameters are inflated [30, 31]. In Vitezica et al. [22], genomic inbreeding was fitted as a covariate and, in the current study, we prove this reasoning by using a parametric genomic model, such as a GBLUP.

Theory and evidence of directional dominance (equivalently, inbreeding depression) suggest that dominance effects of genes (here associated to markers) should have a priori a positive value for traits that exhibit inbreeding depression or heterosis. If we call $\mathbf{d}$ the vector of dominance marker effects, the following prior distribution is plausible:

$$(\mathbf{d}) \sim N\left(\mathbf{1}\mu_d, \mathbf{I}\sigma_d^2\right),$$

where μ_d is the overall mean of dominance effects, which should be positive if there is heterosis due to dominance. A typical model for genomic prediction is that in Toro and Varona [11]:

$$\mathbf{y} = \mathbf{X\beta} + \mathbf{Za} + \mathbf{Wd} + \mathbf{e}, \tag{1}$$

where $\mathbf{y}$ contains phenotypic values; $\mathbf{X\beta}$ stands for fixed effects and random effects other than additive and dominance effects; $\mathbf{a}$ is the vector of "biological" additive SNP effects, $\mathbf{d}$ is the vector of "biological" dominance SNP effects for each of the markers; matrix $\mathbf{Z}$ has entries 1, 0, −1, for SNP genotypes AA, Aa and aa, respectively, while matrix $\mathbf{W}$ has entries 0, 1, 0 for SNP genotypes AA, Aa and aa, respectively. $\mathbf{e}$ is the vector of overall random residual effects.

Typically, genetic models require $\mathbf{a}$ and $\mathbf{d}$ to have zero means, which is not true for $\mathbf{d}$ when directional dominance exist. Defining $\mathbf{d}^* = \mathbf{d} - \mathrm{E}(\mathbf{d})$, then $\mathrm{E}(\mathbf{d}^*) = \mathbf{0}$, and Eq. (1) can be written as:

$$\begin{aligned} \mathbf{y} &= \mathbf{X\beta} + \mathbf{Za} + \mathbf{W}\big(\mathbf{d}^* + \mathrm{E}(\mathbf{d})\big) + \mathbf{e} \\ &= \mathbf{X\beta} + \mathbf{Za} + \mathbf{Wd}^* + \mathbf{W1}\mu_d + \mathbf{e}. \end{aligned}$$

The term $\mathbf{W1}\mu_d$ is actually an average of dominance effects for each individual and is equal to $\mathbf{h}\mu_d$, where $\mathbf{h} = \mathbf{W1}$ contains the row-sums of $\mathbf{W}$, i.e. individual heterozygosities (it should be noted that $\mathbf{W}$ has a value of 1 at heterozygous loci for an individual). Inbreeding coefficients $\mathbf{f}$ can be calculated as:

$$\mathbf{f} = \mathbf{1} - \mathbf{h}/N,$$

where N is the number of SNPs. Then, the prior means $\mathbf{h}\mu_d$ can be rewritten as:

$$\mathbf{h}\mu_d = (\mathbf{1} - \mathbf{f})N\mu_d = \mathbf{1}N\mu_d + \mathbf{f}(-N\mu_d).$$

The term $\mathbf{1}N\mu_d$ is confounded with the overall mean of the model ($\mathbf{\mu}$), while the term $\mathbf{f}(-N\mu_d)$ models the inbreeding depression and $b = (-N\mu_d)$ is the inbreeding depression parameter summed over the SNPs, which has to be estimated. Thus, the linear model including genomic inbreeding is, finally:

$$\mathbf{y} = \mathbf{X\beta} + \mathbf{f}b + \mathbf{Za} + \mathbf{Wd}^* + \mathbf{e}.$$

Thus, we have proven why fitting overall homozygosity for the individual as a measure of inbreeding depression accounts for directional dominance.

Estimating genetic (co)variances of markers with additive and dominance effects

A trivariate model based on "biological" (genotypic) additive and dominance effects of SNPs [22, 32], and including genomic inbreeding as above, was applied considering TNB as a different trait in each population:

$$\begin{aligned} \mathbf{y}_L &= \mathbf{1}\mu_L + \mathbf{f}_L b_L + \mathbf{Z}_L \mathbf{a}_L + \mathbf{W}_L \mathbf{d}_L + \mathbf{e}_L, \\ \mathbf{y}_Y &= \mathbf{1}\mu_Y + \mathbf{f}_Y b_Y + \mathbf{Z}_Y \mathbf{a}_Y + \mathbf{W}_Y \mathbf{d}_Y + \mathbf{e}_Y, \\ \mathbf{y}_{LY} &= \mathbf{1}\mu_{LY} + \mathbf{f}_Y b_Y + \mathbf{Z}_{LY} \mathbf{a}_{LY} + \mathbf{W}_{LY} \mathbf{d}_{LY} + \mathbf{e}_{LY}, \end{aligned} \tag{2}$$

where $\mathbf{y}_L$, $\mathbf{y}_Y$ and $\mathbf{y}_{LY}$ contain corrected phenotypic values for purebred L, purebred Y and crossbred LY sows, respectively; μ_L, μ_Y and μ_{LY} are the respective means; $\mathbf{a}_L$, $\mathbf{a}_Y$ and $\mathbf{a}_{LY}$ are the "biological" additive SNP effects and $\mathbf{d}_L, \mathbf{d}_Y$ and $\mathbf{d}_{LY}$ are the "biological" dominance SNP effects for each of the SNPs for L, Y and LY, respectively; matrices $\mathbf{Z}$ and $\mathbf{W}$ are as above; $\mathbf{f}_L b_L$, $\mathbf{f}_Y b_Y$ and $\mathbf{f}_{LY} b_{LY}$ model the inbreeding depression for L, Y and LY populations; $\mathbf{e}_L, \mathbf{e}_Y$ and $\mathbf{e}_{LY}$ are the overall random residual effects.

Note that "biological" is used here to refer to the genotypic additive and dominance values of the SNPs, to distinguish them from the traditional treatment of quantitative genetics in terms of "statistical" effects (breeding values and dominance deviations) [32].

The above equations can be reformulated to genotypic values of individuals instead of SNPs, in order to be compatible with the classical GBLUP model and animal breeding software, such as BLUPF90 [33] and DMU [34]:

$$
\begin{aligned}
\mathbf{y}_L &= \mathbf{1}\mu_L + \mathbf{f}_L b_L + \mathbf{u}_L + \mathbf{v}_L + \mathbf{e}_L, \\
\mathbf{y}_Y &= \mathbf{1}\mu_Y + \mathbf{f}_Y b_Y + \mathbf{u}_Y + \mathbf{v}_Y + \mathbf{e}_Y, \\
\mathbf{y}_{LY} &= \mathbf{1}\mu_{LY} + \mathbf{f}_{LY} b_{LY} + \mathbf{u}_{LY} + \mathbf{v}_{LY} + \mathbf{e}_{LY}.
\end{aligned} \tag{3}
$$

Note that $\mathbf{u}$ and $\mathbf{v}$ are vectors of genotypic additive and dominance effects and therefore cannot be directly compared to breeding values and dominance deviations in the pedigree-based genetic evaluation. In addition, $\mathbf{f}$ is a vector of genomic inbreeding coefficients and b is a population-specific inbreeding depression parameter per unit of genomic inbreeding, respectively. Note that there is potentially inbreeding depression at the level of the crossbred animals, although, first, the numeric values of the vector $\mathbf{f}$ should be smaller since crossbred animals have a higher level of heterozygosity, and second, the estimates of the inbreeding depression parameters (b) do not need to be identical across the three populations, which thus gives considerable flexibility.

In terms of the genotypic additive effects $\mathbf{u}$, the variances within each breed are:

$$
\begin{aligned}
Var(\mathbf{u}_L) &= var(\mathbf{Z}_L \mathbf{a}_L) = \mathbf{Z}_L \mathbf{Z}_L' \sigma_{a_L}^2, \\
Var(\mathbf{u}_Y) &= var(\mathbf{Z}_Y \mathbf{a}_Y) = \mathbf{Z}_Y \mathbf{Z}_Y' \sigma_{a_Y}^2, \\
Var(\mathbf{u}_{LY}) &= var(\mathbf{Z}_{LY} \mathbf{a}_{LY}) = \mathbf{Z}_{LY} \mathbf{Z}_{LY}' \sigma_{a_{LY}}^2,
\end{aligned}
$$

where $\sigma_{a_L}^2, \sigma_{a_Y}^2$ and $\sigma_{a_{LY}}^2$ are the additive variances of SNP effects in breeds L, Y and LY, respectively. The covariances between the genotypic additive effects $\mathbf{u}$ are:

$$
Cov \begin{pmatrix} \mathbf{u}_L \\ \mathbf{u}_Y \\ \mathbf{u}_{LY} \end{pmatrix} = \begin{pmatrix} \mathbf{Z}_L \mathbf{Z}_L' \sigma_{a_L}^2 & \mathbf{Z}_L \mathbf{Z}_Y' \sigma_{a_{L,Y}} & \mathbf{Z}_L \mathbf{Z}_{LY}' \sigma_{a_{L,LY}} \\ \mathbf{Z}_Y \mathbf{Z}_L' \sigma_{a_{L,Y}} & \mathbf{Z}_Y \mathbf{Z}_Y' \sigma_{a_Y}^2 & \mathbf{Z}_Y \mathbf{Z}_{LY}' \sigma_{a_{Y,LY}} \\ \mathbf{Z}_{LY} \mathbf{Z}_L' \sigma_{a_{L,LY}} & \mathbf{Z}_{LY} \mathbf{Z}_Y' \sigma_{a_{Y,LY}} & \mathbf{Z}_{LY} \mathbf{Z}_{LY}' \sigma_{a_{LY}}^2 \end{pmatrix}, \tag{4}
$$

where $\sigma_{a_{L,Y}}, \sigma_{a_{L,LY}}$ and $\sigma_{a_{Y,LY}}$ are the additive covariances of SNP effects between populations L and Y, populations

L and LY, and populations Y and LY, respectively. Analogous structures exist for dominance genotypic effects:

$$
Cov \begin{pmatrix} \mathbf{v}_L \\ \mathbf{v}_Y \\ \mathbf{v}_{LY} \end{pmatrix} = \begin{pmatrix} \mathbf{W}_L \mathbf{W}_L' \sigma_{d_L}^2 & \mathbf{W}_L \mathbf{W}_Y' \sigma_{d_{L,Y}} & \mathbf{W}_L \mathbf{W}_{LY}' \sigma_{d_{L,LY}} \\ \mathbf{W}_Y \mathbf{W}_L' \sigma_{d_{L,Y}} & \mathbf{W}_Y \mathbf{W}_Y' \sigma_{d_Y}^2 & \mathbf{W}_Y \mathbf{W}_{LY}' \sigma_{d_{Y,LY}} \\ \mathbf{W}_{LY} \mathbf{W}_L' \sigma_{d_{L,LY}} & \mathbf{W}_{LY} \mathbf{W}_Y' \sigma_{d_{Y,LY}} & \mathbf{W}_{LY} \mathbf{W}_{LY}' \sigma_{d_{LY}}^2 \end{pmatrix}.
$$

Estimation of marker-based variance components using an equivalent model

The variance components $\sigma_{a_L}^2, \sigma_{a_Y}^2, \sigma_{a_{LY}}^2$ and $\sigma_{a_{L,Y}}, \sigma_{a_{L,LY}}$, $\sigma_{a_{Y,LY}}$ in Eq. (4) cannot be estimated by regular methods or software (i.e. REML or Gibbs sampling) because they cannot be factorized out from Eq. (4). To fit such a multivariate structure, we used an equivalent model. Additional effects need to be defined, even if they are of no interest per se. For instance, the vectors of hypothetical genotypic additive effects of the genotypes of the L breed on the scale of breed Y ($\mathbf{u}_{L,Y}$) and LY ($\mathbf{u}_{L,LY}$) have variance–covariance matrices $\mathbf{Z}_L \mathbf{Z}_L' \sigma_{a_Y}^2$ and $\mathbf{Z}_L \mathbf{Z}_L' \sigma_{a_{LY}}^2$, respectively. Thus, as a whole, the genetic variance and covariance structure for the genotypic additive effects $\mathbf{u}$ are:

$$
Var(\mathbf{u}) = var \begin{bmatrix} \mathbf{u}_L \\ \mathbf{u}_{L,Y} \\ \mathbf{u}_{L,LY} \\ \mathbf{u}_{Y,L} \\ \mathbf{u}_Y \\ \mathbf{u}_{Y,LY} \\ \mathbf{u}_{LY,L} \\ \mathbf{u}_{LY,Y} \\ \mathbf{u}_{LY} \end{bmatrix} = var \begin{bmatrix} \mathbf{Z}_L \mathbf{a}_L \\ \mathbf{Z}_L \mathbf{a}_Y \\ \mathbf{Z}_L \mathbf{a}_{LY} \\ \mathbf{Z}_Y \mathbf{a}_L \\ \mathbf{Z}_Y \mathbf{a}_Y \\ \mathbf{Z}_Y \mathbf{a}_{LY} \\ \mathbf{Z}_{LY} \mathbf{a}_L \\ \mathbf{Z}_{LY} \mathbf{a}_Y \\ \mathbf{Z}_{LY} \mathbf{a}_{LY} \end{bmatrix}
$$

$$
= \begin{bmatrix} \mathbf{Z}_L \mathbf{Z}_L' & \mathbf{Z}_L \mathbf{Z}_Y' & \mathbf{Z}_L \mathbf{Z}_{LY}' \\ \mathbf{Z}_Y \mathbf{Z}_L' & \mathbf{Z}_Y \mathbf{Z}_Y' & \mathbf{Z}_Y \mathbf{Z}_{LY}' \\ \mathbf{Z}_{LY} \mathbf{Z}_L' & \mathbf{Z}_{LY} \mathbf{Z}_Y' & \mathbf{Z}_{LY} \mathbf{Z}_{LY}' \end{bmatrix} \otimes \begin{bmatrix} \sigma_{a_L}^2 & \sigma_{a_{L,Y}} & \sigma_{a_{L,LY}} \\ \sigma_{a_{Y,L}} & \sigma_{a_Y}^2 & \sigma_{a_{Y,LY}} \\ \sigma_{a_{LY,L}} & \sigma_{a_{LY,Y}} & \sigma_{a_{LY}}^2 \end{bmatrix}
$$

$$
= \mathbf{Z}\mathbf{Z}' \otimes \begin{bmatrix} \sigma_{a_L}^2 & \sigma_{a_{L,Y}} & \sigma_{a_{L,LY}} \\ \sigma_{a_{Y,L}} & \sigma_{a_Y}^2 & \sigma_{a_{Y,LY}} \\ \sigma_{a_{LY,L}} & \sigma_{a_{LY,Y}} & \sigma_{a_{LY}}^2 \end{bmatrix},
$$

where matrix $\mathbf{Z}$ contains elements 1, 0, −1 for the three genotypes, and is defined across the three breeds,

$$
\mathbf{Z} = \begin{bmatrix} \mathbf{Z}_L \\ \mathbf{Z}_Y \\ \mathbf{Z}_{LY} \end{bmatrix}.
$$

To construct a relationship matrix similar to the classical $\mathbf{G}$-matrix of GBLUP [35], Vitezica et al. [22] introduced a normalized genomic relationship matrix $\mathbf{G} = \dfrac{\mathbf{Z}\mathbf{Z}'}{\{\mathrm{tr}[\mathbf{Z}\mathbf{Z}']\}/n}$, where n is the number of animals across the three populations and the division by $\{\mathrm{tr}[\mathbf{Z}\mathbf{Z}']\}/n$ scales the matrix such that the average of the diagonal elements equals 1. This alters the variances across genotypic additive effects $\mathbf{u}$ in the following way:

$$Var(\mathbf{u}) = var \begin{bmatrix} \mathbf{u}_L \\ \mathbf{u}_{L,Y} \\ \mathbf{u}_{L,LY} \\ \mathbf{u}_{Y,L} \\ \mathbf{u}_Y \\ \mathbf{u}_{Y,LY} \\ \mathbf{u}_{LY,L} \\ \mathbf{u}_{LY,Y} \\ \mathbf{u}_{LY} \end{bmatrix} = \mathbf{ZZ'} \otimes \begin{bmatrix} \sigma^2_{a_L} & \sigma_{a_{L,Y}} & \sigma_{a_{L,LY}} \\ \sigma_{a_{Y,L}} & \sigma^2_{a_Y} & \sigma_{a_{Y,LY}} \\ \sigma_{a_{LY,L}} & \sigma_{a_{LY,Y}} & \sigma^2_{a_{LY}} \end{bmatrix}$$

$$= \left(\mathbf{G} \times \frac{\{\mathrm{tr}[\mathbf{ZZ'}]\}}{n} \right) \otimes \begin{bmatrix} \sigma^2_{a_L} & \sigma_{a_{L,Y}} & \sigma_{a_{L,LY}} \\ \sigma_{a_{Y,L}} & \sigma^2_{a_Y} & \sigma_{a_{Y,LY}} \\ \sigma_{a_{LY,L}} & \sigma_{a_{LY,Y}} & \sigma^2_{a_{LY}} \end{bmatrix}$$

$$= \mathbf{G} \otimes \begin{bmatrix} \sigma^2_{A_L} & \sigma_{A_L A_Y} & \sigma_{A_L A_{LY}} \\ \sigma_{A_Y A_L} & \sigma^2_{A_Y} & \sigma_{A_Y A_{LY}} \\ \sigma_{A_L A_{LY}} & \sigma_{A_Y A_{LY}} & \sigma^2_{A_{LY}} \end{bmatrix} = \mathbf{G} \otimes \mathbf{G}_0, \tag{5}$$

where $\mathbf{G}_0$ are variance components associated to the genotypic additive effects $\mathbf{u}$. This structure (a Kronecker product) is compatible with animal breeding software for BLUP and REML and the variance–covariance component $\mathbf{G}_0$ can be estimated in a straightforward manner. Then, the (co)variances of additive genotypic effects of SNPs across populations can be obtained as:

$$\begin{bmatrix} \sigma^2_{a_L} & \sigma_{a_{L,Y}} & \sigma_{a_{L,LY}} \\ \sigma_{a_{Y,L}} & \sigma^2_{a_Y} & \sigma_{a_{Y,LY}} \\ \sigma_{a_{LY,L}} & \sigma_{a_{LY,Y}} & \sigma^2_{a_{LY}} \end{bmatrix} = \mathbf{G}_0 / \{\mathrm{tr}[\mathbf{ZZ'}]\}/n$$

$$= \begin{bmatrix} \sigma^2_{A_L} & \sigma_{A_L A_Y} & \sigma_{A_L A_{LY}} \\ \sigma_{A_Y A_L} & \sigma^2_{A_Y} & \sigma_{A_Y A_{LY}} \\ \sigma_{A_L A_{LY}} & \sigma_{A_Y A_{LY}} & \sigma^2_{A_{LY}} \end{bmatrix} / \{\mathrm{tr}[\mathbf{ZZ'}]\}/n. \tag{6}$$

The variances across genotypic dominance effects $\mathbf{v}$ are altered in a similar way:

$$Var(\mathbf{v}) = var \begin{bmatrix} \mathbf{v}_L \\ \mathbf{v}_{L,Y} \\ \mathbf{v}_{L,LY} \\ \mathbf{v}_{Y,L} \\ \mathbf{v}_Y \\ \mathbf{v}_{Y,LY} \\ \mathbf{v}_{LY,L} \\ \mathbf{v}_{LY,Y} \\ \mathbf{v}_{LY} \end{bmatrix}$$

$$= \mathbf{D} \otimes \begin{bmatrix} \sigma^2_{D_L} & \sigma_{D_L D_Y} & \sigma_{D_L D_{LY}} \\ \sigma_{D_Y D_L} & \sigma^2_{D_Y} & \sigma_{D_Y D_{LY}} \\ \sigma_{D_L D_{LY}} & \sigma_{D_Y D_{LY}} & \sigma^2_{D_{LY}} \end{bmatrix} = \mathbf{D} \otimes \mathbf{D}_0, \tag{7}$$

where $\mathbf{D}_0$ contains variances and covariances associated to the genotypic dominance effects $\mathbf{v}$ and $\mathbf{D} = \frac{\mathbf{WW'}}{\{\mathrm{tr}[\mathbf{WW'}]\}/n}$, where the matrix $\mathbf{W}$ contains elements 0, 1, 0 for the three genotypes, and is defined

across the three breeds $\left(\mathbf{W} = \begin{bmatrix} \mathbf{W}_L \\ \mathbf{W}_Y \\ \mathbf{W}_{LY} \end{bmatrix} \right)$ and $\mathbf{W'} = \begin{bmatrix} \mathbf{W}'_L & \mathbf{W}'_Y & \mathbf{W}'_{LY} \end{bmatrix}$. Then, the (co)variances of dominance genotypic effects of SNPs are:

$$\begin{bmatrix} \sigma^2_{d_L} & \sigma_{d_{L,Y}} & \sigma_{d_{L,LY}} \\ \sigma_{d_{Y,L}} & \sigma^2_{d_Y} & \sigma_{d_{Y,LY}} \\ \sigma_{d_{LY,L}} & \sigma_{d_{LY,Y}} & \sigma^2_{d_{LY}} \end{bmatrix} = \mathbf{D}_0 / \{\mathrm{tr}[\mathbf{WW'}]\}/n$$

$$= \begin{bmatrix} \sigma^2_{D_L} & \sigma_{D_L D_Y} & \sigma_{D_L D_{LY}} \\ \sigma_{D_Y D_L} & \sigma^2_{D_Y} & \sigma_{D_Y D_{LY}} \\ \sigma_{D_L D_{LY}} & \sigma_{D_Y D_{LY}} & \sigma^2_{D_{LY}} \end{bmatrix} / \{\mathrm{tr}[\mathbf{WW'}]\}/n \tag{8}$$

This approach, which is an extension of Vitezica et al. [22], makes it possible to estimate (co)variances of genotypic effects of SNPs in purebred and crossbred populations under a genomic model with additive and non-additive (dominance) inheritance.

Matrices $\mathbf{Z}$ and $\mathbf{W}$, their crossproducts and the inverses of $\mathbf{G}$ and $\mathbf{D}$ were built using own programs. Genetic parameters were estimated by using average information REML with software airemlf90 [33]. Standard errors on functions of genetic parameters (i.e. standard errors on correlations) were estimated from the average information matrix using the REML-MVN method of Houle and Meyer [36].

Additive and dominance variances in purebred and crossbred populations

The additive and dominance (co)variances of genotypic effects of SNPs, either within breed or between breeds, were calculated using Eqs. (6) and (8), respectively. Using these calculated additive and dominance (co)variances of SNPs across all the SNPs, the corresponding traditional, individual-based genetic parameters can be obtained as follows. The genetic parameters obtained are directly comparable to pedigree-based estimates [32].

Consider the allele substitution effect $\alpha = a + (q - p)d$. According to [32], the additive genetic variances for purebred performance (mating animals in the same breed) for breed L ($\sigma^2_{AP_L}$) and Y ($\sigma^2_{AP_Y}$) are:

$$\sigma^2_{AP_L} = \sum \left(2p^L_i q^L_i \right) \sigma^2_{a_L} + \sum \left(2p^L_i q^L_i \left(q^L_i - p^L_i \right)^2 \right) \sigma^2_{d_L},$$

$$\sigma^2_{AP_Y} = \sum \left(2p^Y_i q^Y_i \right) \sigma^2_{a_Y} + \sum \left(2p^Y_i q^Y_i \left(q^Y_i - p^Y_i \right)^2 \right) \sigma^2_{d_Y},$$

where σ^2_a and σ^2_d are the variances of additive and dominance genotypic effects of SNPs in either breed L or Y; p_i and q_i are allele frequencies for SNP i; indices L and Y denote the breeds Landrace and Yorkshire, respectively. For crossbred performance of say, Landrace, the allele substitution effect is $\alpha_{AC_L} = a_{AC_L} + \left(q^Y - p^Y \right) d_{AC_L}$. Thus, the additive genetic variances within purebred L and Y for crossbred performance (due to gametes from the L or Y individuals in the crossbred population) are equal to:

$$\sigma^2_{AC_L} = \sum \left(2p_i^L q_i^L\right)\sigma^2_{a_{LY}} + \sum \left(2p_i^L q_i^L \left(q_i^Y - p_i^Y\right)^2\right)\sigma^2_{d_{LY}},$$

$$\sigma^2_{AC_Y} = \sum \left(2p_i^Y q_i^Y\right)\sigma^2_{a_{LY}} + \sum \left(2p_i^Y q_i^Y \left(q_i^L - p_i^L\right)^2\right)\sigma^2_{d_{LY}},$$

where the $\sigma^2_{AC_L}$ represents the additive genetic variance of animals in breed L when mated to animals in breed Y; the $\sigma^2_{AC_Y}$ represents the additive genetic variance of animals in breed Y when mated to animals in breed L; and $\sigma^2_{a_{LY}}$ and $\sigma^2_{d_{LY}}$ are the variances of additive and dominance genotypic effects of SNPs in the crossbred LY population, respectively. The additive genetic variance for animals in the crossbred LY population ($\sigma^2_{AC_{LY}}$) is the sum of the additive genetic variance of Landrace alleles and that of Yorkshire alleles in the crossbred animals [22] as follows:

$$\sigma^2_{AC_{LY}} = \frac{1}{2}\sigma^2_{AC_L} + \frac{1}{2}\sigma^2_{AC_Y}.$$

Note that this variance is not the additive genetic variance of the crossbred animals acting as reproducers (i.e., creating an F2) [37].

The additive genetic covariances between purebred and crossbred performances within breeds L (σ_{AP_L,AC_L}) and Y (σ_{AP_Y,AC_Y}) are:

Correlations of allele substitution effects between two breeds

The breeding value of an individual includes the allele substitution effects of all genes and the allele frequencies. For purebred performance, the allele substitution effects of one locus for breed L and Y are:

$$\alpha_L = a_L + \left(q_i^L - p_i^L\right)d_L,$$

$$\alpha_Y = a_Y + \left(q_i^Y - p_i^Y\right)d_Y,$$

where a is the additive effect and d is the dominance effect for each SNP; p_i and q_i are allele frequencies for SNP i, with superscripts denoting breeds L or Y. In the case of purely additive gene action, the covariance between α_L and α_Y is $\sigma_{\alpha_{L,Y}}$, which can be interpreted as a genetic correlation among populations [38–40]. Then, the covariance between the allele substitution effects of one locus is:

$$\begin{aligned} cov(\alpha_L, \alpha_Y) &= cov\left(a_L + \left(q_i^L - p_i^L\right)d_L, a_Y + \left(q_i^Y - p_i^Y\right)d_Y\right) \\ &= cov(a_L, a_Y) + \left(q_i^L - p_i^L\right)\left(q_i^Y - p_i^Y\right)cov(d_L, d_Y) \\ &= \sigma_{a_{L,Y}} + \left(q_i^L - p_i^L\right)\left(q_i^Y - p_i^Y\right)\sigma_{d_{L,Y}}, \end{aligned}$$

where $\sigma_{a_{L,Y}}$ and $\sigma_{d_{L,Y}}$ are the additive and dominance covariances of SNP effects between breeds L and Y for additive and dominance, respectively. If we assume that

$$\sigma_{AP_L,AC_L} = \sum \left(2p_i^L q_i^L\right)\sigma_{a_{L,LY}} + \sum \left(2p_i^L q_i^L \left(q_i^L - p_i^L\right)\left(q_i^Y - p_i^Y\right)\right)\sigma_{d_{L,LY}},$$

$$\sigma_{AP_Y,AC_Y} = \sum \left(2p_i^Y q_i^Y\right)\sigma_{a_{Y,LY}} + \sum \left(2p_i^Y q_i^Y \left(q_i^Y - p_i^Y\right)\left(q_i^L - p_i^L\right)\right)\sigma_{d_{Y,LY}},$$

where $\sigma_{a_{L,LY}}$ and $\sigma_{d_{L,LY}}$ are the covariances of SNP effects between purebred L and crossbred LY populations for additive and dominance, respectively; $\sigma_{a_{Y,LY}}$ and $\sigma_{d_{Y,LY}}$ are the covariances of SNP effects between purebred Y and crossbred LY populations for additive and dominance, respectively.

Therefore, the genetic correlations of breeding values between purebred and crossbred performances within L (r_{PC_L}) and Y (r_{PC_Y}) are: $r_{PC_L} = \frac{\sigma_{AP_L,AC_L}}{\sqrt{\sigma^2_{AP_L}\sigma^2_{AC_L}}}$ and $r_{PC_Y} = \frac{\sigma_{AP_Y,AC_Y}}{\sqrt{\sigma^2_{AP_Y}\sigma^2_{AC_Y}}}$.

According to [22], the dominance genetic variances within purebred populations L and Y are $\sigma^2_{D_L} = \sum \left(2p_i^L q_i^L\right)^2 \sigma^2_{d_L}$ and $\sigma^2_{D_Y} = \sum \left(2p_i^Y q_i^Y\right)^2 \sigma^2_{d_Y}$, respectively. The dominance genetic variance in crossbred LY animals is $\sigma^2_{D_{LY}} = \sum \left(4p_i^L q_i^L p_i^Y q_i^Y\right)\sigma^2_d$.

The broad sense heritabilities for purebred performance (H_P^2) were calculated as the ratio of total genetic variances for purebred performance ($\sigma^2_{AP} + \sigma^2_D$) to phenotypic variances ($\sigma^2_{AP} + \sigma^2_D + \sigma^2_e$).

SNP effects (both additive and dominance) are independent across loci, then the covariance between the allele substitution effects across all n loci is:

$$cov(\alpha_L, \alpha_Y) = \sigma_{\alpha_{L,Y}} = \sigma_{a_{L,Y}} + \frac{1}{n}\sum \left(\left(q_i^L - p_i^L\right)\left(q_i^Y - p_i^Y\right)\right)\sigma_{d_{L,Y}}.$$

Also, the variances of allele substitution effects across all n loci for breeds L and Y are:

$$var(\alpha_L) = \sigma^2_{\alpha_L} = \sigma^2_{a_L} + \frac{1}{n}\sum \left(\left(q_i^L - p_i^L\right)^2\right)\sigma^2_{d_L},$$

$$var(\alpha_Y) = \sigma^2_{\alpha_Y} = \sigma^2_{a_Y} + \frac{1}{n}\sum \left(\left(q_i^Y - p_i^Y\right)^2\right)\sigma^2_{d_Y},$$

where σ^2_a and σ^2_d are the additive and dominance variance of SNPs. Then, the correlation of allele substitution effects for purebred performance between populations L and Y is $r_{\alpha P_L,\alpha P_Y} = \frac{\sigma_{\alpha_{L,Y}}}{\sigma_{\alpha_L}\sigma_{\alpha_Y}}$. If there is no dominance variation, the $r_{\alpha P_L,\alpha P_Y}$ relates to additive genetic variances as $r_{\alpha P_L,\alpha P_Y} = \frac{\sigma_{a_{L,Y}}}{\sigma_{a_L}\sigma_{a_Y}}$.

The correlation of allele substitution effects for crossbred performance between populations L and Y is similar to that for purebred performance, but the allele frequencies are swapped, as:

$$r_{\alpha C_L, \alpha C_Y} = \frac{\sigma_{\alpha_{L\,in\,LY}, Y\,in\,LY}}{\sigma_{\alpha_{L\,in\,LY}} \sigma_{\alpha_{Y\,in\,LY}}}$$

$$= \frac{\sigma_{a_{LY}}^2 + \frac{1}{n} \sum \left((q_i^L - p_i^L)(q_i^Y - p_i^Y) \right) \sigma_{d_{LY}}^2}{\sqrt{\sigma_{a_{LY}}^2 + \frac{1}{n} \sum \left((q_i^Y - p_i^Y)^2 \right) \sigma_{d_{LY}}^2} \sqrt{\sigma_{a_{LY}}^2 + \frac{1}{n} \sum \left((q_i^L - p_i^L)^2 \right) \sigma_{d_{LY}}^2}},$$

where $\sigma_{a_{LY}}^2$ and $\sigma_{d_{LY}}^2$ are the additive and dominance variance of SNPs in the crossbred LY population. If there is no dominance variation, the $r_{\alpha C_L, \alpha C_Y}$ is equal to 1, by assumption in the model.

Scenarios

Variance components, genetic correlations of breeding values between purebred and crossbred performances (r_{PC}) within each pure breed and correlations of allele substitution effects for purebred ($r_{\alpha P_L, \alpha P_Y}$) and crossbred ($r_{\alpha C_L, \alpha C_Y}$) performance between two pure breeds were first investigated using the full genomic dataset. To explore the effects of using genomic information and the inclusion of dominance deviation on the genetic evaluation of crossbred performance in the trivariate model, three different scenarios were compared.

Nogen

The statistical model was a trivariate BLUP model, similar to Eq. (3), but the dominance deviation was excluded. Instead of using a genomic relationship matrix, a single relationship matrix **A** was constructed across the three breeds, assuming that they form a single population. Thus, the genetic (co)variances of additive genetic effects **u** were:

$$Var(\mathbf{u}) = \mathbf{A} \otimes \begin{bmatrix} \sigma_{A_L}^2 & \sigma_{A_L A_Y} & \sigma_{A_L A_{LY}} \\ \sigma_{A_Y A_L} & \sigma_{A_Y}^2 & \sigma_{A_Y A_{LY}} \\ \sigma_{A_L A_{LY}} & \sigma_{A_Y A_{LY}} & \sigma_{A_{LY}}^2 \end{bmatrix} = \mathbf{A} \otimes \mathbf{A_0},$$

where $\mathbf{A_0}$ were variance components associated to genetic additive effects and not the genotypic additive effects in Eq. (5). Pedigree-based inbreeding depression was also included in the model. The pedigree-based inbreeding coefficients were calculated as in [41] using the software inbupgf90 [33].

Gen_AM

The statistical model was similar to Eq. (3), but without dominance deviations. Genomic information was used to construct the additive genomic relationship matrix.

Gen_ADM

The statistical model includes additive and dominance effects as in Eq. (3). Genomic information was used to construct the additive and dominance genomic relationship matrices.

To explore the impact of genomic information and dominance effects on genomic evaluation for crossbred performance, the full genomic dataset was split into training and validation populations and the predictive ability for crossbred animals in the validation population was investigated in different scenarios. The farrowing date of January 1, 2013 was used as the cut-off date to divide recorded purebred and crossbred sows into training and validation populations. As a result, 6769 sows (1270 L, 1405 Y and 4094 LY) were included in the training population, while the remaining 2716 sows (854 L, 813Y and 1049 LY) were included in the validation population. Predictive ability of crossbreds was measured as the correlations $cor(y_c, \hat{y})$ in the validation population for each scenario, where y_c is the corrected phenotypic records of TNB for crossbred animals; $\hat{y}$ is the predicted corrected observations of TNB for crossbred animals and is equal to the sum of the estimated population mean ($\hat{\mu}$), inbreeding ($\hat{fb}$) and genotypic values ($\hat{g}$); the genotypic value $\hat{g}$ was calculated as the sum of additive and dominance genetic effects in the scenario Gen_ADM. In the other two scenarios, the genotypic value $\hat{g}$ only included the additive genetic effect. Hotelling–Williams t test at a confidence level of 5% was applied to evaluate the significance of the differences in validation correlations in each scenario. Furthermore, to detect the possible biases in the predictions, the regression coefficients of y_c on $\hat{y}$ were explored. Note that no bias implies that a regression coefficient equals 1. In addition, to measure the uncertainty associated with the predictions, 1000 bootstrap samples [42] was applied to estimate the means and standard errors.

For comparison, the predictive ability of crossbred animals was also investigated in a model without inbreeding depression effects, for all three scenarios. The predictive ability was measured as the correlation $cor(y_c, \hat{y})$, where $\hat{y}$ is the sum of the estimated population mean ($\hat{\mu}$) and genotypic value ($\hat{g}$).

Table 1 Variance components of additive and dominance genetic effects for purebred and crossbred animals

Scenario	Breed	σ^2_{AP}	$\sigma_{AP,AC}$	σ^2_{AC}	σ^2_D	σ^2_e	$\sigma^2_{AC_{LY}}$	$\sigma^2_{D_{LY}}$	$\sigma^2_{e_{LY}}$
Nogen	L	0.99 (0.31)	0.17 (0.07)	0.05 (0.02)	–	10.82 (0.43)	0.05 (0.02)	–	7.35 (0.15)
	Y	1.07 (0.33)	0.15 (0.07)	0.05 (0.02)	–	8.96 (0.38)			
Gen_AM	L	0.87 (0.22)	0.47 (0.10)	0.28 (0.07)	–	10.89 (0.38)	0.28 (0.07)	–	7.11 (0.15)
	Y	0.55 (0.20)	0.17 (0.10)	0.28 (0.07)	–	9.42 (0.33)			
Gen_ADM	L	0.86 (0.21)	0.46 (0.10)	0.28 (0.06)	0.04 (0.03)	10.86 (0.38)	0.28 (0.06)	0.02 (0.01)	7.11 (0.15)
	Y	0.54 (0.18)	0.17 (0.09)	0.28 (0.06)	0.06 (0.05)	9.35 (0.33)			

Numbers in brackets are the standard errors of the corresponding parameters

σ^2_{AP} is the additive genetic variance for purebred performance; $\sigma_{AP,AC}$ is the additive genetic covariance between purebred and crossbred performance; σ^2_{AC} is the additive genetic variance for crossbred performance; σ^2_D is the dominance genetic variance for either purebred animals; σ^2_e is the residual variance for purebred animals; $\sigma^2_{AC_{LY}}$ is the additive genetic variance for the F1 crossbred animals LY; $\sigma^2_{D_{LY}}$ is the dominance genetic variance for the F1 crossbred animals LY; and $\sigma^2_{e_{LY}}$ is the residual variance for the F1 crossbred animals LY

L Landrace, *Y* Yorkshire breeds

Results

Variance components, heritabilities and correlations

Table 1 shows the estimates of variance components for additive genetic effects for purebred (σ^2_{AP}) and crossbred (σ^2_{AC}) performance in different scenarios, and dominance variations (σ^2_D) in the *Gen_ADM* scenario. For all scenarios, the additive genetic variances for purebred performance (σ^2_{AP}) were larger than those for their crossbred performance (σ^2_{AC}). Estimated variance components in the scenarios *Gen_AM* and *Gen_ADM* were very close, but different from those obtained in scenarios without using genomic information. In general, estimates had large standard errors in all scenarios, but no obvious differences in standard errors were detected between different scenarios. Residual variance for purebred animals (σ^2_e) was larger than for crossbred animals ($\sigma^2_{e_{LY}}$) in each scenario. For the scenario *Gen_ADM*, the ratios of dominance genetic variance to additive genetic variance ranged from 5 to 11% for both purebred and crossbred populations.

The broad sense heritabilities for purebred and crossbred animals, genetic correlations between breeding values for purebred and crossbred performances within pure breeds and correlations of allele substitution effects across the two breeds are in Table 2. In different scenarios, the heritabilities of purebred performance (H^2_P) ranged from 0.07 (0.03) to 0.08 (0.03) and from 0.06 (0.03) to 0.10 (0.03) for breeds L and Y, respectively. Standard errors of H^2_P were almost consistent across scenarios. Estimated genetic correlations of breeding values between purebred and crossbred performances (r_{PC}) increased from 0.76 (0.20) (*Nogen*) to 0.95 (0.06) (*Gen_AM*) for breed L and from 0.43 (0.22) (*Gen_ADM*) to 0.54 (0.30) (*Nogen*) for breed Y. The r_{PC} was higher for breed L than for breed Y in all scenarios, but the standard errors of r_{PC} were always higher for breed Y than for breed L. With genomic information, the correlations

Table 2 Heritabilities and genetic correlations between breeding values for purebred and crossbred performances

Scenario	Breed	r_{PC}	H^2_P
Nogen	L	0.76 (0.20)	0.08 (0.03)
	Y	0.54 (0.30)	0.10 (0.03)
Gen_AM	L	0.95 (0.06)	0.07 (0.03)
	Y	0.44 (0.20)	0.06 (0.03)
Gen_ADM	L	0.93 (0.05)	0.08 (0.03)
	Y	0.43 (0.22)	0.06 (0.03)

Numbers between brackets are the standard errors of the corresponding parameters

r_{PC} is the genetic correlation of breeding values between purebred and crossbred performances within the Landrace or Yorkshire breeds; H^2_P is the broad sense heritability for purebred performance for the Landrace and Yorkshire breeds in different scenarios

L Landrace, *Y* Yorkshire

of allele substitution effects between purebred ($r_{\alpha P_L, \alpha P_Y}$) and crossbred ($r_{\alpha C_L, \alpha C_Y}$) performance between breeds L and Y were estimated, as shown in Table 3. For purebred performance, $r_{\alpha P_L, \alpha P_Y}$ was equal to 0.14 and 0.19 in *Gen_AM* and *Gen_AMD*, respectively. However, the standard errors were large, around 0.2 in both scenarios. For crossbred performance, $r_{\alpha C_L, \alpha C_Y}$ was equal to 0.98 in *Gen_ADM*. This high correlation is a byproduct of assuming that additive biological effects in crossbred animals are the same regardless of the Yorkshire or Landrace origin of the allele. However, the same allele has potentially different effects in the respective Landrace or Yorkshire genetic backgrounds, and the difference is modeled through the correlations, hence the low values of $r_{\alpha P_L, \alpha P_Y}$. Without including the dominance effects in the model *Gen_AM*, $r_{\alpha C_L, \alpha C_Y}$ was equal to 1 by definition.

Table 3 Correlations of allele substitution effects for purebred and crossbred performance between Landrace and Yorkshire breeds

Scenario	$r_{\alpha P_L,\alpha P_Y}$	$r_{\alpha C_L,\alpha C_Y}$
Nogen	–	–
Gen_AM	0.14 (0.22)	1
Gen_ADM	0.19 (0.24)	0.98 (0.02)

Numbers between brackets are the standard errors of the corresponding parameters

$r_{\alpha P_L,\alpha P_Y}$ is the correlation of allele substitution effects for purebred performance between the Landrace and Yorkshire breeds; $r_{\alpha C_L,\alpha C_Y}$ is the correlation of allele substitution effects for crossbred performance between the Landrace and Yorkshire breeds. For Gen_AM, $r_{\alpha C_L,\alpha C_Y}$ is equal to 1 by definition

Predictive abilities

Predictive abilities for crossbred pigs in the validation population are in Table 4. The correlation between the corrected phenotypic values and the predicted observations for TNB $(cor(y_c, \hat{y}))$ ranged from 0.010 in the scenario Nogen to 0.056 in scenarios Gen_AM and Gen_ADM. Standard errors of $cor(y_c, \hat{y})$ based on 1000 bootstrap samples were equal to 0.03 across all scenarios. No significant differences in predictive ability between scenarios were detected by the Hotelling–Williams t test at the confidence level of 5%.

The regression coefficients of corrected phenotypic values on the predicted corrected observations for TNB are in the second row of Table 4. Regression coefficients were smaller than 1 for the three scenarios. Among these scenarios, regression coefficients for scenarios with genomic information (Gen_AM and Gen_ADM) were slightly closer to 1 than that for the pedigree-based scenario (Nogen). Except for the Nogen scenario, standard errors of regression coefficients were around 0.39. For the Nogen scenario, the standard error was around 5 times larger than that for other scenarios. Overall, there was no clear trend towards a scenario with less bias.

For comparison, predictive abilities $cor(y_c, \hat{y})$ for crossbred pigs in the validation population for the models without the inbreeding depression effect were equal to −0.08 in scenario Nogen, 0.045 in scenario Gen_AM and 0.046 in scenario Gen_ADM. In all cases, these are lower than the predictive abilities in Table 4, and these differences are statistically significant according to the Hotelling–Williams t test.

Inbreeding depression

Marker-based and pedigree-based inbreeding coefficient (f) for each population and their estimated corresponding inbreeding depression parameters (b) in the different scenarios are in Table 5. Marker-based inbreeding coefficients were almost identical for breeds L and Y, but they were larger than those for LY, which was expected because crossbred animals have a higher level of heterozygosity than purebred animals. However, according to the pedigree-based inbreeding coefficients, the Landrace population was slightly more inbred than the Yorkshire population. In terms of inbreeding depression parameters (b), they were all negative (thus, genomic inbreeding has detrimental effects for TNB even in crossbred animals) but not of the same magnitude across the three populations. Note that for the scenario Nogen, b was estimated based on the pedigree-based inbreeding coefficients. As a whole, breed L had the most negative b, while breed Y had the least negative b, regardless of the scenario. Thus, TNB was more negatively affected by inbreeding in breed L than in breed Y and population LY.

Table 4 Predictive ability for crossbred animals in the validation population

	Nogen	Gen_AM	Gen_ADM
$cor(y_c, \hat{y})$[a]	0.010 (0.031)	0.056 (0.031)	0.056 (0.031)
Regression coefficient[b]	0.703 (2.218)	0.736 (0.386)	0.730 (0.385)

Numbers between brackets are the standard errors of the corresponding parameters

[a] Predictive ability ($cor(y_c, \hat{y})$) is given by the correlation coefficient between the corrected phenotypes (y_c) and their predictions ($\hat{y}$) for total number of piglets born (TNB) in crossbred animals

[b] Regression coefficient of the corrected phenotypes (y_c) on the predicted observations ($\hat{y}$) in crossbred animals

Table 5 Marker-based and pedigree-based inbreeding coefficients f and estimated inbreeding depression parameter b (piglets per 100% of inbreeding) in different scenarios for each breed

	L	Y	LY
Marker-based inbreeding coefficient f[a]	0.695 (0.019)	0.698 (0.020)	0.565 (0.012)
Pedigree-based inbreeding coefficient f[b]	0.111 (0.032)	0.078 (0.031)	0
Nogen (b)	−4.821	−3.561	0
Gen_AM (b)	−9.656	−1.924	−5.122
Gen_ADM (b)	−9.731	−1.878	−5.055

The inbreeding coefficient is the mean inbreeding coefficient across individuals within each breed

Numbers between brackets are the standard deviations of the mean inbreeding coefficient

For Nogen, the inbreeding depression parameter b is the regression of phenotype on pedigree-based inbreeding. For Gen_AM and Gen_ADM, the inbreeding depression parameter b is the regression of phenotype on marker-based inbreeding

[a] Calculated as the proportion of homozygous loci per individual

[b] Calculated as in Meuwissen and Luo [41]

Discussion

This study extended the trivariate GBLUP model of Vite-zica et al. [22] in order to obtain (co)variances of effects of SNPs, genetic correlations of breeding values between purebred and crossbred performances and correlations of allele substitution effects under dominance. We also evaluated this model using different scenarios for the genetic evaluation of crossbred performance in Danish purebred and crossbred pigs. Scenarios that included or not genomic information were studied to estimate the genetic correlations of breeding values between purebred and crossbred performances. To our knowledge, this is the first study to report correlations of allele substitution effects between two breeds in the presence of dominance effects. The results show that the Vitezica model [22] is a tool that can be used for the genomic evaluation of cross-bred performance in genotyped animals. In this study, for TNB, models with dominance deviations did not improve the genomic evaluation of crossbred performance with regard to both predictive ability and unbiasedness, but the inclusion of an inbreeding depression effect in the models significantly improved predictive ability.

Phenotypic variances were larger for purebred animals (11.76 for breed L and 9.99 for breed Y) than for cross-bred animals (7.30 for LY). This could be the reason why the estimated additive genetic variances for purebred performance (σ_{AP}^2) were larger than those for crossbred performance (σ_{AC}^2). However, compared to results in a previous study that used a much larger Danish pure-bred and crossbred dataset [17], both estimated additive genetic variances and phenotypic variances in the cur-rent study were smaller, which is due to three reasons. (1) The dataset in the current study was a genotyped subset of the population used in the previous study. Purebred genotyped individuals were pre-selected and their per-formances were more homogeneous than that of the whole population. The preselection process resulted in a loss of about 15% of the purebred phenotypic varia-tion. However, the genotyped crossbred animals were an almost random sample of the whole population and there was only a small loss of about 5% of phenotypic varia-tion for crossbred animals. (2) The phenotypic values for TNB in the current study were pre-corrected for fixed and non-genetic random effects. This pre-correction led to a loss of about 11 and 17% of phenotypic varia-tion for purebreds and crossbreds, respectively. (3) Dur-ing the pre-correction, some genetic variation may have been allocated to other random effects (e.g. service boar effects), in particular because TNB is a lowly heritable trait.

The estimated heritabilities of TNB for purebred per-formance (H_P^2) were slightly lower than those previously reported (0.11 and 0.09 for breeds L and Y, respectively)

[17, 22, 43]. Large standard errors of H_P^2 implied that the current dataset was not large enough. The consist-ent standard errors across scenarios indicated that even when genomic information was included, the uncertainty of H_P^2 did not decrease. Taking the standard errors into account, the estimated H_P^2 across scenarios were not very different. Compared to the results of [17], the lower H_P^2 found in the current study was due to the sharp decrease in additive genetic variances (σ_{AP}^2).

The ratios of estimated dominance genetic variances to additive genetic variances in the current study (5 to 11%) were generally a little smaller than in other studies on TNB. Vitezica et al. [22] reported that this ratio was equal to about 20% for litter size in both purebred and cross-bred lines by using the same trivariate GBLUP model. Esfandyari et al. [19] stated that, by using purebred genomic information in a univariate Bayesian mixture model at the SNP level, the ratio between dominance variance and additive variance for TNB was equal to 15 and 18% for breeds L and Y, respectively. Based on pedi-gree information, Misztal et al. [10] reported a ratio that reached about 25% for number of piglets born alive in a Yorkshire population. However, there are some stud-ies that did report smaller ratios than those reported here. For instance, Hidalgo [20] reported that, based on genotyped crossbred animals, the dominance variance for TNB accounted for nearly zero of the total genetic variance and concluded that TNB was not affected by dominance effects in the Dutch Landrace and Yorkshire populations. For other traits or species, different ratios of dominance genetic variance to additive genetic variance were also reported. For average daily gain in Duroc pigs, Su et al. [9] estimated a ratio of 15%, but their results were based on genotypic variance components and can-not be directly compared to genetic variance components [32]. For average daily weight gain in Yorkshire and Lan-drace pigs, Lopes et al. [44] reported ratios of 13.8 and 28%, respectively by including genomic information. For Fleckvieh cattle, Ertl et al. [12] calculated ratios that ranged from 3.4% for stature to 69% for protein yield by using a univariate SNP-BLUP model. Overall, these dif-ferent ratios of dominance genetic variance to additive genetic variance may reflect differences in the traits ana-lyzed and in the type of information used for the estima-tion [9], and also uncertainty in the estimates.

The genetic correlation of breeding values between purebred and crossbred performances (r_{PC}) is a key parameter in crossbreeding schemes [2]. In the current study, the estimated r_{PC} was in line with results reviewed by Wei et al. [3]. Lutaaya et al. [5] also reported r_{PC} that ranged from 0.32 to 1. Such differences in r_{PC} may reflect differences in the extent of GxE interactions and the distance across breeds. In our study, estimated r_{PC}

did not vary dramatically across the scenarios, when the standard errors were taken into account. These standard errors were very large, which indicated that the amount of available information was too small to ensure accurate r_{PC} estimates. Across scenarios, standard errors of r_{PC} decreased when genomic information was included, which indicates that including genomic information may reduce the uncertainty of the estimations. r_{PC} was larger for breed L than for breed Y, which was in agreement with a previous study [17] and may be due to the data structure. Among the 5143 crossbred animals, the number of Yorkshire sires (N = 1125) was much smaller than that of Landrace sires (N = 4018). Such a different amount of information affects the accuracy of the estimates, and thus the standard error of r_{PC} was larger for breed Y than for breed L (see Table 2). However, compared to the results reported in [17], the r_{PC} for breed L increased by about 10% while that for breed Y did not change much. Both pre-correction of data and the genotyped subset of original data used may play a role in the differences observed between the current and previous results [17]. In the previous study, a single-step method, which can use pedigree information and genomic information simultaneously, was used. In this study, the use of only phenotypic records on genotyped individuals affected the accuracy of estimates. Our results confirmed the moderate value of the r_{PC} for TNB in breeds L and Y.

To our knowledge, this is the first time that correlations of allele substitution effects for both purebred $(r_{\alpha P_L, \alpha P_Y})$ and crossbred $(r_{\alpha C_L, \alpha C_Y})$ performance between two breeds in the presence of dominance variation are estimated. In genomic selection, SNPs are assumed to be in LD with QTL along the whole genome [45]. The correlation of allele substitution effects between breeds measures the degree of average similarities between SNP effects assuming that the QTL effects are the same in breeds 1 and 2 [38–40]. In practice, the correlation of allele substitution effects between two breeds can be interpreted as indicating "how consistent the SNP substitution effects are across two breeds". For purebred performance, the estimated SNP substitution effects were based on the within-breed allele frequencies. A high $r_{\alpha P_L, \alpha P_Y}$ correlation means that the estimated SNP substitution effects based on allele frequencies from breed L can be used for breed Y and vice versa. However, $r_{\alpha P_L, \alpha P_Y}$ was not significantly different from 0 in the current study, which demonstrates that SNP effects estimated from a reference population that consists of one pure breed (e.g. Landrace) cannot be readily applied to the other breed (e.g. Yorkshire). This was in agreement with the findings of [46] who reported that prediction based on an across-population reference panel was worse than within-population prediction. In other species, estimated correlations

of allele substitution effects between breeds based on models without dominance, oscillate between 0 and 0.8, and are trait-dependent [38, 47]. For crossbred performance, an $r_{\alpha C_L, \alpha C_Y}$ close to 1 was found in the current study, which indicated that the allele substitution effects based on the allele frequencies from the opposite breeds were very similar for the L and Y breeds. In practice, this suggests that SNP substitution effects that are estimated based on a reference population consisting of crossbred animals can be used to estimate crossbred breeding values for both breeds L and Y.

It was expected that genomic evaluations obtained by including dominance deviations in the model would be improved, especially when records of crossbred animals were included [9]. However, our results showed that inclusion of dominance deviations did not increase the predictive ability for crossbreds. This result was in line with conclusions in [9, 12, 20], but was opposite to those in [18, 19, 48, 49]. Theoretically, estimating dominance genetic effects should be useful because ignoring them will result in less accurate estimates of allele substitution effects and consequently less accurate estimated breeding values in genomic prediction [11]. However, regarding the additive genetic variance, estimates were nearly the same in scenarios *Gen_AM* and *Gen_ADM*, which demonstrated that the additive variances were already well captured by the additive model. Thus, the accuracy of the estimated additive genetic effects was not affected when dominance effects were included in the model [12]. Moreover, a simulation study at the level of the gene action showed that when all gene actions were purely additive, including dominance in addition to the additive effects in the model was not advantageous compared to using an additive model. Hidalgo [20] showed that TNB was not affected by dominance in the Dutch crossbred population. In the current study, we also observed similar results, and dominance variation accounted for a small proportion of the total genetic variation (4 to 10%). The lack of change in predictive ability also indicated the difficulty of distinguishing dominance genetic effects from additive genetic effects [9], but it confirmed a previous simulation study that concluded that the use of a dominance model did not negatively affect genomic evaluation even if the trait was purely additive [18].

Scenarios in which genomic information was included (*Gen_AM* and *Gen_ADM*) showed higher predictive abilities than the pedigree-based scenario (*Nogen*). For the *Nogen* scenario, the relationship matrix was constructed based on a base population that was considered as a mixture of L and Y animals, which was not the case. Therefore, the results of the *Gen_AM* and *Gen_ADM* scenarios were more reliable than those of the *Nogen* scenario. Although predictive abilities were not significantly different according to the Hotelling-Williams t-test, the results from 1000 bootstrap samples still showed that the predictive abilities of about 90% of the crossbred animals would be higher when genomic information was available (894 of 1000 bootstrap samples showed higher predictive abilities in scenarios that included genomic information than those in the *Nogen* scenario; results not shown). Comparison of the predictive abilities that were estimated in the current study with those from a previous study [17] indicated that the single-step model [16] might be more robust than the Vitezica model [22] used in this paper in terms of both predictive ability and unbiasedness for the crossbred performance. Our results suggested that using a small set of genotyped animals and pre-corrected data to implement genetic evaluation for crossbred performance was less powerful than using the whole dataset, which is similar to the conclusions for purebred performance [43].

The regression coefficients obtained with the Vitezica model were less than 1, which suggests that variations in total genetic effects could be overestimated (inflated). In terms of unbiasedness, there was no clear trend among the scenarios examined, regardless of whether genomic information was included or not. Overall, unbiasedness was not a problem in the current study because the regression coefficients in all scenarios did not significantly differ from 1.

Inbreeding depression for litter size in pigs is a well-known phenomenon [50, 51], and we found that inclusion of inbreeding effects in the model improved predictive abilities of crossbred animals. Estimates of inbreeding depression effects are rarely reported, but our estimates agree with those previously reported for commercial and Iberian pigs [52]. Inbreeding depression was, for the same amount of marker-based inbreeding, more detrimental in the Landrace than in the Yorkshire breed. There are many possible explanations among which the purging of lethal recessive alleles [53]. We also report an estimate of the inbreeding depression parameter for the crossbred animals, which is between the estimates for the parental breeds. To our knowledge, this estimate has never been reported.

The correlation between breeding values and dominance deviations is of theoretical concern [30]. However, this does not apply to the current marker-based analyses for the following reasons. (1) In a pedigree-based analysis, mating in an inbred population produces deviations from the Hardy–Weinberg equilibrium, which generate correlations between breeding values and dominance deviations [30]. However, in our study, SNPs are in Hardy–Weinberg equilibrium if allele frequencies are considered in the current generation. (2) Such a correlation occurs because the pedigree

information forces the genetic model to refer to the base population, since the state of alleles is not known, i.e. only probabilities of IBD are known. In our study, the states of alleles are known and the model can be described as referring to the current generation instead. (3) The equivalent GBLUP models in Eq. (3) used genotypic additive and dominance values, not breeding values and dominance deviations. A reasonable assumption in the model is that additive and dominance effects are unrelated at each SNP. Thus, covariance between additive and dominance genetic effects was ignored in the current study.

Conclusions

We present for the first time the use of genomic inbreeding in crossbred and purebred genomic evaluation. Estimates are biologically sound and are relevant even for crossbred animals. We also report for the first time, estimated correlations of allele substitution effects in the presence of dominance. For TNB, the dominance genetic variance accounts for only a small proportion of the total genetic variation (4 to 10%). A moderate, positive genetic correlation between breeding values for TNB for purebred and crossbred performances was confirmed. Inclusion of dominance in the GBLUP model did not improve predictive ability for crossbred animals, whereas inclusion of inbreeding depression effects did. An additive GBLUP model is sufficient to capture the additive genetic variances and for genomic evaluation. The GBLUP model [22] was applied successfully for genetic evaluations for crossbred performance in pigs. This model can potentially be a useful tool in genetic evaluation for crossbred performance.

Authors' contributions
TX performed data analysis and wrote the manuscript. All authors participated in the derivation of the theory. AL and OFC coordinated the project, conceived the study, made substantial contributions for the interpretation of results and revised the manuscript. ZGV improved the manuscript and added valuable comments during the study. All authors read and approved the final manuscript.

Author details
[1] Department of Molecular Biology and Genetics, Center for Quantitative Genetics and Genomics, Aarhus University, 8830 Tjele, Denmark. [2] UR1388 GenPhySE, INRA, CS-52627, 31326 Castanet-Tolosan, France. [3] INP, ENSAT, GenPhySE, Université de Toulouse, 31326 Castanet-Tolosan, France.

Acknowledgements
The work was funded through the Green Development and Demonstration Programme (Grant No. 34009-12-0540) by the Danish Ministry of Food, Agriculture and Fisheries, the Pig Research Centre and Aarhus University. The first author benefits from a joint grant from the European Commission and Aarhus University, within the framework of the Erasmus-Mundus joint doctorate "EGS-ABG". AL and ZGV thank financial support from the INRA SelGen metaprogram projects X-Gen and SelDir. OFC acknowledges funding from the GenSAP project. We are grateful to the Genotoul bioinformatics platform Toulouse Midi-Pyrenees for providing computing resources. Discussions with Luis Varona are gratefully acknowledged.

Competing interests

The authors declare that they have no competing interests.

References

1. Wei M, Van der Werf JHJ. Maximizing genetic response in cross-breds using both purebred and crossbred information. Anim Sci. 1994;59:401–13.
2. Bijma P, Bastiaansen JWM. Standard error of the genetic correlation: how much data do we need to estimate a purebred–crossbred genetic correlation? Genet Sel Evol. 2014;46:79.
3. Wei M, Van der Steen HAM. Comparison of reciprocal recurrent selection with pure-line selection systems in animal breeding (a review). Anim Breed Abstr. 1991;59:281–98.
4. Dekkers JCM. Marker-assisted selection for commercial crossbred performance. J Anim Sci. 2007;85:2104–14.
5. Lutaaya E, Misztal I, Mabry JW, Short T, Timm HH, Holzbauer R. Genetic parameter estimates from joint evaluation of purebreds and crossbreds in swine using the crossbred model. J Anim Sci. 2001;79:3002–7.
6. Lo LL, Fernando RL, Grossman M. Genetic evaluation by BLUP in two-breed terminal crossbreeding systems under dominance. J Anim Sci. 1997;75:2877–84.
7. Charlesworth D, Willis JH. The genetics of inbreeding depression. Nat Rev Genet. 2009;10:783–96.
8. Falconer DS, Mackay TFC. Introduction to quantitative genetics. New York: Longman Group Ltd; 1981.
9. Su G, Christensen OF, Ostersen T, Henryon M, Lund MS. Estimating additive and non-additive genetic variances and predicting genetic merits using genome-wide dense single nucleotide polymorphism markers. PLoS One. 2012;7:e45293.
10. Misztal I, Varona L, Culbertson M, Bertrand JK, Mabry J, Lawlor TJ, et al. Studies on the value of incorporating the effect of dominance in genetic evaluations of dairy cattle, beef cattle and swine. Biotechnol Agron Soc Environ. 1998;2:227–33.
11. Toro MA, Varona L. A note on mate allocation for dominance handling in genomic selection. Genet Sel Evol. 2010;42:33.
12. Ertl J, Legarra A, Vitezica ZG, Varona L, Edel C, Emmerling R, et al. Genomic analysis of dominance effects on milk production and conformation traits in Fleckvieh cattle. Genet Sel Evol. 2014;46:40.
13. Fulton JE. Genomic selection for poultry breeding. Anim Front. 2012;2:30–6.
14. Loberg A, Dürr JW. Interbull survey on the use of genomic information. Interbull Bull. 2009;39:3–14.
15. Christensen OF, Legarra A, Lund MS, Su G. Genetic evaluation for three-way crossbreeding. Genet Sel Evol. 2015;47:98.
16. Christensen OF, Madsen P, Nielsen B, Su G. Genomic evaluation of both purebred and crossbred performances. Genet Sel Evol. 2014;46:23.
17. Xiang T, Nielsen B, Su G, Legarra A, Christensen OF. Application of single-step genomic evaluation for crossbred performance in pig. J Anim Sci. 2016;94:936–48.
18. Zeng J, Toosi A, Fernando RL, Dekkers JCM, Garrick DJ. Genomic selection of purebred animals for crossbred performance in the presence of dominant gene action. Genet Sel Evol. 2013;45:11.
19. Esfandyari H, Bijma P, Henryon M, Christensen OF, Sorensen AC. Genomic prediction of crossbred performance based on purebred Landrace and Yorkshire data using a dominance model. Genet Sel Evol. 2016;48:40.
20. Hidalgo AM. Exploiting genomic information on purebred and crossbred pigs. PhD thesis. Swedish University of Agricultural Sciences, Uppsala; 2015.
21. Ibáñez-Escriche N, Fernando RL, Toosi A, Dekkers JCM. Genomic selection of purebreds for crossbred performance. Genet Sel Evol. 2009;41:12.
22. Vitezica ZG, Varona L, Elsen MJ, Misztal I, Herring W, Legarra A. Genomic BLUP including additive and dominant variation in purebreds and F1 crossbreds, with an application in pigs. Genet Sel Evol. 2016;48:6.
23. Madsen P. DMU trace, a program to trace the pedigree for a subset of animals from a large pedigree file, version 2. Tjele: Center for Quantitative Genetics and Genomics, Department of Molecular Biology and Genetics, Aarhus University; 2012.
24. Ramos AM, Crooijmans RP, Affara NA, Amaral AJ, Archibald AL, Beever JE, et al. Design of a high density SNP genotyping assay in the pig using SNPs identified and characterized by next generation sequencing technology. PLoS One. 2009;4:e6524.
25. GeneSeek Company. GGP-for Porcine LD (GeneSeek Genomic Profiler for Porcine Low Density). 2012. http://www.neogen.com/Genomics/pdf/Slicks/GGP_PorcineFlyer.pdf.
26. Xiang T, Ma P, Ostersen T, Legarra A, Christensen OF. Imputation of genotypes in Danish purebred and two-way crossbred pigs using low-density panels. Genet Sel Evol. 2015;47:54.
27. Browning SR. Missing data imputation and haplotype phase inference for genome-wide association studies. Hum Genet. 2008;124:439–50.
28. Silió L, Rodríguez M, Fernández A, Barragán C, Benítez R, Óvilo C, et al. Measuring inbreeding and inbreeding depression on pig growth from pedigree or SNP-derived metrics. J Anim Breed Genet. 2013;130:349–60.
29. Lynch M, Walsh B. Genetics and analysis of quantitative traits. 1st ed. Sunderland: Sinauer Assoc; 1998.
30. de Boer IJM, Hoeschele I. Genetic evaluation methods for populations with dominance and inbreeding. Theor Appl Genet. 1993;86:245–58.
31. Aliloo H, Pryce JE, Gonzalez-Recio O, Cocks BG, Hayes BJ. Accounting for dominance to improve genomic evaluations of dairy cows for fertility and milk production traits. Genet Sel Evol. 2016;48:8.
32. Vitezica ZG, Varona L, Legarra A. On the additive and dominant variance and covariance of individuals within the genomic selection scope. Genetics. 2013;195:1223–30.
33. Misztal I, Tsuruta S, Strabel T, Auvray B, Druet T, Lee DH. BLUPF90 and related programs (BGF90). In Proceedings of the 7th world congress on genetics applied to livestock production, Montpellier; 19–23 August, 2002.
34. Madsen P, Jensen J. A user's guide to DMU, version 6, release 5.2. Tjele: Center for Quantitative Genetics and Genomics, Department of Molecular Biology and Genetics, Aarhus University; 2013.
35. VanRaden PM. Efficient methods to compute genomic predictions. J Dairy Sci. 2008;91:4414–23.
36. Houle D, Meyer K. Estimating sampling error of evolutionary statistics based on genetic covariance matrices using maximum likelihood. J Evol Biol. 2015;28:1542–9.
37. Lo LL, Fernando RL, Grossman M. Covariance between relatives in multi-breed populations: additive model. Theor Appl Genet. 1993;87:423–30.
38. Karoui S, Carabaño MJ, Díaz C, Legarra A. Joint genomic evaluation of French dairy cattle breeds using multiple-trait models. Genet Sel Evol. 2012;44:39.
39. Wientjes YC, Veerkamp RF, Bijma P, Bovenhuis H, Schrooten C, Calus MP. Empirical and deterministic accuracies of across-population genomic prediction. Genet Sel Evol. 2015;47:5.
40. Porto-Neto LR, Barendse W, Henshall JM, McWilliam SM, Lehnert SA, Reverter A. Genomic correlation: harnessing the benefit of combining two unrelated populations for genomic selection. Genet Sel Evol. 2015;47:84.
41. Meuwissen THE, Luo Z. Computing inbreeding coefficients in large populations. Genet Sel Evol. 1992;24:305–13.
42. Mäntysaari EA, Koivula M. GEBV validation test revisited. Interbull Bull. 2012;45:11–6.
43. Guo X, Christensen OF, Ostersen T, Wang Y, Lund MS, Su G. Improving genetic evaluation of litter size and piglet mortality for both genotyped and nongenotyped individuals using a single-step method. J Anim Sci. 2015;93:503–12.
44. Lopes MS, Bastiaansen JWM, Janss L, Knol EF, Bovenhuis H. Estimation of additive, dominance, and imprinting genetic variance using genomic data. G3 (Bethesda). 2015;5:2629–37.
45. Meuwissen THE, Hayes BJ, Goddard ME. Prediction of total genetic value using genome-wide dense marker maps. Genetics. 2001;157:1819–29.
46. Hidalgo AM, Bastiaansen JWM, Lopes MS, Harlizius B, Groenen MAM, de Koning D-J. Accuracy of predicted genomic breeding values in purebred and crossbred pigs. G3 (Bethesda). 2015;5:1575–83.

47. Legarra A, Baloche G, Barillet F, Astruc JM, Soulas C, Aguerre X, et al. Within-and across-breed genomic predictions and genomic relationships for Western Pyrenees dairy sheep breeds Latxa, Manech, and Basco-Béarnaise. J Dairy Sci. 2014;97:3200–12.

48. Moghaddar N, Swan AA, van der Werf JH. Comparing genomic prediction accuracy from purebred, crossbred and combined purebred and cross-bred reference populations in sheep. Genet Sel Evol. 2014;46:58.

49. Sun C, VanRaden PM, Cole JB, O'Connell JR. Improvement of prediction ability for genomic selection of dairy cattle by including dominance effects. PLoS One. 2014;9:e103934.

50. Dickerson GE. Inbreeding and heterosis in animals. In Proceedings of the animal breeding and genetics symposium in honor of Jay L. Lush, 29 July 1972, Blacksburg; 1973. p. 54–77.

51. Leroy G. Inbreeding depression in livestock species: review and meta-analysis. Anim Genet. 2014;45:618–28.

52. Silió L, Barragán C, Fernández AI, García-Casco J, Rodríguez MC. Assessing effective population size, coancestry and inbreeding effects on litter size using the pedigree and SNP data in closed lines of the Iberian pig breed. J Anim Breed Genet. 2016;133:145–54.

53. Hinrichs D, Meuwissen THE, Ødegard J, Holt M, Vangen O, Woolliams JA. Analysis of inbreeding depression in the first litter size of mice in a long-term selection experiment with respect to the age of the inbreeding. Heredity. 2007;99:81–8.

5

Genome-wide association study and accuracy of genomic prediction for teat number in Duroc pigs using genotyping-by-sequencing

Cheng Tan[1,2], Zhenfang Wu[3], Jiangli Ren[1], Zhuolin Huang[1], Dewu Liu[3], Xiaoyan He[3], Dzianis Prakapenka[2], Ran Zhang[1], Ning Li[1], Yang Da[2*] and Xiaoxiang Hu[1*]

Abstract

Background: The number of teats in pigs is related to a sow's ability to rear piglets to weaning age. Several studies have identified genes and genomic regions that affect teat number in swine but few common results were reported. The objective of this study was to identify genetic factors that affect teat number in pigs, evaluate the accuracy of genomic prediction, and evaluate the contribution of significant genes and genomic regions to genomic broad-sense heritability and prediction accuracy using 41,108 autosomal single nucleotide polymorphisms (SNPs) from genotyping-by-sequencing on 2936 Duroc boars.

Results: Narrow-sense heritability and dominance heritability of teat number estimated by genomic restricted maximum likelihood were 0.365 ± 0.030 and 0.035 ± 0.019, respectively. The accuracy of genomic predictions, calculated as the average correlation between the genomic best linear unbiased prediction and phenotype in a tenfold validation study, was 0.437 ± 0.064 for the model with additive and dominance effects and 0.435 ± 0.064 for the model with additive effects only. Genome-wide association studies (GWAS) using three methods of analysis identified 85 significant SNP effects for teat number on chromosomes 1, 6, 7, 10, 11, 12 and 14. The region between 102.9 and 106.0 Mb on chromosome 7, which was reported in several studies, had the most significant SNP effects in or near the *PTGR2*, *FAM161B*, *LIN52*, *VRTN*, *FCF1*, *AREL1* and *LRRC74A* genes. This region accounted for 10.0% of the genomic additive heritability and 8.0% of the accuracy of prediction. The second most significant chromosome region not reported by previous GWAS was the region between 77.7 and 79.7 Mb on chromosome 11, where SNPs in the *FGF14* gene had the most significant effect and accounted for 5.1% of the genomic additive heritability and 5.2% of the accuracy of prediction. The 85 significant SNPs accounted for 28.5 to 28.8% of the genomic additive heritability and 35.8 to 36.8% of the accuracy of prediction.

Conclusions: The three methods used for the GWAS identified 85 significant SNPs with additive effects on teat number, including SNPs in a previously reported chromosomal region and SNPs in novel chromosomal regions. Most significant SNPs with larger estimated effects also had larger contributions to the total genomic heritability and accuracy of prediction than other SNPs.

*Correspondence: yda@umn.edu; huxx@cau.edu.cn
[1] State Key Laboratory for Agrobiotechnology, China Agricultural University, Beijing 100193, China
[2] Department of Animal Science, University of Minnesota, Saint Paul, MN 55108, USA
Full list of author information is available at the end of the article

Background

A sufficient number of teats is necessary for a sow to rear its piglets to weaning age. Many putative QTL (quantitative trait loci) for teat number have been reported on most of the porcine chromosomes, but most of these were detected using microsatellite markers and lacked specific gene targets [1]. Genome-wide association studies (GWAS) using single nucleotide polymorphisms (SNPs) and analyses of candidate genes have identified several specific gene targets that affect teat number in swine. A GWAS using 42,654 SNPs on 936 Large White pigs reported 39 QTL with 211 significant SNP effects on teat number [2]. Among those SNP effects, the region between 102.0 and 105.2 Mb on chromosome 7 had the most significant effects and the percentage of the genetic variance explained by SNPs in this region ranged from 0.04 to 2.51%. Within this region, the *VRTN* and *PROX2* genes were identified as the most convincing candidate genes. The chromosomal locations of the significant SNPs that were detected in this GWAS differed from all previously reported QTL for teat number that have been compiled in the animal QTL database [1]. Another GWAS, using 32,911 SNPs on 1550 Large White pigs, reported 21 QTL with additive effects on chromosomes 6, 7 and 12, one QTL with a dominant effect on chromosome 4, and identified *VRTN* as the most promising candidate gene for teat number [3]. A third GWAS using 41,647 SNPs on 1657 Large White pigs found 65 significant SNPs on chromosomes 1, 2, 7, 8, 12 and 14, including SNPs in the region 102.9 between 105.2 Mb on chromosome 7 [4]. A fourth GWAS using 39,778 SNPs identified the *VRTN* gene with pleiotropic and desirable effects on thoracic vertebral number, teat number and carcass (body) length across four pig populations, and showed that, of all SNPs on chromosome 7, a SNP within the *VRTN* gene had the most significant effect on teat number in Duroc pigs [5]. Among all significant SNPs that have been detected for teat number by GWAS, the significance of the *VRTN* gene on chromosome 7 achieved the widest consensus and has been identified as a strong candidate gene for teat number [2–5]. However, in the literature some discrepancies regarding the most significant location and many SNP effects in other genomic regions have been reported. A GWAS using the porcine 60 K SNP chip on a F2 population from a cross between Landrace and Korean pigs identified highly significant SNPs on chromosome 7 that were more than 40 Mb away from the *VRTN* gene [6], and in another GWAS using 36,588 SNPs and 1024 Duroc pigs, the most significant SNPs on chromosome 7 were found 2 to 3 Mb downstream of the *VRTN* region [7]. However, other than for the *VRTN* region, there is little consensus among the GWAS results on genomic regions that affect teat number [2–4, 6, 7]. Therefore, additional studies are needed to identify the genetic factors that affect swine teat number. Furthermore, it is unclear what the impact of the highly significant SNPs is on the accuracy of genomic prediction for teat number.

The objective of this study was to identify genetic factors that affect teat number in pigs, evaluate the accuracy of genomic prediction, and evaluate the contribution of significant genes and genomic regions to the heritability and accuracy of genomic prediction using 41,108 autosomal SNPs from genotyping-by-sequencing (GBS) on 2936 Duroc boars.

Methods

Animals, phenotyping, and genotyping-by-sequencing

Animal and phenotype data used for this study were provided by Guangdong Wen's Foodstuff Group (Guangdong, China). The study population included 2936 Duroc boars born from September 2011 to September 2013 in 1456 litters from 79 sires with one to three piglets per litter, and all pigs were managed at a single nucleus farm. The left and right teats were counted separately within 48 h after birth and only normal teats were recorded. In this study, the phenotype used for 'teat number' was the total number of teats that was equal to the sum of the left and right normal teats. The 'mean ± (standard deviation)' of teat number was 10.72 ± 1.72. The phenotypic values followed a near bell-shaped distribution (Fig. 1), which was similar to the leptokurtic distribution of teat number that is observed for Landrace and Large White pigs [8], with most animals (1992 out of 2936) having 10 or 11 teats.

Genomic DNA was extracted from ear tissue of all 2936 Duroc boars and quantified using a Qubit 2.0 Fluorometer. DNA concentrations were normalized to 50 ng/ml in 96-well plates. A two-enzyme i.e. *Eco*RI and *Msp*I

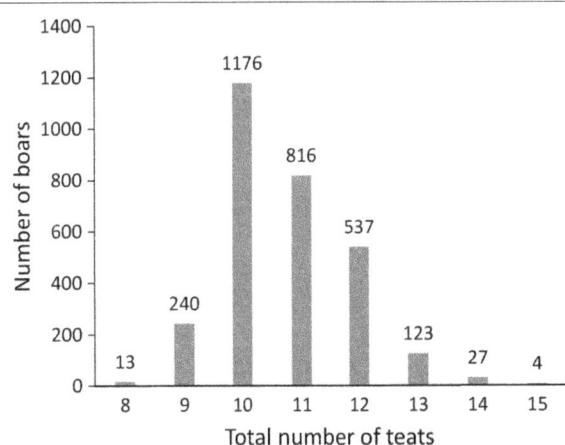

Fig. 1 Phenotypic distribution of total teat number in Duroc boars (N = 2936)

genotyping-by-sequencing (GBS) was used. A set of 96 forward barcoded adapters with an *EcoRI* overhang were designed by the GBS Barcode Generator (http://www. deenabio.com/services/gbs-adapters), and the reverse adapter with a *MspI* overhang was designed according to [9]. DNA samples (150 ng each) were digested with *EcoRI* and *MspI*, then ligated to the designed adapters. Following adapter ligation, samples were pooled in 96-plex and size-selected using two cycles of purification with Agencourt AMPure XP Beads (Beckman Coulter, Pasadena, CA). The purified libraries were amplified by PCR and then sequenced on Illumina NextSeq500 by a 90-bp single-end sequencing. SNP genotypes were called according to the pipeline implemented in Tassel 5.0 with default parameters [10], and Beagle 4.0 was used to impute missing SNP genotypes. A total of 90,051 SNPs were identified for the population used in this study. SNP filtering was based on the following criteria: only SNPs that had a minor allele frequency higher than 5%, for which the frequency of the least frequent homozygous genotype was at least 0.01, and passed the Hardy–Weinberg equilibrium test ($p \geq 10^{-6}$) were retained. Among the autosomal SNPs, 41,108 SNPs satisfied these requirements and were used for analyses.

GWAS analysis

Three single-SNP methods were used for the GWAS analysis: a t test of additive and dominance SNP effects using a generalized least squares (GLS) analysis that takes intraclass correlation of sibs into account and is implemented in the EPISNP2 program [11, 12], and the least squares (LS) analyses of additive effects by PLINK [13] and EPISNP1 [11] with population stratification correction using the first 50 dimensions from multidimensional scaling (MDS) as covariates. We report the PLINK and EPISNP1 results using the first 35 MDS dimensions for stratification correction because the genomic inflation factor [14] and the patterns of Manhattan plots of SNP significance stabilized when fitting the first 35 MDS dimensions (Fig. 2).

The statistical model for the EPISNP2 analysis was:

$$\mathbf{y} = \mathbf{X}_b\mathbf{b} + \mathbf{Xg} + \mathbf{Zf} + \mathbf{e},$$

where $\mathbf{y}$ is the vector of phenotypic values, $\mathbf{b}$ is the vector of fixed year-month effects, $\mathbf{X}_b$ is the incidence matrix for $\mathbf{b}$, $\mathbf{g}$ is the vector of the effects of SNP genotypes, $\mathbf{X}$ is the incidence matrix of $\mathbf{g}$, $\mathbf{f}$ is the vector of random family effects with a common variance σ_f^2 for sibs in the same family, and $\mathbf{Z}$ is the incidence matrix of $\mathbf{f}$. The variance–covariance matrix of the family effects was assumed to be $\mathbf{G} = \mathrm{Var}(\mathbf{f}) = \mathbf{I}\sigma_f^2$, where $\mathbf{I}$ is an identity matrix, and the phenotypic variance–covariance matrix is $\mathrm{Var}(\mathbf{y}) = \mathbf{V} = \mathbf{ZGZ}' + \mathbf{I}\sigma_e^2$ [12].

The statistical model for the PLINK analysis was:

$$\mathbf{y} = \mathbf{X}_b\mathbf{b} + \mathbf{X}_1\mathbf{b}_1 + \alpha\mathbf{x} + \mathbf{e},$$

where $\mathbf{b}_1$ is the vector of fixed effect(s) of the MDS dimension(s), $\mathbf{X}_1$ is the matrix of the MDS dimension(s) as calculated by PLINK from the SNP matrix of identity-by-state [13], α is the additive SNP effect, and $\mathbf{x}$ is a column vector of genotype codes for α created by PLINK.

The statistical model for EPISNP1 analysis was:

$$\mathbf{y} = \mathbf{X}_b\mathbf{b} + \mathbf{X}_1\mathbf{b}_1 + \mathbf{Xg} + \mathbf{e},$$

where the matrices have the same definitions as in the previous two models. Significance tests for additive and dominance SNP effects by EPISNP1 and EPISNP2 were implemented by t tests for the additive and dominance contrasts of the estimated SNP genotypic values [11, 12, 15].

Genomic heritability and accuracy of genomic prediction

Genomic heritability and genomic prediction were estimated by using a mixed model with additive and dominance effects as described previously [16–18]. Briefly, the mixed model for heritability estimation and genomic prediction was:

$$\mathbf{y} = \mathbf{X}_b\mathbf{b} + \mathbf{Za} + \mathbf{Zd} + \mathbf{e},$$

with $\mathrm{Var}(\mathbf{y}) = \mathbf{V} = \mathbf{ZA}_g\mathbf{Z}'\sigma_\alpha^2 + \mathbf{ZD}_g\mathbf{Z}'\sigma_\delta^2 + \mathbf{I}\sigma_e^2$, where $\mathbf{Z}$ is an incidence matrix allocating phenotypic observations to each individual, $\mathbf{a}$ is the vector of genomic additive (breeding) values, $\mathbf{d}$ is the vector of genomic dominance values or dominance deviations, $\mathbf{A}_g$ is a genomic additive relationship matrix calculated from the SNPs, $\mathbf{D}_g$ is a genomic dominance relationship matrix calculated from the SNPs, σ_α^2 is the additive variance, σ_δ^2 is the dominance variance, and σ_e^2 is the residual variance. The $\mathbf{A}_g$ and $\mathbf{D}_g$ matrices were calculated using Definition II of genomic relationships implemented by the GVCBLUP package, and variance components of additive, dominance and random residual values were estimated by genomic restricted maximum likelihood estimation (GREML) using the GREML_CE program in the GVCBLUP package [18]. The genomic heritability was defined as: $h_\alpha^2 = \sigma_\alpha^2/\sigma_y^2$, i.e. the narrow-sense heritability, $h_\delta^2 = \sigma_\delta^2/\sigma_y^2$, i.e. the dominance heritability, and $h_t^2 = h_\alpha^2/h_\delta^2$, i.e. the broad-sense heritability, where $\sigma_y^2 = \sigma_\alpha^2 + \sigma_\delta^2 + \sigma_e^2$ is the phenotypic variance. The genomic best linear unbiased prediction (GBLUP) of additive, dominance and genetic values of individuals in the training and validation samples were calculated at the last iteration of the GREML.

A tenfold validation study was conducted to evaluate the prediction accuracy. The 2936 Duroc boars were randomly divided into 10 validation datasets of 293 individuals except the 10th sample, which included

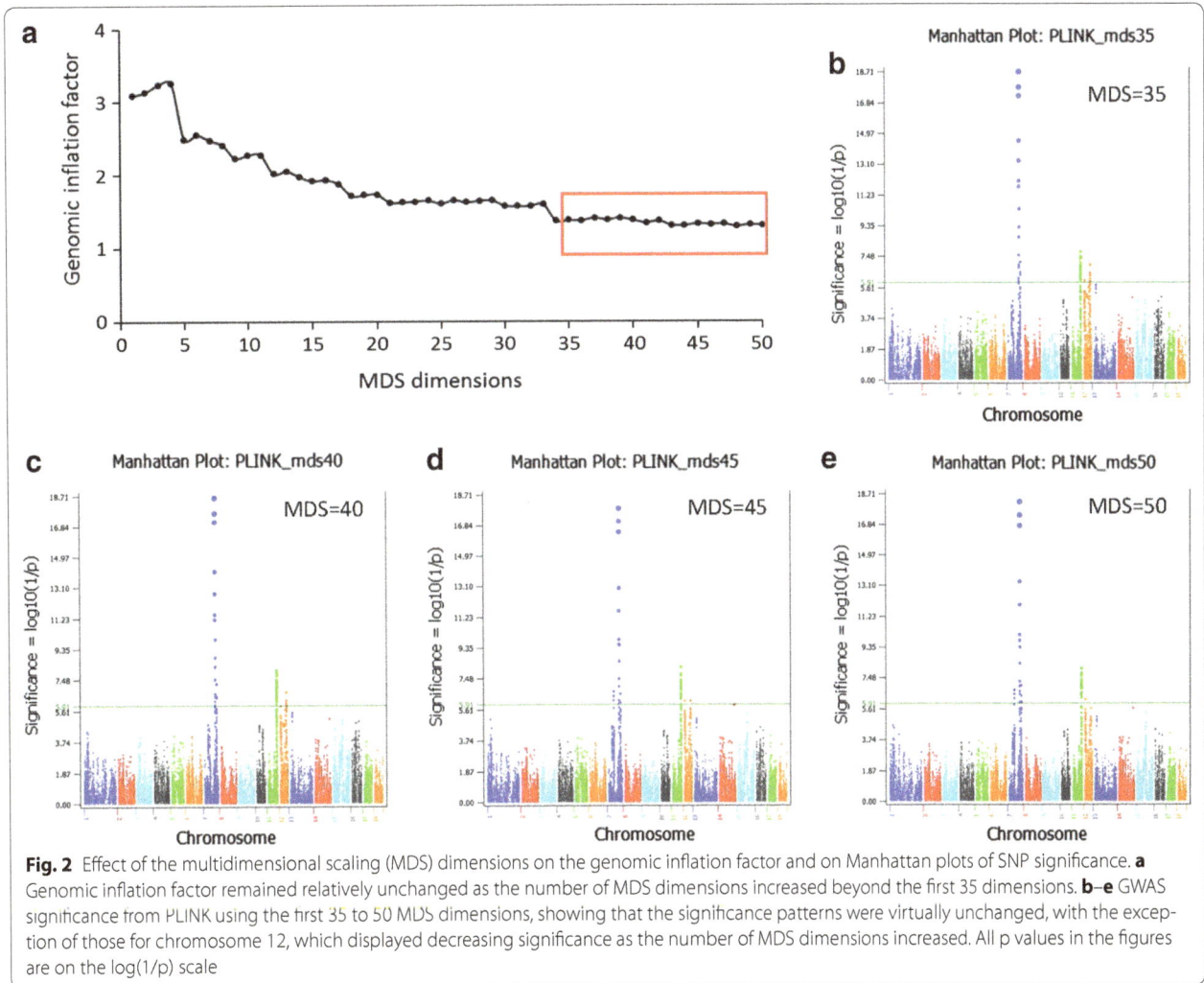

Fig. 2 Effect of the multidimensional scaling (MDS) dimensions on the genomic inflation factor and on Manhattan plots of SNP significance. **a** Genomic inflation factor remained relatively unchanged as the number of MDS dimensions increased beyond the first 35 dimensions. **b–e** GWAS significance from PLINK using the first 35 to 50 MDS dimensions, showing that the significance patterns were virtually unchanged, with the exception of those for chromosome 12, which displayed decreasing significance as the number of MDS dimensions increased. All p values in the figures are on the log(1/p) scale

299 individuals. For each of the 10 validation analyses, phenotypic observations in the validation dataset were omitted in the GBLUP calculation. Three measures of prediction accuracy were calculated and compared: $\hat{R}_{0jp} = \text{corr}\left(\hat{g}_{0j}, y_0\right)$, which is the observed accuracy of predicting the phenotypic values in the validation population and is calculated as the correlation between the estimated genetic values ($\hat{g}_{0j}$) and the phenotypic observations (y_0) of validation individuals, averaged across all validation datasets; $R_{0j} = \text{corr}\left(\hat{g}_{0j}, g_{0j}\right)$, which is the expected accuracy of predicting the true genetic values (g_{0j}) of individuals in the validation population and is calculated as the square root of the reliability estimate for each individual from the GVCBLUP package [18], where 'j = α' indicates additive prediction, 'j = δ' indicates dominance prediction, and 'j = t' indicates prediction of total genetic value; and $R_{0jp} = R_{0j}\sqrt{h_j^2}$, which is the expected accuracy of predicting the phenotypic values, where h_j^2 is the genomic narrow-sense

(j = α), dominance (j = δ), or broad-sense (j = t) heritability. The accuracy of predicting phenotypic values was previously termed as 'predictive ability' [19] to distinguish it from 'expected prediction accuracy' of predicting genetic values (R_{0j} in this study). The formula of the expected accuracy of predicting phenotypic values, $R_{0jp} = R_{0j}\sqrt{h_j^2}$, is a slightly different form of the relationship between 'predictive ability' and 'prediction accuracy' [19]. The mathematical difference between R_{0jp} and R_{0j} is in the denominators of these two measures: the denominator of R_{0jp} is the phenotypic standard deviation, whereas the denominator of R_{0j} is the genetic standard deviation, which is necessarily smaller than the phenotypic standard deviation in the presence of non-zero residual variance. Therefore, R_{0j} is the upper limit of R_{0jp} but this upper limit may not hold for the observed accuracy of predicting phenotypic values ($\hat{R}_{0jp}$) due to unknown variations in the data or genetic mechanisms that are not explained by the statistical model. The observed accuracy of predicting genetic values can only be defined when the true genetic values such as simulated genetic values are known [16] but they could not be defined in this study because the true genetic values were unknown.

Contribution of significant SNPs and genomic regions to total heritability and prediction accuracy

The contributions of each SNP to additive, dominance and total heritability can be estimated [18]. However, as shown by the results of this study, the contribution of each SNP is affected by the number of SNPs in the mixed model: the larger is the number of SNPs, the smaller is the contribution of each SNP. To avoid this dependency in the mixed model for GREML and to estimate each SNP's independent contribution to genomic heritability and prediction accuracy, we used the approach of 'partial heritability' and 'partial accuracy' based on differences in heritability and prediction accuracy between a full model and a reduced model. The full model fits all SNPs as random effects and the reduced model fits the target SNP or SNPs as fixed effects to remove their effects from the phenotypic values. The reduced model was as follows:

$$\mathbf{y} = \mathbf{X}_f\mathbf{b} + \mathbf{X}\mathbf{s} + \mathbf{Z}\mathbf{a} + \mathbf{Z}\mathbf{d} + \mathbf{e},$$

where $\mathbf{s}$ is a column vector of fixed SNP effects and $\mathbf{X}$ is the incidence matrix of $\mathbf{s}$. The phenotypic variance–covariance matrix was assumed to be the same as in the full model, i.e., $\mathrm{Var}(\mathbf{y}) = \mathbf{V} = \mathbf{Z}\mathbf{A}_g\mathbf{Z}'\sigma_\alpha^2 + \mathbf{Z}\mathbf{D}_g\mathbf{Z}'\sigma_\delta^2 + \mathbf{I}\sigma_e^2$, but variance components were estimated under the reduced model. Let $\widehat{h}^2$ ($\widehat{h}_i^2$) be the estimated heritability from the full (reduced) model, and R_0 (R_{0i}) a measure of prediction accuracy for the full (reduced) model. Then, the relative contribution of the ith SNP or the ith set of SNPs to the total heritability was calculated as $c_{hi}^2 = 1 - \widehat{h}_i^2/\widehat{h}^2$, and

the relative contribution of the ith SNP or the ith set of SNPs to the prediction accuracy as $c_{ri} = 1 - R_{0i}/R_0$.

Results

Genomic heritability and prediction accuracy

The estimate of genomic narrow-sense heritability ($\widehat{h}_\alpha^2$) was 0.365 ± 0.030, of dominance heritability ($\widehat{h}_\delta^2$) was 0.035 ± 0.019, of broad-sense heritability (h_t^2) was 0.400 ± 0.034, and the estimate of narrow-sense heritability for the mixed model with additive effects only, was 0.368 ± 0.030, which is slightly higher than the corresponding estimate for the mixed model with additive and dominance effects (Table 1). The observed accuracy of predicting phenotypic values from the tenfold validation study was 0.437 ± 0.064 for the mixed model with additive and dominance SNP effects ($\widehat{R}_{0tp}$, Model 1A in Table 2), and was 0.435 ± 0.064 for the mixed model with additive effects only ($\widehat{R}_{0\alpha p}$, Model 2A in Table 2), which is only 0.46% lower than that from the mixed model with additive and dominance effects. These slight differences in both heritability and accuracy of prediction between the additive model and the model with additive and dominance effects indicates that additive SNP effects were the primary genetic effects that affect teat number and that dominance SNP effects only had a negligible contribution to the prediction accuracy for teat number.

GWAS results

The GWAS that was done with EPISNP2, which accounted for the sib intraclass correlation and was implemented by a GLS analysis [11, 12], identified 73 SNPs on chromosomes 1, 6, 7, 10, 11, 12 and 14 with additive effects but no SNP with dominance effects reached genome-wide significance with the Bonferroni multiple testing correction ($p < 10^{-5.91}$) (Fig. 3a, b; Table 3; Additional file 1: Table S1). LS analysis of PLINK [13] and EPISNP1 [11] with stratification correction using the first

---

Table 1 Estimates of genomic heritabilities for teat number using 41,108 autosomal SNPs on 2936 Duroc boars

Model	All SNPs as random effects	85 significant SNPs removed	85 significant SNPs as fixed effects
Additive and dominance effects	$\widehat{h}_\alpha^2 = 0.365 \pm 0.030$	$\widehat{h}_\alpha^2 = 0.346 \pm 0.030$ $-c_{hi}^2 = -5.20\%$	$\widehat{h}_\alpha^2 = 0.260 \pm 0.030$ $-c_{hi}^2 = -28.77\%$
	$\widehat{h}_\delta^2 = 0.035 \pm 0.019$	$\widehat{h}_\delta^2 = 0.036 \pm 0.020$ $-c_{hi}^2 = +2.86\%$	$\widehat{h}_\delta^2 = 0.037 \pm 0.022$ $-c_{hi}^2 = +5.71\%$
	$\widehat{h}_t^2 = 0.400 \pm 0.034$	$\widehat{h}_t^2 = 0.382 \pm 0.034$ $-c_{hi}^2 = -4.50\%$	$\widehat{h}_t^2 = 0.297 \pm 0.036$ $-c_{hi}^2 = -25.75\%$
Additive effects only	$\widehat{h}_\alpha^2 = 0.368 \pm 0.030$	$\widehat{h}_\alpha^2 = 0.350 \pm 0.030$ $-c_{hi}^2 = -4.89\%$	$\widehat{h}_\alpha^2 = 0.263 \pm 0.030$ $-c_{hi}^2 = -28.53\%$

h_α^2 = narrow-sense heritability. h_δ^2 = dominance heritability. h_t^2 = broad-sense heritability = $h_\alpha^2 + h_\delta^2$. $-c_{hi}^2$ = decrease in heritability relative to the heritability estimated by using all SNPs fitted as random effects

Table 2 Accuracies of genomic prediction for the phenotypic values and true genetic values of teat number using 41,108 autosomal SNPs on 2936 Duroc boars in a tenfold validation study

Model and accuracy change	$\widehat{R}_{0tp} = corr(\widehat{g}_{0j}, y_0)$	$R_{0jp} = R_{0j}\sqrt{h_j^2}$	$R_{0j} = corr(\widehat{g}_{0j}, g_{0j})$
Model 1A	$\widehat{R}_{0tp} = 0.437 \pm 0.064$	$R_{0tp} = 0.460$	$R_{0t} = 0.728 \pm 0.004$
Model 1B	$\widehat{R}_{0tp} = 0.279 \pm 0.076$	$R_{0tp} = 0.360$	$R_{0t} = 0.661 \pm 0.007$
$-c_{ri}$ of 1B relative to 1A	-36.16%	-21.74%	-9.20%
Model 2A	$\widehat{R}_{0\alpha p} = 0.435 \pm 0.064$	$R_{0\alpha p} = 0.425$	$R_{0\alpha} = 0.700 \pm 0.007$
Model 2B	$\widehat{R}_{0\alpha p} = 0.275 \pm 0.074$	$R_{0\alpha p} = 0.320$	$R_{0\alpha} = 0.624 \pm 0.009$
$-c_{ri}$ of 2B relative to 2A	-36.78%	-24.70%	-10.86%
Model 3A	$\widehat{R}_{0tp} = 0.426 \pm 0.066$	$R_{0tp} = 0.446$	$R_{0t} = 0.721 \pm 0.004$
$-c_{ri}$ of 3A relative to 1A	-2.52%	-3.04%	-0.96%
Model 4A	$\widehat{R}_{0\alpha p} = 0.424 \pm 0.066$	$R_{0\alpha p} = 0.409$	$R_{0\alpha} = 0.691 \pm 0.007$
$-c_{ri}$ of 4A relative to 2A	-2.53%	-3.76%	-1.28%

Model 1A has additive and dominance effects and uses all 41,108 autosome SNPs. Model 1B is a modification of Model 1A by using the 85 significant SNPs as fixed non-genetic effects. Model 2A has additive effects only and uses all 41,108 autosome SNPs. Model 2B is a modification of Model 2A by using the 85 significant SNPs as fixed non-genetic effects. Model 3A has additive and dominance effects and uses 41,023 autosomal SNPs after removing the 85 significant SNPs. Model 4A has additive effects only and uses 41,023 autosomal SNPs after removing the 85 significant SNPs. $\widehat{R}_{0jp}$ is the observed accuracy of predicting phenotypic values from tenfold validations. R_{0jp} is the expected accuracy of predicting phenotypic values. R_{0j} is the expected accuracy of predicting genetic values calculated by GVCBLUP from tenfold validations, $j = t$ or α. $h_t^2 = 0.400$ for Model 1A, $= 0.297$ for Model 1B, $= 0.382$ for Model 3. $\widehat{h}_\alpha^2 = 0.368$ for Model 2A, 0.263 for Model 2B, $= 0.350$ for Model 4. $-c_{ri}$ is the decrease in accuracy

35 dimensions of MDS as fixed covariates, identified 54 and 21 significant SNPs, respectively (Fig. 3c, d; Additional file 1: Table S1). Twelve SNPs detected by PLINK and two SNPs detected by EPISNP1 did not overlap with the SNPs detected by EPISNP2. Eighteen SNPs detected by EPISNP1 overlapped with those detected by EPISNP2 and PLINK. For this dataset, EPISNP1 was the most conservative for declaring significance. We report SNPs detected by EPISNP2 because they all had a substantial contribution to the broad-sense genomic heritability (see Additional file 1: Table S1). A graphical view of the GWAS results obtained by EPISNP2, PLINK and EPISNP1 for all autosomes is in Additional file 2: Figure S1.

To evaluate the impact of the significant SNPs on the phenotypic variance, we estimated the decreases in observed genomic narrow-sense heritability and prediction accuracy when the phenotypic values were adjusted for the estimated genotypic values of the significant SNPs. The results showed that the 85 significant SNPs identified by the three methods, i.e. EPISNP2, PLINK and EPISNP1, accounted for 28.5 to 28.8% of the genomic narrow-sense heritability (Table 1) and for 36.2 to 36.8% of the observed prediction accuracy (Model 1A and Model 2A in Table 2). These results show that many SNPs that were deemed insignificant by the GWAS analysis were relevant for genomic prediction of teat number. Each of the 85 SNPs had a relatively large contribution to the genomic heritability and prediction accuracy, with the contribution of each SNP to the observed genomic narrow-sense

heritability ranging from 0.7 to 7.3% and relative contribution of each SNP to the observed prediction accuracy ranging from 0.5 to 5.6% (see Additional file 1: Table S1).

Analysis of the region between 102.9 and 106.0 Mb on chromosome 7

A cluster of 14 SNPs within or near the *PTGR2, FAM161B, LIN52, VRTN, FCF1, AREL1* and *LRRC74A* genes in the region between 102.9 and 106.0 Mb on chromosome 7 had the most significant effects on teat number with genome-wide significance (Fig. 4a). Based on the GLS analysis of EPISNP2, the two SNPs upstream of *PTGR2* had the most significant additive effects, followed by the three SNPs within and upstream of *AREL1*, whereas the LS analysis of PLINK and EPISNP1 with stratification correction ranked the three *AREL1* SNPs as the most significant and the two SNPs upstream of *PTGR2* in the 6th and 7th positions (see Additional file 1: Table S1). The six most significant SNPs in the region between 102.9 and 103.8 Mb on chromosome 7 accounted for 7.4% of the genomic additive heritability and 7.0% of the observed prediction accuracy in the tenfold validation study (Table 3), and all the 14 SNPs in this region with genome-wide significance accounted for 10.0% of the genomic narrow-sense heritability and 8.0% of the observed prediction accuracy. Removal of the genotypic effects of the 14 SNPs by fitting these SNPs as fixed effects in the model for EPISNP2 removed all significant effects in the region between 102.9 and 103.8 Mb on chromosome 7 and also removed the significant effects of seven SNPs

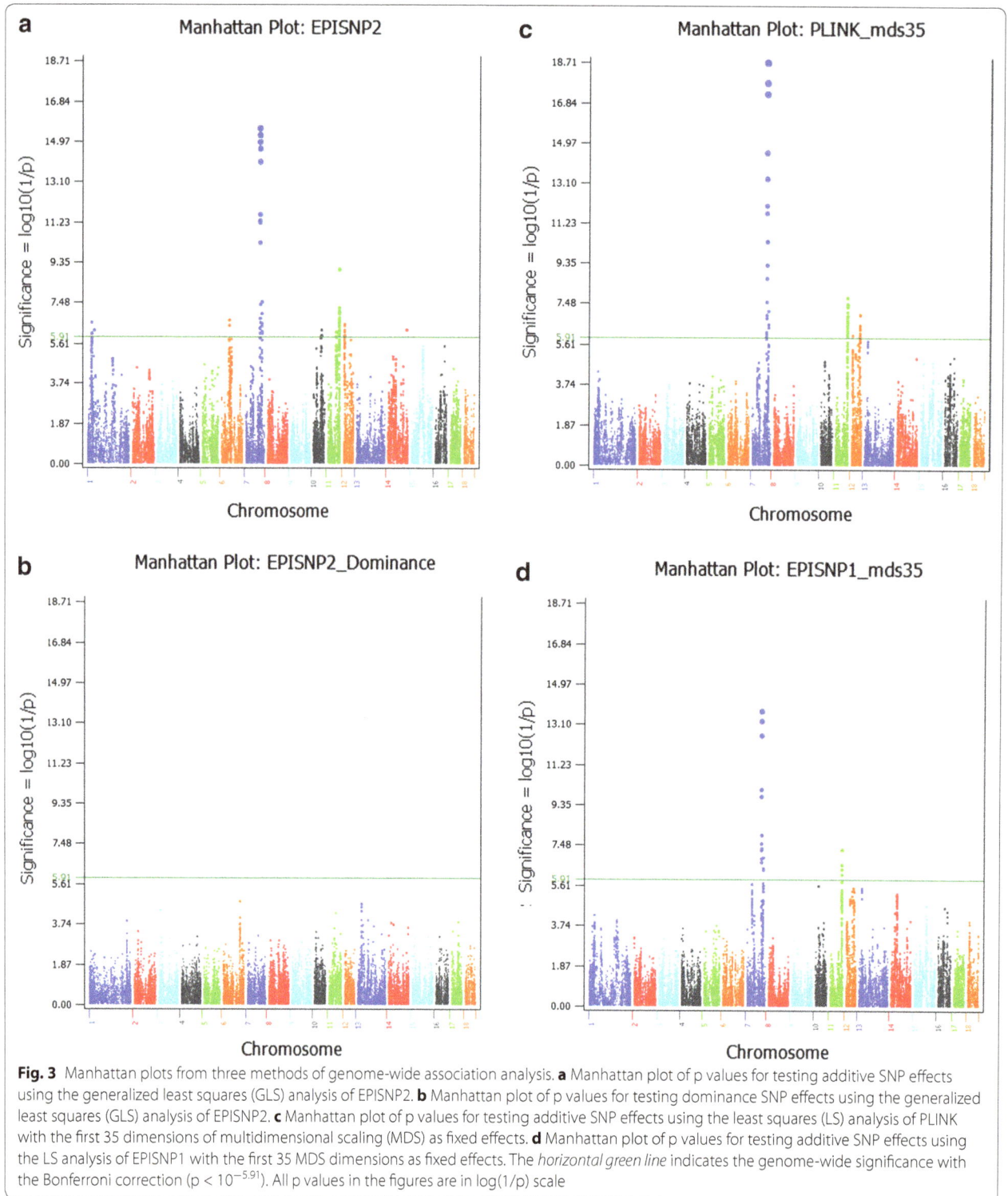

Fig. 3 Manhattan plots from three methods of genome-wide association analysis. **a** Manhattan plot of p values for testing additive SNP effects using the generalized least squares (GLS) analysis of EPISNP2. **b** Manhattan plot of p values for testing dominance SNP effects using the generalized least squares (GLS) analysis of EPISNP2. **c** Manhattan plot of p values for testing additive SNP effects using the least squares (LS) analysis of PLINK with the first 35 dimensions of multidimensional scaling (MDS) as fixed effects. **d** Manhattan plot of p values for testing additive SNP effects using the LS analysis of EPISNP1 with the first 35 MDS dimensions as fixed effects. The *horizontal green line* indicates the genome-wide significance with the Bonferroni correction ($p < 10^{-5.91}$). All p values in the figures are in log(1/p) scale

in the region between 116.1 and 117.4 Mb on chromosome 7 (Fig. 4b). SNPs within or near the *AREL1*, *PTGR2*, *FMA161B*, *LIN52* and *LRRC74A* genes also had the largest contributions to the genomic narrow-sense heritability

and prediction accuracy (Fig. 4c). We did not detect SNPs within the *VRTN* gene but a SNP upstream of and nearest to *VRTN* (S7_103355294) was highly significant, ranking 6th based on EPISNP2 and 5th based on PLINK and

Genome-wide association study and accuracy of genomic prediction for teat number in Duroc pigs...

67

Table 3 Chromosome regions with significant SNP effects on teat number

Chr	Region (Mb)	Size (Mb)	Most significant SNP				Contribution	Gene region
			Name	MAF	p value	% of h_α^2	% of $\widehat{R}_0$	
1	29.63–30.18	0.55	S1_29635241	0.499	2.68 (10^{-07})	2.25	1.51	*TNFAIP3, OLIG3*
7	102.91–103.80	0.89	S7_102911357	0.438	2.49 (10^{-16})	7.35	6.98	*PTGR2* (U), *FAM161B, LIN52, VRTN* (U, D), *FCF1, AREL1* (16 genes)
7	116.07–117.43	1.36	S7_116899295	0.342	2.97 (10^{-08})	3.07	2.13	*GALC, KCNK10, SPATA7, PTPN21, ZC3H14, EML5, TTC8*
11	56.56–58.58*	2.21	S11_58558301	0.422	7.14 (10^{-7})	1.64	2.27	*SPRY2, SNORA70*
11	77.75–79.69	1.94	S11_79009219	0.226	9.16 (10^{-10})	5.07	5.23	*FGF14, BIVM* (U, D), *LOC102167592, LOC102167785* (U)
12	4.53–6.26*	1.73	S12_5615207	0.335	3.16 (10^{-07})	2.59	2.63	*MFSD11-KCTD2* (49 genes)
12	50.54–51.74	1.20	S12_51574540	0.138	1.06 (10^{-07})	0.64	1.16	*OR3A2, ASPA, TRPV3, TRPV1, CTNS, TAX1BP3, EMC6, CAMKK1*

Chr = chromosome; MAF = minor allele frequency. h_α^2 is the additive heritability. $\widehat{R}_0$ is the observed accuracy of prediction. U indicate the significant SNP is located upstream of the gene. D indicates the significant SNP is located downstream of the gene. Contribution was calculated for the six most significant SNPs in each region. * This region has three significant SNPs

EPISNP1. Linkage disequilibrium (LD) analysis using Haploview [20] showed that the two SNPs flanking *VRTN* were in strong LD (D′ = 0.90, Fig. 4d), implying that either of these two SNPs could also be in strong LD with *VRTN*. Therefore, assuming that *VRTN* is a causal gene, the significant effect of S7_103355294 could be a linked effect of *VRTN*. The LD analysis showed that the significant effects of the region between 116.1 and 117.4 Mb on chromosome 7 could also be due to LD with the region between 102.9 and 106.0 Mb since three of the seven SNPs in the former region were in low LD with five significant SNPs in the latter region (D′ = 0.13 to 0.24, Fig. 4d). This did not consider the possibility of multilocus LD between the two regions. The low LD was the only known reason that could explain the disappearance of the significant QTL effects of the region between 116.1 and 117.4 Mb when the 14 SNPs in the region between 102.9 and 106.0 Mb were fitted as fixed effects.

Other chromosomes with significant SNPs

In addition to chromosome 7, SNPs on chromosomes 1, 6, 10, 11, 12 and 14 also had significant effects with genome-wide significance (Table 3; Additional file 1: Table S1). Among these chromosomes, the region between 77.3 and 79.7 Mb on chromosome 11 had the most significant additive effects. The top six SNPs within or near the *FGF14, BIVM, LOC10216759,* and *LOC102167785* genes accounted for 5.1% of the genomic narrow-sense heritability and 5.2% of the observed prediction accuracy (Table 3). The remaining regions on chromosomes 1, 7, 11 and 12 each with at least six SNPs accounted for 0.6 to 3.1% of the genomic narrow-sense heritability and accounted for 1.2 to 2.6% of the observed prediction accuracy (Table 3).

Discussion

Comparison with previous GWAS results

The region between 102.9 and 106.0 Mb on chromosome 7 that was identified in the current study was also reported in several previous GWAS but with varying lengths, e.g., between 102.1 and 105.2 Mb [2], 102.9 and 105.2 Mb [4], 103.0 and 103.6 Mb [3], the *VRTN* gene [5], the *VRTN–PROX2–FOS* region that is equivalent to the region between 103.4 and 104.3 Mb based on a microsatellite study [21], and the region between 106.7 and 106.9 Mb [7]. Five significant SNPs on chromosome 10 were located close to some previously reported significant regions on this chromosome [2], and a significant SNP at 51.6 Mb on chromosome 12 was close to the previously reported chromosome 12 region between 52.9 and 52.6 Mb [2, 4]. In the current study, the region between 77.7 and 79.7 Mb on chromosome 11 was the second most significant chromosome region, which, to our knowledge, has not been reported in previous GWAS. However, the Animal QTLdb [1] has one entry for pig teat number in the region between 79.2 and 85.7 Mb on chromosome 11, which partially overlaps the region between 77.7 and 79.7 Mb that we detected in this study, and this Animal QTLdb result was based on a QTL mapping study using 137 microsatellite markers on 573 F2 females and 530 F2 males from a Meishan × Large White cross [22].

Heritabilities and factors that affect teat number

The genomic narrow-sense heritability estimates reported in the current study are the only ones available for teat number. Estimated narrow-sense heritabilities ranged from 0.346 to 0.350 (Table 1) and were within the range of recently published heritability estimates based on pedigree relationships, e.g., 0.39 in a study using

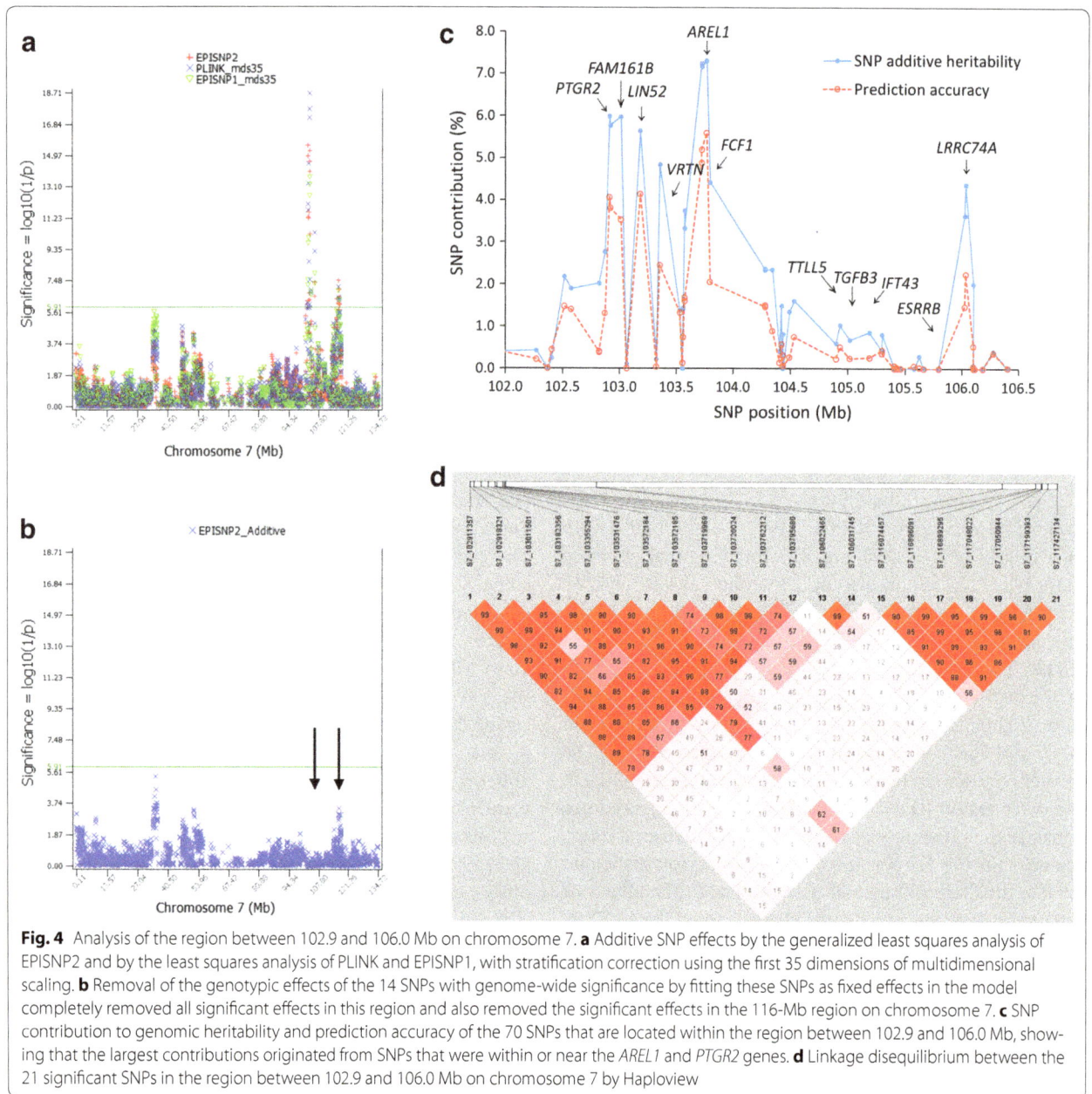

Fig. 4 Analysis of the region between 102.9 and 106.0 Mb on chromosome 7. **a** Additive SNP effects by the generalized least squares analysis of EPISNP2 and by the least squares analysis of PLINK and EPISNP1, with stratification correction using the first 35 dimensions of multidimensional scaling. **b** Removal of the genotypic effects of the 14 SNPs with genome-wide significance by fitting these SNPs as fixed effects in the model completely removed all significant effects in this region and also removed the significant effects in the 116-Mb region on chromosome 7. **c** SNP contribution to genomic heritability and prediction accuracy of the 70 SNPs that are located within the region between 102.9 and 106.0 Mb, showing that the largest contributions originated from SNPs that were within or near the *AREL1* and *PTGR2* genes. **d** Linkage disequilibrium between the 21 significant SNPs in the region between 102.9 and 106.0 Mb on chromosome 7 by Haploview

57,000 Yorkshire pigs [23], and 0.37 in a study using 1550 Landrace pigs [3]. The above genomic and pedigree-based heritability estimates indicate that swine teat number has a strong genetic component, and also that a large portion of the phenotypic variation is not explained by additive genetic effects. Only one GWAS reported dominance effects on chromosome 4 in Landrace pigs [3]. Our GWAS results differ from the reported dominance effects on chromosome 4 because we found that many other chromosomes had more significant dominance effects than chromosome 4, although none of the dominance effects that we observed reached the genome-wide

significance threshold of $p < 10^{-5.91}$ (Fig. 3b). The estimated dominance heritability was low (0.036) and inclusion of dominance effects in the prediction model only had negligible effects on the observed prediction accuracy. Removing dominance effects from the mixed model resulted only in a 0.5% reduction in the observed accuracy of predicting phenotypic values, although the reductions in the expected accuracy of predicting phenotypic values (7.6%) and in the expected accuracy of predicting genetic values (3.9%) were considerably larger for unknown reasons (Table 2). Previously, a maternal effect on teat number was reported [24] but has not been studied by GWAS or genomic prediction. Data error is a source of phenotypic variation, but teat number is easy to measure and any data errors that might have occurred would only be minor, given that our GWAS results on chromosome 7 agreed with those of other studies [2, 3] and that our genomic heritability estimates are consistent with those based on pedigree data [3, 23].

SNP contributions to genomic heritability and prediction accuracy

We analyzed three methods (Methods I, II and III) of estimating the contribution of a set of SNPs to genomic heritability and compared Methods II and III for estimating the contributions of SNPs to both genomic heritability and prediction accuracy. Method I for estimating the contribution of SNPs to genomic heritability consisted of summing together the heritability estimates of the target set of SNPs because the narrow-sense, dominance and broad-sense heritabilities for each SNP can be estimated individually by the GVCBLUP package [18]. However, this method was inappropriate for estimating SNP contributions to the phenotypic variance because the heritability estimate for each SNP decreases as the number of SNPs in the mixed model increases and is approximately proportional to $1/m$, where m is the number of SNPs in the mixed model (see "Appendix" for an approximate proof).

The results in Table 4 show the dependency of the size of the heritability estimate on the number of SNPs fitted as random effects in the mixed model. When dropping every other SNP from the model (i.e. reducing from 41,108 to 20,554 SNPs), the average $h^2_{\alpha i}$ (narrow-sense heritability of the ith SNP) of each SNP for the same 20,554 SNPs when all 41,108 SNPs were fitted in the model nearly doubled ($\bar{h}^2_{\alpha 1}/\bar{h}^2_{\alpha 2} = 1.960$), while the total narrow-sense heritability was nearly unaffected ($\widehat{h}^2_{\alpha 1} = 0.360$ and $\widehat{h}^2_{\alpha 3} = 0.368$), showing that the size of the heritability estimate for each SNP was approximately divided by 2 when the model had twice as many SNPs. Therefore, the heritabilities of the significant SNPs are

not suitable for measuring their contributions to the phenotypic variance due to the dependency of the size of the heritability estimate for each SNP on the number of SNPs fitted as random effects in the mixed model.

Method II for estimating the contribution of a set of target SNPs to genomic heritability and prediction accuracy consisted in calculating the difference between the model with all SNPs and the model without the target SNPs. However, removing the target SNPs from the statistical model may not completely remove their effects because some of them could be explained by other SNPs in LD with the target SNPs. The results in Table 4 supported this expectation, i.e., the total narrow-sense heritability when halving the number of SNPs fitted was nearly unaffected. Using Method II, the 85 significant SNPs accounted for 4.5% of the total genomic heritability (Table 1) and 2.5% of the observed prediction accuracy (Model 3A, Table 2). Due to the partial effects of the removed SNPs that could have been explained by other SNPs, the contributions of SNPs to genomic heritability and observed prediction accuracy using Method II can be considered as the lower bound of the SNP contributions.

Method III for estimating SNP contribution to genomic heritability and to prediction accuracy calculated the difference between the model with all SNPs fitted as random effects and the model with the target SNPs fitted as fixed effects, an approach that we refer to as 'partial heritability' and 'partial accuracy'. The example of the region between 102.9 and 106.3 Mb on chromosome 7 showed that fitting significant SNPs as fixed effects completely removed the significant effects of those SNPs and also removed the effects of the SNPs that are still fitted in the statistical model as random effects but are in LD with the SNPs fitted as fixed effects (Fig. 4b). Figure 5 is a graphical view of a specific chromosome region, showing that the contributions of SNPs to the total narrow-sense heritability from the two models with 41,108 SNPs and 20,554 SNPs estimated using Method III were nearly the same for the region between 102.9 and 106.0 Mb on chromosome 7, i.e., partial heritability estimates were

Table 4 Estimates of SNP additive heritabilities of teat number when using 41,108 autosomal SNPs or every other (20,554) of the 41,108 SNPs

SNP set	Average $\widehat{h}^2_{k\alpha}$ per SNP	Ratio	Total $\widehat{h}^2_{\alpha i}$
20,554 SNPs	$\bar{h}^2_{\alpha 1} = 1.75\ (10^{-5})$	$\bar{h}^2_{\alpha 1}/\bar{h}^2_{\alpha 2} = 1.960$	$\widehat{h}^2_{\alpha 1} = 0.360$
20,554 SNPs with all 41,108 SNPs in the mixed model	$\bar{h}^2_{\alpha 2} = 8.93\ (10^{-6})$	$\bar{h}^2_{\alpha 2}/\bar{h}^2_{\alpha 3} = 0.998$	$\widehat{h}^2_{\alpha 2} = 0.184$
41,108 SNPs	$\bar{h}^2_{\alpha 3} = 8.95\ (10^{-6})$	$\bar{h}^2_{\alpha 1}/\bar{h}^2_{\alpha 3} = 1.955$	$\widehat{h}^2_{\alpha 3} = 0.368$

$\widehat{h}^2_{k\alpha}$ is the heritability of the kth SNP. $\bar{h}^2_{\alpha i} = \sum_{k=1}^{m_i} \widehat{h}^2_{ka}/m_i$ is the average of SNP additive heritability of the ith SNP set, $\widehat{h}^2_{\alpha i} = \sum_{k=1}^{m_i} \widehat{h}^2_{ka}$ is the total additive heritability of all SNPs in the ith SNP set, where m_i is the number of SNPs in the ith SNP set nearly unaffected by the number of SNPs in the model. Using this method, the 85 significant SNPs accounted for 28.5 to 28.8% of the genomic narrow-sense heritability (Table 1) and for 36.2 to 36.8% of the observed prediction accuracy (Table 2). In general, contributions of SNPs to genomic heritability and to the observed prediction accuracy were consistent, i.e., most SNPs with higher contributions to heritability also had greater contributions to prediction accuracies. On average, the contributions of SNPs obtained with Method III were larger than those with Method II by 24.0% (4.6 to 28.6%) for the genomic additive heritability and by 33.3% (2.3 to 35.6%) for the observed prediction accuracy (Table 2). Such large differences could be due to two reasons: overestimation by Method III and underestimation by Method II. Overestimation by Method III is expected since some effects that do not come from the target SNPs that are fitted as fixed effects could also be removed, in addition to removing the effects of the target SNPs. e.g., fitting the 14 SNPs in the region between 102.9 and 106.0 Mb on chromosome 7 as fixed effects also removed the significant effects of SNPs in the region between 116.1 and 117.4 Mb on chromosome 7 (Fig. 4b). Therefore, the contributions of SNPs estimated by Method III could be considered as the upper bound of the true SNP contributions. However, underestimation of Method II is likely the main reason for the large differences between Methods II and III, because a large percentage of the effects of the removed SNPs could have been explained by other SNPs in the

model. For the sample in Table 4, the effects of half of the 41,108 SNPs were almost completely explained by the remaining half of the 41,108 SNPs because estimates of genomic heritabilities from those two sets of SNPs were almost the same, as we discussed above. Based on this analysis, we report contributions of SNPs by Method III in the abstract but also show the results obtained with Method II in the main body of the article (Tables 1, 2).

Observed and expected prediction accuracies

In the current study, we compared three measures of prediction accuracy: the observed prediction accuracy of predicting phenotypic values based on the correlation between predictions from GBLUP and phenotypic observations of the validation individuals ($\widehat{R}_{0jp}$, j = α or t), the expected accuracy of predicting phenotypic values (R_{0jp}), and the expected accuracy of predicting genetic values (R_{0j}). For the models using all 41,108 SNPs fitted as random effects (Model 1A and Model 2A in Table 2), we found excellent consistency between the observed accuracies of predicting phenotypic values ($\widehat{R}_{0tp}$ = 0.437 for Model 1A and $\widehat{R}_{0\alpha p}$ = 0.435 for Model 2A) and the expected accuracies of predicting phenotypic values (R_{0tp} = 0.460 for Model 1A and $R_{0\alpha p}$ = 0.425 for Model 2A). For the models using the 85 SNPs as fixed effects to remove the genetic values of those SNPs from the phenotypic values (Model 1B and Model 2B in Table 2), some differences between the observed accuracies of predicting phenotypic values ($\widehat{R}_{0tp}$ = 0.279 for Model 1B and

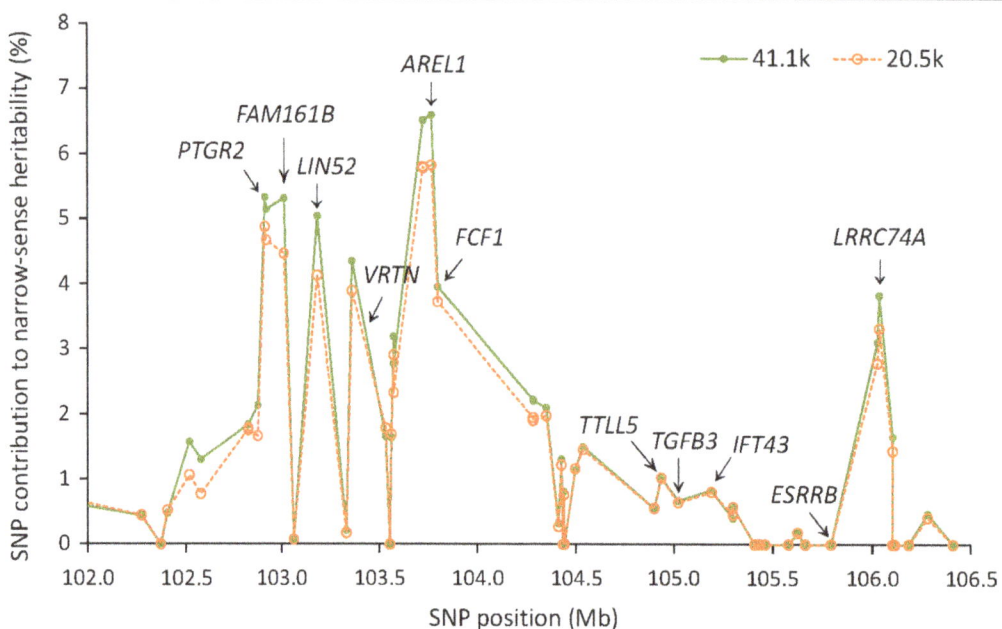

Fig. 5 SNP partial heritability in the region between 102.9 and 106.0 Mb on chromosome 7 from two models with 20.5 and 41 K SNPs. The results show that partial heritability estimates were nearly unaffected by the number of SNPs in the model

$\widehat{R}_{0\alpha p} = 0.275$ for Model 2B) and between the expected accuracies of predicting phenotypic values ($R_{0tp} = 0.360$ for Model 1B and $R_{0\alpha p} = 0.320$ for Model 2B) were larger (Table 2), but those differences were mostly within one standard deviation of the observed accuracies and should be considered as acceptable. As expected, both observed and expected accuracies of predicting phenotypic values ($\widehat{R}_{0jp}$ and R_{0jp}) were lower than the expected accuracies of predicting genetic values (R_{0j}).

Conclusions

Swine teat number has a strong genetic component with narrow-sense heritability estimates of about 0.365. The GWAS results confirmed the previously reported region on chromosome 7 and identified several new regions associated with swine teat number; they also indicated that the additive effects are the primary genetic effects for teat number and indicated consistency between statistical significance of SNP effects and SNP contribution to the genomic heritability. Most SNPs with higher statistical significance also had greater contributions to the genomic broad-sense heritability and prediction accuracy. The 85 significant SNPs accounted for about 28% of the genomic heritability and 36% of the prediction accuracy.

Additional files

Additional file 1: Table S1. Significant SNP effects with Bonferroni significance ($p < 10^{-5.91}$) by three methods of GWAS analysis.

Additional file 2: Figure S1. Manhattan plots of additive SNP effects of all 18 autosomes by three methods of GWAS analysis. All p-values in the figures are in log(1/p) scale.

Authors' contributions
XH, YD, NL and CT designed the experiments. ZW, DL, RZ and XH performed data collection. CT, JR and ZH performed the experiments. CT and YD analyzed the data. DP developed some of the analysis software. YD, CT and XHu wrote the manuscript. All authors read and approved the final manuscript.

Author details
[1] State Key Laboratory for Agrobiotechnology, China Agricultural University, Beijing 100193, China. [2] Department of Animal Science, University of Minnesota, Saint Paul, MN 55108, USA. [3] National Engineering Research Center for Breeding Swine Industry, South China Agricultural University, Guangdong 510642, China.

Funding
This project was supported by the National Basic Research Program of China (2014CB138501), the 948 Program of the Ministry of Agriculture of China (2012-G1(4)), and the National High Technology Research and Development Program of China (2011AA100301).

Competing interests
The authors declare that they have no competing interests.

Appendix: Approximate proof for the decrease in SNP heritability as the number of SNPs increases

The heritability estimate for each SNP decreases as the number of SNPs in the mixed model increases and is approximately proportional to $1/m$, where m is the number of SNPs in the mixed model. An approximate mathematical proof for this result can be derived based on the invariance property of GBLUP and GREML to duplicating SNPs. Assuming a set of m SNPs is duplicated r times in the mixed model, GBLUP of genetic values (additive, dominance and genotypic values) of individuals and SNP genetic variance components, as well as the associated variance estimates by GREML are invariant to the duplication of SNPs, and GBLUP of SNP additive, dominance and genotypic effects differ from those without duplicate SNPs by the square root of r [25]. In the example of additive SNP effects, $\widehat{\alpha}_{ri} = \widehat{\alpha}_i / \sqrt{r}$, where $\widehat{\alpha}_{ri}$ is the additive GBLUP estimate of the ith SNP from the mixed model with m SNPs repeated r times, $\widehat{\alpha}_i$ is the additive GBLUP estimate of the ith SNP from the model with m SNPs. The additive heritability for the ith SNP from the mixed model with m SNPs ($h_{\alpha i}^2$) is: $h_{\alpha i}^2 = \left(\widehat{\alpha}_i^2 / \sum_{i=1}^{m} \widehat{\alpha}_i^2\right) h_\alpha^2$ [18], where $\widehat{\alpha}_i$ is the additive GBLUP of the ith SNP and h_α^2 is the total additive heritability based on all SNPs. Since h_α^2 is unaffected by repeated SNPs, the additive heritability for the ith SNP from the mixed model with m SNPs repeated r times ($h_{\alpha ri}^2$) is:

$$h_{\alpha ri}^2 = \left(\widehat{\alpha}_{ri}^2 \Big/ \sum_{i=1}^{rm} \widehat{\alpha}_{ri}^2\right) h_\alpha^2 = \left(\widehat{\alpha}_i^2 / r\right) \Big/ \left(\sum_{i=1}^{rm} \widehat{\alpha}_i^2 / r\right) h_\alpha^2 = h_{\alpha i}^2 / r,$$

i.e., the heritability for the ith SNP from the model with r times of the m SNPs is $1/r$ of the SNP heritability with m SNPs in the model without repeat. This theoretical result under the simple assumption of repeated SNPs was almost the same as the results obtained with the real data in Table 4, i.e., when the number of SNPs is reduced by half, the heritability of each SNP as an average of all SNP heritability estimates nearly doubled.

References
1. Hu ZL, Park CA, Wu XL, Reecy JM. Animal QTLdb: an improved database tool for livestock animal QTL/association data dissemination in the post-genome era. Nucleic Acids Res. 2013;41:D871–9.
2. Duijvesteijn N, Veltmaat JM, Knol EF, Harlizius B. High-resolution association mapping of number of teats in pigs reveals regions controlling vertebral development. BMC Genomics. 2014;15:542.
3. Lopes MS, Bastiaansen JW, Harlizius B, Knol EF, Bovenhuis H. A genome-wide association study reveals dominance effects on number of teats in pigs. PLoS One. 2014;9:e105867.
4. Verardo LL, Silva FF, Lopes MS, Madsen O, Bastiaansen JW, Knol EF, et al. Revealing new candidate genes for reproductive traits in pigs: combining Bayesian GWAS and functional pathways. Genet Sel Evol. 2016;48:9.
5. Yang J, Huang L, Yang M, Fan Y, Li L, Fang S, et al. Possible introgression of the *VRTN* mutation increasing vertebral number, carcass length and teat number from Chinese pigs into European pigs. Sci Rep. 2016;6:19240.

6. Lee JB, Jung EJ, Park HB, Jin S, Seo DW, Ko MS, et al. Genome-wide association analysis to identify SNP markers affecting teat numbers in an F2 intercross population between Landrace and Korean native pigs. Mol Biol Rep. 2014;41:7167–73.
7. Arakawa A, Okumura N, Taniguchi M, Hayashi T, Hirose K, Fukawa K, et al. Genome-wide association QTL mapping for teat number in a purebred population of Duroc pigs. Anim Genet. 2015;46:571–5.
8. Clayton GA, Powell JC, Hiley PG. Inheritance of teat number and teat inversion in pigs. Anim Prod. 1981;33:299–304.
9. Poland JA, Brown PJ, Sorrells ME, Jannink JL. Development of high-density genetic maps for barley and wheat using a novel two-enzyme genotyping-by-sequencing approach. PLoS One. 2012;7:e32253.
10. Glaubitz JC, Casstevens TM, Lu F, Harriman J, Elshire RJ, Sun Q, et al. TASSEL-GBS: a high capacity genotyping by sequencing analysis pipeline. PLoS One. 2014;9:e90346.
11. Ma L, Runesha HB, Dvorkin D, Garbe J, Da Y. Parallel and serial computing tools for testing single-locus and epistatic SNP effects of quantitative traits in genome-wide association studies. BMC Bioinformatics. 2008;9:315.
12. Ma L, Wiggans GR, Wang S, Sonstegard TS, Yang J, Crooker BA, et al. Effect of sample stratification on dairy GWAS results. BMC Genomics. 2012;13:536.
13. Purcell S, Neale B, Todd-Brown K, Thomas L, Ferreira MA, Bender D, et al. PLINK: a tool set for whole-genome association and population-based linkage analyses. Am J Hum Genet. 2007;81:559–75.
14. Freedman ML, Reich D, Penney KL, McDonald GJ, Mignault AA, Patterson N, et al. Assessing the impact of population stratification on genetic association studies. Nat Genet. 2004;36:388–93.
15. Mao Y, London NR, Ma L, Dvorkin D, Da Y. Detection of SNP epistasis effects of quantitative traits using an extended Kempthorne model. Physiol Genomics. 2006;28:46–52.
16. Da Y, Wang C, Wang S, Hu G. Mixed model methods for genomic prediction and variance component estimation of additive and dominance effects using SNP markers. PLoS One. 2014;9:e87666.
17. Wang C, Da Y. Quantitative genetics model as the unifying model for defining genomic relationship and inbreeding coefficient. PLoS One. 2014;9:e114484.
18. Wang C, Prakapenka D, Wang S, Pulugurta S, Runesha HB, Da Y. GVCBLUP: a computer package for genomic prediction and variance component estimation of additive and dominance effects. BMC Bioinformatics. 2014;15:270.
19. Legarra A, Robert-Granié C, Manfredi E, Elsen JM. Performance of genomic selection in mice. Genetics. 2008;180:611–8.
20. Barrett JC, Fry B, Maller J, Daly M. Haploview: analysis and visualization of LD and haplotype maps. Bioinformatics. 2005;21:263–5.
21. Ren DR, Ren J, Ruan GF, Guo YM, Wu LH, Yang GC, et al. Mapping and fine mapping of quantitative trait loci for the number of vertebrae in a White Duroc × Chinese Erhualian intercross resource population. Anim Genet. 2012;43:545–51.
22. Bidanel J, Rosendo A, Iannuccelli N, Riquet J, Gilbert H, Caritez J, Billon Y, Amigues Y, Prunier A, Milan D. Detection of quantitative trait loci for teat number and female reproductive traits in Meishan × Large White F2 pigs. Animal. 2008;2:813–20.
23. Lundeheim N, Chalkias H, Rydhmer L. Genetic analysis of teat number and litter traits in pigs. Acta Agric Scand A. 2013;63:121–5.
24. Pumfrey RA, Johnson RK, Cunningham PJ, Zimmerman DR. Inheritance of teat number and its relationship to maternal traits in swine. J Anim Sci. 1980;50:1057–60.
25. Da Y, Tan C, Parakapenka D. Joint SNP-haplotype analysis for genomic selection based on the invariance property of GBLUP and GREML to duplicate SNPs. J Anim Sci. 2016;94:161–2.

6

On the performance of tests for the detection of signatures of selection: a case study with the Spanish autochthonous beef cattle populations

Aldemar González-Rodríguez[1], Sebastián Munilla[1,2], Elena F. Mouresan[1], Jhon J. Cañas-Álvarez[3], Clara Díaz[4], Jesús Piedrafita[3], Juan Altarriba[1,5], Jesús Á. Baro[6], Antonio Molina[7] and Luis Varona[1,5]*

Abstract

Background: Procedures for the detection of signatures of selection can be classified according to the source of information they use to reject the null hypothesis of absence of selection. Three main groups of tests can be identified that are based on: (1) the analysis of the site frequency spectrum, (2) the study of the extension of the linkage disequilibrium across the length of the haplotypes that surround the polymorphism, and (3) the differentiation among populations. The aim of this study was to compare the performance of a subset of these procedures by using a dataset on seven Spanish autochthonous beef cattle populations.

Results: Analysis of the correlations between the logarithms of the statistics that were obtained by 11 tests for detecting signatures of selection at each single nucleotide polymorphism confirmed that they can be clustered into the three main groups mentioned above. A factor analysis summarized the results of the 11 tests into three canonical axes that were each associated with one of the three groups. Moreover, the signatures of selection identified with the first and second groups of tests were shared across populations, whereas those with the third group were more breed-specific. Nevertheless, an enrichment analysis identified the metabolic pathways that were associated with each group; they coincided with canonical axes and were related to immune response, muscle development, protein biosynthesis, skin and pigmentation, glucose metabolism, fat metabolism, embryogenesis and morphology, heart and uterine metabolism, regulation of the hypothalamic–pituitary–thyroid axis, hormonal, cellular cycle, cell signaling and extracellular receptors.

Conclusions: We show that the results of the procedures used to identify signals of selection differed substantially between the three groups of tests. However, they can be classified using a factor analysis. Moreover, each canonical factor that coincided with a group of tests identified different signals of selection, which could be attributed to processes of selection that occurred at different evolutionary times. Nevertheless, the metabolic pathways that were associated with each group of tests were similar, which suggests that the selection events that occurred during the evolutionary history of the populations probably affected the same group of traits.

*Correspondence: lvarona@unizar.es
[1] Departamento de Anatomía, Embriología y Genética, Universidad de Zaragoza, 50013 Saragossa, Spain
Full list of author information is available at the end of the article

Background

The evolutionary history of animal populations involves both natural and artificial selection. These processes not only affect the allelic frequencies at causal polymorphisms, but also the surrounding genomic regions due to the so-called "hitchhiking" effect. Thus, they may leave detectable signals on the structure of the genome that can be identified by using appropriate procedures [1, 2].

The vast majority of the procedures used to detect signatures of selection [2] is based on the null hypothesis of absence of selection, which relies on the neutral model of evolution [3]. In fact, these procedures can be classified according to the source of information they use to reject the null hypothesis. Based on the literature [2], three main groups of tests can be identified: the first group is based on the analysis of the site frequency spectrum [4–6], the second group focuses on the study of the extension of the linkage disequilibrium across the length of the haplotypes that surround a polymorphism [7, 8], and the third group is based on several measures of differentiation among populations [9–11]. In addition, the results of all these tests can be affected to some degree by demographic events and by the ascertainment bias caused by the procedure used to select the single nucleotide polymorphisms (SNPs) for the genotyping chip [12]. Thus, the results of each test may not be fully consistent with each other [13], which has led to propose strategies for summarizing results into a single statistic that either does [13] or does not [14, 15] account for the correlations between the results from different methods.

The aim of our study was to compare the performance of a subset of these procedures by using a dataset on seven autochthonous beef cattle populations (Asturiana de los Valles, Avileña-Negra Ibérica, Bruna dels Pirineus, Morucha, Pirenaica, Retinta and Rubia Gallega) which share close genetic relationships between them [16]. A second objective was to identify candidate genes and/or metabolic processes that are associated with the regions involved in the selection processes that occurred during the evolution of these populations.

Methods
Animals and sample size

A total of 171 sire/dam/offspring triplets were collected from seven Spanish beef cattle populations, including Asturiana de los Valles (AV, n = 25), Avileña-Negra Ibérica (ANI, n = 24), Bruna dels Pirineus (BP, n = 25), Morucha (Mo, n = 24), Pirenaica (Pi, n = 24), Retinta (Re, n = 24) and Rubia Gallega (RG, n = 24) breeds. The selected parents were chosen as unrelated as possible to fully represent the diversity of the populations.

SNP genotyping and phasing

Genomic DNA was extracted by standard protocols. High-density SNP genotyping was performed at a commercial laboratory (Xenética Fontao, Lugo, Spain) by using the BovineHD BeadChip (Illumina Inc, USA) according to the manufacturer's protocol; this HD chip is designed to genotype 777,962 SNPs. The SNPs that were retained for our study were located on autosomal chromosomes at a single position. Additional requirements were a Mendelian error rate lower than 0.05, and SNP and individual call rates higher than 0.95. Quality control was performed by using PLINK software [17] and finally, 703,707 SNPs that covered 2,510,606 kb were available for the analyses with on average one SNP per 3.567 kb. Haplotypes for the parental chromosomes were derived with Beagle software [18] using the "TRIO" option.

Detection of signatures of selection

The data were analysed using the following procedures for the detection of signatures of selection.

Tajima

The procedure that was developed by Tajima [4] compares two statistics to estimate the scaled mutation rate. The first statistic (θ_π) is based on the number of segregating sites within a genomic region and the second (θ_κ) is the average heterozygosity at segregating sites in the sample. The standardized difference between these two values, $D = \theta_\pi - \theta_\kappa$, is used to infer departures from neutrality. Theoretically, if $D < 0$ either the population has suffered expansion after a recent bottleneck or a recent selective sweep has taken place; on the contrary if $D > 0$, the population has either experienced a sudden population contraction or is under balancing selection. The analysis was performed over sliding windows of 100 SNPs by using own software.

Fay and Wu

This procedure [6] calculates the following statistic $D = \theta_\pi - \theta_H$, where θ_H depends on the number of sites at which a derived allele is present within a genomic region. In the analysis, the ancestral alleles were extracted from the study of Rocha et al. [19]. This test was computed over sliding windows of 100 SNPs by using own software.

Fu and Li

This procedure [5] is based on counting the number of singletons or alleles present in only one phase. The rationale is that a selection process will extend time to coalescence so that a larger number of mutations may take place in new or external branches of the tree and thus appear only once in the observed sample. As before, the analysis was performed over sliding windows of 100 SNPs by using own software.

iHS

This procedure [8] calculates the ratio of the integrated haplotype score (*iHH*) for the ancestral allele and the derived allele at a given SNP. The *iHH* is the integral (area) of the observed decay of the *EHH* (extended haplotype homozygosity) as defined by Sabeti et al. [7]. As in the previous test, ancestral alleles were extracted from Rocha et al. [19]. The *iHS* was calculated with the *selscan* software [20] using the parameters recommended by the authors. For further calculations, we used the |*iHS*|.

nSL

This procedure was recently presented by Ferrer-Admetlla et al. [21]. The procedure of calculation is similar to *iHS*, but replaces *IHH* by an alternative statistic (*SL*) that measures the length of a segment of haplotype homozygosity in terms of segregating sites. The main advantage of *nSL* over *iHS* is that it uses segregating sites as a measure of distance, while *iHS* needs the recombination distance. Thus, the *iHS* is more sensitive to recombination rate [21]. The analysis used the same parameters as in the *iHS* test and own software. As before, we used |nSL|.

H12

This method was recently proposed [22] with the *H*12 statistic being defined as:

$$H12 = (p_1 + p_2)^2 + \sum_{j>2} p_j^2,$$

where p_j is the frequency of the *j*th most common haplotype in the population. Here, the frequencies of the first and second most common haplotypes were combined into a single frequency. The calculation was performed over sliding windows of 100 SNPs by using own software.

Fixation index (F_{ST})

This procedure was described by Wright [9] and is the most classical approach to study the pattern of differentiation between populations. The fixation index F_{ST} is calculated for each SNP and for each pair of populations as $F_{ST} = (H_O - H_E)/H_E$, where H_O and H_E are the observed and expected heterozygosities, respectively. Estimates for F_{ST} were averaged over sliding windows of 100 SNPs and assigned to the central SNP in each window. The procedure was computed with own software. Finally, the results for each population were computed by averaging the paired F_{ST} estimates with the other six populations.

Selestim

This procedure [10] assumes a hierarchical Bayesian model to distinguish selected polymorphisms from the background of neutral (or almost neutral) polymorphisms and also to estimate the intensity of selection in each population. The model assumes a binomial distribution of the allele counts at each locus and for each population, and the prior distribution of allelic frequencies is modeled under the assumption of a stationary density of the diffusion process [10]. The model is implemented by using a Markov chain Monte Carlo method. SelEstim software (http://www1.montpellier.inra.fr/CBGP/software/selestim/) was used for this purpose with the standard parameters that are recommended by the authors. Among the outputs provided by the Selestim approach, we extracted the σ_{ij} parameter [10], which represents the coefficient of selection for the *i*th subpopulation and the *j*th locus.

XP-CLR

This approach [23] assumes that the allele frequencies of two populations that diverge from an ancestral population follow a Gaussian distribution for which the variance contains information on the history of the populations since they split. Under the assumption that the evolutionary process is reversible, the procedure defines the distribution of allelic frequencies in the first population (reference) given the allele frequencies in the second population (objective). We calculated *XP-CLR* by taking each pair of populations as objective and reference with the software *XPCLR* (http://genetics.med.harvard.edu/reich/Reich_Lab/Software.html). Then, we averaged the six available tests for each population that was treated as an objective population, to infer the signatures of selection for each breed.

XP-EHH

This approach [7] is also computed for each pair of populations. For each population, as in the *iHS* test, it calculates the *EHH* between a core SNP and a set of SNPs within a predefined genomic interval and integrates it with respect to genetic distance to calculate the integrated haplotype score (*IHH*) for populations A and B. Then, the statistic is computed as $XPEHH_{AB} = \ln(IHH_A/IHH_B)$. As previously, we computed this statistic for each SNP and each pair of populations and the results were averaged over the six comparisons to generate a unique result for each population. We used the software *selscan* [20] with the parameters that are recommended by the authors.

VarLD

This procedure [24] evaluates the magnitude of the differences in linkage disequilibrium between a pair of populations. It calculates the linkage disequilibrium as the correlation coefficient between pairs of SNPs within a genomic region and creates a matrix of those correlations for each population. Then, it evaluates the differences between the matrices of both populations as the difference between its eigenvalues. The procedure was computed using the software VarLD [25] over sliding windows of 100 SNPs.

For all the above-described methods, we used the empirical distribution of the results generated along the genome as the null distribution of the test, in order to reduce the possible effects of the demographic history or the ascertainment bias. The underlying hypothesis is that, on average, both demographic events and ascertainment bias affect all the genome in a similar way, and thus, deviations or extreme values of the empirical distribution could be understood as signals of selection events.

Summary of signals of selection

In order to detect communalities and summarize the results of the 11 procedures for ease of interpretation, we normalized these results for each SNP using a logarithm transformation to make the scale of the different results comparable and, then, we calculated the correlation between the logarithms (or the negative of logarithms for the Tajima, Fu and Li and Fay and Wu procedures) for the 703,707 SNPs. In a confirmatory analysis, provided that the methods used to detect signatures of selection were classified into three groups, we performed a factor analysis restricted to a subspace of three axes using a *varimax* rotation [26]. The analysis was done with *R* [27] by using the function *principal()* included in the package *psych.*

Selection of candidate genes

First, we identified candidate genes based on the empirical distribution of the output of the three canonical axes of the factor analysis. Thus, we defined a very strict threshold by selecting the genomic 1-Mb regions with at least 25 SNPs that were in the top 0.1% of the results for each axis. Then, we used the *Ensembl-Biomart* database to identify the genes that were present in those genomic regions and compared our results with those in the literature to identify potential candidate genes for selection in the bovine populations.

Enrichment analysis

Finally, in order to obtain a clearer picture of the metabolic pathways that were affected by the selection processes, we identified the genomic regions that were above the top 5% of each canonical axis. The objective of the relaxation of the empirical threshold was to capture softer signals of selection. With these selected genomic regions for each canonical axis, we used the software WebGestalt [28] (http://bioinfo.vanderbilt.edu/webgestalt/) by setting the *Homo sapiens* genome as the reference genome. In addition, we used a hypergeometric p value to correct for multiple-testing. The results included the top 10 pathways (WikiPathways).

Results and discussion

Summarizing footprints of selection detected by 11 procedures

A large set of procedures is available for the identification of footprints of selection across the genome [2]. Most of these procedures are based on the rejection of the null hypothesis of absence of selection based on the neutral theory of evolution [3]. However, each of these methods calculates a different statistic to test this hypothesis. In addition, they are influenced to varying degrees by demographic history and ascertainment bias caused by the selection of SNPs [12]. Thus, it is expected that each test provides a different output as confirmed by the correlations between the results obtained by the 11 procedures used in this study (Fig. 1) and by the Manhattan plots generated with the results for each test and population (see Additional file 1: Figure S1, Additional file 2: Figure S2, Additional file 3: Figure S3, Additional file 4: Figure S4, Additional file 5: Figure S5, Additional file 6: Figure S6, Additional file 7: Figure S7, Additional file 8: Figure S8, Additional file 9: Figure S9, Additional file 10: Figure S10, Additional file 11: Figure S11).

In order to summarize the signals of selection that were detected by the 11 tests, there are procedures to condense such results into a single statistic by using Bayes factors [14] or a combination of p values [13, 15]. However, these strategies imply that the signals of selection that are captured by the different methods are comparable. Nevertheless, as the definition of the null hypothesis varies between tests, the signals of selection identified by each procedure may correspond to different types of selection events. In fact, some authors [29] pointed out that within-population haplotype length methods [8, 21] can detect only very recent selection processes, because they become ineffective when the selected alleles reach fixation or are very close to fixation. The same authors [29] indicated that signals of selection that are based on a reduction of genetic diversity [4, 5] persist for a longer period of time and these methods can detect older signals of selection, while tests that are based on population differentiation [9, 10] occupy an intermediate position.

In this study, the correlations of the absolute logarithm of the results between the 11 methods used were low or even negative (Fig. 1). However, there are some

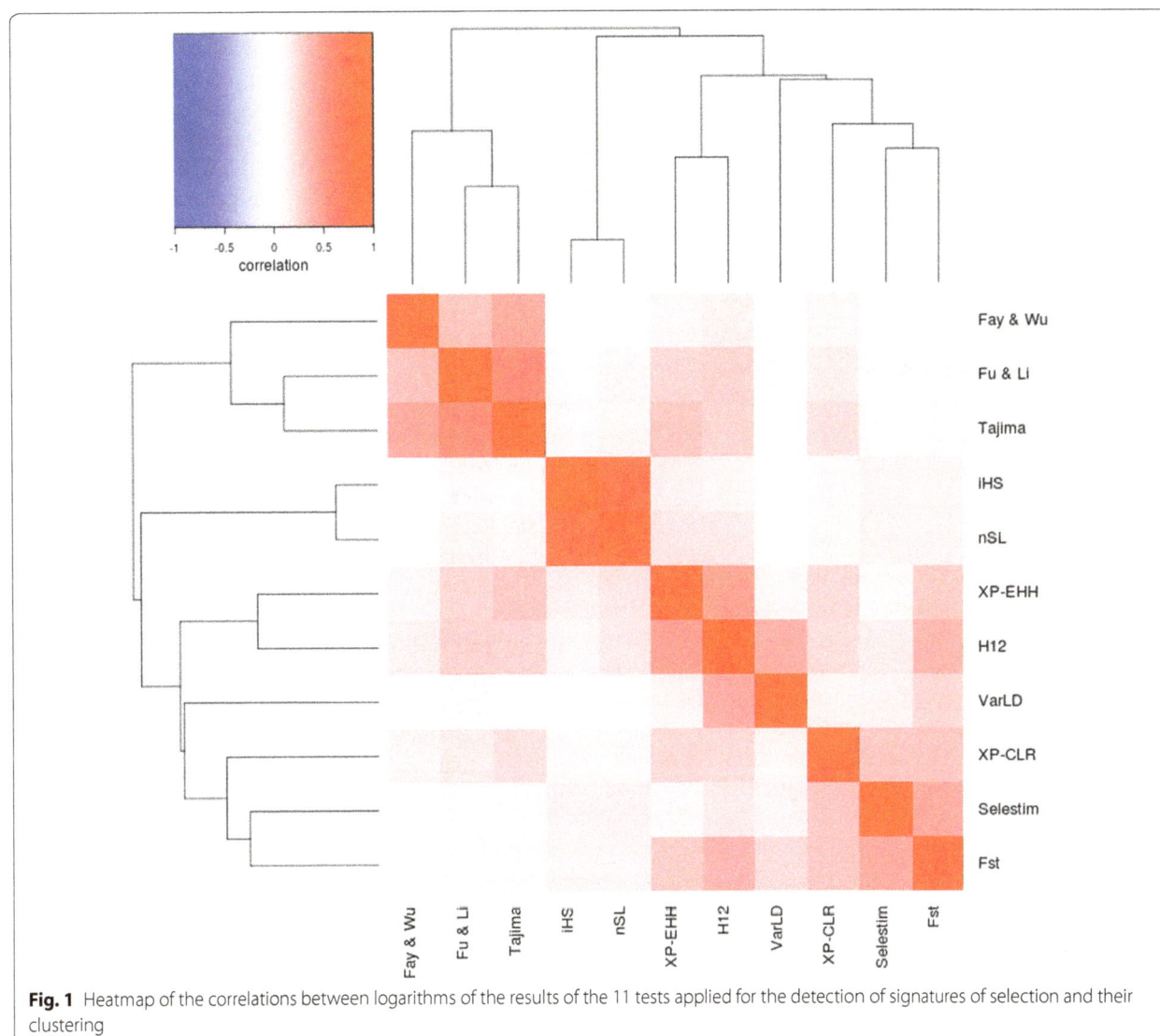

Fig. 1 Heatmap of the correlations between logarithms of the results of the 11 tests applied for the detection of signatures of selection and their clustering

remarkable exceptions such as the correlations between Tajima and Fu and Li, and iHS and nSL tests, that were remarkably high. It should be noted that the Tajima and Fu and Li tests are both based on the analysis of the site frequency spectrum, and that the nSL test is just a modification of iHS where the map distance is replaced by the number of segregating sites [21]. A more detailed analysis of the structure of these correlations allows to identify three main groups of tests, one group based on the site frequency spectrum (Tajima, Fu and Li, Fay and Wu); a second group based on the haplotype length (iHS and nSL), and a third group that focuses on the differentiation between populations (F_{ST}, SelEstim and XP-CLR). The remaining tests (VarLD, H12 and XP-EHH) are in an intermediate position between the latter two, although slightly closer to tests based on population differentiation.

Such a structure of the correlations between the results of these methods indicates that the implementation of a factor analysis, as suggested by Simianer et al. [30], could be appropriate to summarize the results into a few canonical axes. In addition, as described below, each axis was associated with signatures of selection of a different kind. In particular, we applied a factor analysis restricted to three canonical axes using a varimax approximation [26] that explains up to the 56% of variation.

Table 1 shows the loadings for the canonical axes that resulted from the factor analysis for each of the 11 tests used to detect selection signatures. Moreover, Table 1 presents the correlations between the canonical axes and each specific test, which are fully consistent with the results of the correlations presented in Fig. 1. The first axis explains 20% of the variation and shows a high

Table 1 Weights in the factor analysis with, between parentheses, the correlation between the results of each test and the canonical axis, and percentage of variance explained by the three axes

Method	First axis	Second axis	Third axis	% variance
Tajima	−0.07 (0.07)	0.42 (0.85)	−0.01 (0.06)	73
Fu-Li	−0.08 (−0.02)	0.35 (0.68)	−0.05 (−0.04)	47
Fay-Wu	−0.05 (0.09)	0.38 (0.77)	−0.00 (0.07)	61
Selestim	0.28 (0.58)	−0.10 (−0.05)	0.02 (0.13)	36
XPCLR	0.22 (0.51)	0.06 (0.24)	−0.02 (0.06)	32
H12	0.29 (0.67)	0.08 (0.32)	−0.04 (0.06)	55
IHS	−0.06 (0.07)	−0.04 (0.03)	0.53 (0.95)	90
NSL	−0.04 (0.11)	−0.02 (0.07)	0.52 (0.94)	89
F_{ST}	0.38 (0.77)	−0.10 (−0.02)	−0.03 (0.08)	60
XP-EHH	0.17 (0.47)	0.14 (0.40)	0.01 (0.13)	40
VarLD	0.31 (0.59)	−0.11 (−0.09)	−0.10 (−0.08)	36

correlation with the procedures based on the analysis of population differentiation (F_{ST}, SelEstim, XP-EHH, XP-CLR and VarLD) and H12; the second axis explains 19% of the variation and is correlated with the methods based on the site frequency spectrum (Tajima, Fu and Li, and Fay and Wu); and, finally, the third axis is strongly correlated with methods based on the extension of linkage disequilibrium or haplotype length (iHS and nSL)

and explains 17% of the total variation. For each test, the three axes explain between 32 (XP-CLR) and 90% (iHS) of the variation. The Manhattan plots of the results that relate to the three canonical axes are in Figs. 2, 3 and 4. The first two axes presented a higher level of shared signals between populations (see Figs. 2, 3) whereas the results of the third axis were, in general, breed-specific. This statement is supported by the results in Fig. 5, which shows the correlations of the results obtained for the first, second and third canonical axes between populations. An average correlation of 0.50 was found for the first axis [ranging from 0.39 (BP and Re) to 0.71 (AV and RG)]. Furthermore, the second axis also showed high correlations between populations that ranged from 0.37 (Re and Pi) to 0.60 (AV and RG) with an average of 0.49. On the contrary, the correlations for the third axis were lower with an average value of 0.08 and ranged from 0.05 (Pi and Re) to 0.16 (AV and BP). In addition, the structure of the correlations (Fig. 5) confirmed the classification of the populations into two main clusters, one composed by the ANI, Mo and Re populations and the other by the Pi, BP, RG and ANI populations, as previously reported by Cañas-Álvarez et al. [16] based on distance measures and admixture analysis.

As in the study of Sabeti et al. [29], our results may indicate that old selection or adaptation processes that occurred before breed differentiation or during

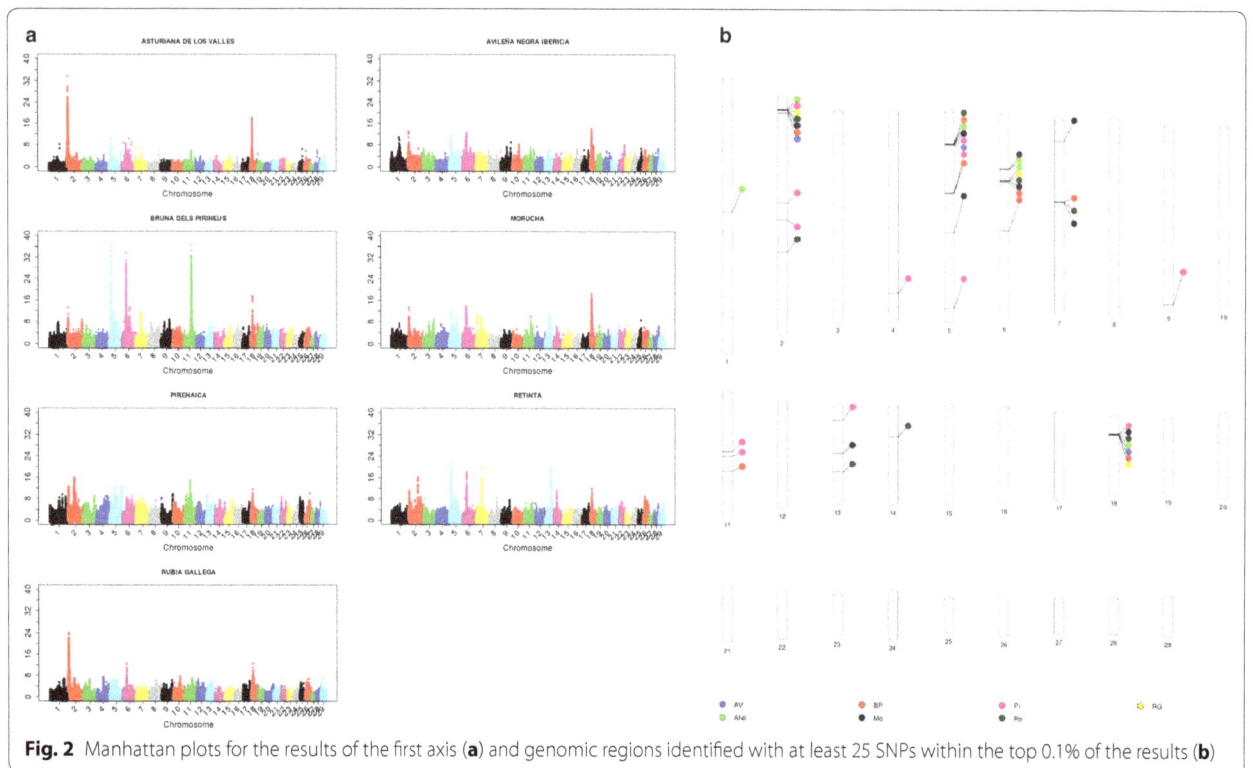

Fig. 2 Manhattan plots for the results of the first axis (**a**) and genomic regions identified with at least 25 SNPs within the top 0.1% of the results (**b**)

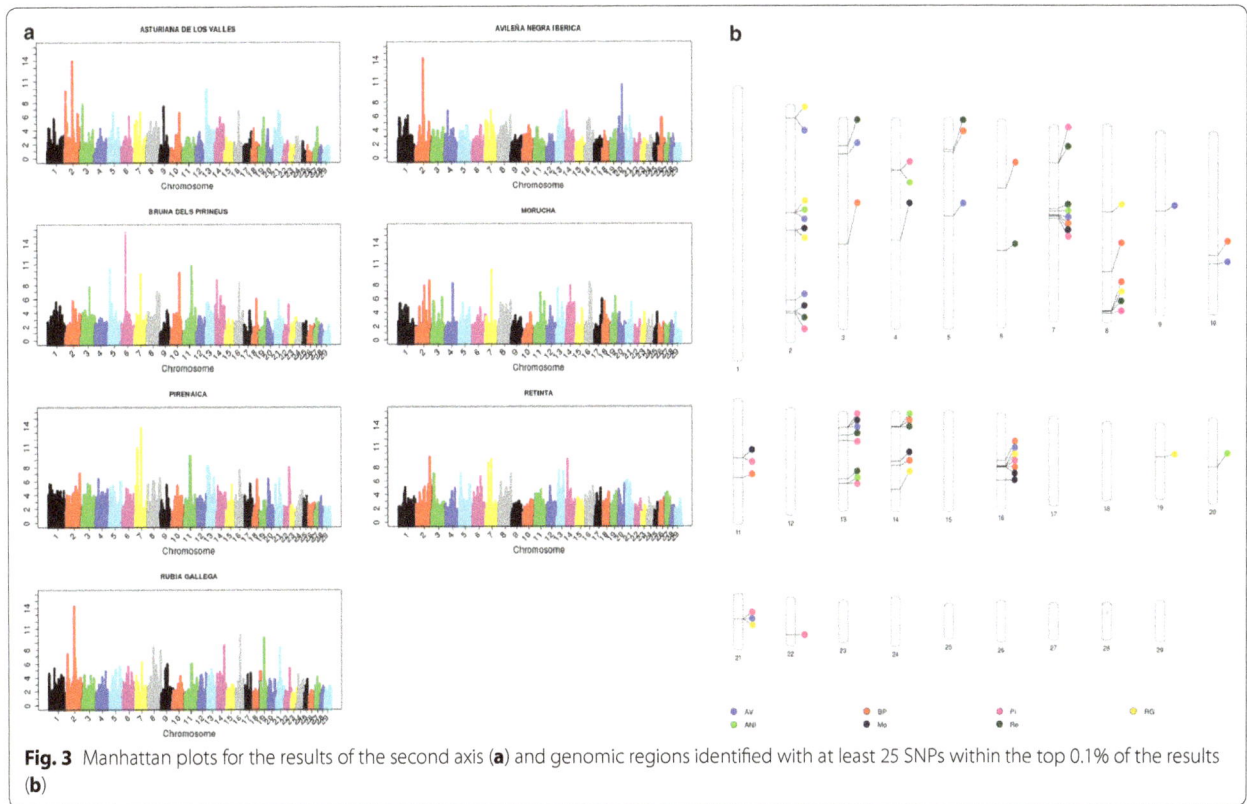

Fig. 3 Manhattan plots for the results of the second axis (**a**) and genomic regions identified with at least 25 SNPs within the top 0.1% of the results (**b**)

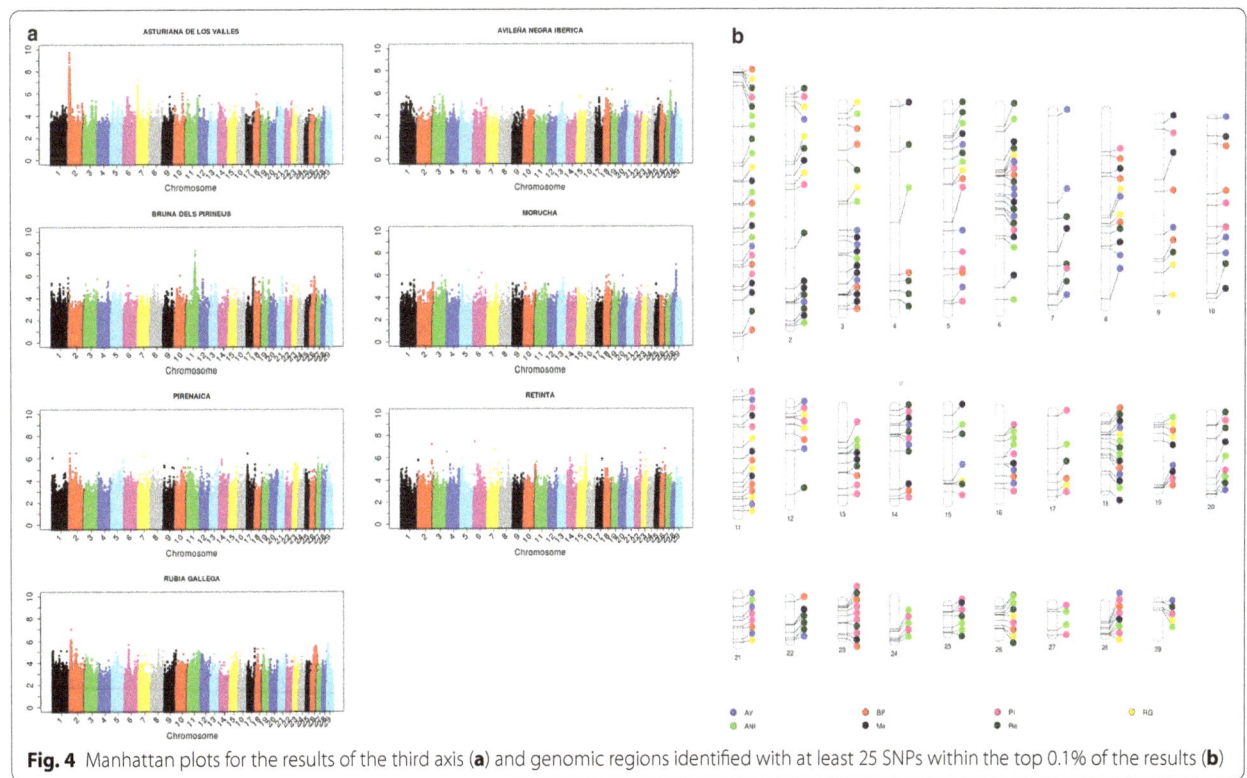

Fig. 4 Manhattan plots for the results of the third axis (**a**) and genomic regions identified with at least 25 SNPs within the top 0.1% of the results (**b**)

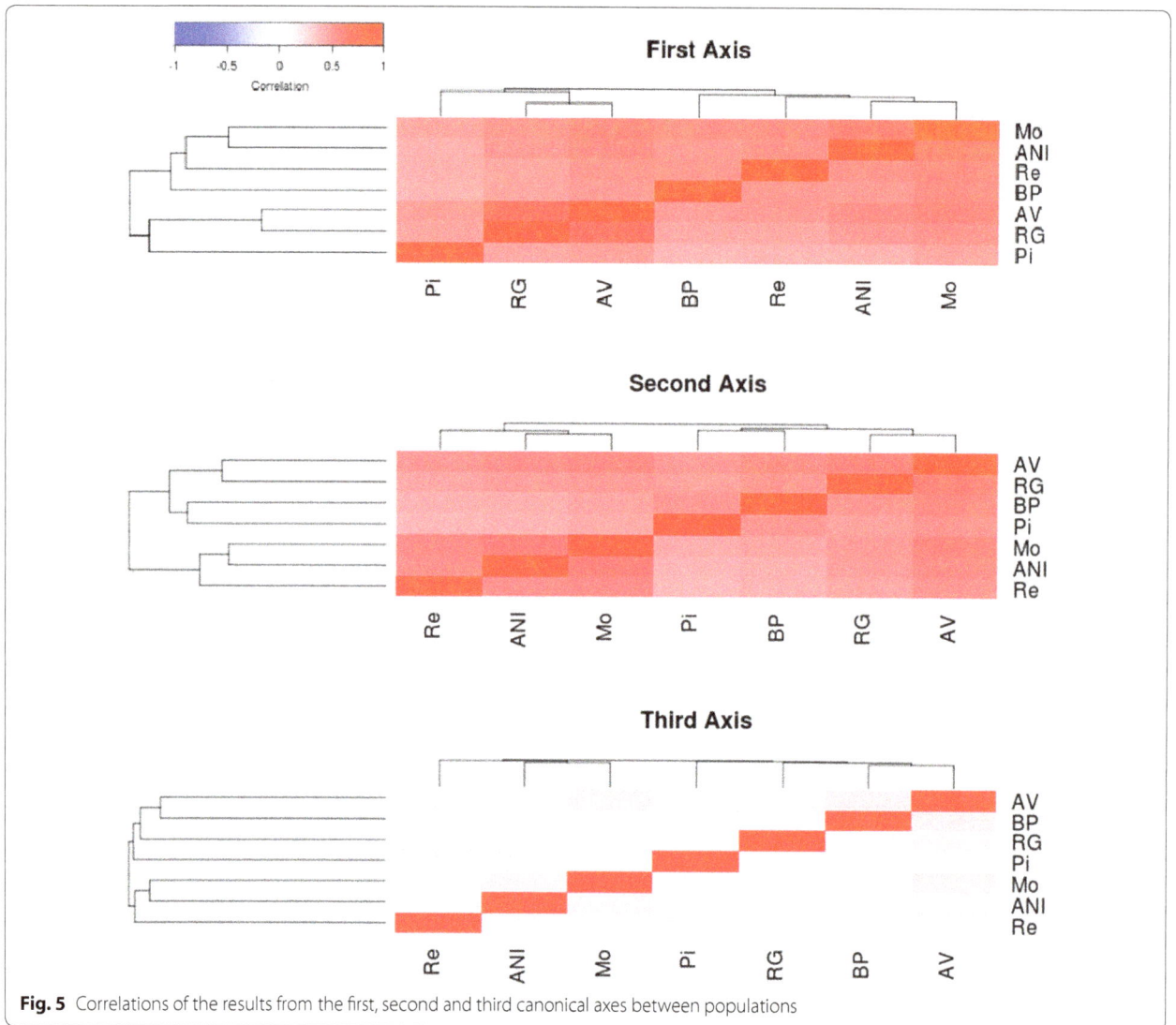

Fig. 5 Correlations of the results from the first, second and third canonical axes between populations

speciation were only detected by the site frequency spectrum methods that are associated with the second canonical axis. The signals of selection that were generated by later isolation and recent selection events within the populations are identified by the differentiation methods, which are linked to the first canonical axis. Finally, the haplotype length methods, summarized in the third canonical axis, identified more recent and, in general, less intense selection events that are mostly specific to each population. The absence of regions with strong recent signals of selection agrees with the postulate that artificial selection processes do not leave relevant signatures of selection [31]. The main reason of this absence can be due to the polygenic nature of most of the traits associated with current selection processes [32–34] or to the effect of epistasis [35].

In fact, the most remarkable signal of selection from this third axis was identified on chromosome 2 around the *myostatin* (*MTSN*) gene (between 6,213,566 and 6,220,196 bp) in the AV population, where double-muscling is included as a criterion of selection in its breeding program. This specific genomic region can be used to illustrate the timing of the signatures of selection that were detected by each group of methods. In the AV population, two large signatures of selection were detected with the first and third canonical axes, respectively. The first axis is related to processes that were involved in the creation of the breed and the third axis to recent selection. In addition, a large signature of selection associated with the first canonical axis in the RG population was observed. However, in this population, there is no relevant signature in the results of the third axis. This result is consistent with several previous studies [36] that reported the presence of haplotypes associated with double-muscling also in the RG population. This may indicate that some degree of selection around this gene may have occurred during the process of breed formation, but the current breeding program does no longer put any selection pressure on double-muscling. Finally, the results from the second canonical axis are less relevant (AV) or even absent (RG), which indicates that, for this group of tests, the selection effects may be diluted because a larger number of generations without selection is considered.

Candidate genes and metabolic paths

The results of the first axis (Fig. 2) allowed us to highlight seven relevant genomic regions on *Bos taurus* chromosome (BTA) 2 (between 1,047,347 and 11,899,039 bp), BTA5 (between 15,920,995 and 20,321,882 bp), BTA6 (between 37,853,912 and 41,160,000 bp), BTA7 (between 47,276,124 and 47,745,164 bp), BTA11 (between 65,077,840 and 72,203,248 bp), BTA13 (between 57,430,392 and 57,754,760 bp) and BTA18 (between 12,675,262 and 16,202,289 bp). In some cases, these genomic regions were extremely large, because of strong signatures of selection such as those on BTA2 for the AV and RG populations or on BTA5, BTA6 and BTA11 for the BP population. Such huge signatures of selection imply that large genomic regions included SNPs that were associated with results above the top 0.1% of the empirical distribution along the genome. However, the localization of the strongest signals within each genomic region and for each population allowed us to narrow down the genomic regions (Fig. 2b), which are similar to those reported in a previous study on the differentiation between populations [37]. These regions included well-known genes that were previously reported as potential candidates of selection signatures in cattle [38], such as *MTSN* (*myostatin*) on BTA2, suggested in several beef cattle populations [39–41], *KIT-LG* (*kit-ligand*) on BTA5 with a very large peak in the BP population, *MC1R* (*melanocortin 1 receptor*) on BTA18, which controls the production of eumelanin (black) or pheomelanin (red) pigments [42] and appears to be relevant in populations with black (AV and Mo) or red (Re) coat color. Moreover, it should be also highlighted that the region on BTA6 that includes *LAP3* (*leucine aminopeptidase3*), *LCORL* (*ligand dependent nuclear receptor corepressor-like*) and *NCAPG* (*non-SMC condensing I complex, subunit G*) and was identified in two meta-analyses [38, 43] as one the genomic regions that is most frequently identified with signatures of selection in the bovine genome. The genomic region identified on BTA7 includes the *CAMLG* (*calcium modulating ligand*) and *TCF* (*transcription*

factor 7) genes, which are close to a strong signature of selection that was reported by Gautier [44] and is associated with the *VDAC1* (*voltage-dependent anion-selective channel protein 1*) gene. A strong signature of selection was observed for the genomic region on BTA13, in the Re population, where is located the *END3* (*endothelin 3*) gene that plays a role in melanocyte development [45] and was recently associated with piebald pattern [44]. Finally, there is a very strong signature of selection on BTA11 for the BP population, where the closest gene to the maximum signal is *BMP10* (*bone morphogenetic protein 10*). Within this genomic region on BTA11, some authors [46, 47] identified signatures of selection and several genes that could be associated with fertility: *PROKR1* (*prokineticin receptor 1*), *GFPT1* (*glutamine-fructose-6-phosphate transaminase 1*), *GMCL1* (*germ cell-less spermatogenesis associated 1*), *PCBP1* (*poly(rC) binding protein 1*) and *EHD3* (*EH-domain containing 3*).

The results of the second canonical axis (Fig. 3) confirmed some of the signals of selection that were detected in the first axis, but also revealed several new ones that are shared by several populations and located on five chromosomes: BTA2 (between 61,684,232 and 62,199,344 and between 72,158,144 and 73,356,296 bp), BTA7 (between 20,612,988 and 21,163,812 bp), BTA13 (between 11,860,881 and 12,062,522 bp), BTA16 (between 44,612,592 and 45,846,144 bp) and BTA21 (between 32,207,264 and 32,414,316 bp). Previously, two meta-analyses [38, 43] showed that these regions were also associated with signatures of selection in other populations. Among the genes included in these regions, some of them may be good candidates for being affected by selection i.e.: (1) genes that are related to energy balance and homeostasis: *R3HGM1* (*R3H domain containing 1*) on BTA2 [48]; *CAMK1D* (*calcium/calmodulin-dependent protein kinase ID*) on BTA13 [49]; and *SLC25A33* (*solute carrier family 25* (*pyrimidine nucleotide carrier*), *member 33*) and *SLC2A5* (*solute carrier family 2* (*facilitated glucose/fructose transporter*), *member 5*) on BTA16 [49, 50]; (2) *PLIN5* (*perilipin 5*) on BTA7, which is involved in the regulation of lipid metabolism in rat and pigs [51, 52]; (3) *SCAPER* (*s-phase cyclin A-associated protein in the endoplasmic reticulum*) on BTA21, which regulates cell cycle progression [53]; and (4) there are two other signatures of selection that are worth noting i.e. one detected for the RG population on BTA19 (between 27,941,270 and 28,571,032 bp) where *ALOX15B* (*arachidonate 15-lipoxygenase, type B*) and *ALOX12B* (*arachidonate 12-lipoxygenase, 12R type*) are located and are involved in immune response [54], and one for the ANI population on BTA20 (between 40,854,136 and 40,996,384 bp) where the *NPR3* (*natriuretic peptide receptor 3*) gene is located, which is related with cattle stature [55].

Although less relevant, the results of the third canonical axis (Fig. 4) confirm the signatures of selection around the *MTSN* gene for the AV, RG and Pi populations, and around the complex *LAP–LCORL–NCAG* for the BP population. There are several other interesting signatures of selection such as those located on BTA27 (between 36,466,580 and 40,862,444 bp) and BTA28 (between 41,643,416 and 45,215,488 bp) for the ANI and Mo populations, respectively. The first genomic region includes the *IKBKB* (*inhibitor of kappa light polypeptide gene enhancer in B-cells, kinase beta*), *DKK4* (*dickkopf WNT signaling pathway inhibitor 4*) and *VDAC3* (*voltage-dependent anion-selective channel protein 3*) genes that are associated with immune response to *trypanosoma* infection in African populations [56], and the second region contains the *ALOX5* (*arachidonate 5-lipoxygenase*) and *RASSF4* (*ras association (RalGDS/AF-6) domain family member 4*) genes, that are related with growth [57] and feed conversion [58], respectively.

Finally, we performed a pathway enrichment analysis [59] for the identified genomic regions by applying a less restrictive criterion (top 5%). The objective of this analysis was to identify a larger number of genomic regions associated with signatures of selection although with less strong signals. The results of the top 10 pathways that were identified by the enrichment analysis are in Table 2. In general, the pathways associated with each canonical axis are coincident, which indicates that the metabolic pathways that were involved in old and recent selection events are similar, although probably with variable intensities and directions [60]. The enrichment analysis identified pathways that are related with immune response (lymphocyte TarBase), muscle development (muscle cell TarBase), protein biosynthesis (translation factors, cytoplasmic ribosomal proteins), skin and pigmentation (epithelium TarBase), glucose metabolism (insulin signaling, integrated pancreatic cancer pathway), fat metabolism (adipogenesis), embryogenesis and morphology (focal adhesion), heart (calcium regulation in the cardiac cell) and uterine metabolism (myometrial relaxation and contraction pathways), regulation of the hypothalamic–pituitary–thyroid axis (TSH signalling pathway), hormonal, cellular cycle (MAPK-signaling pathway, G1 to S cell cycle control, eukaryotic transcription initiation), cell signaling (notch signaling pathway) and extracellular receptors (GPCR, class A rhodopsin-like). Among these, 10 pathways (focal adhesion, integrated pancreatic cancer pathway, adipogenesis, myometrial relaxation and contraction pathways, adipogenesis, lymphocyte TarBase, insulin signaling, MAPK signaling pathway, focal adhesion, epithelium TarBase) had been previously identified in a meta-analysis based on a very large number of studies on selection signatures in cattle [38], which confirmed

Table 2 Top 10 enriched pathways for the three axes

Pathway	Ngenes[a]	Total[b]
First axis		
Focal adhesion	48	185
Integrated pancreatic cancer pathway	46	181
MAPK signalling pathway	43	163
Lymphocite TarBase	96	533
Epithelium TarBase	69	340
TSH signalling pathway	25	70
Adipogenesis	36	130
Cytoplasmic ribosomal proteins	27	88
Muscle cell TarBase	77	424
GPCRs, class A rhodopsin-like	53	259
Second axis		
Epithelium TarBase	51	340
Lymphocyte TarBase	62	533
Translation factors	15	51
Focal adhesion	27	185
Adipogenesis	21	130
Muscle cell TarBase	44	424
Notch signalling pathway	11	45
Integrated pancreatic cancer pathway	23	181
G1 to S cell cycle control	14	77
Eukaryotic transcription initiation	10	41
Third axis		
Lymphocyte TarBase	304	533
MAPK signalling pathway	124	165
Insulin signalling	123	163
Muscle cell TarBase	240	424
Calcium regulation in the cardiac cell	116	151
Focal adhesion	132	185
Integrated pancreatic cancer pathway	130	181
Myometrial relaxation and contraction pathways	116	162
Adipogenesis	99	130
Epithelium TarBase	191	340

[a] Ngenes: number of genes present in the genomic regions

[b] Total: number genes in the pathway

that the metabolic pathways involved in old and recent processes of selection are similar to those detected by using equivalent approaches in other cattle populations.

Conclusions

In this study, we confirm that the results of various procedures used to identify signatures of selection varied largely among groups of tests depending on the source of information they use to reject the null hypothesis of absence of selection. However, we observed some correlations between the results of each test. Accordingly, these tests could be clustered into three groups that matched with the three canonical axes of a factor analysis. Moreover, each canonical factor (or group of tests) identified different signals of selection, which were assigned to selection events that occurred at different evolutionary times. In fact, older selection events generated signatures of selection that presented communalities between populations, whereas more recent selection events were detected specifically for each population. Nevertheless, the enriched metabolic pathways associated to each group of tests showed an important degree of agreement which suggests that the traits involved in the selection events were similar during the evolutionary history of the populations.

Additional files

Additional file 1: Figure S1. Manhattan plots of the results along the autosomal genome obtained with the Tajima procedure.

Additional file 2: Figure S2. Manhattan plots of the results along the autosomal genome obtained with the Fu and Li procedure.

Additional file 3: Figure S3. Manhattan plots of the results along the autosomal genome obtained with the Fay and Wu procedure.

Additional file 4: Figure S4. Manhattan plots of the results along the autosomal genome obtained with the SelEstim procedure.

Additional file 5: Figure S5. Manhattan plots of the results along the autosomal genome obtained with the XP-CLR procedure.

Additional file 6: Figure S6 Manhattan plots of the results along the autosomal genome obtained with the $H12$ procedure.

Additional file 7: Figure S7. Manhattan plots of the results along the autosomal genome obtained with the iHS procedure.

Additional file 8: Figure S8. Manhattan plots of the results along the autosomal genome obtained with the nSL procedure.

Additional file 9: Figure S9. Manhattan plots of the results along the autosomal genome obtained with the F_{ST} procedure.

Additional file 10: Figure S10. Manhattan plots of the results along the autosomal genome obtained with the XP-EHH procedure.

Additional file 11: Figure S11. Manhattan plots of the results along the autosomal genome obtained with the VarLD procedure.

Authors' contributions

LV conceived the study; CD, JP and JA contributed to the study design; AG, SM, EFM, JJC and LV developed and applied the methods; AG, CD, JP and LV drafted the manuscript; CD, JP, JA, AM and JAB contributed with the generation of experimental data. All authors read and approved the final manuscript.

Author details

[1] Departamento de Anatomía, Embriología y Genética, Universidad de Zaragoza, 50013 Saragossa, Spain. [2] Departamento de Producción Animal, Facultad de Agronomía, Universidad de Buenos Aires, 1417 Buenos Aires, Argentina. [3] Grup de Recerca en Remugants, Departament de Ciència Animal i dels Aliments, Universitat Autònoma de Barcelona, 08193 Bellaterra, Barcelona, Spain. [4] Departamento de Mejora Genética Animal, INIA, 28040 Madrid, Spain. [5] Instituto Agroalimentario de Aragón (IA2), 50013 Saragossa, Spain. [6] Departamento de Ciencias Agroforestales, Universidad de Valladolid, 34004 Palencia, Spain. [7] MERAGEM, Universidad de Córdoba, 14071 Córdoba, Spain.

Acknowledgements

The research leading to these results received funding from a Ministerio de Ciencia e Innovación AGL 2010-15903 Grant from the Spanish Government, from the European Union's Seventh Framework Programme for research, technological development and demonstration under Grant Agreement No. 289592—Gene2Farm, from FEDER funds and from the regional government of Aragon (DGA) through its research group A51. The collaboration of Breed Societies in collecting samples and the support of FEAGAS are also acknowledged. The authors wish to thank JJ Arranz (Universidad de Leon) for facilitating the sampling of the Morucha population. J. J. Cañas-Álvarez acknowledges the financial support provided by COLCIENCIAS through the Francisco José de Caldas fellowship 497/2009 and A González-Rodríguez acknowledges the financial support from the Spanish Government BES-2011-045434.

Competing interests

The authors declare that they have no competing interests.

References

1. Nielsen R. Molecular signatures of natural selection. Annu Rev Genet. 2005;39:197–218.
2. Qanbari S, Simianer H. Mapping signatures of positive selection in the genome of livestock. Livest Sci. 2014;166:133–43.
3. Kimura M. The neutral theory of molecular evolution. Cambridge: Cambridge University Press; 1983.
4. Tajima F. Statistical method for testing the neutral mutation hypothesis by DNA polymorphism. Genetics. 1989;123:585–95.
5. Fu YX, Li WH. Statistical tests of neutrality of mutations. Genetics. 1993;133:693–709.
6. Fay JC, Wu CI. Hitchhiking under positive Darwinian selection. Genetics. 2000;155:1405–13.
7. Sabeti PC, Reich DE, Higgins JM, Levine HZP, Richter DJ, Schaffner SF, et al. Detecting recent positive selection in the human genome from haplotype structure. Nature. 2002;419:832–7.
8. Voight BF, Kudaravalli S, Wen X, Pritchard JK. A map of recent positive selection in the human genome. PLoS Biol. 2006;4:e72.
9. Wright S. Isolation by distance. Genetics. 1943;28:114–38.
10. Vitalis R, Gautier M, Dawson KJ, Beaumont MA. Detecting and measuring selection from gene frequency data. Genetics. 2014;196:799–817.
11. Beaumont MA, Balding DJ. Identifying adaptive genetic divergence among populations from genome scans. Mol Ecol. 2004;13:969–80.
12. Nielsen R. Population genetic analysis of ascertained SNP data. Hum Genomics. 2004;1:218–24.
13. Ma Y, Ding X, Qanbari S, Weigend S, Zhang Q, Simianer H. Properties of different selection signature statistics and a new strategy for combining them. Heredity (Edinb). 2015;115:426–36.
14. Grossman SR, Shlyakhter I, Karlsson EK, Byrne EH, Morales S, Frieden G, et al. A composite of multiple signals distinguishes causal variants in regions of positive selection. Science. 2006;327:883–6.
15. Utsunomiya YT, Perez-O'Brien AM, Sonstegard TS, Van Tassell CP, do Carmo AS, Mészáros G, et al. Detecting loci under recent positive selection in dairy and beef cattle by combining different genome-wide scan methods. PLoS One. 2013;8:e64280.
16. Cañas-Álvarez JJ, González-Rodríguez A, Munilla S, Varona L, Díaz C, et al. Genetic diversity and divergence among Spanish beef cattle breeds assessed by a bovine high-density SNP chip. J Anim Sci. 2015;93:5164–74.
17. Purcell S, Neale B, Todd-Brown K, Thomas L, Ferreira MAR, Bender D, et al. PLINK: a tool set for whole-genome association and population-based linkage analyses. Am J Hum Genet. 2007;81:559–75.
18. Browning BL, Browning SR. A unified approach to genotype imputation and haplotype-phase inference for large data sets of trios and unrelated individuals. Am J Hum Genet. 2009;84:210–23.
19. Rocha D, Billerey C, Samson F, Boichard D, Boussaha M. Identification of the putative ancestral allele of bovine single-nucleotide polymorphism. J Anim Breed Genet. 2014;131:483–6.
20. Szpiech ZA, Hernández RD. Selscan: an efficient multi-threaded program to calculate EHH-based scans for positive selection. Mol Biol Evol. 2014;31:2824–7.
21. Ferrer-Admetlla A, Liang M, Korneliussen T, Nielsen R. On detecting incomplete soft or hard selective sweeps using haplotype structure. Mol Biol Evol. 2014;31:1275–91.

22. Garud NR, Messer PW, Buzbas EO, Petrov DA. Recent selective sweeps in North American *Drosophila melanogaster* show signatures of soft sweeps. PLoS Genet. 2015;11:e1005004.

23. Chen H, Patterson N, Reich D. Population differentiation as a test for selective sweeps. Genome Res. 2010;20:393–402.

24. Teo YY, Fry AE, Bhattacharya K, Small KS, Kwiatkowski DP. Genome-wide comparisons of variation in linkage disequilibrium. Genome Res. 2009;19:1849–60.

25. Ong R, Teo YY. VarLD: a program for quantifying variation in linkage disequilibrium patterns between populations. Bioinformatics. 2010;26:1269–70.

26. Kaiser HF. The varimax criterion for analytic rotation in factor analysis. Psychometrika. 1958;23:187–200.

27. R Development Core Team. R: a language and environment for statistical computing. Vienna: R Foundation for Statistical Computing; 2008. ISBN 3-900051-07-0. http://www.R-project.org.

28. Wang J, Duncan D, Shi Z, Zhang B. WEB-based GEne SeT AnaLysis Toolkit (WebGestalt). Nucleic Acids Res. 2013;41:W77–83.

29. Sabeti PC, Schaffner SF, Fry B, Lohmueller J, Varilly P, Shamovsky O, et al. Positive natural selection in human lineage. Science. 2006;312:1614–20.

30. Simianer H, Ma Y, Qanbari S. Statistical problems in livestock population genomics. In: Proceedings, 10th World Congress of genetics applied to livestock production. Vancouver; 2014. https://asas.org/docs/default-source/wcgalp-proceedings-oral/202_paper_10373_manuscript_1346_0.pdf?sfvrsn=2.

31. Kemper KE, Saxton SJ, Bolormaa S, Hayes BJ, Goddard ME. Selection for complex traits leaves little or no classic signatures of selection. BMC Genomics. 2014;15:246.

32. Cole JB, VanRaden PM, O'Connell JR, Van Tassell CP, Sonstegard TS, Schnabel RD, et al. Distribution and location of genetic effects for dairy traits. J Dairy Sci. 2009;92:2931–46.

33. Hayes BJ, Pryce J, Chamberlain AJ, Bowman PJ, Goddard ME. Genetic architecture of complex traits and accuracy of genomic prediction: coat colour, milk-fat percentage and type in Holstein cattle as contrasting model traits. PLoS Genet. 2010;23:e1001139.

34. Pimentel EG, Erbe M, König S, Simianer H. Genome partitioning of genetic variation for milk production and composition traits in Holstein cattle. Front Genet. 2011;2:19.

35. Shao H, Burrage LC, Sinasac DS, Hill AE, Ernest SR, O'Brien W, et al. Genetic architecture of complex traits: large phenotypic effects and pervasive epistasis. Proc Natl Acad Sci USA. 2008;105:19910–4.

36. Dunner S, Miranda ME, Amigues Y, Cañon J, Georges M, Hanset R, et al. Haplotype diversity of the myostatin gene among beef cattle breeds. Genet Sel Evol. 2003;35:103–18.

37. González-Rodríguez A, Toro MA, Varona L, Carabaño MJ, Cañas-Álvarez JJ, Altarriba J, et al. Genome-wide analysis of genetic diversity in Autochthonous Spanish populations of beef cattle. In: Proceedings of the 10th World Congress of genetics applied to livestock production. Vancouver; 2014. https://asas.org/docs/default-source/wcgalp-proceedings-oral/255_paper_9607_manuscript_1524_0.pdf?sfvrsn=2.

38. Gutierrez-Gil B, Arranz JJ, Wiener P. An interpretive review of selective sweep studies in *Bos taurus* cattle populations: identification of unique and shared selection signals across breeds. Front Genet. 2015;6:167.

39. Wiener P, Gutierrez-Gil B. Assessment of selection mapping near the *myostatin* gene (*GDF-8*) in cattle. Anim Genet. 2009;40:598–608.

40. Boitard S, Rocha D. Detection of signatures of selective sweeps in the Blonde d'Aquitaine cattle breed. Anim Genet. 2013;44:579–83.

41. Druet T, Pérez-Pardal L, Charlier C, Gautier M. Identification of large selective sweeps associated with major genes in cattle. Anim Genet. 2013;44:758–62.

42. Werth LA, Hawkins GA, Eggen A, Petit E, Elduque C, Kreigesmann B, et al. Rapid communication: *melanocyte stimulating hormone receptor* (*MC1R*) maps to bovine chromosome 18. J Anim Sci. 1996;74:262.

43. Randhawa IAS, Khatkar MS, Thomson PC, Raadsma HW. A meta-assembly of selection signatures in cCattle. PLoS One. 2016;11:e0152013.

44. Gautier M. Genome-wide scan for adaptive divergence and association with population-specific covariates. Genetics. 2015;201:1555–79.

45. Saldana-Caboverde A, Kos L. Roles of endothelin signaling in melanocyte development and melanoma. Pigment Cell Melanoma Res. 2010;23:160–70.

46. Stella A, Ajmone-Marsan P, Lazzari B, Boettcher P. Identification of selection signatures in cattle breeds selected for dairy production. Genetics. 2010;185:1451–61.

47. Rothammer S, Seichter D, Förster M, Medugorac I. A genomic scan for signatures of differential artificial selection in ten cattle breeds. BMC Genomics. 2013;14:908.

48. Burt DW. The cattle genome reveals its secrets. J Biol. 2009;8:36.

49. Randhawa IAS, Khatkar MS, Thomson PC, Raadsma HW. Composite selection signals can localize the trait specific genomic regions in multi-breed populations of cattle and sheep. BMC Genet. 2014;15:34.

50. Bomba L, Nicolazzi EL, Milanesi M, Negrini R, Mancini G, Biscarini F, et al. Relative extended haplotype homozygosity signals across breeds reveal dairy and beef specific signatures of selection. Genet Sel Evol. 2015;47:25.

51. Bosma M, Sparks LM, Hooiveld GJ, Jorgensen JA, Houten SM, Schrauwen P, et al. Overexpression of *PLIN5* in skeletal muscle promotes oxidative gene expression and intramyocellular lipid content without compromising insulin sensitivity. Biochim Biophys Acta. 2013;1831:844–52.

52. Puig-Oliveras A, Ramayo-Caldas Y, Corominas J, Estellé J, Pérez-Montarelo D, Hudson NJ, et al. Differences in muscle composition transcriptome among pigs phenotypically extreme for fatty acid composition. PLoS One. 2014;9:e99720.

53. Tsang WY, Wang L, Chen Z, Sanchez I, Dynlacht BD. SCAPER, a novel cyclin A-interacting protein that regulates cell cycle progression. J Cell Biol. 2007;178:621–33.

54. Mashima R, Okuyama T. The role of lipoxygenases in pathophysiology; new insights and future perspectives. Redox Biol. 2015;6:297–310.

55. Randhawa IAS, Khatkar MS, Thomson PC, Raadsma HW. Composite selection signals for complex traits exemplified through bovine stature using multibreed cohorts of European and African *Bos taurus*. G3 (Bethesda). 2015;5:1391–401.

56. Noyes H, Brass A, Obara I, Anderson S, Archibald AL, Bradley DG, et al. Genetic and expression analysis of cattle identifies candidate genes in pathways responding to *Trypanosoma congolense* infection. Proc Natl Acad Sci USA. 2011;108:9304–9.

57. Pareek CS, Michno J, Smoczynski R, Tyburski J, Golebiewski M, Piechocki K, et al. Identification of predicted genes expressed differentially in pituitary gland tissue of young growing bulls revealed by the cDNA-AFLP technique. Czech J Anim Sci. 2013;58:147–58.

58. Yao C, Spurlock DM, Armentano LE, Page CD Jr, VandeHaar MJ, Bickhart DM, et al. Random Forests approach for identifying additive and epistatic single nucleotide polymorphisms associated with residual feed intake in dairy cattle. J Dairy Sci. 2013;96:6716–29.

59. Flori L, Fritz S, Jaffrézic F, Boussaha M, Gut I, Heath S, et al. The genome response to artificial selection: a case study in dairy cattle. PLoS One. 2009;4:e6595.

60. Fellius M, Beerling ML, Buchanan DS, Theunissen B, Kollmees PA, Lenstra JA. On the history of cattle genetic resources. Diversity. 2014;6:705–50.

7

Novel optimum contribution selection methods accounting for conflicting objectives in breeding programs for livestock breeds with historical migration

Yu Wang*[iD], Jörn Bennewitz and Robin Wellmann

Abstract

Background: Optimum contribution selection (OCS) is effective for increasing genetic gain, controlling the rate of inbreeding and enables maintenance of genetic diversity. However, this diversity may be caused by high migrant contributions (MC) in the population due to introgression of genetic material from other breeds, which can threaten the conservation of small local populations. Therefore, breeding objectives should not only focus on increasing genetic gains but also on maintaining genetic originality and diversity of native alleles. This study aimed at investigating whether OCS was improved by including MC and modified kinships that account for breed origin of alleles. Three objective functions were considered for minimizing kinship, minimizing MC and maximizing genetic gain in the offspring generation, and we investigated their effects on German Angler and Vorderwald cattle.

Results: In most scenarios, the results were similar for Angler and Vorderwald cattle. A significant positive correlation between MC and estimated breeding values of the selection candidates was observed for both breeds, thus traditional OCS would increase MC. Optimization was performed under the condition that the rate of inbreeding did not exceed 1% and at least 30% of the maximum progress was achieved for all other criteria. Although traditional OCS provided the highest breeding values under restriction of classical kinship, the magnitude of MC in the progeny generation was not controlled. When MC were constrained or minimized, the kinship at native alleles increased compared to the reference scenario. Thus, in addition to constraining MC, constraining kinship at native alleles is required to ensure that native genetic diversity is maintained. When kinship at native alleles was constrained, the classical kinship was automatically lowered in most cases and more sires were selected. However, the average breeding value in the next generation was also lower than that obtained with traditional OCS.

Conclusions: For local breeds with historical introgressions, current breeding programs should focus on increasing genetic gain and controlling inbreeding, as well as maintaining the genetic originality of the breeds and the diversity of native alleles via the inclusion of MC and kinship at native alleles in the OCS process.

Background

In recent decades, the widespread use of artificial insemination and other reproductive technologies has resulted in substantial genetic gains in livestock populations. However, another consequence is that only a limited number of animals with high estimated breeding values (EBV) have been intensively used in breeding programs, which can result in increasing rates of inbreeding to undesired levels. A high rate of inbreeding not only leads to considerable reduction in genetic variation but also more deleterious recessive alleles become homozygous, which may threaten the entire future of the population [1]. Thus, there is a conflict between maximizing genetic gain and managing the rate of inbreeding.

*Correspondence: yu.wang@uni-hohenheim.de
Institute of Animal Science, University of Hohenheim, 70593 Stuttgart, Germany

Crossbreeding has been demonstrated to be an efficient method to reduce the threat of inbreeding depression and increase the level of genetic diversity [2]. In addition, local breeds are often crossed with breeds of high economic value to improve performance. However, such introgressions of genetic material can be a threat for maintaining local breeds. Amador et al. [3] confirmed that, after several generations without management, even a small introduction of foreign genetic material will rapidly disperse throughout the original population, and that this material is difficult to remove. Therefore, foreign introgressions present a large risk for the conservation of local breeds, which leads to a conflict in current breeding programs between increasing the contribution of foreign genetic material and conserving local breeds.

Optimum contribution selection (OCS) is a selection method that is effective at achieving a balance between rate of inbreeding and genetic gain. This selection process maximizes genetic gain in the next generation while constraining the rate of inbreeding via restriction of relatedness among offspring [4–6]. The superiority of OCS has been demonstrated with both simulated [7, 8] and real data [9–11]. The objective function for OCS has been optimized using Lagrange multipliers [4, 8, 12], evolutionary algorithms [7, 13, 14], and semidefinite programming algorithms [9, 15, 16]. A similar related optimization problem was expressed as a mixed-integer quadratically constrained optimization problem and solved with branch-and-bound algorithms [17]. In this paper, we applied the algorithm described in [18] for solving cone-constrained convex problems by using R package *optiSel*.

OCS is efficient for controlling the level of kinship among progeny and the rate of inbreeding in future generations and can ultimately maintain genetic diversity [12, 16, 19, 20]. However, a high level of genetic diversity can be achieved by a large genetic contribution from migrant breeds, which is undesirable for the conservation of local breeds, because it reduces their genetic uniqueness, as well as the genetic diversity between breeds [21]. Thus, conflicting objectives are observed with regards to maintaining genetic diversity and conserving genetic uniqueness of local small breeds with historical migrations.

Instead of focusing on genetic gain and rate of inbreeding only, a reasonable breeding objective would be to also include recovery of genetic originality by reducing migrant contributions (MC). The diversity of native alleles may also be important for conservation. Thus, to conserve breeds with historical migrations, Wellmann et al. [22] recommended that approaches should not only constrain MC, but also aim at increasing the probability that alleles originating from native founders are not identical by descent (IBD).

Our aim was to investigate whether including MC and modified kinship matrices that account for breed origin of alleles as additional constraints in OCS can improve breeding programs in local breeds. Both conservation progress and genetic gain were evaluated. The following scenarios based on different objective functions were considered: (1) maximizing the diversity of native alleles while restricting MC and/or the average breeding value of the progeny generation at desired levels; (2) minimizing MC while restricting the loss of diversity of native alleles and/or the average breeding value of the progeny generation at desired levels; and (3) maximizing the average breeding value of the progeny generation while restricting MC and/or the loss of diversity of native alleles at desired levels. The traditional pedigree-based kinship was constrained in all optimization scenarios.

Methods
Data
Data from two local German cattle breeds, Angler and Vorderwald, were analyzed. The Angler breed is mainly located in the northern part of Germany and represents a dual-purpose breed, although the primary emphasis is on milk production. With the introduction of other breeds to improve milk yield, the Angler breed has experienced a considerable amount of migrant breed introgressions [23]. The Angler dataset was provided by the VIT (Vereinigte Informationssysteme Tierhaltung w.V., Verden), Germany. The Vorderwald breed is a dual-purpose breed located in the black forest region of southwest Germany. Similarly, due to their frequent crossing with high-yield breeds, the genetic originality of Vorderwald cattle has decreased dramatically [24, 25]. The Vorderwald dataset was provided by the Institute for Animal Breeding, Bavarian State Research Center for Agriculture in Grub, Germany. Both datasets consist of pedigrees with information on sex, breed, birth year and estimated breeding values for milk production obtained from routine genetic evaluations. Animals with an unknown pedigree born before 1970 were classified as purebred. Animals from other breeds and animals with an unknown pedigree born after 1970 were considered as migrants, although some may have purebred ancestors. The Angler dataset included 109,109 animals born between 1906 and 2015, of which 86,269 (79.1%) were classified as Angler. The Vorderwald dataset included 200,468 animals born between 1906 and 2010, of which 180,646 (90.1%) were classified as Vorderwald. MC for each animal was calculated and expressed as the proportion of migrant breed alleles based on pedigree information.

Selection candidates

Selection candidates were chosen among animals that were classified as purebred in the herdbook in order to compute their optimum contributions with different approaches. Sires that had progeny born in 2005 and 2006 were set as male selection candidates and selected males were mated to 1000 randomly chosen dams, which are called female selection candidates. For the Angler breed, 1199 selection candidates were available and 15,370 animals were involved in the pedigree that included all selection candidates and their ancestors. For the Vorderwald breed, 1123 selection candidates were available and 12,934 animals were involved in the pedigree. For a better comparison of results between the two breeds, EBV were normalized across all selection candidates of each breed, with a mean of 0 and a standard deviation of 1.

Optimum contribution selection strategies

The output of the optimum contribution selection procedure is a vector $\mathbf{c}$ with individual genetic contributions. The genetic contribution c_i of animal i is the fraction of genes in the next generation that originate from this individual. Genetic contributions cannot be negative, i.e. $c_i \geq 0$, which is denoted as constraint (a) in the following. The total genetic contribution of each sex must be equal to 0.5 for diploid species, i.e. $\mathbf{c}'\mathbf{s} = 0.5$ and $\mathbf{c}'\mathbf{d} = 0.5$ (constraint b), where $\mathbf{s}$ and $\mathbf{d}$ are vectors of the indicators (0/1) of a candidate's sex. Because cows can produce only a limited number of calves, all female selection candidates were used for breeding and the genetic contributions were forced to be equal, i.e. $c_{d_1} = c_{d_2} = \cdots = c_{d_n}$ (constraint c). Thus, optimization was only performed for bulls. For male selection candidates, the number of offspring is not limited, thus the maximum genetic contribution is 0.5, i.e. $c_{s_i} \leq 0.5$. To calculate the proportion of sires with non-zero genetic contributions, a sire i is considered to have a non-zero genetic contribution only if $c_{s_i} \geq 0.00025$ to account for possible numerical inaccuracies of the algorithm.

Four kinships that are involved in the calculation of the OCS procedure were applied. The diversity parameters described in [22] are complementary to the kinships used here, i.e. these kinship values are equal to 1 minus the corresponding diversity denoted as $\varphi_A, \ldots, \varphi_D$ in [22]. The relevant derivations of the formulas for calculating the diversity parameters are provided in detail in [22].

The classic kinship f_A between individuals i and j (element of matrix $\mathbf{f_A}$), which describes the probability that two alleles, X_i and X_j, at a locus that are randomly selected from individuals i and j are IBD

(i.e. $\mathbf{f_A}(i,j) = \mathbf{P}\left(X_i \underset{\text{IBD}}{=} X_j\right)$), was restricted in all scenarios. For breeds with historical migrations and

foreign introgressions, Wellmann et al. [22] proposed that the breed origin of the alleles should be considered to preserve the local breed. Thus, we considered different approaches that account for the origin of alleles, denoted as f_B, f_C and f_D. Kinship matrix $\mathbf{f_B}$ contains the probabilities that two alleles randomly chosen from two individuals at a locus are IBD or that at least one allele is from a migrant breed ($\mathcal{M}$):

$$\mathbf{f_B}(i,j) = \mathbf{P}\left(X_i \underset{\text{IBD}}{=} X_j \text{ or } X_i \in \mathcal{M} \text{ or } X_j \in \mathcal{M}\right).$$

Note that this is equal to the probability that both alleles are IBD and native plus the probability that at least one allele is from a migrant.

Kinship matrix $\mathbf{f_C}$ contains the probabilities that two alleles randomly chosen from two individuals at a locus are IBD or both alleles are from migrant breeds:

$$\mathbf{f_C}(i,j) = \mathbf{P}\left(X_i \underset{\text{IBD}}{=} X_j \text{ or } X_i, X_j \in \mathcal{M}\right).$$

This is equal to $\mathbf{f_B}(i,j) = \mathbf{f_C}(i,j) + P(either X_i \in \mathcal{M}$ or $X_j \in \mathcal{M})$. The probability that at least one of the two randomly chosen alleles is from a migrant breed is higher than the probability that both are from migrant breeds. Thus, $\mathbf{f_B}$ is greater than $\mathbf{f_C}$. In general, $\mathbf{f_A} \leq \mathbf{f_C} \leq \mathbf{f_B}$ (element-wise). The kinship at native alleles f_D is defined as the conditional probability that two alleles X and Y at a locus that are randomly chosen from the offspring population are IBD, given that both descended from native founders ($\mathcal{F}$):

$$f_D(\mathbf{c}) = \mathbf{P}\left(X \underset{\text{IBD}}{=} Y \mid X, Y \in \mathcal{F}\right).$$

Note that this value says nothing about the kinship at loci that originate from migrants or about the MC. The mean kinships for the offspring generation are $\mathbf{c}'\mathbf{f_A}\mathbf{c}$, $\mathbf{c}'\mathbf{f_B}\mathbf{c}$ and $\mathbf{c}'\mathbf{f_C}\mathbf{c}$, respectively. Mean kinship f_D in the offspring population was calculated as $f_D(\mathbf{c}) = 1 - \frac{1 - \mathbf{c}'\mathbf{f_B}\mathbf{c}}{\mathbf{c}'\mathbf{f_N}\mathbf{c}}$, where $\mathbf{f_N}$ is a matrix containing the probabilities that both randomly chosen alleles at a locus originated from native founders.

Our aim was to identify the best method of accounting for the conflicting objectives of a breeding program, which are to increase breeding values, to maintain genetic diversity, and to maintain genetic originality of the breed. Since $1 - f_D(\mathbf{c}) = \mathbf{P}(X \underset{\text{IBD}}{\neq} Y | X, Y \in \mathcal{F})$ is the genetic diversity at native alleles, the constraint on f_D is used to maintain or increase genetic diversity at native alleles and is a parameter of interest. Kinship f_B and f_C were considered because minimizing or constraining f_D is in general not a convex problem, so minimizing f_B and f_C could result in lower f_D values than minimizing f_D itself.

In the different scenarios, an upper bound for MC (ub.MC) and/or a lower bound for the average EBV (lb.EBV) were set as additional constraints. The expectation of the average EBV in the next generation is $\mathbf{c}'\mathbf{EBV}$, where $\mathbf{EBV}$ is a vector of the EBV of each selection candidate. The expectation of the average MC of the next generation is $\mathbf{c}'\mathbf{MC}$, where $\mathbf{MC}$ is a vector of the MC of each selection candidate.

For all optimization problems, constraints a, b, and c were applied to limit the solution for c_i to within a reasonable range. Solver "cccp" [18], which was called from the R package optiSel [26], was used to solve the optimization problems. This solver contains routines for solving cone constrained convex problems using interior-point methods that are partially ported from Python's CVXOPT and based on Nesterov-Todd scaling [27]. The solver uses a primal–dual path following algorithms for linear and quadratic cone constrained programming.

Scenarios were categorized based on three main objective functions: minimizing kinships, minimizing MC and maximizing genetic gain in the next generation. For minimizing kinships, three sub-scenarios were considered, which involved minimizing f_B, f_C and f_D, respectively. Parameters ub.f_A, ub.f_B, ub.f_C, ub.f_D and ub.MC were defined as the upper bound values of the corresponding parameters in the next generation, whereas lb.EBV was set as the lower bound of the mean EBV for the next generation. One or several of the following constraints were used to define the optimization problems for each breed:

$$\mathbf{c}'\mathbf{f_A}\mathbf{c} \leq \text{ub.}f_A,$$

$$\mathbf{c}'\mathbf{f_B}\mathbf{c} \leq \text{ub.}f_B,$$

$$\mathbf{c}'\mathbf{f_C}\mathbf{c} \leq \text{ub.}f_C,$$

$$f_D(\mathbf{c}) \leq \text{ub.}f_D,$$

$$\mathbf{c}'\mathbf{MC} \leq \text{ub.MC},$$

$$\mathbf{c}'\mathbf{EBV} \geq \text{lb.EBV}.$$

The OCS scenarios considered are listed in Table 1. The name of each optimization scenario consists of a prefix that indicates the objective function and a suffix that indicates the constraint settings. For example, scenario *maxEBV.A.B.MC* indicates a scenario that maximizes the average EBV in the next generation, while constraining

Table 1 Names of the OCS scenarios based on different objective functions

Objective function	Name of the scenario[a]
Minimizing f_B	*minfB.A; minfB.A.MC; minfB.A.MC.EBV*
Minimizing f_C	*minfC.A; minfC.A.MC; minfC.A.MC.EBV*
Minimizing f_D	*minfD.A; minfD.A.MC; minfD.A.MC.EBV*
Minimizing MC	*minMC.A; minMC.A.EBV; minMC.A.B.EBV; minMC.A.C.EBV; minMC.A.D.EBV*
Maximizing EBV	*maxEBV.A; maxEBV.A.MC; maxEBV.A.B.MC; maxEBV.A.C.MC; maxEBV.A.D.MC*

[a] The name of each optimization scenario consists of a prefix that indicates the objective function and a suffix that indicates the constraint settings. For example, scenario *minfB.A* indicates that the objective function is to minimize the average f_B value in the following generation with a constraint on f_A

f_A, f_B, and MC. The vector of genetic contributions for this scenario is denoted as $\mathbf{c}_{\text{maxEBV.A.B.MC}}$.

Criteria for comparing scenarios included not only the result of the objective function, but also the other parameters obtained in the scenario, in particular EBV, MC, classic kinship, and kinship at native alleles. To evaluate the effectiveness of the OCS scenarios, the results were compared with the output from a reference scenario (*REF*) and the output from a truncation selection scenario (*TS*). In scenario *REF* all selection candidates were used as parents and had equal contributions to the offspring generation. For endangered breeds, an effective population size (N_e) of 50 is often considered as sufficient [28]. Based on the equation in [1], $\frac{1}{N_e} = \frac{1}{4*N_{\text{sire}}} + \frac{1}{4*N_{\text{dam}}}$, the 13 sires with the highest EBV were selected as male selection candidates in the *TS* scenario, and mated to the 1000 dams. All parents had equal contributions to the offspring generation in this scenario.

To ensure that optimal solutions exist in all scenarios for each breed, feasible threshold values must be set for the constraints. To restrict the rate of inbreeding, the upper bound (ub.f_A) was defined as follows. When N_e is equal to 50, the rate of inbreeding ΔF, which can be calculated from $\Delta F = \frac{1}{2N_e}$, is 1% per generation. Based on this, the threshold for f_A was calculated as ub.$f_A = \overline{f_A} + \left(1 - \overline{f_A}\right)\Delta F$, where $\overline{f_A}$ is the average kinship of the selection candidates.

To calculate the constraint setting for the other parameters, we used the results from the scenario that optimizes the corresponding parameter with restriction only on f_A and the *REF* scenario, using the following calculations:

$$\text{ub.}f_B = \lambda \mathbf{c}'_{\text{minfB.A}}\mathbf{f_B}\mathbf{c}_{\text{minfB.A}} + (1-\lambda)\mathbf{c}'_{\text{REF}}\mathbf{f_B}\mathbf{c}_{\text{REF}},$$

$$\text{ub.}f_C = \lambda \mathbf{c}'_{\text{minfC.A}}\mathbf{f_C}\mathbf{c}_{\text{minfC.A}} + (1-\lambda)\mathbf{c}'_{\text{REF}}\mathbf{f_C}\mathbf{c}_{\text{REF}},$$

$$\text{ub.}f_D = \lambda \mathbf{f_D}(\mathbf{c}_{\text{minfD.A}}) + (1-\lambda)\mathbf{f_D}(\mathbf{c}_{\text{REF}}),$$

$$ub.MC = \lambda \mathbf{c}'_{minMC.A}\mathbf{MC} + (1 - \lambda)\mathbf{c}'_{REF}\mathbf{MC},$$

$$lb.EBV = \lambda \mathbf{c}'_{maxEBV.A}\mathbf{EBV} + (1 - \lambda)\mathbf{c}'_{REF}\mathbf{EBV},$$

where λ is a parameter that indicates the proportion of progress to be accomplished for each constrained parameter relative to the scenario with a restriction only on f_A. The value of λ can be determined by the breeding organization. A higher λ value indicates a stricter setting for all constraints. We set λ at 0.3 to ensure that optimized solutions were found for all scenarios and for both breeds. The specific values used for all constraints for each breed are in Additional file 1: Table S1.

Results

Results of the basic statistical analyses for average kinship, MC and EBV of the parent generation are in Table 2 for both breeds. Average kinship f_A was lower for the Angler population than for the Vorderwald population (0.020 vs. 0.025) but f_B (0.910 vs. 0.853) and f_C levels (0.488 vs. 0.381) were higher. On average, 69.5 and 60.7% of the genetic material of the Angler and Vorderwald cattle, respectively, originated from migrant breeds. Native effective population sizes of 86 and 49 were estimated from six previous generations for Angler and Vorderwald cattle, respectively. Native effective population size is a parameter that quantifies the decrease in native allele diversity and is defined in [22]. If the native effective size is high, then native allele diversity decreases slowly. Thus, the diversity of native alleles decreased more rapidly in Vorderwald cattle than in Angler cattle, whereas MC were higher in Angler cattle. Average EBV for both breeds were below the current population mean, which is 100 for Angler and 0 for Vorderwald because selection candidates were sampled from old age cohorts. A positive correlation between EBV and MC was found for both breeds (Figs. 1, 2).

Minimizing average kinship

Genetic contributions of the selection candidates were optimized to minimize f_B, f_C and f_D with restrictions on

MC and/or average EBV in the offspring generation for each breed, (see Tables 3, 4, 5, respectively). Compared to the *REF* scenario, all OCS scenarios showed superior results for the optimized criteria as expected.

Table 3 shows the results obtained when minimizing f_B in the offspring generation under the different constraints for each breed. The lowest f_B for Angler cattle was 0.827 when the upper bound for f_A in the next generation was set to 0.030. MC was lower than the constraint value setting (0.570 vs. 0.677). Thus, the minimum f_B did not change after adding the constraint on MC (*minfB.A.MC*). When the restriction on average EBV was set to 0.516, the average kinship f_B increased to 0.866, which was still lower than the f_B obtained in the *REF* scenario (0.926). Similar results were obtained for Vorderwald cattle. When the upper bound for f_A in the progeny generation was set to 0.035, the minimum f_B level in the progeny generation was 0.789. Again, f_B did not change after adding an upper bound for MC (0.528 vs. 0.582). f_B increased to 0.813 when the EBV constraint was set to 0.550, although it was lower than the f_B obtained in the *REF* scenario (0.852).

Results when minimizing f_C were similar to minimizing f_B (see Table 4). The f_C of the progeny generation decreased to 0.345 for Angler cattle when the upper bound for f_A was set to 0.030. When f_C was minimized, MC decreased to a value lower than the constraint level setting (0.570 vs. 0.677). Thus, minimizing f_C gave the same results for scenarios *minfC.A* and *minfC.A.MC*. After adding an EBV constraint of 0.516, f_C increased to 0.404 but was lower than the f_C obtained in the *REF* scenario (0.527). For Vorderwald cattle, the minimum average f_C in the progeny generation was 0.300 when f_A was restricted to 0.035, even after adding a higher constraint on MC (0.582 vs. 0.528). In scenario *minfC.A.MC.EBV*, f_C reached 0.327 after adding an EBV constraint of 0.550, although this was lower than the f_C obtained in the *REF* scenario (0.380).

When the kinship at native alleles, f_D, was minimized, the average kinship f_A was automatically lowered in most cases (Table 5); in Angler cattle, f_A reached 0.020, which was lower than the constraint level (0.030). In this case, the minimum f_D was 0.040. When MC was restricted to 0.677, the minimum f_D increased to 0.044. When an EBV constraint of 0.516 was added, the minimum f_D increased to 0.047, which was still lower than the f_D obtained in the *REF* scenario (0.049). For Vorderwald cattle, when f_A was restricted to 0.035 in the progeny generation, the lowest f_D was 0.057. When the maximum MC was set to 0.582, f_D increased to 0.058. When adding an EBV constraint of 0.550, the lowest f_D was 0.064, which was still lower than the f_D obtained in the *REF* scenario (0.072).

Table 2 Descriptive statistics for the active breeding population in the Angler and Vorderwald breeds

	Angler (N = 1199)		Vorderwald (N = 1123)	
	Mean	SD	Mean	SD
f_A	0.020	0.027	0.025	0.027
f_B	0.910	0.055	0.853	0.084
f_C	0.488	0.123	0.381	0.128
MC	0.695	0.126	0.607	0.153
EBV	86.868	13.901	−512.020	502.465

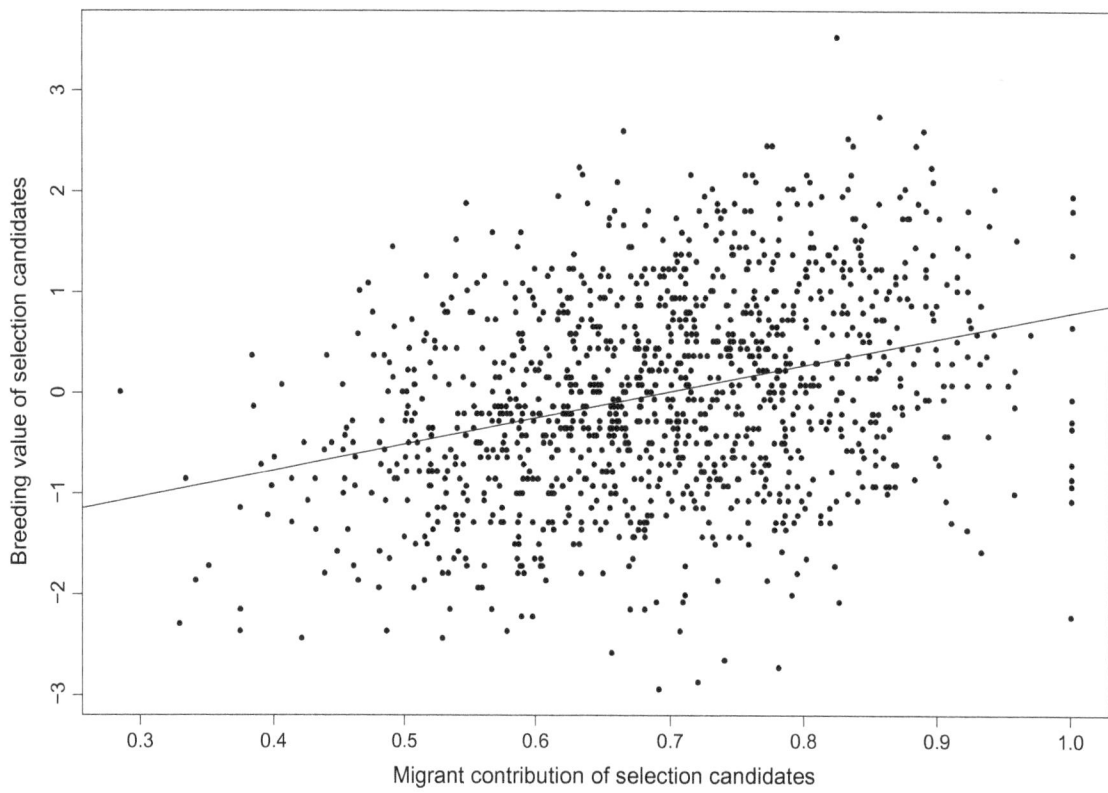

Fig. 1 Relationship between migrant contribution and the estimated breeding value of selection candidates in the Angler cattle population. The correlation between the EBV and MC is 0.328 and the regression coefficient between the EBV and MC is 2.614

Minimizing migrant contribution

Table 6 shows the results of minimizing MC under various constraints. When f_A was restricted to 0.030 in the progeny generation for Angler cattle, MC was equal to 0.570. When constraining the EBV to at least 0.516, MC in scenario *minMC.A.EBV* increased to 0.622 and f_B and f_C were lower than their constraint settings (0.866 vs. 0.896 and 0.404 vs. 0.472, respectively). Thus, adding constraints for f_B or f_C did not change the results. When the upper bound for f_D was set to 0.046, MC increased to 0.683, which was less than that achieved in the *REF* scenario (0.722). Results were similar for Vorderwald cattle. The minimum MC achieved in the next generation was 0.527 when the upper bound for f_A was 0.035. When the lower bound for EBV was set to 0.550, the minimal MC increased to 0.555. Adding a lower constraint for f_B (0.813 vs. 0.833) or f_C (0.327 vs. 0.356) did not change results. When the upper bound for f_D was set to 0.067 as an additional constraint, the minimum MC was 0.571, which was less than that obtained in the *REF* scenario (0.605).

Maximizing the average EBV

Results for maximizing the average EBV in the progeny generation under various constraints are in Table 7.

For both breeds, the *REF* scenario achieved the lowest average EBV in the offspring generation. This value was not zero because male and female selection candidates had different mean EBV. For Angler cattle, scenario *maxEBV.A* achieved a higher EBV (1.226 vs. 1.184) than the TS scenario, although the average kinship f_A was restricted (0.030 vs. 0.031). The average EBV decreased when adding the MC restriction, and f_B and f_C decreased to a level lower than their upper bound settings. Restricting f_D also lowered f_A. The EBV dropped to its lowest value of 0.449 when restricting f_A, f_D and MC, although this was still around twice that obtained in the *REF* scenario (0.211). Similar results were observed for the Vorderwald cattle population. Scenario *maxEBV.A* achieved a similar EBV as the *TS* scenario (1.164 vs. 1.161) but the average kinship f_A was much lower (0.035 vs. 0.043). When adding restrictions on f_D and MC, the maximum EBV decreased to 0.636, which was more than twice that obtained in the *REF* scenario (0.287).

The number of selected sires with non-zero genetic contributions was calculated in each scenario, as well as the standard deviation of the genetic contribution of all male selection candidates. Among all scenarios, TS selected the smallest number of sires. Adding a constraint

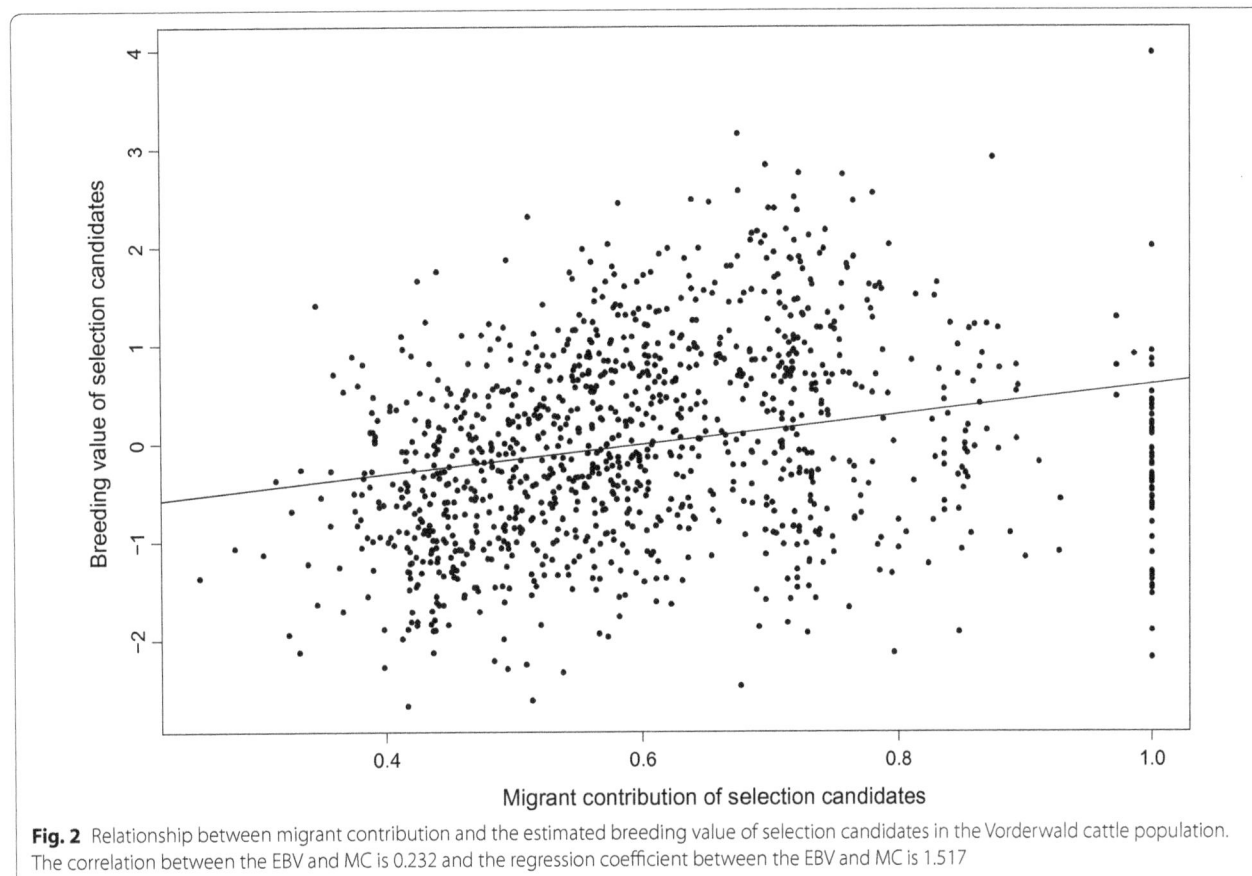

Fig. 2 Relationship between migrant contribution and the estimated breeding value of selection candidates in the Vorderwald cattle population. The correlation between the EBV and MC is 0.232 and the regression coefficient between the EBV and MC is 1.517

Table 3 Optimization of the genetic contributions when minimizing kinship f_B with a restriction on migrant contribution and/or mean estimated breeding values

Scenario[a]	Parameter[b]							
	f_A	f_B[c]	f_C	f_D	MC	EBV	Selected[d]	SD of c_s[e]
Angler								
REF	0.022	0.926	0.527	0.049	0.722	0.211	–	–
TS	0.031	0.939	0.565	0.067	0.722	1.184	0.065	0
minfB.A	*0.030*	0.827	0.345	0.082	0.570	−0.295	0.106	0.012
minfB.A.MC	*0.030*	0.827	0.345	0.082	**0.570**	−0.295	0.111	0.012
minfB.A.MC.EBV	*0.030*	0.866	0.404	0.083	**0.623**	*0.516*	0.081	0.012
Vorderwald								
REF	0.030	0.852	0.380	0.072	0.605	0.287	–	–
TS	0.043	0.882	0.432	0.093	0.645	1.161	0.106	0
minfB.A	*0.035*	0.789	0.300	0.074	0.528	−0.111	0.260	0.011
minfB.A.MC	*0.035*	0.789	0.300	0.074	**0.528**	−0.111	0.260	0.011
minfB.A.MC.EBV	*0.035*	0.813	0.327	0.075	**0.555**	*0.550*	0.228	0.010

[a] The name of each optimization scenario consists of a prefix that indicates the objective function and a suffix that indicates the constraint settings

[b] The parameter used as a constraint is marked in italics in the scenario. Bold italic values indicate that the actual value obtained does not reach the limit of the corresponding constraint (value higher than the lower limit or lower than the upper limit)

[c] Objective function

[d] Proportion of selected sires with non-zero genetic contributions; a c_{s_i} value lower than 0.00025 is treated as zero

[e] Standard deviation of the genetic contributions of all male selection candidates

Table 4 Optimization of the genetic contribution when minimizing kinship f_C with a restriction on migrant contribution and/or mean estimated breeding values

Scenario[a]	Parameter[b]							
	f_A	f_B	f_C[c]	f_D	MC	EBV	Selected[d]	SD of c_s[e]
Angler								
REF	0.022	0.926	0.527	0.049	0.722	0.211	–	–
TS	0.031	0.939	0.565	0.067	0.722	1.184	0.065	0
minfC.A	*0.030*	0.827	0.345	0.082	0.570	−0.299	0.111	0.012
minfC.A.MC	*0.030*	0.827	0.345	0.082	***0.570***	−0.299	0.111	0.012
minfC.A.MC.EBV	*0.030*	0.866	0.404	0.083	***0.623***	*0.516*	0.091	0.012
Vorderwald								
REF	0.030	0.852	0.380	0.072	0.605	0.287	–	–
TS	0.043	0.882	0.432	0.093	0.645	1.161	0.106	0
minfC.A	*0.035*	0.789	0.300	0.074	0.528	−0.109	0.276	0.010
minfC.A.MC	*0.035*	0.789	0.300	0.074	***0.528***	−0.109	0.276	0.010
minfC.A.MC.EBV	*0.035*	0.813	0.327	0.075	***0.555***	*0.550*	0.228	0.010

[a] The name of each optimization scenario consists of a prefix indicating the objective function and a suffix indicating the constraint settings

[b] The parameter used as a constraint is marked in italic in the scenario. Bold italic values show that the actual value obtained does not reach the limit of the corresponding constraint in this scenario (value higher than the lower limit or lower than the upper limit)

[c] Objective function

[d] Proportion of selected sires with non-zero genetic contributions; a c_{s_i} value lower than 0.00025 is treated as zero

[e] Standard deviation of the genetic contributions of all male selection candidates

Table 5 Optimization of genetic contribution when minimizing kinship f_D with restriction on migrant contribution and/or mean estimated breeding values

Scenario[a]	Parameter[b]							
	f_A	f_B	f_C	f_D[c]	MC	EBV	Selected[d]	SD of c_s[e]
Angler								
REF	0.022	0.926	0.527	0.049	0.722	0.211	–	–
TS	0.031	0.939	0.565	0.067	0.722	1.184	0.065	0
minfD.A	***0.020***	0.954	0.614	0.040	0.782	0.078	0.434	0.009
minfD.A.MC	***0.019***	0.899	0.464	0.044	*0.677*	0.090	0.414	0.004
minfD.A.MC.EBV	***0.020***	0.899	0.464	0.047	*0.677*	*0.516*	0.333	0.005
Vorderwald								
REF	0.030	0.852	0.380	0.072	0.605	0.287	–	–
TS	0.043	0.882	0.432	0.093	0.645	1.161	0.106	0
minfD.A	***0.035***	0.895	0.456	0.057	0.669	0.759	0.398	0.015
minfD.A.MC	***0.027***	0.833	0.352	0.058	*0.582*	0.145	0.472	0.006
minfD.A.MC.EBV	***0.029***	0.833	0.353	0.064	*0.582*	*0.550*	0.358	0.007

[a] The name of each optimization scenario consists of a prefix indicating the objective function and a suffix indicating the constraint settings

[b] The parameter used as a constraint is marked in italic in the scenario. Bold italic values show that the actual value obtained does not reach the limit of the corresponding constraint in this scenario (value higher than the lower limit or lower than the upper limit)

[c] Objective function

[d] Proportion of selected sires with non-zero genetic contributions; a c_{s_i} value lower than 0.00025 is treated as zero

[e] Standard deviation of the genetic contributions of all male selection candidates

Table 6 Optimization of the genetic contribution when minimizing the migrant contribution with restricted kinship and/or mean estimated breeding values

Scenario[a]	Parameter[b]							
	f_A	f_B	f_C	f_D	MC^c	EBV	Selected[d]	SD of c_s[e]
Angler								
REF	0.022	0.926	0.527	0.049	0.722	0.211	–	–
TS	0.031	0.939	0.565	0.067	0.722	1.184	0.065	0
minMC.A	*0.030*	0.827	0.345	0.083	0.570	−0.289	0.106	0.012
minMC.A.EBV	*0.030*	0.866	0.404	0.084	0.622	*0.516*	0.091	0.012
minMC.A.B.EBV	*0.030*	**0.866**	0.404	0.084	0.622	*0.516*	0.091	0.012
minMC.A.C.EBV	*0.030*	0.866	**0.404**	0.084	0.622	*0.516*	0.091	0.012
minMC.A.D.EBV	**0.020**	0.903	0.472	*0.046*	0.683	*0.516*	0.342	0.005
Vorderwald								
REF	0.030	0.852	0.380	0.072	0.605	0.287	–	–
TS	0.043	0.882	0.432	0.093	0.645	1.161	0.106	0
minMC.A	*0.035*	0.789	0.300	0.074	0.527	−0.111	0.276	0.011
minMC.A.EBV	*0.035*	0.813	0.327	0.075	0.555	*0.550*	0.220	0.010
minMC.A.B.EBV	*0.035*	**0.813**	0.327	0.075	0.555	*0.550*	0.220	0.010
minMC.A.C.EBV	*0.035*	0.813	**0.327**	0.075	0.555	*0.550*	0.211	0.010
minMC.A.D.EBV	**0.031**	0.825	0.342	*0.067*	0.571	*0.550*	0.317	0.008

[a] The name of each optimization scenario consists of a prefix indicating the objective function and a suffix indicating the constraint settings

[b] The parameter used as a constraint is marked in italic in the scenario. Bold italic values show that the actual value obtained does not reach the limit of the corresponding constraint in this scenario (value higher than the lower limit or lower than the upper limit)

[c] Objective function

[d] Proportion of selected sires with non-zero genetic contributions; a c_{s_i} value lower than 0.00025 is treated as zero

[e] Standard deviation of the genetic contributions of all male selection candidates

on f_D resulted in all cases in more selected sires and a lower standard deviation.

Discussion

For the breeding schemes of the two breeds considered in this study, two conflicts must be addressed: (1) the conflict between increasing genetic gain while managing inbreeding and (2) the conflict between maintaining genetic diversity while controlling loss of genetic uniqueness. The purpose of this study was to determine whether OCS with additional constraints that involve modified kinship matrices and MC was more efficient at conserving genetic diversity and originality while also ensuring genetic improvement than traditional OCS. Using data on German Angler and Vorderwald cattle, various scenarios were compared. Both breeds have been frequently crossed with high-yielding breeds to improve performance. We found that diversity of native alleles decreased more rapidly in Vorderwald cattle than in Angler cattle, whereas MC was higher in Angler cattle. The consequences of the scenarios were similar for both breeds. Compared to traditional OCS, constraining kinship f_D and MC promoted recovery of genetic originality

in the breeds and diversity of native alleles but reduced response to selection.

Traditional OCS achieved the highest average EBV in the progeny generation among all scenarios with a restriction on rate of inbreeding, which, in our study, is represented by scenario *maxEBV.A*. Compared to the *TS* scenario, average EBV was higher in the traditional OCS scenario for both breeds, while the average relatedness was lower. Probably, the average EBV in TS was smaller because the TS scenario assumed equal contributions for selected sires, whereas OCS optimizes their contributions. Because MC and EBV were positively correlated, traditional OCS increased the average MC, which is undesirable when the aim is to conserve the genetic originality of local breeds.

Different kinship estimates

Both f_B and f_C take probabilities of IBD and probabilities of alleles originating from migrant breeds into account, i.e. they account for both level of inbreeding and level of genetic originality. Although theoretically, MC affects f_B more than f_C, results from minimizing f_B and f_C were almost identical for the two breeds considered. Wellmann

Table 7 Optimization of the genetic contribution when maximizing the breeding value with restricted kinship and/or mean estimated migrant contributions

Scenario[a]	Parameter[b]							
	f_A	f_B	f_C	f_D	MC	EBV[c]	Selected[d]	SD of c_s[e]
Angler								
REF	0.022	0.926	0.527	0.049	0.722	0.211	–	–
TS	0.031	0.939	0.565	0.067	0.722	1.184	0.065	0
maxEBV.A	*0.030*	0.937	0.560	0.082	0.743	1.226	0.085	0.012
maxEBV.A.MC	*0.030*	0.901	0.471	0.082	*0.677*	0.979	0.070	0.012
maxEBV.A.B.MC	*0.030*	***0.893***	0.454	0.082	***0.664***	0.884	0.075	0.012
maxEBV.A.C.MC	*0.030*	0.901	***0.471***	0.082	*0.677*	0.979	0.070	0.012
maxEBV.A.D.MC	***0.020***	0.899	0.464	*0.046*	*0.677*	0.449	0.347	0.005
Vorderwald								
REF	0.030	0.852	0.380	0.072	0.605	0.287	–	–
TS	0.043	0.882	0.432	0.093	0.645	1.161	0.106	0
maxEBV.A	*0.035*	0.895	0.456	0.077	0.666	1.164	0.203	0.013
maxEBV.A.MC	*0.035*	0.835	0.357	0.079	*0.582*	0.812	0.220	0.011
maxEBV.A.B.MC	*0.035*	***0.832***	0.353	0.078	***0.579***	0.787	0.220	0.011
maxEBV.A.C.MC	*0.035*	0.835	*0.356*	0.078	***0.581***	0.808	0.220	0.011
maxEBV.A.D.MC	***0.031***	0.834	0.354	*0.067*	*0.582*	0.636	0.317	0.008

[a] The name of each optimization scenario consists of a prefix indicating the objective function and a suffix indicating the constraint settings

[b] The parameter used as a constraint is marked in italic in the scenario. Bold italic values show that the actual value obtained does not reach the limit of the corresponding constraint in this scenario (value higher than the lower limit or lower than the upper limit)

[c] Objective function

[d] Proportion of selected sires with non-zero genetic contributions; a c_{s_i} value lower than 0.00025 is treated as zero

[e] Standard deviation of the genetic contributions of all male selection candidates

et al. [22] reported a larger difference between these two methods, which is probably because contributions of both sexes were optimized in their work. Minimizing neither f_B nor f_C reduced the kinship at native alleles, f_D, thus these two criteria are not an alternative for controlling the kinship at native alleles directly. Results from minimizing f_B and f_C were very similar to the results from minimizing MC. Hence, instead of minimizing or constraining f_B or f_C, it is recommended to control MC. To control the diversity at native alleles, f_D must be constrained or minimized directly, although this optimization problem may be not convex. However, because minimizing f_D did not reduce MC, a constraint on MC is needed for all optimizations that involve f_D. Minimizing f_D is different from minimizing f_A with an additional constraint on MC because minimizing f_A resulted in a larger f_D than minimizing f_D when MC is constrained to the same level (results not shown). Similarly, when including kinship f_D as an additional constraint in the OCS, the level of kinship f_A decreased in all scenarios. Thus, if f_D is constrained, then MC must be constrained as well and the constraint for f_A can be omitted.

Among all the scenarios, TS used the smallest number of sires and resulted in the highest average genetic contribution of selected sires. Including kinship f_D as an additional constraint in the OCS scenarios resulted in a larger number of selected sires than including f_B or f_C. Therefore, including f_D is an efficient method to avoid overuse of sires with high EBV and limits the rate of inbreeding in the long run. Compared with the inclusion of f_B or f_C, inclusion of f_D resulted in a lower average EBV in the progeny generation, depending on the constraint level setting. In most cases, OC was negatively correlated with MC and positively correlated with the average EBV, as illustrated in Additional file 2: Table S2, which represents a desirable result for future selection and breeding programs.

Scenarios with optimizations of both male and female contributions were also evaluated (results not shown), using the same calculation methods to obtain the constraint value settings. For all scenarios and both breeds, the constraint settings were stricter than in the scenarios that optimized male contributions. The performance of all scenarios improved when both male and female selection were optimized, which is consistent with Sánchez-Molano et al. [8], who used OCS to improve fitness and productivity traits. To achieve these improvements, however, additional reproductive techniques must be applied due to the limited reproduction rate of female animals.

Considering the migrant contribution

Previous OCS approaches for maximizing genetic gain while limiting rate of inbreeding did not consider MC. Introgression of migrant breed alleles must be managed to maintain genetic uniqueness and conserve local breeds. As expected, the average EBV obtained with and without MC as a constraint showed that controlling MC restricts increases in genetic gain. Interestingly, kinship at native alleles increased compared to the *REF* scenario when MC was constrained or minimized. Hence, the individuals with the lowest MC may not carry some native alleles that are still present in individuals with higher MC. Thus, in this case, constraining f_D is required to ensure that native genetic diversity is maintained.

However, maximum genetic gains can only be achieved by allowing for the introgression of foreign genetic material. Therefore, the two main purposes in a breeding program, i.e. conserving local breeds and improving genetic gain, are contradictory and must be balanced by the breeding organization. In this study, we set the proportion of breeding progress to be achieved at $\lambda = 0.3$ to determine the constraint level required for achieving optimal solutions for both breeds. Depending on the situation, the breeding organization could select an appropriate value of λ to emphasize conservation of local breeds or genetic improvement, thus facilitating both purposes.

Future improvements

Because of advances in molecular genetics, genome-wide dense marker genotype data are increasingly available, even for some endangered breeds and have shown promise in capturing genetic variation due to Mendelian sampling [29]. The application of genomic data provides a more accurate method of calculating relationships between individuals compared with the use of estimates from pedigree data [30]. Breeding values estimated by genomic approaches are also more accurate and show more within-family variation compared with breeding values estimated via traditional approaches [31]. Furthermore, compared to the use of pedigree kinship, the use of genomic kinship is substantially more efficient in maintaining genetic diversity when optimizing genetic contributions [8, 12, 16, 32]. Moreover, new methods to estimate kinship at native alleles, i.e. f_D, can be developed based on genomic data and the use of genomic data may enable estimation of MC for selection candidates without using pedigree data.

Conclusions

Maintaining genetic originality is essential for conserving local breeds. It was shown that using an OCS approach as developed in this study can effectively maintain the diversity of native alleles and genetic originality, while ensuring genetic improvement. The most promising

approach involved the inclusion of additional constraints for migrant contributions and kinship at native alleles f_D. When a constraint for f_D was included, the classical kinship f_A in the offspring was lower than the constraint level, so the constraint on f_A could be removed. More sires were selected when f_D was constrained than when f_D was not constrained and the standard deviation of the contributions was lower, i.e., the optimum contributions of the selected sires were more similar.

Authors' contributions

YW did the statistical analyses and wrote the first draft of the paper. RW developed the OCS approaches and wrote the software. RW and JB initiated and oversaw the project and refined the manuscript. All authors read and approved the final manuscript.

Acknowledgements

The study was supported by the German Research Foundation (Deutsche Forschungsgemeinschaft, DFG). The authors thank the VIT (Vereinigte Informationssysteme Tierhaltung w.V., Verden), Germany for providing the Angler data and the Institute of Animal Breeding of the Bavarian State Research Center for Agriculture for providing the Vorderwald data. In addition, we wish to thank the editor JCM Dekkers as well as three anonymous reviewers for their constructive suggestions to improve the final manuscript.

Competing interests

The authors declare that they have no competing interests.

References

1. Falconer DS, Mackay TF. Introduction to quantitative genetics. 4th ed. Harlow: Pearson Education Limited; 1996.
2. Frankham R, Ballou JD, Briscoe DA 2nd, editors. Introduction to conservation genetics. Cambridge: Cambridge University Press; 2002.
3. Amador C, Toro MÁ, Fernández J. Removing exogenous information using pedigree data. Conserv Genet. 2011;12:1565–73.
4. Meuwissen THE. Maximizing the response of selection with a predefined rate of inbreeding. J Anim Sci. 1997;75:934–40.
5. Grundy B, Villanueva B, Woolliams JA. Dynamic selection procedures for constrained inbreeding and their consequences for pedigree development. Genet Res. 1998;72:159–68.
6. Woolliams JA, Berg P, Dagnachew BS, Meuwissen THE. Genetic contributions and their optimization. J Anim Breed Genet. 2015;132:89–99.
7. Gourdine JL, Sørensen AC, Rydhmer L. There is room for selection in a small local pig breed when using optimum contribution selection: a simulation study. J Anim Sci. 2012;90:76–84.
8. Sánchez-Molano E, Pong-Wong R, Banos G. Genomic-based optimum contribution in conservation and genetic improvement programs with antagonistic fitness and productivity traits. Front Genet. 2016;7:25.
9. Schierenbeck S, Pimentel ECG, Tietze M, Körte J, Reents R, Reinhardt F, et al. Controlling inbreeding and maximizing genetic gain using semidefinite programming with pedigree-based and genomic relationships. J Dairy Sci. 2011;94:6143–52.

10. Howard DM, Pong-Wong R, Knap PW, Kremer VD, Woolliams JA. The structural impact of implementing optimal contribution selection in a commercial pig breeding population. In: Proceedings of the 10th world congress on genetics applied to livestock production: 17–22 August 2014; Vancouver. 2014.

11. Dagnachew BS, Meuwissen THE. A fast Newton-Raphson based iterative algorithm for large scale optimal contribution selection. Genet Sel Evol. 2016;48:70.

12. Eynard SE, Windig JJ, Hiemstra SJ, Calus MPL. Whole-genome sequence data uncover loss of genetic diversity due to selection. Genet Sel Evol. 2016;48:33.

13. Sørensen MK, Sørensen AC, Borchersen S, Berg P. Consequences of using EVA software as a tool for optimal genetic contribution selection in Danish Holstein. In: Proceedings of the 8th world congress on genetics applied to livestock production: 13–18 August 2006; Belo Horizonte. 2006.

14. Sørensen MK, Sørensen AC, Baumung R, Borchersen S, Berg P. Optimal genetic contribution selection in Danish Holstein depends on pedigree quality. Livest Sci. 2008;118:212–22.

15. Pong-Wong R, Woolliams JA. Optimisation of contribution of candidate parents to maximise genetic gain and restricting inbreeding using semidefinite programming. Genet Sel Evol. 2007;39:3–25.

16. Gómez-Romano F, Villanueva B, Fernández J, Woolliams JA, Pong-Wong R. The use of genomic coancestry matrices in the optimisation of contributions to maintain genetic diversity at specific regions of the genome. Genet Sel Evol. 2016;48:2.

17. Mullin TJ, Belotti P. Using branch-and-bound algorithms to optimize selection of a fixed-size breeding population under a relatedness constraint. Tree Genet Genomes. 2016;12:4.

18. Pfaff B. The R package cccp: Design for solving cone constrained convex programs. R Finance. 2014. http://www.pfaffikus.de/files/conf/rif/rif2014.pdf.

19. Ducro B, Windig J. Genetic diversity and measures to reduce inbreeding in Friesian horses. In: Proceedings of the 10th world congress on genetics applied to livestock production: 17–22 August 2014; Vancouver; 2014.

20. Stachowicz K, Sørensen AC, Berg P. Optimum contribution selection conserves genetic diversity better than random selection in small populations with overlapping generations. In: Proceedings of the 55th annual meeting of the European Association for Animal Production (EAAP): 5–9 September 2004; Bled. 2004.

21. Bennewitz J, Simianer H, Meuwissen THE. Investigations on merging breeds in genetic conservation schemes. J Dairy Sci. 2008;91:2512–9.

22. Wellmann R, Hartwig S, Bennewitz J. Optimum contribution selection for conserved populations with historic migration. Genet Sel Evol. 2012;44:34.

23. Bennewitz J, Meuwissen THE. Estimation of extinction probabilities of five German cattle breeds by population viability analysis. J Dairy Sci. 2005;88:2949–61.

24. Hartwig S, Wellmann R, Hamann H, Bennewitz J. The contribution of migrant breeds to the genetic gain of beef traits of German Vorderwald and Hinterwald cattle. J Anim Breed Genet. 2014;131:496–503.

25. Hartwig S, Wellmann R, Emmerling R, Hamann H, Bennewitz J. Short communication: importance of introgression for milk traits in the German Vorderwald and Hinterwald cattle. J Dairy Sci. 2015;98:2033–8.

26. Wellmann R. optiSel: optimum contribution selection and population genetics. R package version 0.7.1. 2017. https://cran.r-project.org/web/packages/optiSel/optiSel.pdf.

27. Vandenberghe L. The CVXOPT linear and quadratic cone program solvers. 2010. http://abel.ee.ucla.edu/cvxopt/documentation/coneprog.pdf.

28. Meuwissen THE. Genetic management of small populations: a review. Acta Agric Scand A-AN. 2009;59:71–9.

29. Avendaño S, Woolliams JA, Villanueva B. Mendelian sampling terms as a selective advantage in optimum breeding schemes with restrictions on the rate of inbreeding. Genet Res. 2004;83:55–64.

30. Sonesson AK, Woolliams JA, Meuwissen THE. Genomic selection requires genomic control of inbreeding. Genet Sel Evol. 2012;44:27.

31. Hill WG. On estimation of genetic variance within families using genome-wide identity-by-descent sharing. Genet Sel Evol. 2013;45:32.

32. Clark SA, Kinghorn BP, Hickey JM, van der Werf J. The effect of genomic information on optimal contribution selection in livestock breeding programs. Genet Sel Evol. 2013;45:44.

8

Estimation of breeding values for uniformity of growth in Atlantic salmon (*Salmo salar*) using pedigree relationships or single-step genomic evaluation

Panya Sae-Lim[1*], Antti Kause[2], Marie Lillehammer[1] and Han A. Mulder[3]

Abstract

Background: In farmed Atlantic salmon, heritability for uniformity of body weight is low, indicating that the accuracy of estimated breeding values (EBV) may be low. The use of genomic information could be one way to increase accuracy and, hence, obtain greater response to selection. Genomic information can be merged with pedigree information to construct a combined relationship matrix (**H** matrix) for a single-step genomic evaluation (ssGBLUP), allowing realized relationships of the genotyped animals to be exploited, in addition to numerator pedigree relationships (**A** matrix). We compared the predictive ability of EBV for uniformity of body weight in Atlantic salmon, when implementing either the **A** or **H** matrix in the genetic evaluation. We used double hierarchical generalized linear models (DHGLM) based either on a sire-dam (sire-dam DHGLM) or an animal model (animal DHGLM) for both body weight and its uniformity.

Results: With the animal DHGLM, the use of **H** instead of **A** significantly increased the correlation between the predicted EBV and adjusted phenotypes, which is a measure of predictive ability, for both body weight and its uniformity (41.1 to 78.1%). When log-transformed body weights were used to account for a scale effect, the use of **H** instead of **A** produced a small and non-significant increase (1.3 to 13.9%) in predictive ability. The sire-dam DHGLM had lower predictive ability for uniformity compared to the animal DHGLM.

Conclusions: Use of the combined numerator and genomic relationship matrix (**H**) significantly increased the predictive ability of EBV for uniformity when using the animal DHGLM for untransformed body weight. The increase was only minor when using log-transformed body weights, which may be due to the lower heritability of scaled uniformity, the lower genetic correlation of transformed body weight with its uniformity compared to the untransformed traits, and the small number of genotyped animals in the reference population. This study shows that ssGBLUP increases the accuracy of EBV for uniformity of body weight and is expected to increase response to selection in uniformity.

Background

In aquaculture, selection to increase economically important traits such as growth is one of the main breeding goals. However, fish producers show interest to improve not only the mean but also the variance of traits [1]. Uniformity of growth is preferable because more uniform growth allows a more uniform product, harvest of a larger proportion of the population at market size, and reduction of size grading and multiple harvests [2–4]. More uniform growth may also reduce competitive interactions between animals, which contributes to reduce feed monopolization and dominant behaviour, and thus improve well-being of fish [5]. Uniformity is also important for traits that have an intermediate optimal trait value [6], such as fillet lipid%, body shape, and condition

*Correspondence: panya.sae-lim@nofima.no
[1] Nofima Ås, Osloveien 1, P.O. Box 210, 1431 Ås, Norway
Full list of author information is available at the end of the article

factor in the aquaculture industry. A fish whose growth is sensitive to non-measurable environmental factors, known as micro-environments, shows micro-environmental sensitivity, which results in high environmental variance and consequently contributes to increased phenotypic variation, leading to increased size variation within a group of fish. A number of empirical studies in terrestrial and aquatic species show that uniformity is partly determined by genetic factors [4, 7–16]. Thus, selective breeding can open up one avenue to improve uniformity of fish traits.

Atlantic salmon (*Salmo salar* L.) is a farmed fish that is of major economic importance. Heritability for uniformity of body weight has been estimated in Atlantic salmon [14], rainbow trout (*Oncorhynchus mykiss* Walbaum) [4, 8], and Nile tilapia (*Oreochromis niloticus*) [15, 16]. In general, heritability for uniformity (h_v^2) is low in livestock and aquaculture species ($h_v^2 < 0.05$), indicating that the prediction accuracy of breeding values for uniformity may be low [17, 18]. However, the coefficient of genetic variation (*GCV*) of uniformity of body weight is high in fish species (median *GCV* = 34.0%: min = 17.4% and max = 64.0%), which indicates high potential for response to selection [4, 8, 14, 16, 19]. One way to increase response to selection for uniformity is to increase the accuracy of estimated breeding values (EBV) for uniformity [20].

In aquaculture, full- and half-sib family sizes are usually large and thus the accuracy of EBV based on full-sibs, half-sibs and own performance is high for body weight, but not for uniformity due to its low heritability [8]. One approach to increase the accuracy of EBV is to use genomic information [21]. With genomic selection, genomic estimated breeding values (GEBV) can be obtained for the selection candidates that are genotyped, even when they have no phenotype records. One reason why genomic selection results in higher accuracy of selection is the more accurate estimation of the Mendelian sampling genetic effects through realized additive genetic relationships among animals [22]. Consequently, individual squared residuals, which is the phenotype for uniformity in a double hierarchical generalized linear model (DHGLM), may also be more accurately estimated when using genomic information.

In many cases, combining numerator pedigree and genomic information in genomic evaluations is implemented in multiple steps, which may introduce bias and need some calculations to combine with pedigree-based EBV [23, 24]. Single step genomic best linear unbiased prediction (ssGBLUP) avoids this, and genomic and pedigree information are combined in one step [23], which may lead to less bias and is less prone to double counting of information compared to genomic evaluation methods

that are performed in multiple steps. The ssGBLUP augments the numerator relationship (**A**) matrix by the genomic relationship (**G**) matrix in conventional genetic evaluation using BLUP [24]. This combined numerator and genomic relationship matrix is known as the **H** matrix [25]. In fish breeding, combining pedigree and genomic information allows exploiting the large full- and half-sib families and the more accurate relationships of the genotyped animals, and may yield a higher accuracy of selection for uniformity than the use of the **A** matrix.

To date, the use of ssGBLUP for uniformity has not been studied. Furthermore, according to a previous study, the sire-dam model, but not the animal model, implemented within the framework of DHGLM provided unbiased (co)variance component estimates [14]. However, an animal DHGLM is expected to perform better than a sire-dam DHGLM for genetic evaluation because the animal DHGLM uses full relationships between animals rather than only among sires and dams. This is particularly important for uniformity, which is quantified by the residuals of individuals, which in the animal model do not contain the Mendelian sampling term. Moreover, for genetic evaluation, the animal DHGLM uses all phenotypic information and, for most breeding programs, at least part of the selection candidates, e.g. females for sex-linked traits, have phenotypes available at the time of selection. Use of the animal DHGLM with ssGBLUP for uniformity has not been tested.

In this study, we implemented ssGBLUP for predicting GEBV for uniformity in Atlantic salmon. Specifically, our aim was to compare the predictive ability of EBV for uniformity of body weight when implementing either BLUP with the **A** matrix or ssGBLUP with the **H** matrix. The (co)variance components were estimated from the sire-dam DHGLM with either **A** or **H** and compared prior to genetic evaluation.

Methods
Data

The data used in this study originated from the experiment conducted by Nofima AS and the breeding company SalmoBreed in Norway. The experiment followed all the regulations of animal ethical practice and was approved by the Norwegian Research Animal Committee (ID 6489). In 2013, 234 full-sib families were established from the mating of 131 sires to 234 dams (Table 1) during four weeks. Forty-seven percent of the parents were from year class 2009 and the rest from year class 2010. After hatching, fingerlings from each family were held in a 180-L family tank until tagging size (at mean body weight of 50 g). Each animal was tagged using passive integrated transponder (PIT) tags (Satpos AS, Norway). During tagging, a fin sample for genotyping was collected

Table 1 Population structure of Atlantic salmon

Population structure	
Sires, dams	131, 234
Sires per dam, mean (range)	1.0 (1.0)
Dams per sire, mean (range)	1.78 (1–3)
Full-sib families	234
Fish per full-sib family, mean (range)	15.4 (4–54)
Number of fish with records	3595
Genotyped animals	
Full-sib families	50
Fish per full-sib family, mean (range)	28.3 (21–38)
Number of fish with records	1416

from 21 to 38 sibs of each of 50 full-sib families. Thereafter, all fish were randomly allocated to three experiment tanks and grown for 11 months. At the average age of 16 months, all fish were challenged with sea lice using a co-habitat challenge, and at the end of the challenge test, final body weight (g) was measured for all 3595 fish with an electronic balance. A total of 1416 offspring (39% of all offspring) and the 131 sires and 234 dams were genotyped using the 31 K Affymetrix single nucleotide polymorphism (SNP) chip for Atlantic salmon developed by Nofima. Quality control of SNPs was performed in PLINK v1.9 [26] based on the following criteria: SNPs were removed if (1) their call rate was lower than 90%, (2) they deviated from Hardy–Weinberg equilibrium with a P value cut-off of 10^{-15}, and (3) their minor allele frequency (MAF) was lower than 0.01. After quality control, 921 of 31,013 SNPs were removed (2.9%) and, thus 30,092 SNPs remained to create the genomic relationship.

Relationship matrix

The numerator relationship (**A**) matrix with 814 ancestors in four generations was prepared based on pedigree information using ASReml [27]. The combined numerator and genomic relationship (**H**) matrix was defined as [23]:

$$\mathbf{H} = \begin{bmatrix} \mathbf{A}_{11} + \mathbf{A}_{12} + \mathbf{A}_{22}^{-1}(\mathbf{G} - \mathbf{A}_{22})\mathbf{A}_{22}^{-1}\mathbf{A}_{21} & \mathbf{A}_{12}\mathbf{A}_{22}^{-1}\mathbf{G} \\ \mathbf{G}\mathbf{A}_{22}^{-1}\mathbf{A}_{21} & \mathbf{G} \end{bmatrix},$$

where $\mathbf{A}_{11}$ is the pedigree relationship matrix between non-genotyped animals, $\mathbf{A}_{12}$ and $\mathbf{A}_{21}$ are pedigree relationship matrices between genotyped and non-genotyped animals, $\mathbf{A}_{22}$ is the pedigree relationship matrix between genotyped animals, and **G** is the genomic relationship matrix between genotyped animals. The **G** matrix was computed as [28]: $\mathbf{G} = \frac{\mathbf{WW'}}{N}$, where **W** is the matrix of the scaled SNP genotypes for all loci and N is the total number of SNPs (30,092). The elements of **W** were calculated as:

$$w_{ij} = \frac{(x_{ij} - 2p_j)}{\sqrt{2p_j(1 - p_j)}},$$

where x_{ij} is the SNP genotype (coded 0, 1, or 2) for the ith individual at SNP j and p_j is the allele frequency of the homozygous genotype coded as 2.

However, Aguilar et al. [29] and Christensen and Lund [30] showed that the inverse of the **H** matrix can be computed as:

$$\mathbf{H}^{-1} = \mathbf{A}^{-1} + \begin{bmatrix} \mathbf{0} & \mathbf{0} \\ \mathbf{0} & \mathbf{G}^{-1} - \mathbf{A}_{22}^{-1} \end{bmatrix},$$

which is less computational demanding and more simple than preparing and subsequently inverting the **H** matrix. The $\mathbf{H}^{-1}$ was prepared by using the Calc_grm computer software [31], which prepares both $\mathbf{A}^{-1}$ and $\mathbf{G}^{-1}$ internally before computing $\mathbf{H}^{-1}$.

Statistical analysis

Analysis of residuals

Uniformity can be quantified by squared residuals from a BLUP mixed model equation [32]. The use of genomic information to construct realised relationships between animals, especially for full-sibs, is expected to increase the accuracy of residual estimates due to a greater accuracy of EBV for body weight. Therefore, we investigated the effect of ssGBLUP and traditional BLUP on individual residual estimation. Furthermore, we investigated sire–dam and animal models because residual estimates from a sire-dam model contain not only the unexplained environmental effects but also Mendelian sampling genetic effects. Residual estimates from an animal model do not contain the latter when EBV are estimated with an accuracy of 1. In total, residuals from four models were compared, i.e. the sire-dam or animal model with either **A** or **H**.

The animal mixed model was:

$$y_{iklmn} = \mu + \beta age_k + t_l + yc_m + a_i + c_n + e_{iklmn}, \quad (1)$$

where y_{iklmn} is the observation (body weight) of the ith individual, μ is the overall mean, *age* is the fixed covariate effect due to different levels of age of the fish, calculated from the start feeding date until the date of measurement (day), β is the fixed linear regression coefficient on age, t is the lth fixed communal tank effect, yc is the mth fixed effect of year class of the parents, a_i is the random additive genetic effect, $\boldsymbol{a} \sim [0, \mathbf{A}\sigma_a^2]$, where **A** is the numerator relationship matrix, or $\boldsymbol{a} \sim N[0, \mathbf{H}\sigma_a^2]$, where **H** is the combined genomic and pedigree relationship matrix, N is the normal distribution, and σ_a^2 is the additive genetic variance for body weight, c_n is the random common effect for full-sibs, $\boldsymbol{c} \sim N[0, \mathbf{I}\sigma_c^2]$, where **I** is the identity matrix and σ_c^2 is the common environmental variance of body weight, and e_{iklmn} is the random residual effect, $e \sim N\left[0, \mathbf{I}\sigma_e^2\right]$, where σ_e^2 is the residual variance of body weight assumed to be homogeneous. For the sire-dam model, the term a_i in Eq. (1) was replaced by the random sire-dam (u_i) effect, $\boldsymbol{u} \sim N\left[0, \mathbf{A}\sigma_u^2\right]$ or $\boldsymbol{u} \sim N\left[0, \mathbf{H}\sigma_u^2\right]$. The same $\mathbf{A}$ and $\mathbf{H}$ matrices were used for the sire-dam and the animal models.

Estimation of genetic parameters for uniformity

To estimate genetic parameters for body weight and its uniformity, the sire-dam DHGLM was used [33, 34] because it is expected to provide unbiased (co)variance components for uniformity [14]. Body weight records were treated in two different ways. First, observed body weight was standardized to a mean of 0 and variance of 1, which facilitates convergence of the DHGLM. Second, we used either the natural log or the Box–Cox transformation to account for possible scale effects, because variances typically increase with increasing trait means [35, 36]. For the Box–Cox transformation, each observation was computed as $\frac{y_i^\lambda - 1}{\lambda}$, where λ is the transformation parameter, which was estimated based on Eq. (1) without the random effects [37] by maximum likelihood using the MASS package in R software [38]. The estimate of λ was close to 0 (0.076), indicating that the Box–Cox transformation is very similar to log-transformation, which sets λ equal to 0. Therefore, the Box–Cox transformed body weight was not used further. The standardized body weight and natural logarithm body weight are abbreviated as stdWT and lnWT, respectively.

To estimate genetic parameters, standardized and transformed body weights were modelled using sire-dam DHGLM in ASReml [32]:

$$\begin{bmatrix} \mathbf{y} \\ \boldsymbol{\Psi} \end{bmatrix} = \begin{bmatrix} \mathbf{X} & \mathbf{0} \\ \mathbf{0} & \mathbf{X}_v \end{bmatrix} \begin{bmatrix} \mathbf{b} \\ \mathbf{b}_v \end{bmatrix} + \begin{bmatrix} (\mathbf{Z}_s + \mathbf{Z}_d) & \mathbf{0} \\ \mathbf{0} & (\mathbf{Z}_s + \mathbf{Z}_d)_v \end{bmatrix}$$
$$\times \begin{bmatrix} \mathbf{u} \\ \mathbf{u}_v \end{bmatrix} + \begin{bmatrix} \mathbf{Q} & \mathbf{0} \\ \mathbf{0} & \mathbf{Q}_v \end{bmatrix} \begin{bmatrix} \mathbf{c} \\ \mathbf{c}_v \end{bmatrix} + \begin{bmatrix} \mathbf{e} \\ \mathbf{e}_v \end{bmatrix}, \quad (2)$$

where $\mathbf{y}$ is the vector of stdWT or lnWT records for the ith individual; $\boldsymbol{\Psi}$ is the vector of response variables for the residual variance, where $\psi_i = \log\left(\hat{\sigma}_{e_i}^2\right) + \frac{\hat{e}_i^2}{1-h_i} - \hat{\sigma}_{e_i}^2$, which was linearized using a Taylor series approximation in ASReml [34], $\hat{e}_i^2$ is the squared residual of the ith body weight record, h_i is the diagonal element in the hat-matrix of $\mathbf{y}$ [39], and $\hat{\sigma}_{e_i}^2$ is the estimated residual variance of the ith observation in the previous iteration of ASReml; $\mathbf{X}$ and $\mathbf{X}_v$ are incidence matrices of the fixed effects described in Eq. (1) for the trait mean and its uniformity, respectively; $\mathbf{b}$ ($\mathbf{b}_v$) is the solution vector for the corresponding fixed

effects; $\mathbf{Z}_s$ and $\mathbf{Z}_d$ are incidence matrices for the random sire (s) and dam (d) effects; $\mathbf{u}$ ($\mathbf{u}_v$) is the vector of additive genetic effects of sire-dam on the weight (uniformity), which was assumed to follow a normal distribution for the $\mathbf{A}$ matrix:

$$\begin{bmatrix} \mathbf{u} \\ \mathbf{u}_v \end{bmatrix} \sim N\left(\begin{bmatrix} \mathbf{0} \\ \mathbf{0} \end{bmatrix}, \frac{1}{4} \begin{bmatrix} \sigma_a^2 & \sigma_{a,a_v,exp} \\ \sigma_{a,a_v,exp} & \sigma_{a_v,exp}^2 \end{bmatrix} \otimes \mathbf{A} \right),$$

and for the $\mathbf{H}$ matrix:

$$\begin{bmatrix} \mathbf{u} \\ \mathbf{u}_v \end{bmatrix} \sim N\left(\begin{bmatrix} \mathbf{0} \\ \mathbf{0} \end{bmatrix}, \frac{1}{4} \begin{bmatrix} \sigma_a^2 & \sigma_{a,a_v,exp} \\ \sigma_{a,a_v,exp} & \sigma_{a_v,exp}^2 \end{bmatrix} \otimes \mathbf{H} \right),$$

where the $\frac{1}{4}$ accounts for the fact that the sire and dam each explain only a quarter of the additive genetic variance; $\mathbf{Q}$ ($\mathbf{Q}_v$) is the incidence matrix for the random common effects to full-sibs; $\mathbf{c}$ ($\mathbf{c}_v$) is the vector of common effects to full-sibs:

$$\begin{bmatrix} \mathbf{c} \\ \mathbf{c}_v \end{bmatrix} \sim N\left(\begin{bmatrix} \mathbf{0} \\ \mathbf{0} \end{bmatrix}, \begin{bmatrix} \sigma_c^2 & \sigma_{c,c_v,exp} \\ \sigma_{c,c_v,exp} & \sigma_{c_v,exp}^2 \end{bmatrix} \otimes \mathbf{I} \right).$$

The residuals of $\mathbf{y}$ ($\mathbf{e}$) and $\boldsymbol{\Psi}$ ($\mathbf{e}_v$) were assumed to be independently normally distributed as follows:

$$\begin{bmatrix} \mathbf{e} \\ \mathbf{e}_v \end{bmatrix} \sim N\left(\begin{bmatrix} \mathbf{0} \\ \mathbf{0} \end{bmatrix}, \begin{bmatrix} \mathbf{W}^{-1}\sigma_\epsilon^2 & \mathbf{0} \\ \mathbf{0} & \mathbf{W}_v^{-1}\sigma_{\epsilon\ominus_v}^2 \end{bmatrix} \right),$$

where $\mathbf{W} = \text{diag}(\hat{\boldsymbol{\Psi}}^{-1})$ and $\mathbf{W}_v = \text{diag}\left(\frac{1-\mathbf{h}}{2}\right)$, and σ_ϵ^2 ($\sigma_{\epsilon_v}^2$) is a scaled variance that was expected to be 1. The sire-dam DHGLM was fitted iteratively to update $\boldsymbol{\Psi}$, $\text{diag}(\mathbf{W})$ and $\text{diag}(\mathbf{W}_v)$ until the log-likelihood converged [34].

Calculation of genetic parameters

In the sire-dam DHGLM, the estimated variance for sires was set equal to the estimated genetic variance for dams and equal to one quarter of the additive genetic variance. Hence, the additive genetic variance for body weight (σ_a^2) and its uniformity ($\sigma_{a_v,exp}^2$) were equal to $4\sigma_u^2$ and $4\sigma_{u_v,exp}^2$, respectively. Estimates for $\sigma_{u_v,exp}^2$ and $\sigma_{c_v,exp}^2$ for uniformity of body weight were on the exponential scale (exp) and were converted to an additive scale ($\sigma_{u_v}^2$ and $\sigma_{c_v}^2$) using the extension of the equations of Mulder et al. [17], as derived by Sae-Lim et al. [8]. The additive genetic variance for uniformity of body weight on the additive scale was equal to $4\sigma_{u_v}^2$. Phenotypic variance (σ_P^2) of body weight was equal to $2\sigma_u^2 + \sigma_c^2 + \sigma_e^2$, where σ_c^2 is the variance component for the effect common to full-sibs and σ_e^2 is the residual variance of body weight. Heritability for body weight (h^2) was calculated as σ_a^2/σ_P^2. Heritability for uniformity of body weight (h_v^2) on the additive scale was calculated as $\frac{\sigma_{a_v}^2}{2\sigma_P^4 + 3\left(\sigma_{a_v}^2 + \sigma_{c_v}^2\right)}$ [8, 40]. Similarly, the common environmental effect was calculated as $c^2 = \sigma_c^2/\sigma_P^2$ for body weight and as $c_v^2 = \frac{\sigma_{c_v}^2}{2\sigma_P^4 + 3\left(\sigma_{a_v}^2 + \sigma_{c_v}^2\right)}$ for uniformity of body weight

[8]. The genetic coefficient of variation for uniformity of body weight (*GCV*) was calculated as $\sqrt{\sigma^2_{a_v,exp}}$. Standard errors of h^2_v and *GCV* were approximated using the equations presented by Mulder et al. [41].

Genetic evaluation and cross-validation

Two genetic evaluations, i.e., BLUP with **A** and ssGBLUP with **H**, were performed in a 10-fold cross-validation using the genetic parameters estimated based on the sire-dam DHGLM and !BLUP option in ASReml. In total, four models were used in the 10-fold cross-validation, i.e. animal DHGLM with either **A** or **H** on stdWT and lnWT.

The 10-fold cross-validation was performed on standardized and transformed body weight data as follows:

1. Adjusted phenotypes for body weight (y^*_i) and its uniformity (ψ^*_i) were calculated as $y^*_i = \hat{a}_i + \hat{c}_i + \hat{e}_i$ and $\psi^*_i = \hat{a}_{v_i} + \hat{c}_{v_i} + \hat{e}_{v_i}$, using the solutions from the analysis with Eq. (2) on the full dataset.
2. In a modified dataset, approximately 10% of observed phenotypes (y_i) of animals from each family were masked (=10% of the full dataset). All phenotypes had an equal chance to be masked, but the animals that were masked in the previous fold were not masked again in the next fold.
3. The genetic analysis with Eq. (2) was run on the modified dataset using the **A** and **H** matrices and EBV for body weight and its uniformity were predicted for the masked animals.
4. For each fold, two measurements were computed:
   (a) The predictive ability of EBV was calculated as the Pearson correlation of adjusted phenotypes (step 1) with the corresponding EBV ($\hat{a}^*$) (step 3) for the masked animals that were genotyped, i.e., cor(y^*_i, $\hat{a}^*_i$) for body weight and cor(ψ^*_i, $\hat{a}^*_{v_i}$) for uniformity. Kendall and Spearman correlations were also calculated for uniformity because ψ^*_i was exponentially rather than normally distributed.
   (b) To measure the degree of bias and accuracy of EBV or GEBV of the masked records, the mean square error prediction (MSEP) was calculated as $\frac{\sum^n_i (\hat{a}^*_i - y^*_i)^2}{n}$ for body weight and $\frac{\sum^n_i (\hat{a}^*_{v_i} - \psi^*_i)^2}{n}$ for uniformity of body weight, where n is the number of masked records in each fold. The MSEP was scaled by the variance of the adjusted phenotypes of the corresponding trait.
5. Steps (1) to (4) were repeated for each of the 10 folds.

Finally, average Pearson, Kendall, and Spearman correlations, MSEP and their standard error (SE) over the 10 folds were calculated. A 95% confidence interval of the difference (d) in the predictive ability from different models with either **A** or **H** was constructed using $d \pm 1.96 \times SE_d$, where the $SE_d = \sqrt{\frac{SD^2_{animal\ DHGLM} + SD^2_{sire-dam\ DHGLM}}{number\ of\ folds}}$. When 0 was not within the 95% confidence interval, the predictive abilities of two models were considered statistically different ($P < 0.05$).

Results

Residual estimates

Individual residuals estimated from using the **A** (BLUP) and **H** matrices (ssGBLUP) were plotted against each other to examine their relationship. As expected, the range of residual estimates from the animal models was lower than that from the sire-dam model since residual estimates from the sire-dam model included the entire Mendelian sampling term (Fig. 1).

For the sire-dam model, the use of **H** instead of **A** did not affect estimated residuals of genotyped animals since the regression coefficient of the estimated residuals using **H** on the estimated residuals using **A** and the Pearson correlation between the two were equal to 0.999, which was very similar to the regression coefficient of non-genotyped animals (0.998). The Pearson correlations between estimated residuals using **H** and **A** were the same as regression coefficients of estimated residuals using **A** on estimated residuals using **H** for genotyped animals (0.999) and non-genotyped animals (0.998).

In contrast, the use of **H** in the animal model affected residual estimates of genotyped animals since their distribution was much more scattered (Fig. 1). The slope of estimated residuals using **A** on estimated residuals using **H** was lower than 1 and slightly steeper for genotyped animals (regression coefficient = 0.7025) than for non-genotyped animals (regression coefficient = 0.6798). The Pearson correlations between estimated residuals using **H** and **A** were equal to 0.922 for genotyped animals and 0.966 for non-genotyped animals.

When using the sire-dam model, the difference in estimated residuals with **H** and **A** was small and ranged from −10.8 to 10.0. When using the animal model, this difference was larger and ranged from −95.3 to 104.5.

Genetic parameters of body weight and its uniformity

For body weight, estimates of additive genetic variances from the sire-dam DHGLM with either **A** or **H** were similar (Table 2). Likewise, estimates of h^2 were similar with **A** and **H** for both traits: 0.266 and 0.296, respectively for stdWT and 0.325 and 0.346, respectively for lnWT.

When using **A**, the estimate of h^2_v was higher for uniformity of stdWT (0.036) than for uniformity of lnWT (0.015), while the use of **H** did not affect the magnitude

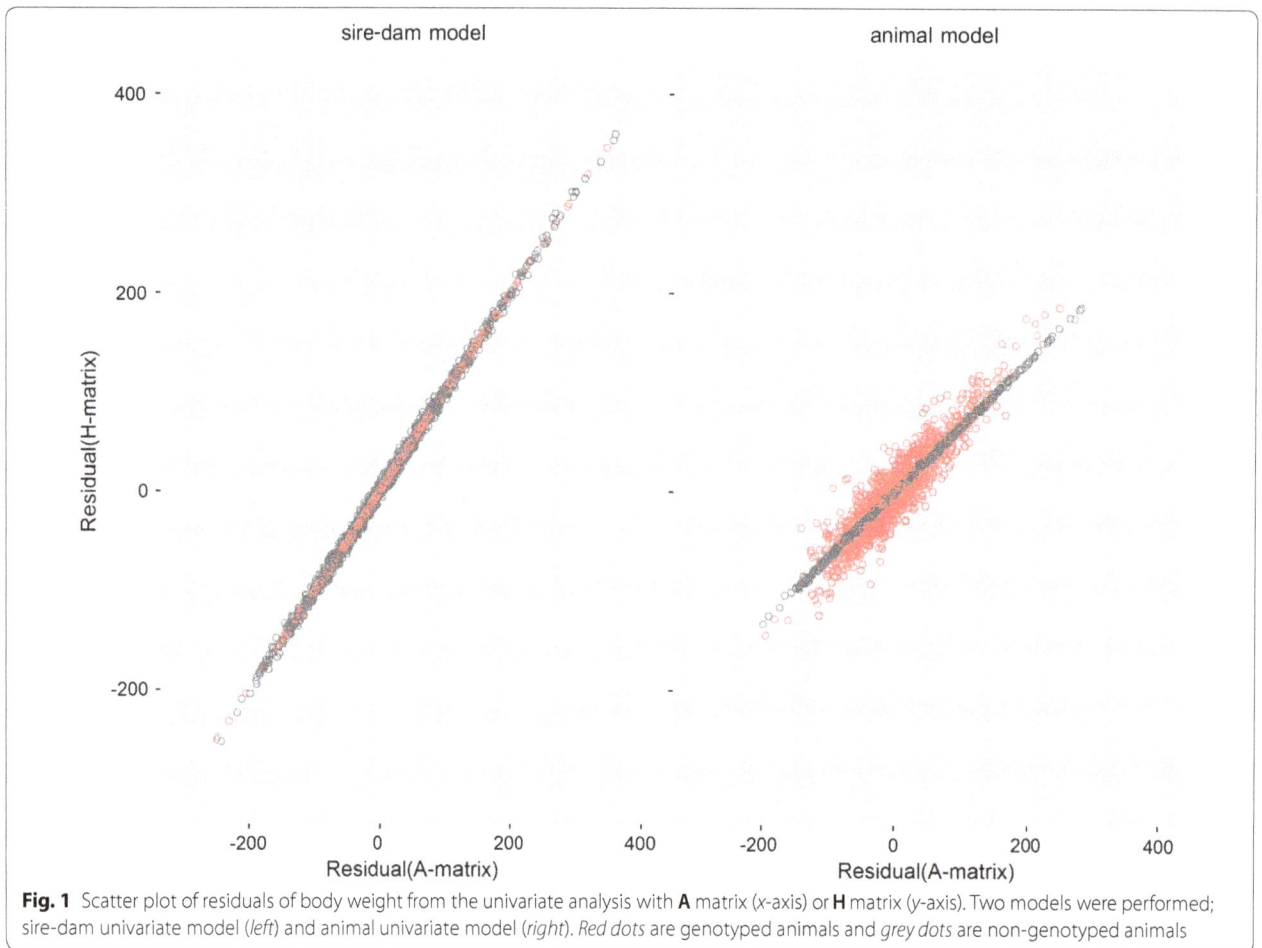

Fig. 1 Scatter plot of residuals of body weight from the univariate analysis with **A** matrix (*x*-axis) or **H** matrix (*y*-axis). Two models were performed; sire-dam univariate model (*left*) and animal univariate model (*right*). *Red dots* are genotyped animals and *grey dots* are non-genotyped animals

of h_v^2 for uniformity. Standard errors of h_v^2 estimates were, however, high (Table 2).

Although the estimates of h_v^2 were low, estimates of *GCV* were high for uniformity of stdWT (48.0% for **A** and 52.3% for **H**), which indicates substantial genetic potential for response to selection. After accounting for scale effects, estimates of *GCV* for uniformity of lnWT were reduced to 30% (for both **A** and **H**), which supports the existence of genetic variation for uniformity beyond the scale effects.

Estimates of c^2 for stdWT and lnWT were moderate and similar for **A** (0.103 to 0.117) and **H** (0.103 to 0.111), which suggests that part of the phenotypic variation was explained by non-genetic effects that are common to full-sibs. Instead, the estimates of c_v^2 for uniformity of stdWT and lnWT were very low and ranged from 0.001 to 0.022 for **A** and 0.002 to 0.019 for **H** (Table 2).

The estimate of the genetic correlation between stdWT and its uniformity was close to 1, using either **A** (0.952) or **H** (0.951), which shows the high dependency between mean and variance of body weight. However, the estimate of the genetic correlation between

lnWT and its uniformity was reduced to −0.093 with **A** and to 0.024 with **H**, which suggests that after accounting for the scale effects, the mean and variance became independent.

Cross-validation

The use of **H** instead of **A** with the animal DHGLM resulted in more variation of the within-family GEBV for stdWT and its uniformity, compared to within-family EBV (Fig. 2; Additional file 1: Figure S1), for sire-dam DHGLM).

The average correlation of adjusted phenotypes with predicted breeding values for stdWT and its uniformity was significantly higher with **H** (stdWT = 0.443; uniformity = 0.217 to 0.317) than with **A** (stdWT = 0.372; uniformity = 0.128 to 0.192). However, after accounting for scale effects using log-transformation, the average Pearson, Kendall and Spearman correlations of adjusted phenotypes with predicted breeding values for lnWT and their uniformity were only slightly higher with **H** than with **A**, and not significantly different from each other ($P > 0.05$).

Table 2 Estimates of variance components and genetic parameters of body weight and its uniformity based on the sire-dam double hierarchical generalized linear model when using pedigree (A) or combined pedigree and genomic relationships (H) and standard or log-transformed phenotypes

Trait/parameter	Standardization		Logarithm	
	A	H	A	H
Body weight				
σ_p^2	0.843	0.856	0.131	0.132
σ_a^2	0.216	0.243	0.043	0.046
σ_c^2	0.095	0.091	0.013	0.014
h^2	$0.266^{0.095}$	$0.296^{0.102}$	$0.325^{0.102}$	$0.346^{0.107}$
c^2	$0.117^{0.037}$	$0.111^{0.037}$	$0.103^{0.038}$	$0.103^{0.038}$
Uniformity of body weight				
$\sigma_{a_v,exp}^2$	$0.2303^{0.1094}$	$0.2732^{0.1211}$	$0.0896^{0.0569}$	$0.0885^{0.0598}$
$\sigma_{a_v}^2$	0.0612	0.0677	0.0005	0.0005
$\sigma_{c_v}^2$	0.0360	0.0306	0.0000	0.0001
GCV	$0.480^{0.114}$	$0.523^{0.116}$	$0.299^{0.095}$	$0.298^{0.100}$
h_v^2	$0.036^{0.019}$	$0.038^{0.020}$	$0.015^{0.014}$	$0.014^{0.013}$
c_v^2	0.022	0.019	0.001	0.002

A = pedigree based relationship matrix; **H** = combined genotyped and non-genotyped relationship matrix; σ_p^2 = phenotypic variance ($2\sigma_u^2 + \sigma_c^2 + \sigma_e^2$), where σ_e^2 is the residual variance for body weight; σ_a^2 and $\sigma_{a_v}^2$ = additive genetic variance for body weight and its uniformity, respectively; σ_c^2 = common environmental variance; GCV = coefficient of additive genetic variance for uniformity ($\sqrt{\sigma_{a_v,exp}^2}$); h^2 = heritability for body weight; c^2 = common environmental effect due to full-sib tanks; h_v^2 = heritability for uniformity; c_v^2 = same as c^2 but for uniformity of body weight. Superscripts are SE of the estimates

The average MSEP for uniformity from the animal DHGLM (0.608 to 0.944) were lower than those from the sire-dam DHGLM (0.973 to 1.112), suggesting that the use of an animal DHGLM increases the accuracy and may reduce bias in predicting breeding values for uniformity (Table 3; Additional file 2: Table S1). However, the average MSEP for uniformity of stdWT and lnWT obtained with **H** (0.608 to 0.944) were not notably different from those obtained with **A** (0.625 to 0.936).

The predictive ability of EBV of uniformity was sensitive to the type of correlation used, i.e. Pearson, Kendall and Spearman (Table 3). Spearman correlations were 39.1 to 49.0% higher than Kendall correlations. Predictive abilities of EBV and GEBV for uniformity of lnWT differed more from each other based on Kendall and Spearman correlations, albeit not significant at $P < 0.05$, than based on Pearson correlations. However, the SE of Kendall correlations were approximately 50% lower than the SE of Pearson and Spearman correlations, suggesting that Kendall correlations provide a more reliable estimate of predictive ability than Pearson and Spearman correlations.

Discussion

To the best of our knowledge, this is the first study that compares the use of the numerator relationship (**A**) and a combined genomics and numerator relationship (**H**) matrix for estimating genetic parameters and predicting breeding values for body weight and its uniformity. The use of the animal DHGLM with **H** significantly improved the predictive ability of GEBV for uniformity of body weight (stdWT) but not for scale-adjusted uniformity.

Genetic parameters

The estimate of heritability for uniformity of stdWT from sire-dam DHGLM with **A** was low ($h_v^2 = 0.036$) but higher than estimates of h_v^2 obtained in previous studies on rainbow trout [4, 8] and Nile tilapia [15, 16] ($\bar{h}_v^2 = 0.016$: min = 0.010: max = 0.024). However, after accounting for scale effects by logarithm transformations, the estimate of h_v^2 decreased to 0.014 to 0.015, which is in line with the previous reports that also used transformations [4, 8, 14–16].

Estimates of h_v^2 for stdWT and lnWT using the sire-dam DHGLM with **H** did not differ from those with **A**, which is in line with estimates of h_v^2 for uniformity of piglet birth weight obtained using either **A** or only the genomic relationship matrix (**G**) [42], while lower estimates were reported for environmental variance of somatic cell score in dairy cattle when using **G** compared to **A** [43]. The similarity of the estimates of h_v^2 obtained by using **A** or **H** in this study can be explained by the very similar estimated residuals (proxy of uniformity) between non-genotyped and genotyped animals when using the sire-dam model with **A** and **H**. The sire-dam model only exploits relationships between sires and dams and does not exploit the full potential of the genotype-based relationships between animals, and especially between full-sibs. In contrast, residuals of genotyped animals estimated by using the animal model were more differentiated when either **A** or **H** was used, and likely more accurate than estimates of residuals for non-genotyped animals. However, in a DHGLM analysis, the sire-dam model provides less biased (co)variance components than the animal model [14], likely because of the dependence between estimates of the breeding value and residual of an individual, which are obtained from the same phenotype of body weight. The use of genomic relationships combined with numerator relationships is expected to reduce the dependency between EBV and estimated residuals because the EBV are more accurate. Therefore, we performed the animal DHGLM with **H** but the model did not converge when the variance components were estimated, which may be due to (1) the dependency between EBV and estimated residuals for body weight remaining high, or (2) the difficulty to disentangle genetic

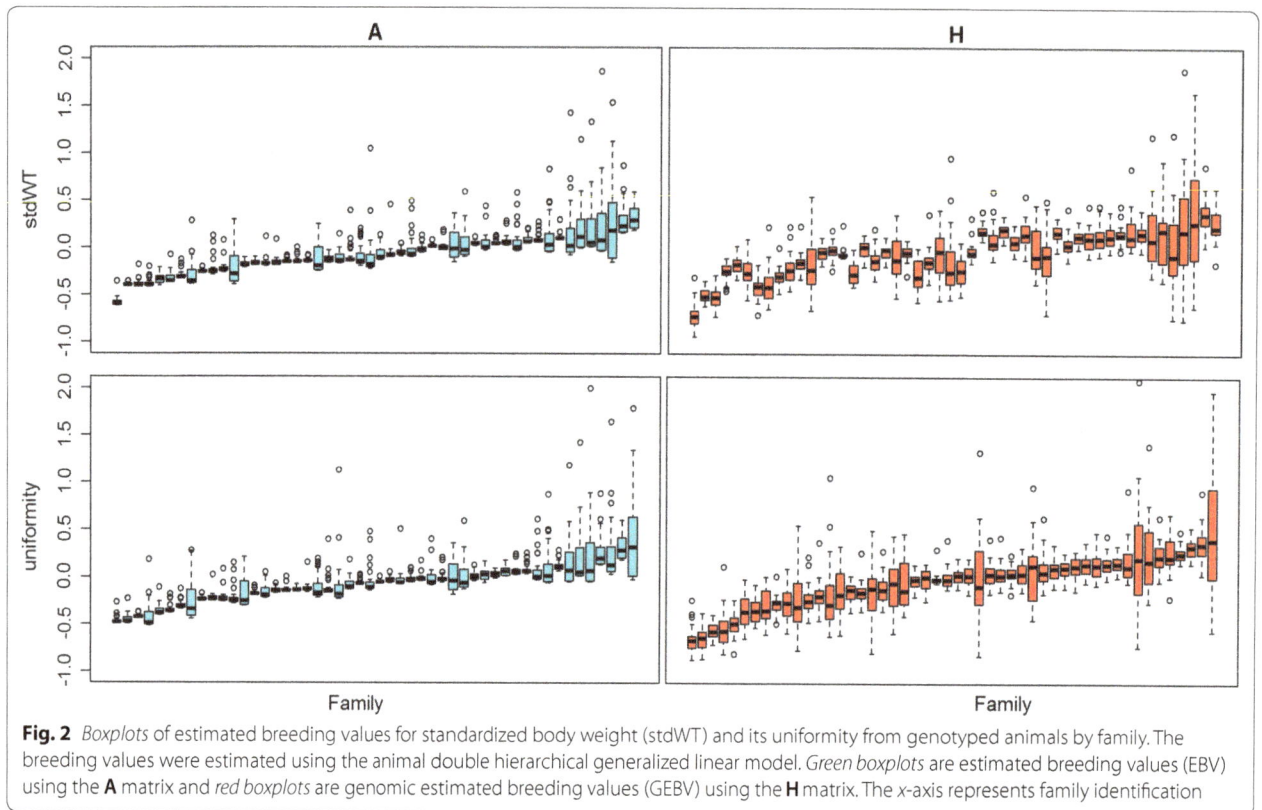

Fig. 2 *Boxplots* of estimated breeding values for standardized body weight (stdWT) and its uniformity from genotyped animals by family. The breeding values were estimated using the animal double hierarchical generalized linear model. *Green boxplots* are estimated breeding values (EBV) using the **A** matrix and *red boxplots* are genomic estimated breeding values (GEBV) using the **H** matrix. The *x*-axis represents family identification

Table 3 **Average Pearson, Kendall and Spearman correlations and mean square error prediction from a 10-fold cross-validation based on the sire-dam double hierarchical generalized linear model[a] when using pedigree (A) or combined pedigree and genomic relationships (H) and standard or log-transformed phenotypes**

Transformation	Relationship	Body weight		Uniformity of body weight			
		Pearson	MSEP	Pearson	Kendall	Spearman	MSEP
Standardized	**A** matrix	$0.372^{0.013}$	$0.724^{0.021}$	$0.192^{0.033}$	$0.128^{0.021}$	$0.178^{0.030}$	$0.625^{0.086}$
	H matrix	$0.443^{0.017}$	$0.682^{0.021}$	$0.271^{0.018}$	$0.217^{0.017}$	$0.317^{0.025}$	$0.608^{0.082}$
Logarithm	**A** matrix	$0.396^{0.019}$	$0.823^{0.029}$	$0.378^{0.032}$	$0.182^{0.016}$	$0.263^{0.023}$	$0.936^{0.085}$
	H matrix	$0.440^{0.016}$	$0.813^{0.028}$	$0.383^{0.026}$	$0.203^{0.014}$	$0.294^{0.020}$	$0.944^{0.085}$

[a] The variance components from the sire-dam double hierarchical generalized linear model were converted to the animal double hierarchical generalized linear model and were used in the 10-fold cross-validation. Relationship = relationship matrix, where **A** refers to pedigree-based relationship matrix and **H** refers to combined genotyped and non-genotyped relationship matrix. The predictability was calculated as the Pearson, Kendall and Spearman correlations between marked phenotype and predicted breeding value. MSEP was scaled by the phenotypic variance of corresponding traits

effects from the common environmental effects for uniformity of body weight.

The standard errors of h_v^2 estimates were high, which may be due to the large variation in family size (4 to 54). According to Hill and Mulder [30], large family sizes or repeated measurements are recommended for estimating the genetic heteroscedasticity of traits. The optimal full-sib family size is 39 with a *GCV* of 39% and an h^2 of 0.36 [18]. The use of **H** did not affect the standard errors of the h_v^2 estimates, which does not agree with previous studies, for example Veerkamp et al. [44] reported lower standard

errors of h^2 estimates for dry matter intake, milk yield, and body weight of heifers when using genomic relationships with an animal model. One possible explanation could be that the benefit of using the genomic relationship matrix may be limited when the sire-dam DHGLM is applied, since the variance of genomic relationships between sires (0.02) was very similar to the variance of numerator relationships between sires (0.01). In contrast, the variance in genomic relationships between animals (0.01) was much larger than the variance of numerator relationships between animals (0.004). Hence, the

SE of h_v^2 estimates may be lower when using an animal DHGLM with a genomic relationship matrix, compared to the numerator relationship matrix. Nevertheless, in our study, it was not possible to investigate this phenomenon since the log-likelihood did not converge for the animal DHGLM with **H**.

The GCV of uniformity for stdWT was substantial (48.0%), which indicates high potential for response to selection. This result is in the upper range of previous findings in fish species (17.4 to 64.0%) [4, 8, 14–16, 19] and in terrestrial animals (10.0 to 58.0%) [6, 9–13, 42]. After accounting for scale effects by logarithm transformations, the GCV for uniformity was reduced but still substantial (29.5 to 29.9%), which was also reported in previous studies on Atlantic salmon [14], rainbow trout [8], rabbit, and pig [45]. Thus, scale effects affect estimates of genetic parameters for uniformity of body weight considerably, but there is genetic variation for uniformity beyond the scale effect.

Genomic information slightly increased the GCV of uniformity of stdWT (from 48.0 to 52.3%). In contrast, genomic information did not influence the GCV for uniformity of lnWT (29.8%). Since estimates of genetic parameters obtained with **A** and **H** were similar, the GCV for body weight remained similar, which is in agreement with the previous comparison between **A** ($GCV = 11.0$ to 12.0%) and **G** ($GCV = 10.0$ to 11.0%) for uniformity of birth weight of piglets using a dam model [42].

Genetic and genomic predictions

In this study, we used the Pearson correlation of EBV and GEBV with adjusted phenotype as the measure of predictive ability. The use of **H** instead of **A** in the animal DHGLM significantly improved the ability to predict breeding values for stdWT (19%) and its uniformity (41.1 to 78.1%). Furthermore, the use of the animal DHGLM instead of the sire-dam DHGLM significantly increased the predictive ability of EBV and GEBV for uniformity (see Additional file 2: Table S1), as expected.

Our findings indicate that ssGBLUP with an animal DHGLM can increase the accuracy of EBV for uniformity substantially compared to pedigree-based BLUP. However, after accounting for scaling effects by using log transformations, the use of **H** compared to **A** only slightly improved the correlation (1.6 to 13.9%) and MSEP between GEBV and adjusted phenotypes, and the improvement was not significant. There are two main reasons why these results differed between uniformity of stdWT and lnWT. First, log-transformation substantially reduced the genetic correlation of stdWT with its uniformity. As a result, any increases in predictive ability of GEBV of lnWT when using **H** instead of **A** (15.7%) did not positively influence the predictive ability of GEBV

of its uniformity. Second, the lower additive genetic variance and h_v^2 for uniformity after accounting for scale effects reduces the accuracy of EBV for uniformity. Consequently, MSEP increased from 0.63 (stdWT) to 0.94 (lnWT) with **A** and from 0.61 (stdWT) to 0.94 (lnWT) with **H** in the animal DHGLM.

The accuracy of genomic selection is expected to increase when the number of genotyped animals in the reference population increases for any trait [46] but in particular for lowly heritable traits, such as uniformity as shown by Sell-Kubiak et al. [42] and somatic cell score by Mulder et al. [43]. In this study, uniformity of lnWT had an even lower h_v^2 than uniformity of stdWT. The number of genotyped animals in the reference population used for cross-validation was on average equal to 1274, which may have limited the benefit of using genomic information in ssGBLUP. A future empirical study should investigate the effect of the number of genotyped animals in the reference population on the ability to predict breeding values for uniformity to validate our findings and conclusions.

Pearson or rank correlations?

Squared residuals or adjusted phenotype for uniformity (ψ_i^*) are exponentially rather than normally distributed, which may not justify quantifying predictive ability using a Pearson correlation. Thus, we also calculated distribution-free rank correlations (Kendall and Spearman) and indeed found estimation of the predictive ability for uniformity to be sensitive to the type of correlation used.

Although not significantly different, Kendall and Spearman correlations explained differences in predictive ability of EBV and GEBV for uniformity of lnWT slightly better than the Pearson correlation. Hence, the conclusion that the benefit of using genomic relationship for computing EBV for uniformity of logarithm transformations is limited remained the same when using Kendall and Spearman correlations. Colwel and Gillett [47] showed that, in general, estimates of Kendall correlations are similar to estimates of Spearman correlations, but in some cases, the magnitude of Spearman correlations can be 50% greater than the magnitude of Kendall correlations [47]. This is in line with our findings since Spearman correlations were 42.2 to 49.0% and 39.1 to 46.1% greater than Kendall correlations for the sire-dam DHGLM (see Additional file 2: Table S1) and the animal DHGLM, respectively. Nevertheless, the SE of Kendall correlations were notably lower by approximately 50% than the SE of Pearson and Spearman correlations, which indicates that the Kendall correlation may be a more reliable estimate of predictive ability than the Pearson and Spearman correlations. Hence, it is recommended to use Kendall instead of Pearson correlations when studying predictive ability for uniformity.

Selection for uniformity

For fish breeding, major goals are to increase mean body weight and reduce variability (more uniformity) of body weight. Nevertheless, definitions of uniformity of stdWT and lnWT are not the same. From a biological point of view, genetic variation for environmental canalization can be quantified after the scale effect is accounted for. However, from an animal breeding point of view, uniformity on the observed scale explains the actual range of fish sizes that are processed by aquaculture industries.

Selection for body weight and uniformity may be challenging because the genetic correlation between body weight and its variability is high and positive, and sometimes approaches 1. A general observation is that the genetic correlation between log-transformed body weight and its variability is zero or even negative, allowing selection to simultaneously increase transformed body weight and reduce variability. Therefore log-transformed body weight and its variability could be included in a selection index. This would require knowledge of the genetic correlation between variability of stdWT and variability of lnWT, which is not available. Hence, we used the sire-EBV and sire-GEBV, obtained from BLUP and ssGBLUP with the animal DHGLM, to calculate the Pearson correlations between EBV for the trait and its variability. Pearson correlations between EBV for variability of stdWT and EBV for variability of lnWT were positively moderate with BLUP (0.48) and ssGBLUP (0.68). The Pearson correlation between EBV for stdWT and EBV for its variability was close to 1 with either BLUP (0.96) or ssGBLUP (0.97), while the Pearson correlation between EBV of lnWT and its variability was −0.24 for BLUP and -0.005 for ssGBLUP. Not surprisingly, the Pearson correlation between EBV for lnWT and EBV for stdWT was highly positive (0.82). These Pearson correlations suggest that variability of lnWT should be included in the selection index because the GEBV for variability of lnWT are positively correlated with GEBV for variability of stdWT, which indicates that selection against variability of lnWT will indirectly reduce variability on the observed scale. Furthermore, GEBV for variability of lnWT are not correlated to GEBV for lnWT and thus selection against variability of lnWT will not indirectly reduce lnWT.

To elucidate responses to selection for body weight and its variability, we performed truncation selection on sires based on their GEBV from the animal DHGLM with ssGBLUP. The breeding goal could include body weight and variability on the observed scale and their economic values (v): $v_1 \mathrm{BV}_{\mathrm{stdWT}} - v_2 \mathrm{BV}_{\mathrm{variability}_{\mathrm{stdWT}}}$, while, based on the Pearson correlations discussed above, the selection index (I) could include lnWT and its variability with their relative weighting factors (b):

$$\mathrm{I} = b_1 \mathrm{GEBV}_{\mathrm{lnWT}} - b_2 \mathrm{GEBV}_{\mathrm{variability}_{\mathrm{lnWT}}}.$$

Selecting the 10% best sires on an index with b_1 of 0.3, i.e., $\mathrm{I} = 0.3 * \mathrm{GEBV}_{\mathrm{lnWT}} - 0.7 * \mathrm{GEBV}_{\mathrm{variability}_{\mathrm{lnWT}}}$ provides almost no genetic gain in variability of stdWT (-0.001) but positive genetic gain in stdWT (3.62% of mean body weight in g). In contrast, a selection index, based on breeding goal traits: $\mathrm{I} = 0.52 * \mathrm{GEBV}_{\mathrm{stdWT}} - 0.48 * \mathrm{GEBV}_{\mathrm{variability}_{\mathrm{stdWT}}}$, provides zero genetic gains for both stdWT and its variability, showing no possibility to achieve genetic gain on body weight while maintaining stable phenotypic variability. Nevertheless, the genetic gain in stdWT was much greater (17.32% of mean body weight in g) when variability was not included in the selection index. Therefore, although it is possible to increase body weight while keeping variability constant, there is a trade-off in genetic gain for body weight when selecting for reduced variability.

Conclusions

The use of the animal DHGLM instead of the sire-dam DHGLM substantially increased the predictive ability for breeding values of uniformity, because the animal DHGLM fully exploits the relationships between full- and half-sibs. When using the animal DHGLM, the use of a combined numerator and genomic relationship matrix significantly increased the predictive ability for breeding values of uniformity of body weight, but only a slight and non-significant increase was observed after accounting for the scale effects by using transformed body weights. The small increase in predictive ability with transformed body weights may be due to lower heritability for uniformity of transformed body weight, a lower genetic correlation between transformed body weights and their uniformities, and/or a small number of genotyped animals in the reference population. The use of a Kendall correlation provided the lowest SE of predictive ability for uniformity and provided a more accurate estimate of predictive ability for uniformity over Pearson and Spearman correlations. In conclusion, the use of ssGBLUP increases the accuracy of breeding values for uniformity of harvest weight, which is expected to increase response to selection in uniformity.

Additional files

Additional file 1: Figure S1. Estimated breeding values for standardized body weight and its uniformity based on sire-dam DHGLM. The data provided represent the boxplot of estimated (genomic) breeding values of genotyped animals based on sire-dam DHGLM when using pedigree (**A**) or combined pedigree and genomic relationships (**H**).

Additional file 2: Table S1. Cross-validation based on sire-dam DHGLM. The data provided represent the results from 10-fold cross-validations based on sire-dam DHGLM.

Authors' contributions
 PSL analyzed the data. HAM, AK, and ML provided theoretical support for genomic DHGLM and cross-validation. HAM, AK, and ML contributed to the discussion of the results. PSL drafted the manuscript. HAM, AK, and ML improved the manuscript. All authors read and approved the final manuscript.

Author details
[1] Nofima Ås, Osloveien 1, P.O. Box 210, 1431 Ås, Norway. [2] Biometrical Genetics, Natural Resources Institute Finland, 31600 Jokioinen, Finland. [3] Animal Breeding and Genomics Centre, Wageningen University and Research, P.O. Box 338, 6700 AH Wageningen, The Netherlands.

Acknowledgements
This study is a part of the research project entitled STABLEFISH funded by Norwegian Research Council (NRC: 234144/E49). We would like to thank SalmoBreed for providing data for this study. Matthew Baranski genotyped the animals and generated genotype data file for this study. Arthur Gilmour is acknowledged for help in implementing genomic DHGLM in ASReml v4. PSL would like to thank Mario Calus for his guidance on Cal_grm computer software, Solomon Antwi Boison and Sergio Vela Avitúa for a fruitful discussion during the drafting of this manuscript.

Competing interests
The authors declare that they have no competing interests.

References
1. Sae-Lim P, Komen H, Kause A, van Arendonk JAM, Barfoot AJ, Martin KE, et al. Defining desired genetic gains for rainbow trout breeding objective using analytic hierarchy process. J Anim Sci. 2012;90:1766–76.
2. Gilmour KM, DiBattista JD, Thomas JB. Physiological causes and consequences of social status in salmonid fish. Integr Comp Biol. 2005;45:263–73.
3. Janhunen M, Kause A, Järvisalo O. Costs of being extreme - Do body size deviations from population or sire means decrease vitality in rainbow trout? Aquaculture. 2012;370–371:123–9.
4. Janhunen M, Kause A, Vehviläinen H, Jarvisalo O. Genetics of microenvironmental sensitivity of body weight in rainbow trout (Oncorhynchus mykiss) selected for improved growth. PLoS One. 2012;7:e38766.
5. Baras E, Jobling M. Dynamics of intracohort cannibalism in cultured fish. Aquacult Res. 2002;33:461–79.
6. Mulder HA, Bijma P, Hill WG. Selection for uniformity in livestock by exploiting genetic heterogeneity of residual variance. Genet Sel Evol. 2008;40:37–59.
7. Mulder H, Hill W, Vereijken A, Veerkamp R. Estimation of genetic variation in residual variance in female and male broiler chickens. Animal. 2009;3:1673–80.
8. Sae-Lim P, Kause A, Janhunen M, Vehviläinen H, Koskinen H, Gjerde B, et al. Genetic (co) variance of rainbow trout (Oncorhynchus mykiss) body weight and its uniformity across production environments. Genet Sel Evol. 2015;47:46.
9. Ros M, Sorensen D, Waagepetersen R, Dupont-Nivet M, SanCristobal M, Bonnet JC, et al. Evidence for genetic control of adult weight plasticity in the snail Helix aspersa. Genetics. 2004;168:2089–97.
10. Rowe S, White IM, Avendano S, Hill WG. Genetic heterogeneity of residual variance in broiler chickens. Genet Sel Evol. 2006;38:617–35.
11. Wolc A, White IM, Avendano S, Hill WG. Genetic variability in residual variation of body weight and conformation scores in broiler chickens. Poult Sci. 2009;88:1156–61.
12. Ibáñez-Escriche N, Moreno A, Nieto B, Piqueras P, Salgado C, Gutiérrez JP. Genetic parameters related to environmental variability of weight traits in a selection experiment for weight gain in mice; signs of correlated canalised response. Genet Sel Evol. 2008;40:279–93.
13. Ibáñez-Escriche N, Varona L, Sorensen D, Noguera JL. A study of heterogeneity of environmental variance for slaughter weight in pigs. Animal. 2008;2:19–26.
14. Sonesson A, Ødegård J, Ronnegard L. Genetic heterogeneity of within-family variance of body weight in Atlantic salmon (Salmo salar). Genet Sel Evol. 2013;45:41.
15. Marjanovic J, Mulder H, Khaw H, Bijma P. Genetic parameters for uniformity of harvest weight in the gift strain of nile tilapia estimated using double hierarchical generalized linear models. In: Proceedings of the international symposium on genetics in aquaculture XII, 21–27 June 2015; Santiago de Compostela; 2015. http://isga2015.acuigen.es/isga-2015-Abstract-Book.pdf.
16. Khaw HL, Ponzoni RW, Yee HY, bin Aziz MA, Mulder HA, Marjanovic J, et al. Genetic variance for uniformity of harvest weight in Nile tilapia (Oreochromis niloticus). Aquaculture. 2016;451:113–20.
17. Mulder HA, Bijma P, Hill WG. Prediction of breeding values and selection response with genetic heterogeneity of environmental variance. Genetics. 2007;175:1895–910.
18. Sae-Lim P, Gjerde B, Nielsen HM, Mulder H, Kause A. A review of genotype-by-environment interaction and micro-environmental sensitivity in aquaculture species. Rev Aquacult. 2015;8:369–93.
19. Marjanovic J, Mulder HA, Khaw HL, Bijma P. Genetic parameters for uniformity of harvest weight and body size traits in the GIFT strain of Nile tilapia. Genet Sel Evol. 2016;48:41.
20. Falconer DS, Mackay TFC. Introduction to quantitative genetics. 4th ed. London: Pearson; 1996.
21. Meuwissen THE, Hayes BJ, Goddard ME. Prediction of total genetic value using genome-wide dense marker maps. Genetics. 2001;157:1819–29.
22. Goddard M, Hayes B, Meuwissen THE. Genomic selection in farm animal species-lessons learnt and future perspectives. In: Proceedings of the 9th world congress on genetics applied to livestock production, 1–6 August 2010; Leipzig. 2010.
23. Misztal I, Aggrey SE, Muir WM. Experiences with a single-step genome evaluation. Poult Sci. 2013;92:2530–4.
24. Misztal I, Legarra A, Aguilar I. Computing procedures for genetic evaluation including phenotypic, full pedigree, and genomic information. J Dairy Sci. 2009;92:4648–55.
25. Legarra A, Aguilar I, Misztal I. A relationship matrix including full pedigree and genomic information. J Dairy Sci. 2009;92:4656–63.
26. Purcell S, Neale B, Todd-Brown K, Thomas L, Ferreira MA, Bender D, et al. PLINK: a tool set for whole-genome association and population-based linkage analyses. Am J Hum Genet. 2007;81:559–75.
27. Gilmour AR, Gogel BJ, Cullis BR, Thompson R. ASReml User Guide Release 4.0. Hemel Hempstead: VSN International Ltd; 2012.
28. VanRaden PM. Efficient methods to compute genomic predictions. J Dairy Sci. 2008;91:4414–23.
29. Aguilar I, Misztal I, Johnson DL, Legarra A, Tsuruta S, Lawlor TJ. Hot topic: a unified approach to utilize phenotypic, full pedigree, and genomic information for genetic evaluation of Holstein final score1. J Dairy Sci. 2010;93:743–52.
30. Christensen OF, Lund MS. Genomic prediction when some animals are not genotyped. Genet Sel Evol. 2010;42:2.
31. Calus MPL, Vandenplas J. calc_grm—a program to compute pedigree, genomic, and combined relationship matrices. Animal Breeding and Genomics Centre: Wageningen; 2015.
32. Hill WG, Mulder HA. Genetic analysis of environmental variation. Genet Res (Camb). 2010;92:381–95.
33. Rönnegård L, Felleki M, Fikse F, Mulder H, Strandberg E. Genetic heterogeneity of residual variance—estimation of variance components using double hierarchical generalized linear models. Genet Sel Evol. 2010;42:8.
34. Felleki M, Lee D, Lee Y, Gilmour AR, Rönnegård L. Estimation of breeding values for mean and dispersion, their variance and correlation using double hierarchical generalized linear models. Genet Res (Camb). 2012;94:307–17.
35. Lande R. On comparing coefficients of variation. Syst Zool. 1977;26:214–7.
36. Box GE, Cox DR. An analysis of transformations. J R Stat Soc Ser B Stat Methodol. 1964;26:211–52.
37. Sakia R. The Box–Cox transformation technique: a review. Statistician. 1992;41:169–78.
38. R Development Core Team. R: a language and environment for statistical computing. Vienna: The R Foundation for Statistical Computing; 2011.
39. Hoaglin DC, Welsch RE. The hat matrix in regression and ANOVA. Am Stat. 1978;32:17–22.

40. Felleki M, Lundeheim N. Genetic control of residual variance for teat number in pigs. Proc Assoc Advmt Anim Breed Genet. 2013;20:538–41.

41. Mulder HA, Visscher J, Fablet J. Estimating the purebred–crossbred genetic correlation for uniformity of eggshell color in laying hens. Genet Sel Evol. 2016;48:39.

42. Sell-Kubiak E, Wang S, Knol EF, Mulder HA. Genetic analysis of within-litter variation in piglets' birth weight using genomic or pedigree relationship matrices. J Anim Sci. 2015;93:1471–80.

43. Mulder HA, Crump RE, Calus MPL, Veerkamp RF. Unraveling the genetic architecture of environmental variance of somatic cell score using high-density single nucleotide polymorphism and cow data from experimental farms. J Dairy Sci. 2013;96:7306–17.

44. Veerkamp RF, Mulder HA, Thompson R, Calus MPL. Genomic and pedigree-based genetic parameters for scarcely recorded traits when some animals are genotyped. J Dairy Sci. 2011;94:4189–97.

45. Yang Y, Christensen O, Sorensen D. Analysis of a genetically structured variance heterogeneity model using the Box–Cox transformation. Genet Res (Camb). 2011;93:33–46.

46. Daetwyler HD, Villanueva B, Woolliams JA. Accuracy of predicting the genetic risk of disease using a genome-wide approach. PLoS One. 2008;3:e3395.

47. Colwell DJ, Gillett JR. 66.49 Spearman versus Kendall. Math Gaz. 1982;66:307–9.

9

Selection for environmental variance of litter size in rabbits

Agustín Blasco[1]* , Marina Martínez-Álvaro[1], Maria-Luz García[2], Noelia Ibáñez-Escriche[3] and María-José Argente[2]

Abstract

Background: In recent years, there has been an increasing interest in the genetic determination of environmental variance. In the case of litter size, environmental variance can be related to the capacity of animals to adapt to new environmental conditions, which can improve animal welfare.

Results: We developed a ten-generation divergent selection experiment on environmental variance. We selected one line of rabbits for litter size homogeneity and one line for litter size heterogeneity by measuring intra-doe phenotypic variance. We proved that environmental variance of litter size is genetically determined and can be modified by selection. Response to selection was 4.5% of the original environmental variance per generation. Litter size was consistently higher in the Low line than in the High line during the entire experiment.

Conclusions: We conclude that environmental variance of litter size is genetically determined based on the results of our divergent selection experiment. This has implications for animal welfare, since animals that cope better with their environment have better welfare than more sensitive animals. We also conclude that selection for reduced environmental variance of litter size does not depress litter size.

Background

In recent years, there has been increasing interest in the genetic determination of environmental variance. The reasons are summarized by Morgante et al. [1] and Sørensen et al. [2]. In evolutionary genetics, how phenotypic variance is maintained under several models of selection is a key issue. For example, Zhang and Hill [3] examined models for maintenance of environmental variance under stabilizing selection, in which phenotypes near the optimum are selected and, consequently, less variable genotypes are favored. In medical genetics, there are several foci of interest, such as differences in the penetrance of risk alleles [1] or the evolution of health indicators over time [2]. In animal and plant genetics, selection to reduce environmental variance can lead to more uniform products without compromising future genetic progress, since genetic variance of the trait is not

affected [4]. In addition, genetic uniformity can be useful for production traits; for example, homogeneity of birth weight within litters in rabbits is related to higher viability of the kits [5].

In the case of litter size, which is a trait directly related to fitness, environmental variance can be related to the capacity of animals to cope with new environmental conditions. Females with less adaptable genotypes are more sensitive to diseases and to stress and show a higher degree of variability in litter size [6–8]. Selection to reduce environmental variance would produce animals that cope better with their environment, which is a definition of animal welfare [9].

There is evidence that environmental variance is under genetic control in several species. Most of this evidence is indirect, because it comes from analyses of databases and not from experiments designed to assess the genetic determination of environmental variance (litter size in sheep [10] and pigs [11], birth weight and stillbirth in pigs [12], weight in snails [13], birth weight in mice [14], uterine capacity in rabbits [15], weight in poultry [16] and beef cattle [17], milk yield of dairy cattle [18], and

*Correspondence: ablasco@dca.upv.es
[1] Institute for Animal Science and Technology, Universitat Politècnica de València, Valencia, Spain
Full list of author information is available at the end of the article

weight in trout [19] and salmon [20]). Other evidence of the existence of a genetic component for environmental variance comes from a few experiments on inbred lines of *Drosophila melanogaster* [1] and from only two selection experiments on birth weight, in rabbits [21] and mice [22]. Models used to analyze environmental variance were reviewed by Hill and Mulder [23]. They are highly parametrized and not robust; for example, Yang et al. [24] showed that small deviations from normality in the residuals can substantially change estimates of genetic parameters.

In the experiment reported in this paper, we avoided the use of complex models of environmental variance by directly selecting for this trait as an observed trait. Environmental variance of litter size can be directly recorded by computing the intra-doe variance of litter size. Since the genetic determination of litter size is approximately the same for all parities of a rabbit doe and permanent effects are the same along parities [25], the intra-doe phenotypic variance represents the environmental variance if no other systematic environmental effects are acting. We developed a divergent selection experiment on intra-doe phenotypic variance as a measure of environmental variance of litter size.

Methods

Animals

The rabbits used in this study came from a maternal synthetic line created from commercial crossbred animals [26]. The rabbits were bred at the farm of the Universidad Miguel Hernández of Elche. Reproduction was organized in discrete generations. Does were first mated at 18 weeks of age and thereafter 10 days after parturition. They were under a constant photoperiod of 16:8 h and controlled ventilation. The animals were fed a standard commercial diet. All experimental procedures were approved by the Committee of Ethics and Animal Welfare of the Miguel Hernández University, according to Council Directives 98/58/EC and 2010/63/EU.

Selection for environmental variance

A divergent selection experiment on environmental variance of litter size was carried out across 10 generations. Each divergent line had approximately 125 females and 25 males per generation. Data from 12,174 litters from 2769 does were used in the experiment. The average number of litters per doe was 4.5, ranging from 2 to 9 (Fig. 1).

Selection was based on environmental variance of litter size, V_e, which was calculated as the within-doe variance of litter size after litter size was pre-corrected for year-season and three levels of parity-lactation status: first

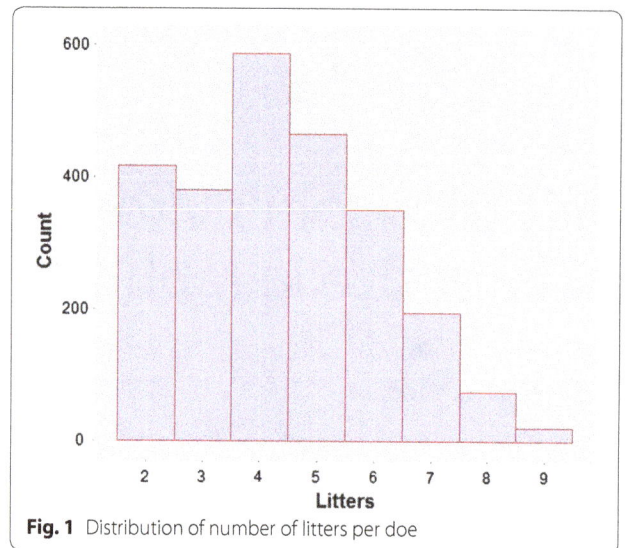

Fig. 1 Distribution of number of litters per doe

parity, other parities for lactating females, and other parities for non-lactating females, to avoid systematic effects that could affect the variance. The intra-doe phenotypic variance represents the environmental variance for litter size under the assumption that the genetic determination is approximately the same for all parities of a rabbit doe and the permanent effects are the same across parities [25]. Variance V_e for each doe was calculated using the minimum quadratic risk estimator:

$$V_e = \frac{1}{n+1} \sum_{1}^{n} (x_i - \bar{x})^2,$$

where x_i is the pre-corrected litter size of a doe's parity i, and n is the total number of parities of the doe (n varying from 2 to 9). This estimator has lower risk (lower expected mean square error) than that of maximum likelihood (ML) or restricted maximum likelihood (REML) [27]. The environmental variance of litter size without pre-correction was also calculated.

All dams were ranked based on their estimate of intra-doe variance of litter size, without using pedigree information for genetic evaluation. Only dams with four or more parities were considered for selection. Selection candidates came from females that had four or five parities, with some exceptions. The best 20% dams were used to breed the next generation. Each sire was mated with five dams and one male progeny from the best dam that a sire was mated to was selected to breed the next generation. This within-male family selection was performed in order to reduce inbreeding. Selection was based on the individual record of each female.

Statistical analysis

Response to selection was estimated as the difference between lines in each generation. These differences between lines were analyzed using a simple linear model with a line-generation effect:

$$\mathbf{y} = \mathbf{Xb} + \varepsilon,$$

where $\mathbf{y}$ is a vector with one record per doe, i.e. its environmental variance V_e, and $\mathbf{b}$ is a vector of the line-generation effect. This linear model has heterogeneous variances, because not all does had the same number of litter size records, so V_e is calculated based on different numbers of parities. The weights for taking this into account were calculated as [28]:

$$\frac{2(n-1)}{(n+1)^2}\sigma_\varepsilon^2,$$

where n is the number of parities of each doe and σ_ε^2 the residual variance. To check the robustness of the model, the same analysis was performed with homogeneous variances, which led to the same results with small variations in the confidence intervals.

Response to selection was also estimated as the average of the genetic values in each generation by using a mixed model with generation as a fixed effect and the breeding value of each doe as a random effect. Breeding values were assumed normally distributed with variance $\mathbf{A}\sigma_u^2$, where $\mathbf{A}$ is the pedigree-based relationship matrix and σ_u^2 is the variance of the breeding values. In this model, the generation effect captures systematic changes in environment over generations.

Correlated response in litter size was estimated as the differences in litter size between lines in each generation. It was analyzed using a standard mixed model with fixed effects of line-generation, parity-lactation status (first parity, and lactating or not at mating in other parities) and year-season, along with a random permanent environmental effect across parities for each doe, which was assumed normally distributed.

Bayesian analyses were performed to fit the above models, with bounded flat priors for all unknowns. Features of the marginal posterior distributions were estimated using Gibbs sampling. After some exploratory analyses, we used a chain of 60,000 samples for differences between lines with a burn-in period of 10,000; only one of every 10 samples was saved for inferences. For the genetic analyses, we used a chain of 1,000,000 samples, with a burn-in of 500,000; only one of every 100 samples was saved for inferences. Convergence was tested using the Z criterion of Geweke [29], and Monte Carlo sampling errors were computed using time-series procedures, as described in [30]. In all Bayesian analyses, the Monte Carlo standard errors were small and lack of convergence was not detected by the Geweke test. Special software code was developed for analyses of differences between lines and the program TM was used for the genetic analyses [31].

Results

Descriptive results

Table 1 summarizes the descriptive features of the traits in the base population. We estimated intra-doe phenotypic variance by pre-correcting for the effects of year-season and parity-lactation status (first parity, and lactating or not at mating in all subsequent parities). Pre-correction had little effect with environmental variance before (V_r) and after pre-correction (V_e) being practically the same. In both cases, environmental variances were highly variable, with a large standard deviation and high coefficient of variation, which helps explain the large response to selection, which will be presented below. The median of the environmental variance differs from its mean, showing that its distribution is asymmetrical, as expected (Fig. 2a). Although normality is not required for comparison of means when the sample size is moderate or large, we applied a normalizing transformation to the environmental variance. We chose the square root because it has a biological interpretation, i.e. environmental standard deviation (SD_e). For this trait, the mean and median were similar (Table 1; Fig. 2b). The distribution of the number of parities per doe is in Fig. 1.

Response to selection

Response to selection was high and equal to approximately 4.5% of the mean of the environmental variance per generation. In generation 10, response to selection was 1.67 kits^2, which is 45% of the original mean, with a 95% confidence interval of [0.85, 2.53]. In a Bayesian context, several confidence intervals can be easily estimated. We can provide intervals [k, $+\infty$), where k can be interpreted as a guaranteed value with a determined probability [32]. The guaranteed value of the environmental variance at 80% probability was 1.32 kits^2, which means that the response was at least 1.32 kits^2 with 80%

Table 1 Descriptive statistics of the evaluated traits in the base population

	Mean	Median	SD	CV
V_e	3.73	2.72	3.36	0.90
V_r	3.96	3.13	3.55	0.90
SD_e	1.74	1.65	0.84	0.48
LS	8.71	9.00	3.01	0.35

SD, standard deviation; CV, coefficient of variation; V_e, environmental variance of litter size based on pre-corrected data; V_r, environmental variance of litter size based on uncorrected data; SD_e, environmental standard deviation of litter size based on pre-corrected data; LS, litter size

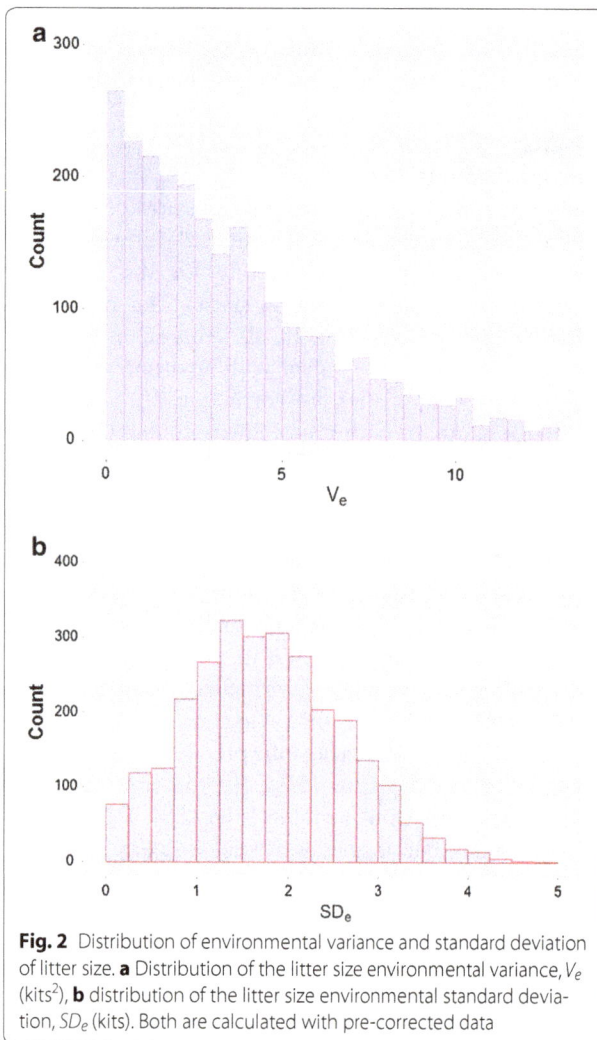

Fig. 2 Distribution of environmental variance and standard deviation of litter size. **a** Distribution of the litter size environmental variance, V_e (kits²), **b** distribution of the litter size environmental standard deviation, SD_e (kits). Both are calculated with pre-corrected data

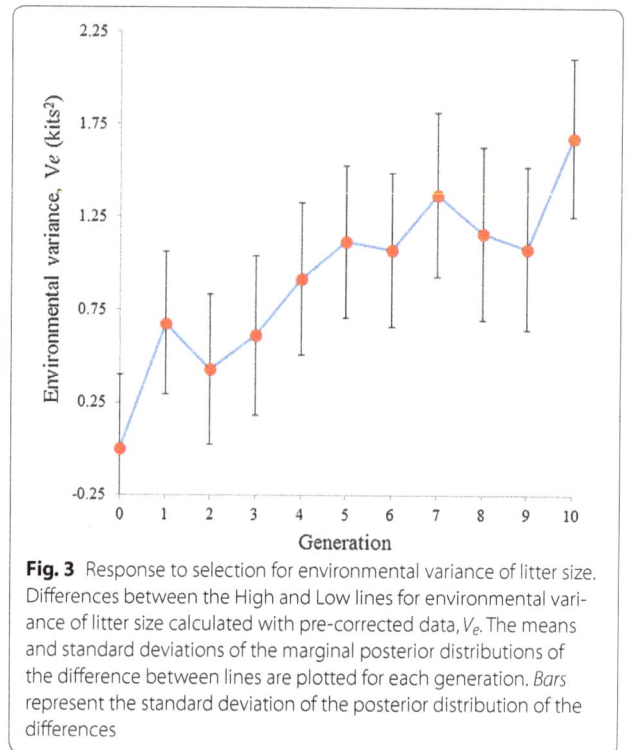

Fig. 3 Response to selection for environmental variance of litter size. Differences between the High and Low lines for environmental variance of litter size calculated with pre-corrected data, V_e. The means and standard deviations of the marginal posterior distributions of the difference between lines are plotted for each generation. *Bars* represent the standard deviation of the posterior distribution of the differences

probability. When the environmental variance was taken without pre-correcting data, the response in generation 10 was 1.74 kits², with a 95% confidence interval of [0.88, 2.61], and a guaranteed value at 80% probability of 1.36 kits², showing that pre-correction had a small effect. The average standard deviation (SD_e) had a response of 0.46 kits in generation 10, with a guaranteed value of 0.36 kits at 80% probability.

For each generation, the mean and standard deviation of the marginal posterior distributions of the differences between the High and Low lines are plotted in Fig. 3. Response to selection was higher in the first generation, likely due to the higher selection pressure applied (Table 2). In divergent selection experiments, the number of animals in the base generation is twice the size of each divergent line, and greater selection pressures can be applied. Response to selection in each line, which is derived from the estimated genetic means in each generation, is shown in Fig. 4, with the corresponding standard

deviations of the posterior distributions. Selection appeared to be more successful in increasing environmental variance than in decreasing it, which agrees with the lower selection differentials that could be applied in the Low line (Table 2). The differences in genetic means between lines are consistent with the phenotypic differences found in Fig. 3, which corroborates the model used.

Correlated response in litter size
Litter size was consistently larger in the Low line than in the High line throughout the experiment (Fig. 5). In the last generation of selection, the difference in litter size between the Low and High lines was 0.80 kits, with a 95% confidence interval of [0.34, 1.26] and a guaranteed value of 0.60 kits at 80% probability and 0.41 kits at 95% probability.

Genetic parameters
Heritabilities and genetic correlations of V_r and LS with V_e are in Table 3. The heritability of LS was low, as expected, but the heritability of V_e was also low; thus, the response to selection in V_e that was obtained should be attributed to its large variability (Table 1). The genetic correlation between litter size variance before and after pre-correction was near 1, which indicates that the impact of pre-correction on the genetic determination of this trait was small. The genetic correlation between V_e

Table 2 Weighted selection differentials for V_e (kits²) by generation

	High line	Low line
Base population	3.0	1.5
Generation 1	1.5	0.2
Generation 2	1.7	0.3
Generation 3	2.9	0.6
Generation 4	1.8	0.2
Generation 5	2.0	0.9
Generation 6	2.4	1.0
Generation 7	2.9	0.8
Generation 8	1.7	0.2
Generation 9	2.4	0.4

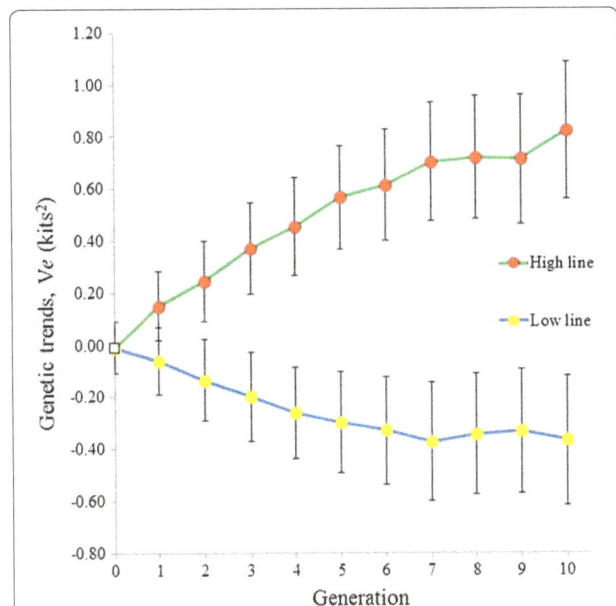

Fig. 5 Correlated response to selection in litter size. Differences in litter size between the High and Low lines. Means and standard deviations of the marginal posterior distributions of the difference between lines are plotted for each generation. *Bars* represent the standard deviation of the posterior distribution of the differences

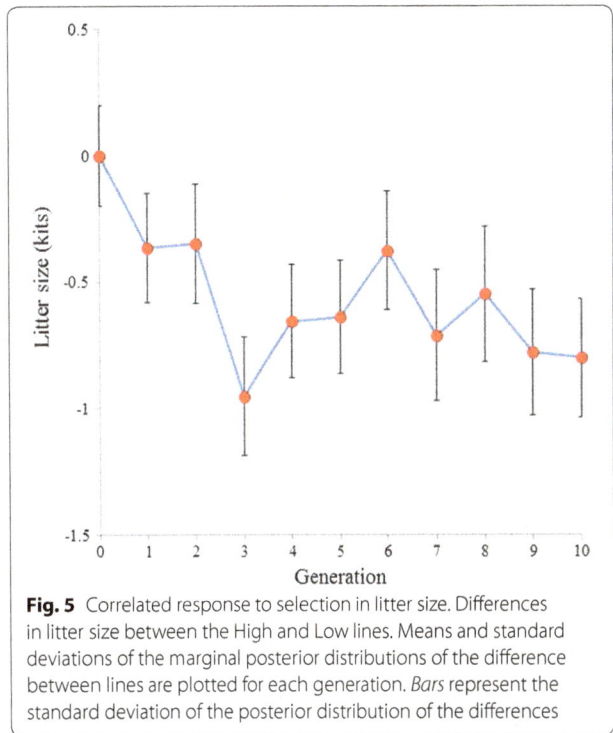

Fig. 4 Response to selection for environmental variance of litter size in the High and Low lines. Genetic means per generation of the litter size environmental variance calculated with pre-corrected data, V_e. *Bars* represent the standard deviation of the posterior distribution of the genetic means

and *LS* was almost null, which indicates that selection for homogeneity does not reduce litter size.

Discussion

There is some evidence in several species that environmental variance can be under genetic control, although only two selection experiments to investigate this have been performed, both using selection for birth weight, in mice [22] and rabbits [21]. A major problem in analyzing environmental variance comes from the complexity of the models that are often used, which are highly parametrized and have nested effects and parameters. Double

hierarchical generalized linear models [18, 33, 34] using residual maximum likelihood and Bayesian nested models [11, 15] have been proposed to analyze this problem. These models are not robust, as shown by Yang et al. [24], who compared genetic parameters after a Box–Cox transformation to normalize the residuals in litter size in pigs and uterine capacity in rabbits. These authors showed that the coefficient of correlation between the trait and its residual variance changed dramatically as a result of transformation, when compared to the results on the untransformed scale, even changing sign in the case of pig litter size. Here, we used a more straightforward and robust criterion for selection, the intra-doe phenotypic variance for litter size, which was considered as the observed environmental variance for litter size. Models as simple as those currently used for other observed traits can then be used to analyze response to selection.

Environmental variance of litter size was estimated as intra-doe phenotypic variance for litter size after pre-correction for season and parity-lactation status. This pre-correction was made under the hypothesis that systematic effects can affect environmental litter size variance of does; for example, a doe that has more parities during one season could have a smaller environmental variance than a doe that has parities across several seasons. The same could occur with the parity-lactation effect; it is well known that there is an effect of first parity

Table 3 Genetic parameters

	h^2	HPD95%	r_g	HPD95%	r_p	HPD95%
V_e	0.08	0.05, 0.11				
V_r	0.09	0.05, 0.13	0.99	0.97, 1.00	0.97	0.967, 0.972
LS	0.10	0.08, 0.13	−0.06	−0.31, 0.21	−0.09	−0.14, −0.03

h^2, heritability; HPD95%, high posterior density interval at 95%; r_g, genetic correlation with V_e; r_p, phenotypic correlation with V_e; V_e, environmental variance of litter size based on pre-corrected data; V_r, environmental variance of litter size based on uncorrected data; LS, litter size

on litter size when compared with other parities (Fig. 6); failing to consider this would cause overestimation of the environmental variance of females that have few parities. Nevertheless, in our data, these effects were so small that we would have obtained almost the same genetic response if these corrections had not been considered, since the genetic correlation between environmental variance with and without pre-corrected data was almost 1 (Table 3). Since the number of parities per doe was not large, variance estimators did not give the same result. We decided to estimate intra-doe variance using the best quadratic estimator; i.e. the one with the smallest risk.

Response to selection was estimated in two ways: as phenotypic differences between lines in each generation and as genetic trends from the estimated genetic means. All methods that are based on genetic means (best linear unbiased prediction—restricted maximum likelihood or Bayesian methods) are model-dependent, and the genetic trends depend directly on the genetic parameters used [35, 36]; for example, if the narrow-sense heritability is overestimated because dominant and epistatic components are not considered in the model, a higher genetic trend and a decreasing environmental trend will

be observed. The advantage of the simple phenotypic difference between the High and Low lines is that they are independent of any model; whether there are major genes, dominance or other effects, the difference between lines is only due to genetic causes, since they were bred and raised in the same environment. When the phenotypic differences are coincident with the estimates based on a genetic model, the genetic model is corroborated (in the Popper sense [37], i.e. the model has more support for the results obtained). Conversely, the advantage of using genetic means is that we can observe the evolution of the genetic means in each line separately. Resulting responses to selection by line (Fig. 4) indicated some asymmetry in responses, with selection appearing less successful in the Low line than in the High line. There are many reasons that can explain asymmetrical response to selection (for example, Falconer and MacKay [38] list eight different reasons); here, the trend towards more homogeneity in litter size tends to reduce the possibility of high selective pressure.

The line selected for low environmental variance of litter size resulted in larger litter size in all generations than the High line. Estimating the correlation between the mean and the variance of a trait has been the goal of several studies, with various results. A negative relationship between the mean of a trait and its environmental variance was detected for litter size in pigs [11, 34], for litter size and litter weight at birth in mice [14, 39], for weight gain in mice [40], for uterine capacity in rabbits [15], and for body weight in broiler chickens [41]. By contrast, no relationship between mean and environmental variance was found for slaughter weight in pigs [42] or for birth weight in rabbits [21, 43], and a positive correlation between mean and environmental variance was found for body weight in snails [13] and broiler chickens [16] and for body conformation in broiler chickens [16]. There has been some controversy about the validity of the analyses of genetic parameters when environmental variance is estimated with highly parametrized models, such as the model of San Cristobal et al. [44]. Yang et al. [24] showed that the negative genetic correlation between uterine capacity in rabbits and its residual variance reported by Ibáñez-Escriche et al. [15] became almost null when the residuals were normalized. In our case, the estimate of

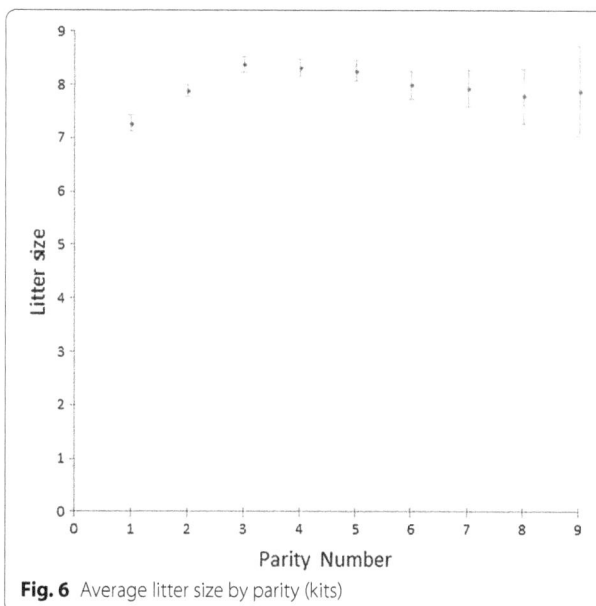

Fig. 6 Average litter size by parity (kits)

the genetic correlation between V_e and LS was almost null, which agrees with the results of Yang et al. [24] for uterine capacity in rabbits, a trait that is closely related to litter size [45]. However, as we have seen, litter size was consistently larger in the Low line than in the High line throughout the experiment, which is compatible with a low negative genetic correlation within the limits of the high posterior density interval at 95% (HPD95%). The important result is that selection for homogeneity does not seem to reduce litter size of does.

The line selected for litter size homogeneity also tolerated external stressors more effectively than the line selected for litter size heterogeneity. The High line had a higher subclinical immune response, which is related to a greater sensitivity to diseases or to less tolerance to common microorganisms in the farm microenvironment [46], and after vaccination, the Low line had a quicker and higher response to invading agents [6, 7]. Response to stress was also better in the Low line; after injection of the stressing agent adrenocorticotropic hormone, the High line had a higher cortisol level, thus a higher level of stress than the Low line. The High line also showed higher hepatic activity [8]. Thus, in general, the High line was more sensitive to stress and had a lower immune response to infections. This has consequences on disease resistance but also on animal welfare, since animals that cope more effectively with their environment have better welfare than animals that are more sensitive.

Conclusions

This is the first experiment of selection on the environmental variance of litter size and the first experiment in which selection has been directly performed on environmental variance as an observed trait. We conclude that the environmental variance of litter size is genetically determined, based on the result of our divergent selection experiment. This has consequences on animal welfare, since animals that cope better with their environment have better welfare than more sensitive animals. We also conclude that selection for reduced litter size variability does not depress litter size.

Authors' contributions
AB designed the selection experiment and wrote the paper; MJA and MLG carried out the selection experiment; MMA, NI, MLG and MJA analyzed the results; and all authors contributed to the discussion and interpretation of the results. All authors read and approved the final manuscript.

Author details
[1] Institute for Animal Science and Technology, Universitat Politècnica de València, Valencia, Spain. [2] Departamento de Tecnología Agroalimentaria, Universidad Miguel Hernández de Elche, Orihuela, Spain. [3] Genètica i Millora Animal, Institut de Recerca i Tecnologia Agroalimentàries, Caldes de Montbui, Spain.

Acknowledgements
We are grateful to Lois Bodin and Daniel Sorensen for useful comments in the design of the experiment and to Luis Varona for his help in the analyses.

Competing interests
The authors declare that they have no competing interests.

Funding
This research was funded by the Ministerio de Economía y Competitividad (Spain), Projects AGL2014-55921, C2-1-P and C2-2-P. Marina Martínez-Alvaro has a Grant from the same funding source, BES-2012-052655.

References
1. Morgante F, Sørensen P, Sorensen DA, Maltecca C, Mackay TFC. Genetic architecture of micro-environmental plasticity in Drosophila melanogaster. Sci Rep. 2015;5:9785.
2. Sørensen P, de los Campos G, Morgante F, Mackay TFC, Sorensen D. Genetic control of environmental variation of two quantitative traits of Drosophila melanogaster revealed by whole-genome sequencing. Genetics. 2015;201:487–97.
3. Zhang XS, Hill WG. Evolution of the environmental component of the phenotypic variance: stabilizing selection in changing environments and the homogeneity cost. Evolution. 2005;59:1237–44.
4. Mulder HA, Bijma P, Hill WG. Selection for uniformity in livestock by exploiting genetic heterogeneity of residual variance. Genet Sel Evol. 2008;40:37–59.
5. Bodin L, Bolet G, Garcia M, Garreau H, Larzul C, David I. Robustesse et canalisation, vision de généticiens. INRA Prod Anim. 2010;23:11–22.
6. García ML, Argente MJ, Muelas R, Birlanga V, Blasco A. Effect of divergent selection for residual variance of litter size on health status and welfare. In: Proceedings of the 10th World Rabbit Congress. Sharm El-Sheikh; 2012. p. 103–6.
7. Argente MJ, García ML, Zbynovska K, Petruska P, Capcarova M, Blasco A. Effect of selection for residual variance of litter size on hematology parameters as immunology indicators in rabbits. In: Proceedings of the 10th World Congress on genetics applied to livestock production. Vancouver; 2014.
8. García ML, Zbynovska K, Petruska P, Bovdisová I, Kalafová A, Capcarova M, et al. Effect of selection for residual variance of litter size on biochemical parameters in rabbits. In: Proceedings of the 67th annual meeting of the European Federation of Animal Science. Belfast; 2016.
9. Broom DM. Welfare assessment and relevant ethical decisions: key concepts. Annu Rev Biomed Sci. 2008;20:79–90.
10. SanCristobal-Gaudy M, Bodin L, Elsen JM, Chevalet C. Genetic components of litter size variability in sheep. Genet Sel Evol. 2001;33:249–71.
11. Sorensen D, Waagepetersen R. Normal linear models with genetically structured residual variance heterogeneity: a case study. Genet Res. 2003;82:207–22.
12. Mulder HA, Hill WG, Knol EF. Heritable environmental variance causes nonlinear relationships between traits: application to birth weight and stillbirth of pigs. Genetics. 2015;199:1255–69.
13. Ros M, Sorensen D, Waagepetersen R, Dupont-Nivet M, San Cristobal M, Bonnet JC. Evidence for genetic control of adult weight plasticity in the snail Helix aspersa. Genetics. 2004;168:2089–97.
14. Gutiérrez JP, Nieto B, Piqueras P, Ibáñez N, Salgado C. Genetic parameters for components analysis of litter size and litter weight traits at birth in mice. Genet Sel Evol. 2006;38:445–62.
15. Ibáñez-Escriche N, Sorensen D, Waagepetersen R, Blasco A. Selection for environmental variation: a statistical analysis and power calculations to detect response. Genetics. 2008;180:2209–26.
16. Wolc A, White IM, Avendano S, Hill WG. Genetic variability in residual variation of body weight and conformation scores in broiler chickens. Poult Sci. 2009;88:1156–61.
17. Fina M, Ibáñez-Escriche N, Piedrafita J, Casellas J. Canalization analysis of birth weight in Bruna dels Pirineus beef cattle. J Anim Sci. 2013;91:3070–8.
18. Mulder HA, Rönnegård L, Fikse WF, Veerkamp RF, Strandberg E. Estimation of genetic variance for macro- and micro-environmental sensitivity using double hierarchical generalized linear models. Genet Sel Evol. 2013;45:23.

19. Janhunen M, Kause A, Vehviläinen H, Järvisalom O. Genetics of microenvironmental sensitivity of body weight in rainbow trout (*Oncorhynchus mykiss*) selected for improved growth. PLoS One. 2012;7:e38766.

20. Sonesson AK, Ødegård J, Rönnegård L. Genetic heterogeneity of within-family variance of body weight in Atlantic salmon (*Salmo salar*). Genet Sel Evol. 2013;45:41.

21. Garreau H, Bolet G, Larzul C, Robert-Granie C, Saleil G, SanCristobal M, et al. Results of four generations of a canalising selection for rabbit birth weight. Livest Sci. 2008;119:55–62.

22. Pun A, Cervantes I, Nieto B, Salgado C, Pérez-Cabal MA, Ibáñez-Escriche N, et al. Genetic parameters for birth weight environmental variability in mice. J Anim Breed Genet. 2012;130:404–14.

23. Hill WG, Mulder HA. Genetic analysis of environmental variation. Genet Res (Camb). 2010;92:381–95.

24. Yang Y, Christensen OF, Sorensen D. Analysis of a genetically structured variance heterogeneity model using the Box–Cox transformation. Genet Res (Camb). 2011;93:33–46.

25. Piles M, Garcia ML, Rafel O, Ramon J, Baselga M. Genetics of litter size in three maternal lines of rabbits: repeatability versus multiple-trait models. J Anim Sci. 2006;84:2309–15.

26. Estany J, Baselga M, Blasco A, Camacho J. Mixed model methodology for the estimation of genetic response to selection in litter size of rabbits. Livest Prod Sci. 1989;21:67–75.

27. Box GEP, Tiao GC. Bayesian inference in statistical analysis. New York: Wiley; 1973.

28. Searle SR. Matrix algebra useful for statistics. Toronto: Wiley; 1982.

29. Sorensen D, Gianola D. Likelihood, Bayesian and MCMC methods in quantitative genetics. New York: Springer; 2002.

30. Geyer CM. Practical Markov chain Monte Carlo (with discussion). Stat Sci. 1992;7:467–511.

31. Legarra A. TM threshold model. 2008. http://genoweb.toulouse.inra.fr/~alegarra/tm_folder/. Accessed 02 May 2017.

32. Blasco A. Bayesian data analysis for animal scientists. New York: Springer; 2017.

33. Rönnegård L, Felleki M, Fikse F, Mulder HA, Strandberg E. Genetic heterogeneity of residual variance—estimation of variance components using double hierarchical generalized linear models. Genet Sel Evol. 2010;42:8.

34. Felleki M, Lee D, Lee Y, Gilmour AR, Rönnegård L. Estimation of breeding values for mean and dispersion, their variance and correlation using double hierarchical generalized linear models. Genet Res (Camb). 2012;94:307–17.

35. Thompson R. Estimation of realized heritability in a selected population using mixed model methods. Genet Sel Evol. 1986;18:475–84.

36. Sorensen DA, Johansson K. Estimation of direct and correlated responses to selection using univariate animal models. J Anim Sci. 1992;70:2038–44.

37. Popper K. The logic of scientific discovery. London: Hutchinson & Co; 1959.

38. Falconer DS, MacKay TFC. An introduction to quantitative genetics. 4th ed. Harlow: Longman Group Ltd; 1996.

39. Formoso-Rafferty N, Cervantes I, Ibáñez-Escriche N, Gutiérrez JP. Correlated genetic trends for production and welfare traits in a mouse population divergently selected for birth weight environmental variability. Animal. 2016;10:1770–7.

40. Ibáñez-Escriche N, Moreno A, Nieto B, Piqueras P, Salgado C, Gutiérrez JP. Genetic parameters related to environmental variability of weight traits in a selection experiment for weight gain in mice; signs of correlated canalised response. Genet Sel Evol. 2008;40:279–93.

41. Mulder HA, Hill WG, Vereijken A, Veerkamp RF. Estimation of genetic variation in residual variance in female and male broiler chickens. Animal. 2009;3:1673–80.

42. Ibáñez-Escriche N, Varona L, Sorensen D, Noguera JL. A study of heterogeneity of environmental variance for slaughter weight in pigs. Animal. 2008;2:19–26.

43. Bolet G, Garreau H, Hurtaud J, Saleil G, Esparbié J, Falieres J. Canalising selection on within litter variability of birth weight in rabbits: responses to selection and characteristics of the uterus of the does. In: Proceedings of the 9th World Rabbit Congress. Verona; 2008. p. 51–6.

44. San Cristobal-Gaudy M, Elsen JM, Bodin L, Chevalet C. Prediction of the response to a selection for canalisation of a continuous trait in animal breeding. Genet Sel Evol. 1998;30:423–51.

45. Argente MJ, Santacreu MA, Climent A, Blasco A. Genetic correlations between litter size and uterine capacity. In: Proceeding of the 8th World Rabbit Congress. Valencia; 2000. p. 333–38.

46. Rauw WM. Immune response from a resource allocation perspective. Front Genet. 2012;3:267.

10

Insight into the genetic composition of South African Sanga cattle using SNP data from cattle breeds worldwide

Sithembile O. Makina[1*], Lindsey K. Whitacre[2], Jared E. Decker[2], Jeremy F. Taylor[2], Michael D. MacNeil[1,3,4], Michiel M. Scholtz[1,3], Este van Marle-Köster[5], Farai C. Muchadeyi[6], Mahlako L. Makgahlela[1] and Azwihangwisi Maiwashe[1,3]

Abstract

Background: Understanding the history of cattle breeds is important because it provides the basis for developing appropriate selection and breed improvement programs. In this study, patterns of ancestry and admixture in Afrikaner, Nguni, Drakensberger and Bonsmara cattle of South Africa were investigated. We used 50 K single nucleotide polymorphism genotypes that were previously generated for the Afrikaner (n = 36), Nguni (n = 50), Drakensberger (n = 47) and Bonsmara (n = 44) breeds, and for 394 reference animals representing European taurine, African taurine, African zebu and *Bos indicus*.

Results and discussion: Our findings support previous conclusions that Sanga cattle breeds are composites between African taurine and *Bos indicus*. Among these breeds, the Afrikaner breed has significantly diverged from its ancestral forebears, probably due to genetic drift and selection to meet breeding objectives of the breed society that enable registration. The Nguni, Drakensberger and Bonsmara breeds are admixed, perhaps unintentionally in the case of Nguni and Drakensberger, but certainly by design in the case of Bonsmara, which was developed through cross-breeding between the Afrikaner, Hereford and Shorthorn breeds.

Conclusions: We established patterns of admixture and ancestry for South African Sanga cattle breeds, which provide a basis for developing appropriate strategies for their genetic improvement.

Background

South Africa is richly endowed with indigenous cattle breeds, among which are the Afrikaner, Nguni and Drakensberger breeds. These breeds played important roles in the social, cultural and economic development of the country [1]. Previously, Makina et al. [2] described these breeds as being genetically distinct from the European *Bos taurus* breeds (Angus and Holstein) and as having genomic regions associated with tropical adaptation [3]. Therefore, they may hold potential for production in harsh and fluctuating South African environments based on their adaptation to the nutritional, parasitic, and

pathogenic challenges they are faced with. These breeds are not endangered and have reasonable effective population sizes [4–6]. Given their adaptive characteristics, they are potentially valuable to breeding programs in other regions that face biological stresses such as famine, drought or disease epidemics [7]. Furthermore, there is a worldwide drive for the effective management of indigenous genetic resources, which includes these breeds [7].

Afrikaner and Nguni cattle were brought to Southern Africa by the Khoi-Khoi people who migrated southwards from the African Great Lakes region between 600 and 700 AD [1]. Summers [8] postulated that ancestors of Afrikaner cattle migrated very quickly along the eastern side of Southern Africa to the current Western Cape and western parts of the Northern Cape. Ancestors of Nguni cattle are believed to have moved southward in the

*Correspondence: somakina@gmail.com
[1] Agricultural Research Council-Animal Production Institute, Private Bag X 2, Irene 0062, South Africa
Full list of author information is available at the end of the article

African continent at a much slower pace [8]. Afrikaner, Nguni and Drakensberger are classified as Sanga cattle and are thought to result from crossbreeding between thoracic-humped Lateral Horned zebu and humpless Egyptian Longhorn cattle [9–11]. The initial admixture probably occurred when African taurine cattle migrated south from Egypt and the Sudan, and indicine cattle migrated to the eastern seaboard of Africa from Arabia and India [9, 11]. However, Bisschop [12] suggested that Sanga cattle originated from crosses between humpless Egyptian Longhorn and short-horned *B. taurus brachyceros*. The cross-section of the horns of Egyptian Longhorn are oval, which is similar to those of Afrikaner cattle, while those of *B. taurus brachyceros* are round as in Nguni cattle [12].

Based on analyses with microsatellite markers, Hanotte et al. [11] and Freeman et al. [13] predicted that Sanga cattle resulted from the crossbreeding of African taurine and zebu cattle around 700 AD, which was confirmed by studies based on single nucleotide polymorphisms (SNPs) [14–18]. However, in spite of these studies, the genetic composition of South African Sanga cattle remains uncertain [19]. The genetic distance between cattle breeds appears, at least in part, inversely related to the geographic proximity of their origin [20]. Hanotte et al. [11] and Freeman et al. [13] also found that the extent of genetic introgression of zebu cattle across the African continent decreases from eastern to western Africa. MacHugh et al. [21] reported that the cattle breeds from the tsetse-infested areas of West and Central Africa had limited or no zebu ancestry, which concurs with their susceptibility to trypanosomiasis [21].

Genomic characterisation of South African Sanga cattle is a first step towards the development of appropriate breeding and selection strategies for these breeds. Makina et al. [2] characterized the relationships between the Afrikaner, Nguni, Drakensberger and Bonsmara breeds using Angus and Holstein as reference breeds, without including any other indicine or African taurine cattle. The limited number of breeds analyzed precluded detection of patterns of co-ancestry or admixture in the South African Sanga breeds. Thus, the aim of our study was to provide a more precise analysis of patterns of admixture and ancestry in the Afrikaner, Nguni, Drakensberger and Bonsmara cattle of South Africa using a subset of data that were generated for cattle breeds worldwide [14–17].

Methods
Description of samples and quality control
Genotypes for four South African Sanga cattle breeds [Afrikaner—AFR (n = 36), Nguni—NGU (n = 50), Drakensberger—DRA (n = 47) and Bonsmara—BON

(n = 44)] originated from previous studies [2, 3, 6]. They were generated using the Illumina BovineSNP50 BeadChip v2, which features 54,609 SNPs distributed across the bovine genome with an average spacing of 49.9 kb [16]. These data were combined with genotypes from an additional 394 reference animals representing European taurine cattle i.e. Shorthorn (SH), Hereford (HFD), Simmental (SM), Limousin (LM), Angus (AN) and Holstein (HOL), African taurines i.e. N'Dama (NDAM), Somba (SOM), Kuri (KUR), Lagune (LAG) and Baoule (BAO), African zebu i.e. Ankole-Watusi (ANKW), Boran (BOR) and Sheko (SHK), East African zebu i.e. short-horned zebu (ZEB) and zebu Bororo (ZBO), and *Bos indicus* i.e. Brahman (BR), Nelore (NEL), Bhagnari (BAG) and Gir (GIR). Samples from these reference individuals were obtained with permission (see [22]) and were selected based on their land of origin and previous characterization. These samples originated from the following studies: Gautier et al. [14]; The Bovine HapMap Consortium [15]; Matukumalli et al. [16]; Decker et al. [17]; Gautier et al. [23]; and Decker et al. [24]. Additional file 1: Table S1 provides breed names and acronyms, number of individuals per breed, sampling area, land of origin and references to the original studies from which the samples came from.

These data were merged in PLINK [25] and autosomal SNPs that were common to all datasets were retained. This resulted in 35,155 SNPs and 548 individuals after removing SNPs with a MAF lower than 0.005, a call rate lower than 0.98 and individuals with more than 5% missing genotypes.

Genetic relationships and population structure
Patterns of admixture and relationships among South African Sanga cattle in relation to the 20 reference breeds were determined using principal component analysis [26] implemented in the SNP Variation suite (SVS 8.1; Golden Helix Inc., Bozeman, Montana) and variational Bayesian inference as implemented in fastSTRUCTURE [27]. The data were evaluated for K values ranging from 2 to 20 to evaluate ancestry proportions from K ancestral populations assuming a simple non-informative prior. The K_ε^* and $K_{\emptyset}c^*$ metrics from fastSTRUCTURE were used to determine the appropriate values of K for the population structure explained by the dataset. The K_ε^* metric is the value of K, which maximizes the log marginal likelihood lower bound and the $K_{\emptyset}c^*$ metric is the minimum value of K that explains almost all of the ancestry in the dataset. Outputs from fastSTRUCTURE [27] were plotted using the GENESIS software [28]. To further test for evidence of admixture in South African Sanga cattle, ancestry graph [29], three-population (f_3) [30, 31] and four-population (f_4) tests

[30, 32] implemented in TreeMix [29] were also used. The maximum likelihood tree (ancestry graph) [29] was first built for all 24 populations (see Additional file 2: Figure S1), after which, migration events were sequentially added to the tree until no more meaningful increases in the proportion of variance explained were observed (see Additional file 3: Table S2).

Results

Principal component analysis

The principal component assessment agreed with previous findings, which partitioned bovine breeds into three distinct groups representing European taurines, African taurines and indicines [14–18, 23, 24] (Fig. 1). Afrikaner and Nguni cattle were situated on the gradient between the indicine and African taurine breeds, but more towards the latter. The Bonsmara and Drakensberger breeds clustered towards the centre of the triangle, which suggests that they have three ancestries (European taurine, African taurine and indicine).

Population structure analysis

Allowing for three ancestral populations ($K = 3$) (Fig. 2) supported the classification of bovine populations into three distinct groups i.e. European taurine, African taurine and *Bos indicus*. This analysis predicted that the composition of Afrikaner cattle was approximately 70% African taurine and 28%, indicine, while that of Nguni was 60% African taurine, 30% indicine, and 10%

European taurine. Predicted compositions of Bonsmara and Drakensberger were 41 and 46% European taurine, 42 and 38% African taurine, and 16 and 15% indicine, respectively.

Increasing K from 3 to 5 assigned Afrikaner individuals into a single cluster and suggested that 97% of the Afrikaner genome was not shared with any of the reference breeds. The remaining 3% ancestral portion of the Afrikaner genome was shared with the African zebu breeds (1.6%) and with the African taurine and indicine reference breeds (<1%). Also at $K = 5$, Nguni, Drakensberger and Bonsmara remained admixed with a distinct genome component that was shared with African zebus (ZBO, ZEB, ANKW, SHK and BOR) and Kuri (a hybrid between African taurine and indicine populations [14]), but absent from indicines (BR, NEL, GIR, with the exception of BAG < 1%) and African taurines (NDAM, LAG, SOM and BAO). We note that the distinct component in BAG was only observed in a few individuals, which suggests that it may have been introduced through unsupervised crossbreeding.

Increasing K to 7 separated European taurines into breed clusters and confirmed that the genetic composition of Bonsmara included Afrikaner (40%), Shorthorn (33%), African zebu (19%) and Hereford (5%). Drakensberger shared ancestry with African zebu (31%), Holstein (32%), Afrikaner (20%), and Shorthorn, (13%). Nguni was predicted to be a hybrid between African zebu (60%) and Afrikaner (30%), with traces of indicine (3%), European

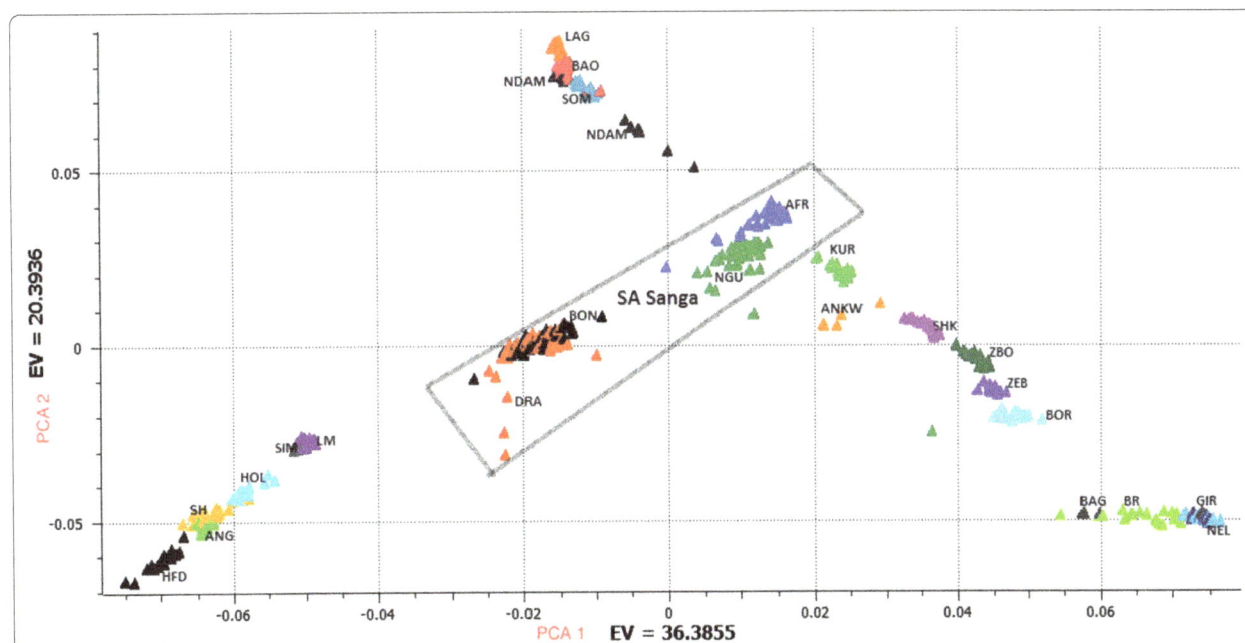

Fig. 1 Principal component analysis incorporating South African Sanga breeds into a set of 20 worldwide cattle breeds. For full definitions of breeds (see Additional file 1 Table S1). *EV* explained variance

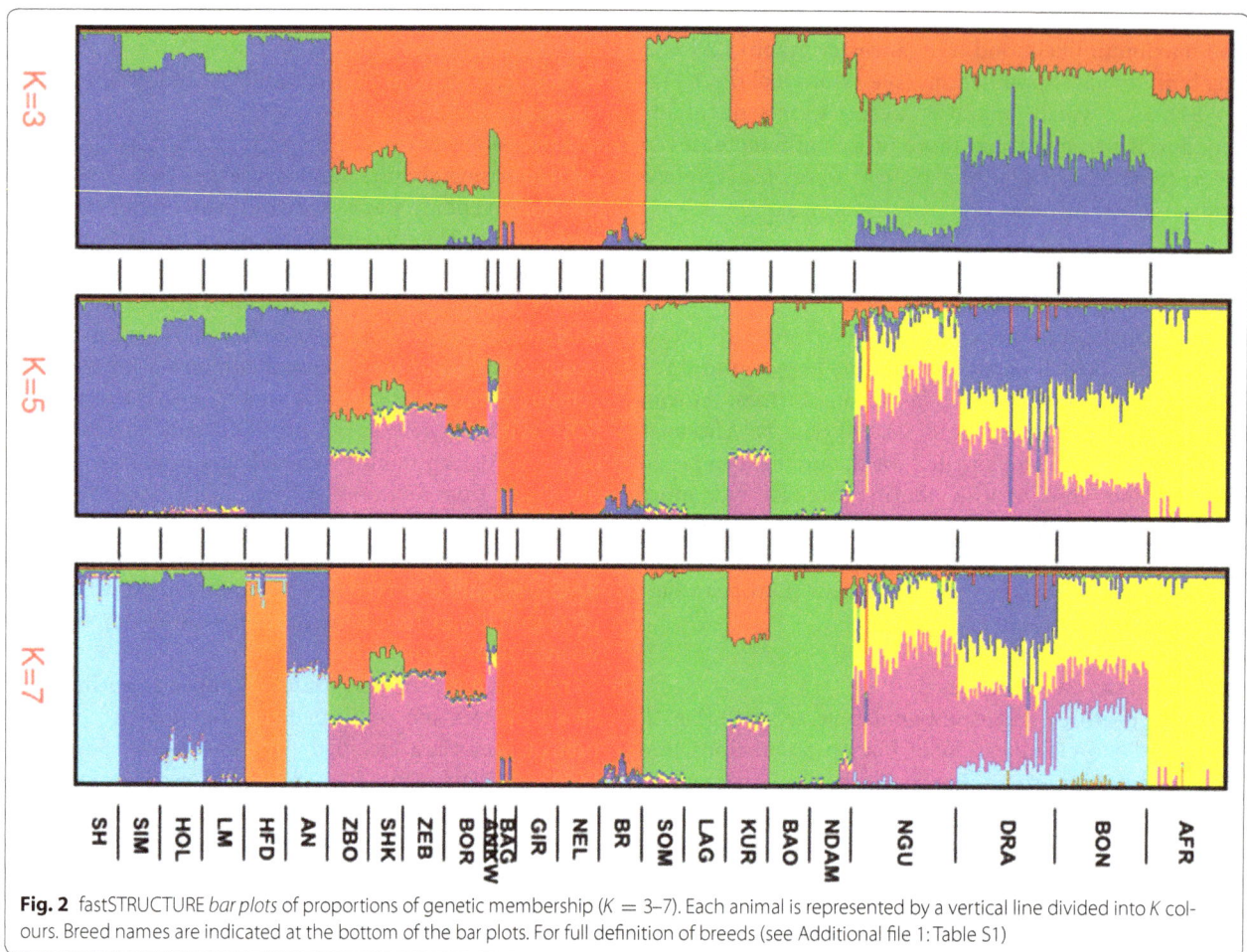

Fig. 2 fastSTRUCTURE *bar plots* of proportions of genetic membership ($K = 3$–7). Each animal is represented by a vertical line divided into K colours. Breed names are indicated at the bottom of the bar plots. For full definition of breeds (see Additional file 1: Table S1)

taurine (3%) and African taurine (2%) ancestry. Increasing K further towards the number of populations studied assigned Drakensberger ($K = 9$) and Nguni ($K = 11$) to discrete clusters.

Ancestry graph

The ancestry graph with 10 migration edges as developed using TreeMix [29] is in Fig. 3. This graph revealed the introduction of NDAM or NDAM relatives into the Nguni and Afrikaner cattle. This finding agrees with the results from the cluster analyses ($K = 3$), which indicated that the Afrikaner and Nguni cattle received approximately 60 to 70% of their ancestry from African taurine cattle (Fig. 2). In addition, we observed an admixture edge between Shorthorn and Bonsmara that was consistent with the history of the development of this breed. The eight other admixture processes modelled by network edges were previously characterized by Gautier et al. [14]; The Bovine HapMap Consortium [15]; Matukumalli et al. [16]; Decker et al. [17]; Gautier et al. [23]; and Decker et al. [24].

Formal tests of admixture

The three-population statistic f_3(A;B,C) [30, 31] tests for bifurcating tree-like relationships in the evolution of populations and significant negative values of the f_3 statistic imply that population A is admixed and is a mixture of populations related to B and C [30, 31]. In agreement with results from the cluster analyses, we detected strong evidence of admixture in Drakensberger, Bonsmara and Nguni cattle. Examining Nguni in conjunction with any of the populations related to African taurines, indicines, African zebus, European taurines or Afrikaner yielded significant tests, for example, f_3[NGU;NEL,LAG] (Z-score $= -18.00$); f_3[NGU;BOR,LAG] (Z-score $= -7.15$); f_3[NGU;AN,AFR] (Z-score $= -2.81$) and f_3[NGU;SIM,AFR] (Z-score $= -2.11$). Similarly, significant negative values were detected for f_3 statistics for trios of f_3[BON;AFR,SH] (Z-score $= -26.89$) and f_3 [BON;HFD,AFR] (Z-score $= -14.93$). The three population test for Drakensberger revealed that the most significant Z-scores were f_3[DRA;AFR,SH] (Z-score $= -15.28$); f_3[DRA;HOL,AFR] (Z-score $= -12.98$) and f_3

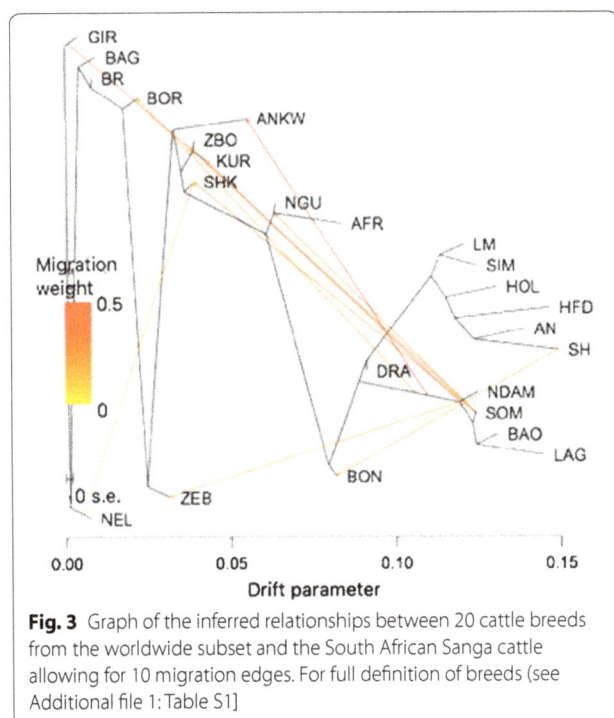

Fig. 3 Graph of the inferred relationships between 20 cattle breeds from the worldwide subset and the South African Sanga cattle allowing for 10 migration edges. For full definition of breeds (see Additional file 1: Table S1)

[DRA;ANG,AFR] (Z-score = −12.02). The f_3 statistic failed to detect admixture within the Afrikaner cattle and agreed with the cluster analyses at $K = 5$ as described above.

The four population test f_4(A,B;C,D) [30, 32] tests whether (A, B) and (D, C) represent distinct clades in a population tree. In this test, significant non-zero values indicate the presence of gene flow between the (A, B) and (C, D) groups in the tree [30, 32]. In this test, combining Afrikaner and any of the African zebu or African taurine breeds yielded the most significant values (see Additional file 4: Table S3), which suggested gene flow from African zebu breeds or African taurines into the Afrikaner cattle. For Nguni, Bonsmara and Drakensberger, the most significant non-zero values were obtained when they were combined with any of the indicine, European taurine, African taurine and African zebu breeds (see Additional file 4: Table S3), which indicates the presence of gene flow between these breeds.

Discussion

This study aimed at unravelling patterns of ancestry and admixture in South African Sanga cattle. Cluster analysis ($K = 3$) and the principal component analysis were mutually supportive and highlighted the presence of three genetic backgrounds for the populations studied. The classification of the bovine populations into a triangle-like structure is well-established [14–18, 23, 24]. As already mentioned, the Nguni and Afrikaner cattle were observed on the gradient between the indicines and African taurines, but more towards the African taurines, which indicates that the genomes of these breeds include more African taurine genetic background than indicine genetic background. This was supported by the proportions inferred by the admixture analysis at $K = 3$, which indicated that about 70 and 60% of the Afrikaner and Nguni ancestries were derived from African taurines. The detection of a migration edge from NDAM or NDAM relatives into Nguni and Afrikaner cattle in the TreeMix analysis provided further support. This larger proportion of ancestry shared between African taurines and South African Sanga (Afrikaner and Nguni) at $K = 3$ was consistent with the hypothesis of selection against susceptibility to trypanosomiasis, which may have led to a reduction in the extent of indicine ancestry in the surviving cattle, since indicine cattle are susceptible to trypanosomiasis [21, 33, 34].

As expected, Bonsmara and Drakensberger clustered towards the centre of the triangle in the principal component analysis, which suggests that these breeds have three ancestries (European taurine, African taurine and indicine). This was supported by the results of the cluster analysis at $K = 3$, which indicated that the Bonsmara and Drakensberger breeds were 41 and 46% European taurine, 42 and 38% African taurine, and 16 and 15% indicine, respectively. This was further confirmed by the detection of strong evidence for admixture by the three and four population tests (f_3 [30, 31] and f_4 [30, 32]) when these breeds were examined in conjunction with European taurines, African taurines and indicines. These results are consistent with the histories of the development of these breeds. The Bonsmara breed was developed at the Mara and Messina research stations between 1937 and 1963 under the guidance of the late Professor Jan C. Bonsma [35]. The aim was to produce a local breed that was well adapted to sub-tropical environments and had superior production compared to Afrikaner cattle. Five types of crosses were tested with Afrikaner cattle that included Red Aberdeen Angus, Hereford, Red Poll, Shorthorn and Sussex breeds. Pilot trials revealed that breed compositions including 5/8 Afrikaner and 3/8 Hereford or Shorthorn ancestries resulted in increased calving percentages and weaning weights with reduced calf mortalities relative to the purebreds [36]. Introgression of European taurine into the Drakensberger breed could have occurred due to the association of this breed with European settlers [1]. In 1837, several "Voortrekker" families (settlers) left Cape Province and traveled north with herds of similar black oxen that were then called "Vaderland" cattle. Most of these trekkers settled along the Drakensberg Mountain range and among them, the Uys family is believed to have played a significant role in the development of the "Uys-cattle" through strong selection within their closed herd; the "Uys-cattle" was later referred to as the Drakensberger breed by the Department of Agriculture owing to their prevalence in the pastures of the Drakensberg Mountains [36].

Allowing for five ancestral populations partitioned the African taurine and indicine admixture ancestry that was observed in the Nguni, Drakensberger and Bonsmara breeds ($K = 3$). With the exception of the African zebus (ZBO, ZEB, ANKW, SHK and BOR) and the Kuri (hybrid between African taurine and indicine populations [14]), this component was unique to the Nguni, Drakensberger and Bonsmara breeds and absent in indicine (BR, NEL, GIR) and African taurine (NDAM, LAG, SOM and BAO) cattle. Thus, the indicine component of African zebus and Sanga cattle seems to differ from that observed in other modern indicine cattle (NEL, BR and GIR). We hypothesize that strong selection within the African context caused the significant divergence between the indicine genome present in African admixed cattle and the genome of other modern indicine cattle. Alternatively, the founders of indicine cattle that migrated into Africa may have differed from those of the modern indicines used in the analysis.

Afrikaner cattle appear to have diverged from their ancestral populations and are recognized as a distinct breed ($K \geq 5$), which is likely due to the effects of genetic drift after admixture and strong selection of animals to conform to the standards and breeding objectives of the breed society. Afrikaner is the oldest indigenous cattle breed in South Africa and was the first indigenous breed to form a breed society in 1902 [1]. These results are consistent with the higher levels of inbreeding postulated by Coetzer and Van Marle [37] and detected by Makina et al. [2].

In agreement with the f_3 statistics, the Nguni cattle were predicted to be admixed ($K = 5$ and 7), their genetic makeup being predominantly African zebu with traces of indicines, African taurines and European taurines ancestries. The higher proportion of African zebu ancestry within the Nguni cattle is in agreement with the previous report by Makina et al. [3], who detected shared signatures of selection between Nguni and African zebu cattle. Historically, Nguni cattle were reared in extensive communal grazing systems in the presence of numerous other cattle representing various breeds and their crosses [38], which may also explain their admixture. The production potential of Nguni was only recognized in the early 1980s after the introduction of beef cattle recording schemes and the publication of results on the characterization of their productivity [39]. The Nguni breed society was established in 1986 and prior to this date, Nguni cattle were bred for various practical

purposes and mated at random, which potentially led to admixture due to their close association with the indigenous people of South Africa and the communal husbandry they practiced [38].

In summary, our analyses support the view that Sanga cattle are composites of African taurine and *Bos indicus* [9, 11]. The Afrikaner breed clearly diverged from its ancestral forebears, probably due to genetic drift and alternative breeding objectives. The Nguni, Drakensberger and Bonsmara breeds are admixed, which was perhaps unintentional for Nguni and Drakensberger, but was certainly done by design in the case of Bonsmara that was developed through crossbreeding of Afrikaner, Hereford and Shorthorn.

Conclusions

This study presents a comprehensive genome-wide characterization of South African Sanga cattle and confirms that South African Sanga cattle originated from African taurine and *Bos indicus*. The hybrid origin of Bonsmara cattle was confirmed and is consistent with the history of its development. Thus, genome-wide characterization of these populations has accurately recapitulated the history of the breeds' formation [16]. These results improve our understanding of the composition and origins of South African Sanga cattle.

Additional files

Additional file 1: Table S1. Description of samples included in the analyses. This table provides information about the samples used in this study such as the sample size, samples origin, land of origin and the references where the sample were previously characterized.

Additional file 2: Figure S1 Graph of the inferred relationships between 24 cattle breeds without adding migration edges. For full definition of breeds see Additional file 1: Table S1. This graph shows the genetic relationship of South African cattle breeds in relation to the cattle breeds of the world before migration edges were sequentially added.

Additional file 3: Table S2. Explained covariance between populations for the ancestral graph with 10 migration edges. This data shows the changes in explained covariance between the populations as the number of migration edges added to the ancentral graph increases. It shows that after adding 10 migration edges, no significant changes were oberved in explained covariance.

Additional file 4: Table S3. Some significant f_4 statistics for South African Sanga cattle. This data shows some of the significant f_4 statistics and indicates the presence of gene flow between these breeds.

Authors' contributions

MDM, SOM, and JED conceived the study in response to a comment on a previous paper [2]. The study was refined with additional inputs from LKW, JED and JFT who together with SOM conducted the statistical analyses and drafted the initial manuscript. EVM and MMS provided historical background on the Sanga breeds in South Africa. All authors provided insights leading to improved interpretation of the results and added critically important scientific content to the manuscript. All authors read and approved the final manuscript.

Author details
[1] Agricultural Research Council-Animal Production Institute, Private Bag X 2, Irene 0062, South Africa. [2] Division of Animal Sciences, University of Missouri, Columbia, MO 65211, USA. [3] Department of Animal, Wildlife and Grassland Sciences, University of Free State, Bloemfontein 9300, South Africa. [4] Delta G, Miles City, MT 59301, USA. [5] Department of Animal and Wildlife Sciences, University of Pretoria, Private Bag X 20, Hatfield 0028, South Africa. [6] Agricultural Research Council-Biotechnology Platform, Private Bag X 5, Onderstepoort 0110, South Africa.

Acknowledgements
The authors thank the South African breeders and research institutions for providing blood and hair samples from their animals. We thank Ms Pienaar who provided hair samples for Afrikaner cattle. Financial support from the ARC, Red Meat Research and Development Trust and National Research Council –THRIP of South Africa is greatly appreciated. The authors would like to acknowledge the reviewers for constructive inputs to this work.

Competing interests
The authors declare that they have no competing interests.

References

1. Scholtz MM. Beef breeding in South Africa. 2nd ed. Pretoria: Agricultural Research Council; 2010.
2. Makina SO, Muchadeyi FC, van Marle-Köster E, MacNeil MD, Maiwashe A. Genetic diversity and population structure among six cattle breeds in South Africa using a whole genome SNP panel. Front Genet. 2014;5:333.
3. Makina SO, Muchadeyi FC, van Marle-Köster E, Taylor JF, Makgahlela L, Maiwashe A. Genome wide scan for signatures of selection among six cattle breeds in South Africa. Genet Sel Evol. 2015;47:92.
4. Tada O, Muchenje V, Dzama K. Effective population size and inbreeding rate of indigenous Nguni cattle under in situ conservation in the low-input communal production system. S Afr J Anim Sci. 2013;43:137–42.
5. Pienaar L, Neser FWC, Grobler JP, Scholtz MM, MacNeil MD. Pedigree analysis of the Afrikaner cattle breed. Anim Genet Resour. 2015;57:51–6.
6. Makina SO, Taylor JF, van Marle-Köster E, Muchadeyi FC, Makgahlela ML, MacNeil MD, et al. Extent of linkage disequilibrium and effective population size in four South African Sanga cattle breeds. Front Genet. 2015;6:337.
7. Hanotte O, Jianlin H. Genetic characterization of livestock populations and its use in conservation decision making. In: Ruane J, Sonnino A, editors. The role of biotechnology in exploring and protecting agricultural genetic resources. Rome: FAO; 2005. p. 89–96.
8. Summers R. Environment and culture in southern Rhodesia: a study in the "personality" of a land-locked country. Proc Am Philos Soc. 1960;3:266–92.
9. Curson HH, Thornton RW. A contribution to the study of African native cattle. Onderstepoort J Vet Sci Anim Ind. 1936;7:613–739.
10. Epstein H. The origin of the domestic animals of Africa. Vol. 1: dog, cattle, buffalo. New York: Africana Publishing Corporation; 1971.
11. Hanotte O, Bradley DG, Ochieng JW, Verjee Y, Hill EW, Rege JEO. African pastoralism: genetic imprints of origins and migrations. Science. 2002;296:336–9.
12. Bisschop JHR. Parent stocks and derived types of African cattle, with particular reference to the importance of conformational characteristics in their study of origin. S Afr J Sci. 1937;33:853–70.
13. Freeman AR, Bradley DG, Nagda S, Gibson JP, Hanotte O. Combination of multiple microsatellite data sets to investigate genetic diversity and admixture of domestic cattle. Anim Genet. 2006;37:1–9.
14. Gautier M, Flori L, Riebler A, Jaffrézic F, Laloé D, Gut I, et al. A whole genome Bayesian scan for adaptive genetic divergence in West African cattle. BMC Genomics. 2009;10:550.
15. Bovine HapMap Consortium, Gibbs RA, Taylor JF, Van Tassel CP, Barendse W, Eversole KA, et al. Genome-wide survey of SNP variation uncovers the genetic structure of cattle breeds. Science. 2009;324:528–32.

16. Matukumalli LK, Lawley CT, Schnabel RD, et al. Development and characterization of a high density SNP genotyping assay for cattle. PLoS One. 2009;4:e5350.
17. Decker JE, Pires JC, Conant GC, McKay SD, Heaton MP, Chen K, et al. Resolving the evolution of extant and extinct ruminants with high-throughput phylogenomics. Proc Natl Acad Sci USA. 2009;106:18644–9.
18. Mbole-Kariuki MN, Sonstegard T, Orth A, Thumbi SM, Bronsvoort BM, Kiara H, et al. Genome-wide analysis reveals the ancient and recent admixture history of East African Shorthorn Zebu from Western Kenya. Heredity (Edinb). 2014;113:297–305.
19. Scholtz MM, Gertenbach W, Hallowell G. History and origin of Nguni cattle. In: Scholtz MM, Gertenbach W, Hallowell G, editors. The Nguni breed of cattle: past, present and future. Pretoria: The Nguni Cattle Breeders' Society; 2011. p. 29–36.
20. McKay SD, Schnabel RD, Murdoch BM, Matukumalli LK, Aerts J, Coppieters W, et al. An assessment of population structure in eight breeds of cattle using a whole genome SNP panel. BMC Genet. 2008;9:37.
21. MacHugh DE, Shriver MD, Loftus RT, Cunningham P, Bradley DG. Microsatellite DNA variation and the evolution, domestication and phylogeography of taurine and Zebu cattle (Bos taurus and Bos indicus). Genetics. 1997;146:1071–86.
22. Decker JE, McKay SD, Rolf MM, Kim JW, Molina Alcalá A, Sonstegard TS, et al. Data from: Worldwide patterns of ancestry, divergence, and admixture in domesticated cattle. Dryad Digital Repository; 2014. doi:10.5061/dryad.th092. Accessed 20 Oct 2015.
23. Gautier M, Laloë D, Moazami-Goudarzi K. Insights into the genetic history of French cattle from dense SNP data on 47 worldwide breeds. PLoS One. 2010;5:e13038.
24. Decker JE, McKay SD, Rolf MM, Kim J, Molina Alcalá A, Sonstegard TS, et al. Worldwide patterns of ancestry, divergence, and admixture in domesticated cattle. PLoS Genet. 2014;10:e1004254.
25. Purcell S, Neale B, Todd-Brown K, Thomas L, Ferreira MA, Bender D, et al. PLINK: a tool set for whole-genome association and population-based linkage analyses. Am J Hum Genet. 2007;81:559–75.
26. Patterson N, Price AL, Reich D. Population structure and eigenanalysis. PLoS Genet. 2006;2:e190.
27. Raj A, Stephens M, Pritchard JK. fastSTRUCTURE: variational inference of population structure in large SNP data sets. Genetics. 2014;197:573–89.
28. Buchmann R, Hazelhurst S. Genesis manual. Technical Report 2015. University of the Witwatersrand. http://www.bioinf.wits.ac.za/software/genesis/Genesis.pdf.
29. Pickrell JK, Pritchard JK. Inference of population split and admixture from genome wide allele frequency data. PLoS Genet. 2012;8:e1002967.
30. Patterson N, Moorjani P, Luo Y, Mallick S, Rohland N, Zhan Y, et al. Ancient admixture in human history. Genetics. 2012;192:1065–93.
31. Reich D, Thangaraj K, Patterson N, Price AL, Singh L. Reconstructing Indian population history. Nature. 2009;461:489–94.
32. Keinan A, Mullikin JC, Patterson N, Reich D. Measurement of the human allele frequency spectrum demonstrates greater genetic drift in East Asians than in Europeans. Nat Genet. 2007;39:1251–5.
33. Gifford-Gonzalez D, Hanotte O. Domesticating animals in Africa: implications of genetic and archaeological findings. J World Prehist. 2011;24:1–23.
34. Mwai O, Hanotte O, Kwon YJ, Cho S. African indigenous cattle: unique genetic resources in a rapidly changing world. Asian Australas J Anim Sci. 2015;28:911–21.
35. Bonsma JC. Cross-breeding, breed creation and the genesis of the Bonsmara. In: Livestock production: a global approach. Tafelberg, Cape Town: Publishers Ltd; 1980. p. 126–36.
36. Dreyer CJ. The breed structure of the Drakensberger cattle breed and factors that influence the efficiency of production. Pretoria: University of Pretoria; 1982.
37. Coetzer WA, Van Marle J. Die voorkoms van puberteit en daaropvolgende estrusperiode by vleisrasverse. S Afr J Anim Sci. 1972;2:17–8.
38. Hofmeyr JH. Findings of the committee on a gene bank for livestock. In: Proceedings of the conference on conservation of early domesticated animals of southern Africa: 3–4 March 1994; Pretoria. 1994.
39. Scholtz MM, Ramsey KA. Establishing a herd book for the Nguni breed in South Africa. In: Scholtz MM, Gertenbach W, Hallowell G, editors. The Nguni breed of cattle: past, present, and future. Pretoria: Nguni Cattle Breeders' Society; 2011. p. 22–5.

11

The 1.78-kb insertion in the 3′-untranslated region of *RXFP2* does not segregate with horn status in sheep breeds with variable horn status

Gesine Lühken[1], Stefan Krebs[2], Sophie Rothammer[3], Julia Küpper[1], Boro Mioč[4], Ingolf Russ[5] and Ivica Medugorac[3*] ●iD

Abstract

Background: The mode of inheritance of horn status in sheep is far more complex than a superficial analysis might suggest. Observations, which were mostly based on crossbreeding experiments, indicated that the allele that results in horns is dominant in males and recessive in females, and some authors even speculated about the involvement of more than two alleles. However, all recent genome-wide association analyses point towards a very strong effect of a single autosomal locus on ovine chromosome 10, which was narrowed down to a putatively causal insertion polymorphism in the 3′-untranslated region of the *relaxin/insulin-like family peptide receptor 2* gene (*RXFP2*). The main objective of this study was to test this insertion polymorphism as the causal mutation in diverse sheep breeds, including breeds with a variable and/or sex-dependent horn status.

Results: After re-sequencing a region of about 246 kb that covered the *RFXP2* gene and its flanking regions for 24 sheep from six completely horned and six completely polled breeds, we identified the same insertion polymorphism that was previously published as segregating with horn status in these breeds. Multiplex PCR genotyping of 489 sheep from 34 breeds and some crosses between sheep breeds showed a nearly perfect segregation of the insertion polymorphism with horn status in sheep breeds of Central and Western European origin. In these breeds and their crossings, heterozygous males were horned and heterozygous females were polled. However, this segregation pattern was not, or at least not completely, reproducible in breeds with sex-dependent and/or variable horn status, especially in sheep that originated from even more southern European regions and from Africa. In such breeds, we observed almost all possible combinations of genotype, sex and horn status phenotype.

Conclusions: The 1.78-kb insertion polymorphism in the 3′-untranslated region of *RXFP2* and SNPs in the 3′-UTR, exon 14 and intron 11 of this gene that we analyzed in this study cannot be considered as the only cause of polledness in sheep and are not useful as a universal marker to define the genetic horn status in sheep.

Background

In sheep, horn status is influenced by sex and varies between breeds. Castle [1] categorized sheep breeds into three types, i.e. (1) both sexes carry horns but those of the females are much smaller (similar to the horn status of wild sheep in Central Asia); (2) males have well-developed horns, females are polled (similar to the horn status of most mouflons that originate from Sardinia [2]); and (3) both sexes are polled (this is the case for the majority of domestic sheep breeds). However, regarding the horn status, many sheep breeds do not fall into these three categories. For example, in several breeds such as Soay, Bündner Oberland and Sakiz, and also in the mouflons

*Correspondence: ivica.medjugorac@gen.vetmed.uni-muenchen.de
[3] Chair of Animal Genetics and Husbandry, LMU Munich, Veterinaerstrasse 13, 80539 Munich, Germany
Full list of author information is available at the end of the article

that originate from Corsica [2, 3], males are strictly horned and females may or may not be horned, while the reverse is observed in other breeds, e.g. Altamurana and Red Karaman, with females being strictly polled and males having or not having horns. Finally, there are some breeds in which the occurrence of horns varies both in males and females, for example in some strains of Steinschaf and of Pramenka such as Travnička Pramenka [4].

Another important and complicating feature of the horn phenotype in some breeds of sheep is the development of knobs and scurs, which is sex-dependent and breed-dependent. Warwick and Dunkle [5] describe knobs as protrusions from the scull that resemble horn cores, except that they are usually less than 2.5 cm (1 in.) high and covered with skin. Scurs have a horn-like covering but are smaller than normal horns and irregular in shape. According to these authors, knobs and scurs are observed in females of Merino-type breeds (including the Rambouillet breed) in which males are horned and females are polled. In contrast, in breeds in which both sexes are polled (e.g. Shropshire, Southdown, and Suffolk), depressions in the skull instead of horn cores are observed in both males and females. It should be mentioned at this point that, in some publications, animals with knobs or scurs are referred to as "horned", whereas in others, they are referred to as "polled".

The mode of inheritance of the horn phenotype in sheep is far more complex than a superficial analysis might suggest. Already more than 100 years ago, several studies showed that in crosses between Dorset Horn (a breed in which both sexes are strictly horned) and completely hornless (polled) breeds [6–8], only the male offspring inherited the horned phenotype. Based on these observations, Wood [7] stated that horns are dominant in male sheep and recessive in female sheep. Results from subsequent studies in the same and other breeds also suggested that the effects of the horned and polled alleles differed between male and female sheep, e.g. [9–13]. More recently, Johnston et al. [14] concluded that in the Soay breed, the mode of inheritance of horns was additive in ewes and dominant in rams.

In 1912, Arkell and Davenport [15] suggested the existence of a sex chromosome-linked inhibitor of horn development, but this hypothesis did not gain common acceptance. In a back-cross between one F_1 Dorset × Rambouillet ram and Merino and Rambouillet ewes that carried knobs, Warwick and Dunkle [5] observed eight individuals with horns and 14 with knobs among the female progeny, but no individual showed depressions in the skull instead of horn cores. Because of this absence of polled sheep with depressions, they concluded that the three genes responsible for the absence of horns or polledness i.e. H, for Dorset horns i.e. H'

(responsible for horns in strictly horned breeds), and for Merino (and Rambouillet) horns i.e. h (responsible for sex-dependent horns), are not independent genes but three alleles at one locus. In the Merino breed, polledness was observed to be produced by an incompletely dominant gene, P. Polled Merinos were supposed to be PP or Pp, while Merino rams with horns and ewes with knobs or short scurs were supposed to be pp [15]. Dolling [16] suggested that a third gene, P', either an allele of P/p or closely linked to P and p, was also involved in the horn phenotype since Peppin Merino ewes carry horns although in Merino breeds ewes are polled. Based on the results from a series of crosses between Boder Leicester and Australian Merino sheep, Dolling [9] concluded that the genes that cause polledness in these two breeds were either closely-linked alleles of two loci or were identical and favored the last interpretation.

In 1996, Montgomery et al. [17] showed that the horn status in sheep is controlled by a locus on the proximal end of sheep chromosome OAR10 (OAR for *Ovis aries* chromosome), and since then several mapping studies have confirmed and gradually narrowed down the location of the responsible region [18–20]. As a preliminary result, Pickering et al. [20] reported that an approximately 3-kb retrotransposed insertion in the 3' untranslated region (UTR) of an unnamed candidate gene was present only in polled animals. Several whole-genome association analyses using the Illumina ovine 50K SNP (single nucleotide polymorphism) chip to genotype different sheep breeds identified several SNPs that were strongly associated with horn status and located between positions 29.36 and 29.51 Mb on OAR10 (the positions refer to the sheep genome assembly Oar_v3.1 and correspond with a region on chromosome 10 (NC_019467.2) between 29.34 and 29.49 Mb of the sheep genome assembly Oar_v4.0); some of these SNPs were immediately adjacent to or even located within the *relaxin/insulin-like family peptide receptor 2* gene (*RXFP2*) [14, 21, 22]. SNP OAR10_29511510.1 was the third most strongly associated SNP in the study of Johnston et al. [14] and is located in intron 11 of *RXFP2*.

Sequence analyses of the promoter region and of all the exons and their boundaries of the *RXFP2* gene on DNA samples from male and female Tan sheep (that include horned males and horned, scurred or polled females) and from males and females of the strictly polled Suffolk breed allowed the detection of several additional novel SNPs [23]. Four of these SNPs were located in the coding sequence of *RXFP2* and two of these caused amino acid substitutions. None of the SNPs segregated with horn status through a Mendelian mode of inheritance in either of these breeds. However, one synonymous SNP in exon 14 (c.1125A>G) was found to be a potential indicator of the presence or absence of horns in Tan sheep [23].

Initially, the objective of our study was to test the association between horn status in sheep and the 3-kb insertion polymorphism that was previously identified by Pickering et al. [20] but not described in more detail and to prove the need for developing a mapping design across a wide range of sheep breeds. In order to retrieve the 'Pickering' insertion and to test for its postulated mode of inheritance and the possible existence of additional polymorphisms in the target region, we re-sequenced a 246-kb region that included the *RFXP2* gene and its flanking regions in sheep from strictly horned and strictly polled breeds. During the course of our work, Wiedemar and Drögemüller [24] in a study on several Swiss sheep breeds rediscovered most possibly the same insertion and showed that it was associated with horn status. Once our sequencing results had confirmed the results of Pickering et al. [20] and of Wiedemar and Drögemüller [24], our primary aim was to test the segregation of this insertion polymorphism in several sheep breeds and in crosses between breeds with different horn statuses, in particular breeds with a variable and/or sex-dependent horn status. Additional SNPs within and near the *RXFP2* gene were tested on a reduced animal dataset to demonstrate the absence of any direct causal relationship of this gene with polledness in a wide range of sheep breeds.

Methods
Animal samples and DNA extraction
Collection of the blood samples used in this study was performed exclusively by local veterinarians during regular quality control of breeding records (e.g. paternity testing) on the farms, thus no authorization from the ethics committee was required.

Extraction of genomic DNA from peripheral blood was done by using either a modified high salt method [25] or a commercial spin column kit (QIAamp DNA blood mini, Qiagen, Hilden). DNA was obtained from 489 pure and crossbred sheep, which were sampled in Germany except for three breeds or strains that originated from Croatia (Cres sheep, Krk sheep and Travnička Pramenka). However, many of the sheep bred and sampled in Germany have their genetic origin in various regions of Europe. Figure 1 shows an overview of the sample sets used and the number of samples per set analyzed with different molecular genetic methods.

The first sample set included 208 samples from 17 completely polled and 84 samples from eight completely horned breeds (see details in Table 1). Sample set 2 included samples from 18 sheep that originated either from one cross between a completely polled breed and a completely horned breed or from multiple crosses between completely polled and completely horned breeds (details in Table 2). The third sample set consisted

of 179 samples from nine breeds that have heterogeneous horn statuses, i.e. that ranged from breeds in which all males are horned and all females are (usually) polled to breeds in which both males and females can be polled or horned (details in Table 3). Some breeds also included individuals with scurs or horn rudiments. All sheep of sample sets 2 and 3 were phenotyped for horn status at an age at which horns are usually developed. In general, the names of the sheep breeds mentioned in this article refer to Mason and Porter [3], except for the breeds Travnička Pramenka [4], Alpines Steinschaf and Cameroon sheep (dad.fao.org).

Sequence capture and sequencing of the *RXFP2* gene and its flanking regions
A region of about 246 kb (NC_019467.2:29,331,000-29,577,000, Oar_v4.0) on OAR10 that included the complete *RXFP2* gene (about 60 kb), the region flanking exon 1 and its 3′-UTR (about 185 kb) was sequenced. For this purpose, 24 DNA samples that included two individuals from six completely polled breeds (Bentheimer, East Friesian, German White Mountain, Pomeranian Coarsewool, Rhön, and White Polled Heidschnucke) and six completely horned breeds (Grey Horned Heidschnucke, Roux du Valais, Scottish Blackface, Valais Blacknose, White Horned Heidschnucke, and Wiltshire Horn) were selected from sample set 1 (Fig. 1).

For the generation of sequencing libraries, 1 μg of genomic DNA was sonicated (Bioruptor, Diagenode, Liege) for 25 cycles (30 s on/off at "low" intensity) and processed with the Accel-DNA 1S kit (Swift Biosciences, Ann Arbor) according to the manufacturer's instructions. The resulting barcoded whole-genome libraries were pooled in equimolar amounts and hybridized to a genomic tiling array (Agilent 244 k capture Array, Agilent, Santa Clara; custom designed by e-array, repeat-masked, 2-bp tiling). Briefly, the libraries were hybridized for 65 h at 65 °C, washed and eluted with nuclease-free water for 10 min at 95 °C. The eluted DNA was concentrated in a vacuum centrifuge, amplified by PCR (10 cycles at 98 °C for 15 s, 65 °C for 30 s, and 72 °C for 30 s) and purified on Ampure XP beads. The target-enriched libraries were sequenced on a Hiseq 1500 instrument (Illumina, SanDiego) in paired-end mode with a read length of 100 nt for each read. Reads were de-multiplexed and mapped to the reference sheep genome (Oar_v4.0) using the BWA software package [26]. After removal of PCR duplicates, variants were called by using the software VarScan, which is optimized to detect SNPs and short indels, and then searched for potential candidate variants that co-segregated with the horned phenotype but were absent in polled animals. The insertion polymorphism was identified manually using the Integrative

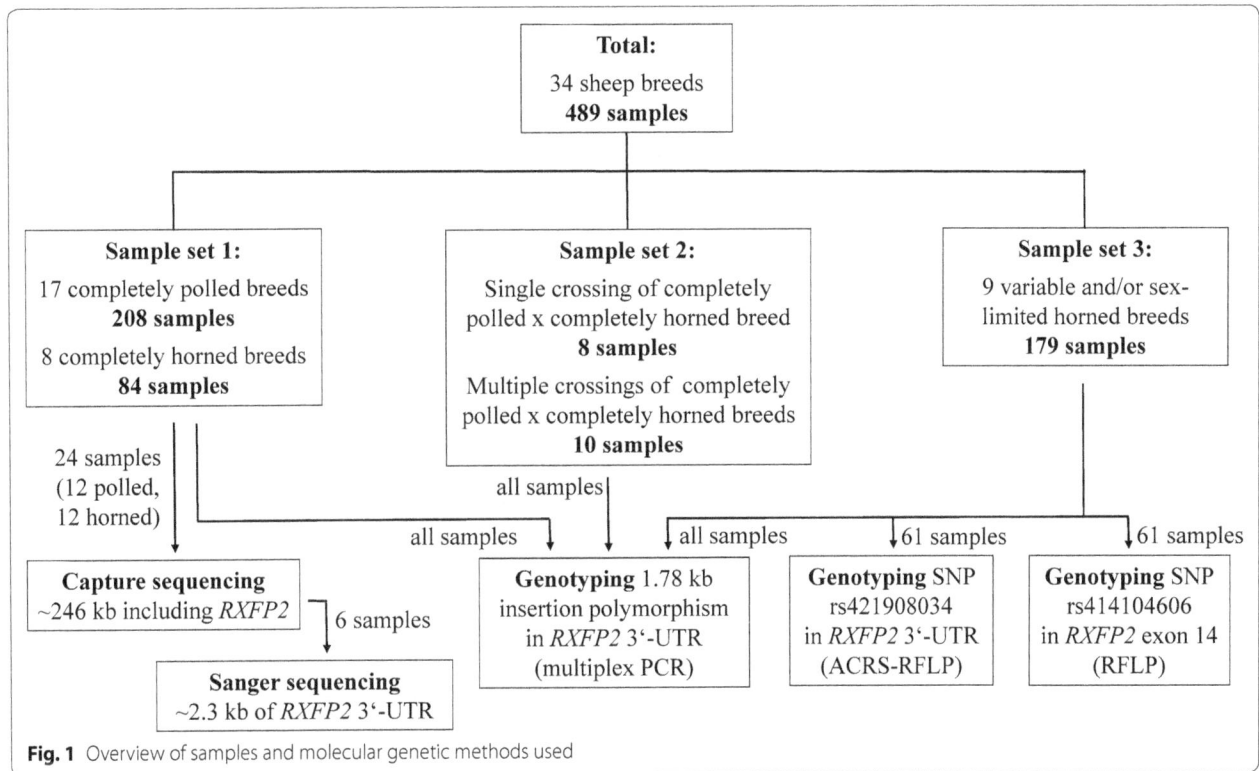

Fig. 1 Overview of samples and molecular genetic methods used

Genomics Viewer software (IGV, http://www.software. broadinstitute.org/software/igv/). Because of the manual process used for the IGV analyses, they were very time-consuming, thus we focused mainly on the sequence data of some previously published regions within the 246-kb region.

To identify the chromosomal segment that segregates with polledness in sheep, all informative SNPs [with a minor allele frequency (MAF) higher than 0.05] were filtered out from the data obtained in the VarScan process. For each SNP, the major allele that was present in the 12 polled animals, i.e. the case group, was identified and their average frequencies were estimated separately for the polled and horned groups using 20-kb sliding windows. Finally, the F_{ST}-values were estimated based on the allele frequencies of the same 20-kb windows and graphically represented. The identical by descent (IBD) region was then defined as the region showing F_{ST}-values higher than 0.3159 (mean F_{ST}-value plus one standard deviation).

Determination of the exact sequences of the identified variations in the 3′-UTR of *RXFP2* by Sanger sequencing

In order to determine the exact sequence of two variants that had been identified in the 3′-UTR of *RXFP2* by targeted sequencing, primers F1 and R1 (see Additional file 1: Table S1) and standard PCR conditions were used to amplify two fragments of about 500 bp from the DNA of two individuals from completely horned breeds (Roux du Valais and Valais Blacknose) and of about 2.3 kb from the DNA of two individuals from completely polled breeds (Bentheimer and Rhön). These PCR products were sequenced (Eurofins Genomics, Ebersberg) on both strands using the same PCR primers. In order to obtain the complete sequence of the insertion, additional sequencing reactions with primers F3 and F4 (see Additional file 1: Table S1) were performed on PCR products from polled sheep only. The resulting chromatograms were analyzed with the software *ChromasPro* version 1.33 (Technelysium Pty Ltd, Tewantin) and compared with the GenBank reference sequence of ovine *RXFP2*, NC_019467.2 (based on the sheep genome assembly Oar_v4.0).

To distinguish between ancestral and derived alleles, we performed multiple sequence alignment of the region that surrounds the variable site in the 3′-UTR of *RXFP2* of whole-genome sequences of seven *Bovidae* using NCBI's blast server (http://www.blast.ncbi.nlm.nih.gov/Blast.cgi) with default settings (Megablast).

Genotyping of ancestral and derived alleles in the 3′-UTR of *RXFP2* by multiplex PCR

To determine the ancestral and derived alleles of the 3′-UTR of *RXFP2* that segregated with horn status in the 24 sequenced samples of sample set 1, a multiplex PCR was designed and applied to all 489 samples of sample sets 1, 2 and 3: if the insertion was present, primer pairs F1/R1, F1/R2 and F2/R1 (see Additional file 1: Table S1) were expected to produce fragments of 2286, 389 and 676 bp, respectively, but if the individual carried the ancestral allele (without the insertion), primer pair F1/R1 was expected to amplify a fragment of only 506 bp. PCR amplifications were carried out according to the manufacturer's standard protocol for the Promega Go Taq polymerase (Promega, Mannheim) in a 15-µL reaction volume that contained the four primers (6 pmol of primer F2, 4 pmol of each of the other three primers), 1.0 mM $MgCl_2$ and 10 to 50 ng of genomic DNA. PCR conditions were as follows: 2 min at 95 °C followed by 33 cycles at 95 °C for 30 s, 60 °C for 45 s and 72 °C for 1 min, and a final elongation step at 72 °C for 10 min. After electrophoresis in a 1.5 % agarose gel stained with Midori Green Advance as recommended by the manufacturer (Biozym Scientific, Hessisch Oldendorf), PCR products were detected using UV light.

Genotyping of a SNP in the 3′-UTR of RXFP2 by amplification-created restriction site-restriction fragment length polymorphism (ACRS-RFLP) analysis

A 104-bp fragment that included an A/G substitution in the 3′-UTR of RXFP2 at position NC_019467.2:29,432,846 (rs421908034) was amplified using the forward primer 5′-CAAGCCAAAAAGGTGAATGG-3′ and the reverse primer 5′-GTGGAGCAGCAGCTTTGAAAT-3′. The reverse primer included a mismatch nucleotide (underlined) to create a restriction site for the restriction enzyme Hpy188I in the presence of allele G. PCR amplifications were carried out according to the manufacturer's standard protocol for the Promega Go Taq polymerase (Promega, Mannheim) in a 25-µL reaction volume that included forward and reverse primers (10 pmol of each), 2.0 mM $MgCl_2$ and 10–50 ng of genomic DNA. PCR conditions were as follows: 1.5 min at 95 °C followed by 35 cycles at 95 °C for 15 s, 56 °C for 20 s and 72 °C for 30 s, and a final elongation step at 72 °C for 5 min. After electrophoresis on a 3.0 % agarose gel stained with Midori Green Advance (as above), PCR products were detected using UV light. PCR products of 61 samples from sample set 3 (that comprised the Wallachenschaf, Dorper, and Cameroon breeds) were incubated with Hpy188I as recommended by the manufacturer (New England Biolabs, Ipswich, MA). Expected sizes of the fragments after digestion were 104 bp if allele A was present and 82 and 22 bp if allele G was present.

Genotyping of a SNP in exon 14 of RXFP2 by restriction fragment length polymorphism (RFLP) analysis

A fragment that included the synonymous SNP c.1125A>G in exon 14 of RXFP2 was amplified as described by Wang et al. [23]. However, based on the ovine genome sequence, the resulting fragment was expected to include 754 bp instead of the 765 bp that were indicated by Wang et al. [23]. PCR products of 61 samples from sample set 3 (that included the Walachenschaf, Dorper and Cameroon breeds) were incubated with Hpy188I (as above). Expected sizes of the fragments after digestion were 726 and 28 bp if allele A was present and 429, 297 and 28 bp if allele G was present.

Analysis of genotyping data for SNP OAR10_29511510.1 in breeds with a fixed horn status

The International Sheep Genomics Consortium (ISGC) [22] produced genotyping data for 49,034 SNPs, including SNP OAR10_29511510.1 (rs413264476) that is located within intron 11 of RXFP2, on 2819 individuals that represent a diverse collection of 74 sheep breeds. However, the horn status of each individual is not available for this data. Thus, among the genotyped breeds, we selected those with a fixed, breed-specific horn status, which included breeds that are completely polled or completely horned as well as breeds with a fixed sex-dependent horn status (all males horned, all females polled). The horn status of the breeds was retrieved from relevant publications such as Mason and Porter [3], online breed databases (e.g. DAD-IS, dad.fao.org) and information provided by breeding organizations. Breed names were those indicated by the ISGC.

Results

Sequence capture and Sanger sequencing results

Alignment of sequence data generated from a 246-kb region that included the RXFP2 gene and part of the flanking regions from 24 sheep samples of sample set 1 showed efficient enrichment of the target region (>100× coverage) and revealed an insertion polymorphism of about 2 kb in the 3′-UTR of RXFP2. Besides this manually-detected insertion polymorphism, only one SNP (A/G) at position NC_019467.2:29,432,846 (rs421908034) passed the automated filtering process and remained as a putative causal candidate based on the VarScan results. This SNP is located 214 bp away from the insertion polymorphism and Fig. 2 shows that it is positioned close to the beginning of the estimated IBD region. The insertion was absent in the 12 sheep from horned breeds, whereas among the sheep from completely polled breeds, 11 were homozygous and only a single female Pomeranian Coarsewool individual was heterozygous for the insertion. The nearby 3'-UTR SNP rs421908034 showed exactly the same genotype distribution (12 horned sheep *GG*, 11 polled sheep *AA*, and one polled Pomeranian Coarsewool sheep *AG*). Sequence data of all 24 samples were deposited in the SRA archive (http://www.ncbi.nlm.nih.gov/Traces/sra/; accession number SRP057491).

Sanger sequencing of a 2286-bp PCR fragment obtained with primer pair F1/R1 and DNA from completely horned and completely polled breeds revealed that the exact size of the insertion segregating with polledness is 1780 bp (Fig. 3a). This insertion is between positions 29, 433, 060 and 29, 434, 923 bp in the reference sequence NC_019467.2 (Oar_v4.0). This part of the ovine *RXFP2* reference sequence includes a stretch of 83 bp (NC_019467.2:29,434,159-29,434,241) that was absent from the sequence of all polled sheep analyzed in this study. In addition, all sequenced polled sheep were homozygous for 13 single nucleotide substitutions and one single nucleotide deletion compared to the reference sequence. Another homozygous nucleotide substitution that was located upstream of the 1.78-kb insertion was identified in all sequenced sheep. Both the 83- and the 1-bp deletion were also observed in polled sheep by Wiedemar and Drögemüller [24]. Sequences with and without the 1.78-kb insertion including all substitutions and deletions not present in the NC_019467.2 (Oar_v4.0) reference sequence were submitted to GenBank (http://www.ncbi.nlm.nih.gov/nucleotide; accession numbers KX084522 and KX084523).

Multiple sequence alignment of the region that surrounds the variable site in the 3'-UTR of *RXFP2* to whole-genome sequences of seven *Bovidae* showed that the 1.78-kb insertion is detected only in the sheep reference genome (which originates from a polled Texel sheep) and not in any of the reference genomes of horned *Bovidae* (see Additional file 2: Figure S1). This clearly confirmed that the 1.78-kb insertion is a derived allele. In the following sections and tables, we refer to the two alleles of the 3'-UTR of *RXFP2* as derived (*der*, with the 1.78-kb insertion) and ancestral (*anc*, without the insertion) alleles.

Genotyping results of the 1.78-kb insertion polymorphism in the 3'-UTR *RXFP2*

The established multiplex PCR proved to be a useful tool to determine the three possible genotypes *anc/anc*, *anc/der* and *der/der* (Figs. 3b, c) and was used to genotype all individuals of the three sample sets. In sample set 1, 204 sheep from completely polled breeds were homozygous *der/der*, whereas one Pomeranian Coarsewool female and three Barbados Black Belly females

were heterozygous *anc/der*. In contrast, all 84 sheep from completely horned breeds were homozygous *anc/anc*. There was no disagreement between the multiplex PCR genotypes and the 24 sequencing results (Table 1).

In sample set 2 (Table 2), the eight single crossbred sheep were the progeny of *der/der* sires from completely polled breeds (East Friesian or Merinolandschaf) mated with *anc/anc* females from the completely horned breed

Table 1 Genotypes at the insertion polymorphism in the 3'-UTR of *RXFP2* for sheep from completely polled and completely horned breeds (sample set 1)

Horn status of breeds	Breed	Genotypes (n)		
		anc/anc	anc/der	der/der
Completely polled	Barbados Black Belly		3[a]	7
	Bentheimer[b]			4
	Charollais			3
	Coburger			18
	East Friesian (White)[b]			20
	German Black-headed Mutton			20
	German Brown Mountain			4
	German White Mountain[b]			20
	Ile de France			3
	Merinolandschaf			20
	Pomeranian Coarsewool[b]		1[a]	18
	Rhön[b]			20
	Rouge du Roussillon			4
	Shropshire			4
	Suffolk			17
	Texel			20
	White Polled Heidschnucke[b]			2
Completely horned	Grey Horned Heidschnucke[b]	26		
	Racka	7		
	Roux du Valais[b]	2		
	Scottish Blackface[b]	18		
	Soay[c]	9		
	Valais Blacknose[b]	18		
	White Horned Heidschnucke[b]	2		
	Wiltshire Horn[b]	2		

anc ancestral allele, *der* derived allele (with the 1.78-kb insertion), *n* number

[a] Female(s)

[b] Two animals of this breed were also used for sequencing

[c] Originating from a completely horned population

Table 2 Genotypes at the insertion polymorphism in the 3′-UTR of *RXFP2* for crossbreds between completely polled and completely horned sheep breeds (sample set 2)

Cross	Crossed breeds	Horn status of the animals	Sex	Genotypes (n)		
				anc/anc	anc/der	der/der
Single cross	East Friesian (white) × Grey Horned Heidschnucke	Horned	Male		3	
		Polled	Female		3	
	Merinolandschaf × Grey Horned Heidschnucke	Horned	Male		2	
Multi cross	Completely polled breeds × completely horned breeds	Horned	Female	3		
		Polled	Female		4	2
		Horned[a]	Female		1	

anc ancestral allele, *der* = derived allele (with the 1.78-kb insertion), *n* number

[a] Horn rudiments

Grey Horned Heidschnucke. As expected, all single crossbred sheep were heterozygous *anc/der*. All males were horned and all females were polled. Among the 10 female progenies from multiple crosses between completely polled breeds and completely horned breeds (Table 2), all three possible genotypes were observed. Horns were observed only in sheep with the genotype *anc/anc*, whereas polled sheep had genotypes *der/der* or *anc/der*. However, horn rudiments were detected in a single heterozygous *anc/der* individual.

In contrast to our observations for completely polled and completely horned breeds and their crosses, the ancestral and derived alleles of the 3′UTR of *RXFP2* did not segregate with horn status in most of the sheep breeds of sample set 3 with sex-dependent or variable horn status (Table 3). Although all nine horned Cameroon rams were homozygous *anc/anc*, all 11 polled ewes of this breed were also *anc/anc*. Similarly, 13 polled Travnička Pramenka ewes and one polled Cres ewe were homozygous *anc/anc*. However, all 47 genotyped sheep of the breeds Dorper and Bovec-like were monomorphic for the derived allele, although 18 individuals (males as well as Bovec-like females) were horned and 12 additional sheep of both sexes had scurs or horn rudiments. In the Walachenschaf, Alpines Steinschaf and Bavarian Forest breeds, discrepancies regarding the segregation of ancestral and derived alleles with horn status were not as obvious as in Cameroon, Travnička Pramenka, Dorper and Bovec-like sheep. The discrepancies mostly concerned heterozygous individuals and/or the occurrence of scurs or horn rudiments in males and females (Table 3).

Genotyping results of the SNP in the 3′-UTR *RXFP2*

Besides the 24 sheep for which sequence capture and sequencing were available, 61 sheep from sample set 3 from a breed with sex-dependent horns (Cameroon) and from two breeds with variable horn status (Walachenschaf and Dorper) were also genotyped for the SNP in the 3′-UTR of *RXFP2* (rs421908034) (see Additional file 3: Table S2). In most (59) of the 61 analyzed sheep, allele *A* segregated with the derived allele and allele *G* with the ancestral allele. Only two Cameroon sheep (one polled female and one horned male) that were heterozygous (*AG*) for this SNP but homozygous *anc/anc* were in contradiction with perfect LD between both mutations in the 3′-UTR of *RXFP2*. The presence of all three genotypes at the rs421908034 SNP in horned male sheep excluded a direct causal relationship with polledness in the overall sheep population. Therefore, genotyping of the SNP in the 3′-UTR of *RXFP2* was not pursued further in this study.

Genotyping results for the SNP in exon 14 of *RXFP2*

The 61 sheep of the same sample set 3 were also genotyped for the SNP in exon 14 of *RXFP2* (rs414104606). As for the insertion polymorphism and the SNP in the 3′-UTR of *RXFP2*, all 24 Dorper sheep were homozygous *AA* for this SNP, although this group contained horned, polled and scurred male sheep as well as scurred female sheep (see Additional file 3: Table S2). In the Cameroon breed, the most common genotype was *AA* with some heterozygous *AG* sheep. These two genotypes were found in horned males as well as in polled females. In 13 horned Walachenschaf individuals, all three possible genotypes (*AA*, *GG* and *AG*) at SNP rs414104606 were observed. For the 24 sheep for which sequence capture and sequencing were performed, all polled animals were homozygous *AA*, 11 horned sheep were homozygous *GG*, and one horned female was heterozygous *AG*. Therefore, as for the SNP in the 3′-UTR of *RXFP2*, genotyping of the SNP in exon 14 was not pursued further.

Genotype distribution of SNP OAR10_29511510.1 in intron 11 of *RXFP2*

Among the 74 international sheep breeds that were genotyped for the SNP OAR10_29511510.1 in intron 11 of

Table 3 Genotypes at the insertion polymorphism in the 3'-UTR of *RXFP2* for sheep from breeds with variable horn status (sample set 3)

Horn status of breed	Breed	Sheep (n)	Horn status of animals	Sex	Genotypes (n)		
					anc/anc	anc/der	der/der
Variable horn status of males and females	Bavarian forest	9	Horned	Male	6	3	
		3	Horned	Female	1	2	
		1	Horned[a]	Female		1	
		1	Polled	Male			1
		4	Polled	Female		3	1
		2	Scurred	Male		2	
		3	Scurred	Female		3	
Variable horn status of males and females	Alpines Steinschaf	6	Horned	Male	6		
		6	Horned	Female	6		
		2	Horned[b]	Male		2	
		1	Polled	Male		1	
		8	Polled	Female		7	1
		1	Scurred	Male			1
Males horned, females variable	Walachenschaf	5	Horned	Male	4	1	
		5	Horned	Female	5		
		3	Horned[b]	Female	2	1	
Males horned, females usually polled	Cres sheep	2	Polled	Female	1	1	
Males horned, females usually polled	Krk sheep	3	Polled	Female			3
Variable horn status of males and females	Travnička Pramenka	2	Horned	Male	2		
		12	Horned	Female	12		
		7	Horned[b]	Female	6	1	
		5	Polled	Male		4	1
		15	Polled	Female	13	2	
		2	Scurred	Female	2		
Variable horn status of males and females	Bovec-like (Krainer Steinschaf)	5	Horned	Male			5
		5	Horned	Female			5
		1	Horned[b]	Male			1
		3	Polled	Male			3
		8	Polled	Female			8
		1	Scurred	Female			1
Variable horn status of males and females	Dorper	8	Horned	Male			8
		6	Polled	Male			6
		3	Scurred	Male			3
		7	Scurred	Female			7
Males horned, females polled	Cameroon sheep	9	Horned	Male	11		
		1	Polled	Male		1	
		13	Polled	Female	10	2	

anc ancestral allele, *der* = derived allele (with 1780 bp-insertion), n number

[a] Right horned, left scurred

[b] Horn rudiments

RXFP2, reliable information about the horn status was available for 38 breeds (28 breeds completely polled, 5 breeds with horned males and polled females, 5 breeds completely horned). Figure 4 shows the assignment to one of the three horn status groups, the number of genotyped sheep and genotype frequencies for each breed.

Among the 28 completely polled breeds, all except seven breeds were monomorphic for genotype *GG*. For the seven non-monomorphic breeds, we estimated frequencies of genotype *AG* that ranged from 0.01 (Australian Suffolk) to 0.08 (Scottish Texel), but no homozygous *AA* sheep were detected. The distribution of SNP

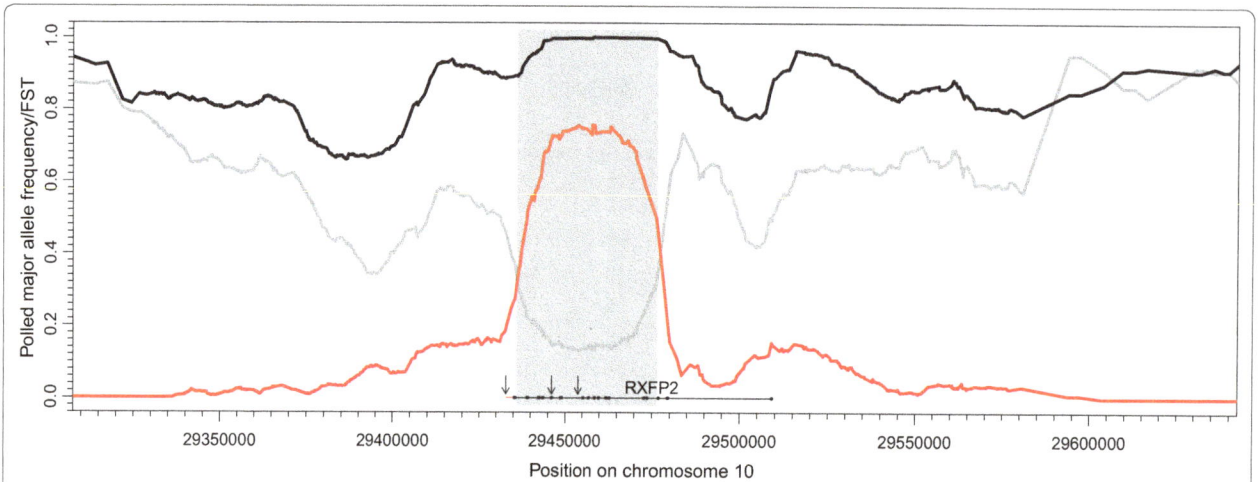

Fig. 2 F_{ST}-values and frequencies of the major allele of the sequence of the captured region. In order to define the chromosomal segment that segregates with the polled phenotype within the sequenced region obtained by sequence capture, frequencies of the polled major allele were estimated in both groups (polled vs. horned) separately and averaged over 20-kb sliding windows (*black*: polled and *gray*: horned). F_{ST}-values (*red curve*) were estimated based on the allele frequencies of the same 20-kb windows. The *gray box* outlines F_{ST}-values that are higher than 0.3159 (mean F_{ST}-value plus one standard deviation). In addition, the position of the *RXFP2* gene is indicated as a *black line* below the curves (*thin line*: introns and *thick line*: exons). The insertion polymorphism is indicated as a *red line* on the edge of *RXFP2*. From *left* to *right*, the positions of SNPs rs421908034 (3′-UTR), rs414104606 (exon14) and OAR_29511510.1/rs413264476 (intron 11) are indicated with *arrows*

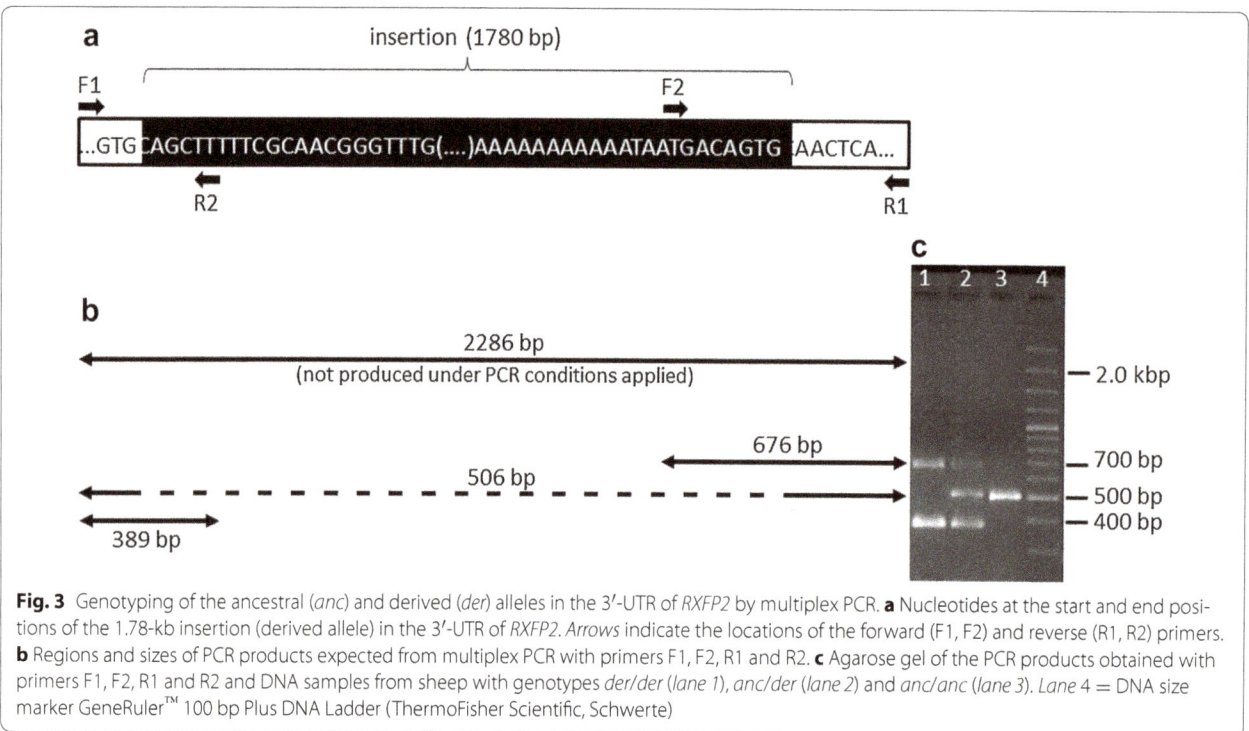

Fig. 3 Genotyping of the ancestral (*anc*) and derived (*der*) alleles in the 3′-UTR of *RXFP2* by multiplex PCR. **a** Nucleotides at the start and end positions of the 1.78-kb insertion (derived allele) in the 3′-UTR of *RXFP2*. *Arrows* indicate the locations of the forward (F1, F2) and reverse (R1, R2) primers. **b** Regions and sizes of PCR products expected from multiplex PCR with primers F1, F2, R1 and R2. **c** Agarose gel of the PCR products obtained with primers F1, F2, R1 and R2 and DNA samples from sheep with genotypes *der/der* (*lane 1*), *anc/der* (*lane 2*) and *anc/anc* (*lane 3*). *Lane 4* = DNA size marker GeneRuler™ 100 bp Plus DNA Ladder (ThermoFisher Scientific, Schwerte)

OAR10_29511510.1 genotypes was similar for breeds with horned males and polled females with three breeds having genotype *GG* and two breeds that included some individuals with genotype *AG* (*AG* frequencies of 0.01 and 0.12 were estimated in the Rambouillet and Ethiopian Menz breeds, respectively). In contrast, three of the five completely horned breeds were homozygous *AA* and no homozygous *GG* sheep were detected (Fig. 4). In the

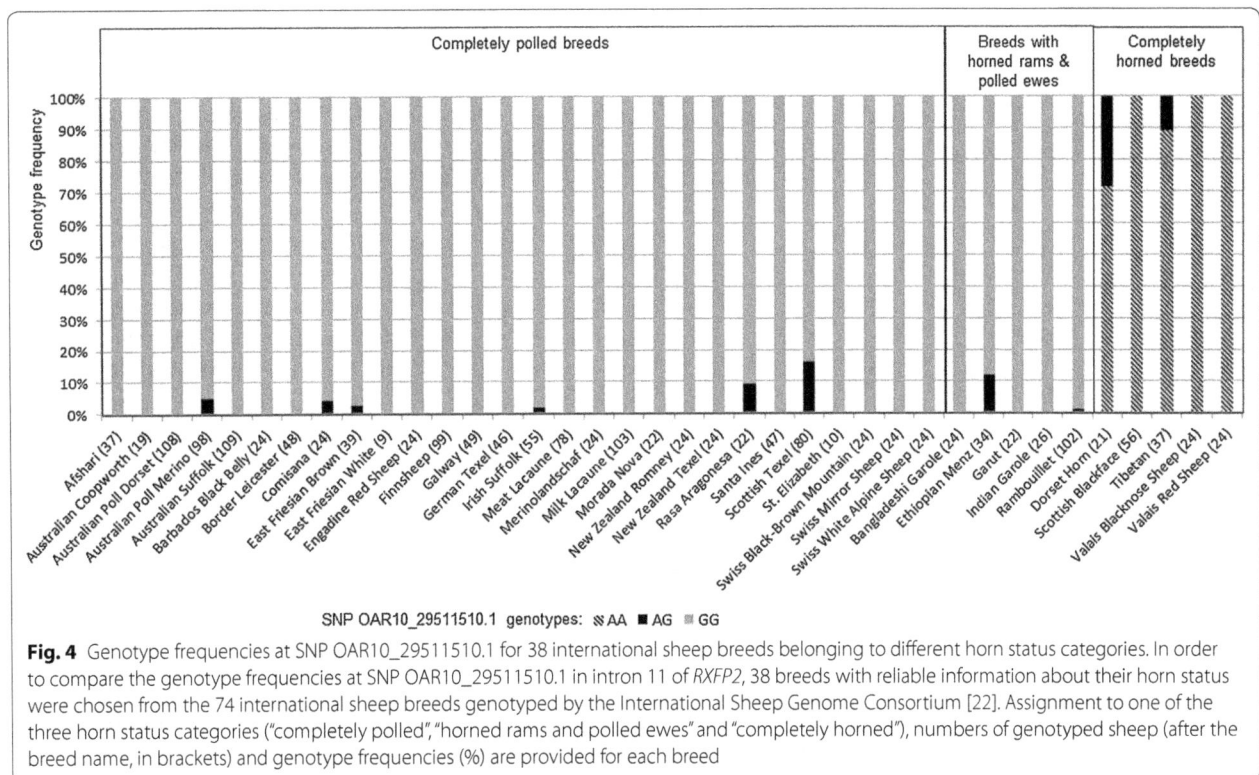

Fig. 4 Genotype frequencies at SNP OAR10_29511510.1 for 38 international sheep breeds belonging to different horn status categories. In order to compare the genotype frequencies at SNP OAR10_29511510.1 in intron 11 of *RXFP2*, 38 breeds with reliable information about their horn status were chosen from the 74 international sheep breeds genotyped by the International Sheep Genome Consortium [22]. Assignment to one of the three horn status categories ("completely polled", "horned rams and polled ewes" and "completely horned"), numbers of genotyped sheep (after the breed name, in brackets) and genotype frequencies (%) are provided for each breed

Tibetan and the Dorset Horn breeds, the frequencies of genotype *AG* were 0.11 and 0.29, respectively.

Table 4 provides the sex of the sheep that were genotyped for SNP OAR10_29511510.1 from completely polled and completely horned breeds that were not monomorphic and from all five breeds with horned males and polled females. Animals with genotype *AG* occurred in both sexes and 17 males from completely polled breeds and six females from completely horned breeds were also heterozygous *AG*. Moreover, Table 4 shows that nearly all males from breeds with horned rams and polled ewes were homozygous *GG*.

Discussion

Segregation of the 1.78-kb insertion polymorphism in the 3'-UTR of *RXFP2* with horn status

Targeted sequencing of the *RXFP2* gene and its flanking regions in 24 sheep from 12 different either completely polled or completely horned breeds allowed us to identify the same insertion polymorphism within the 3'-UTR of *RXFP2* that was recently shown to segregate with horn status in Swiss breeds by Wiedemar and Drögemüller [24]. Sequence comparisons with other horned *Bovidae* confirmed this insertion as the derived allele. In both studies, completely horned breeds were shown to be fixed for the ancestral allele without the insertion in the 3'-UTR of *RXFP2*. In contrast to the Swiss study

in which all sheep from completely polled breeds were monomorphic for the insertion, we detected four ewes from completely polled breeds (Barbados Black Belly and Pomeranian Coarsewool) that carried the insertion on only one chromosome. However, under the hypothesis that horns are dominant in the male and recessive in the female [7], heterozygous ewes are expected to be polled and hence in agreement with the breed's standard horn status. The relative high proportion of heterozygous sheep (3 out of 10) among the Barbados Black Belly sheep may not be representative since all investigated samples of this breed originated from a single small breeding flock and thus the ewes were related to each other.

A sex-limited effect of the derived allele became even clearer after genotyping sheep that originated from single or multiple crosses between completely polled breeds and completely horned breeds: male sheep with the genotype *anc/der* were horned, whereas female sheep with this genotype were polled. This is in agreement with previous studies on crossed sheep [6–8]. However, the fact that we identified a single female sheep with horn rudiments, which originated from multiple crosses between polled and horned breeds, and a polled Cameroon ram (atypical for this breed) that both had the genotype *anc/der* disputed complete association (Tables 2, 3).

Table 4 Genotypes at SNP OAR10_29511510.1 for sheep from completely polled, sex-dependent horned, and completely horned sheep breeds

Horn status of breed	Breed	Sex	Genotypes (n)		
			AA	AG	GG
Completely polled	Australian Poll Merino	Male	0	5	93
	Australian Suffolk	Male	0	1	108
	Comisana	Female	0	5	93
	East Friesian Brown	Male	0	0	15
		Female	0	1	23
	Irish Suffolk	Male	0	1	38
		Female	0	0	16
	Rasa Aragonesa	Male	0	1	3
		Female	0	1	17
	Scottish Texel	Male	0	9	31
		Female	0	4	36
Males horned, females polled	Ethiopian Menz	Male	0	1	17
		Female	0	3	13
	Bangladeshi Garole	Male	0	0	6
		Female	0	0	18
	Garut	Male	0	0	8
		Female	0	0	14
	Indian Garole	Male	0	0	4
		Female	0	0	22
	Rambouillet	Male	0	1	75
		Female	0	0	26
Completely horned	Dorset horn	Male	3	1	0
		Female	12	5	0
	Tibetan	Male	19	3	0
		Female	14	1	0

n number

Similarly, two sheep breeds that originated from Southern Germany (Alpines Steinschaf and Bavarian Forest) further challenged the segregation of the insertion in the 3′-UTR of *RXFP2* with the polled phenotype (Table 3). In summary, for these breeds both the effect of the *anc/der* genotype on the one hand and the scurs/horn rudiments phenotype on the other hand showed no regular pattern. In addition, we observed one male Alpines Steinschaf with scurs that was homozygous for the *der/der* genotype.

Among our samples, those that originated from even more southern European regions (Bovec-like sheep, Cres sheep, Travnička Pramenka and Walachenschaf) as well as from Africa (Dorper and Cameroon sheep) provided additional evidence that challenged a causal relationship or segregation of the insertion polymorphism in the 3′-UTR of *RXFP2* with horn status. In these breeds, we observed almost all possible combinations of genotypes, sex and

horn status phenotypes (Table 3). Only polled males with the *anc/anc* genotype were not observed, which could be due to the lower proportion of sampled polled males (10 %) in the three relevant Southern European sheep breeds.

Based on our results, the insertion polymorphism in the 3′-UTR of *RXFP2* appears to segregate with horn status only in certain sheep breeds, such as the completely polled and completely horned breeds, and (up to a certain degree) in some breeds with variable horn status. In other breeds with variable horn status in both sexes or with a sex-dependent horn status, this segregation is not observed. Furthermore, our results suggest that the perfect LD between the insertion polymorphism and the horn phenotype is gradually eroded in the Southeast European and African breeds as the spatial distance to the North and Central European breeds with fixed horn status increases. However, the absence of *anc/anc* polled males should not be considered as trivial because without this combination, we cannot definitely exclude the 1.78-kb insertion as one of maybe several causal contributors to the complex polled phenotype in sheep.

Segregation of the SNPs in the 3′-UTR and exon 14 of *RXFP2* with horn status

In addition to the genotypes that were obtained by sequence capture and sequencing of 24 sheep, both candidate SNPs (rs421908034 and rs414104606) were also genotyped in 61 sheep from breeds with variable and sex-dependent horn statuses. As shown in Table S2 (see Additional file 3: Table S2), LD between these two SNPs and the 1.78-kb insertion is nearly perfect. Allele G of SNP rs421908034 in the 3′-UTR of *RXFP2* forms a haplotype with the ancestral allele (without the insertion), whereas allele A is linked with the derived allele (with the insertion). Results for only two Cameroon sheep disagree with a perfect LD between these two closely located (214 bp) mutations. The synonymous SNP (rs414104606) in exon 14 of *RXFP2* is located further away (11.1 kb) from the insertion in the 3′UTR of this gene. Consequently, genotyping results suggest an additional deviation from perfect LD between both candidate mutations. Genotyping results showed that Dorper sheep were monomorphic for genotype *AA*, although horned, scurred and polled phenotypes were involved. Moreover, segregation of SNP rs414104606 with horn status was not observed in the two other genotyped breeds with variable (Walachenschaf) and sex-dependent (Cameroon sheep) horn statuses. In addition, the alleles of SNP rs414104606 and the insertion polymorphism in the 3′-UTR of *RXFP2* segregated predominantly in opposite phases in these two breeds (see Additional file 3: Table S2). In accordance with our results, Wang et al. [23] also observed that sheep from a completely polled breed (Suffolk) carried genotype *AA* at SNP rs414104606. They also identified a polled female Han sheep with genotype *GG*, which adds another sheep breed with variable horn status to the group of breeds in which this SNP does not segregate with horn status.

Segregation of SNP OAR10_29511510.1 with horn status

By re-analyzing SNP genotyping data provided by the ISGC, we made the same observations for SNP OAR10_29511510.1 in the intron of *RXFP2*. On the one hand, this SNP segregated perfectly with horn status in 21 completely polled and in three completely horned breeds (polled: *GG*, horned: *AA*). According to our observation that horn status is sex-dependent in polled and completely horned sheep breeds that are heterozygous for the insertion polymorphism in the 3′-UTR of *RXFP2*, sheep that are heterozygous for SNP OAR10_29511510.1 would fit their breed's standard horn status only if all heterozygous animals were either female (polled breeds) or male (horned breeds). However, among the completely polled breeds, five breeds included males with genotype *AG* and two completely horned breeds included females with genotype *AG* (Table 4). Even more significant was the identification of 110 male sheep with genotype *GG* at SNP OAR10_29511510.1 from five breeds with horned males and polled females (Table 4). If the genotype *GG* segregates with polledness, those males would be expected to be polled. This clearly contrasts with their breed standard that defines males as consistently horned. Admittedly, the individual horn status of these sheep was not known. Thus, it is possible that some of the sheep included in the data that we used did not comply with the standard horn status of their respective breeds (e.g. it is known that in polled breeds, sheep with knobs, scurs or even normal horns occasionally occur). However, this would not explain the large number of cases for which the genotype at OAR10_29511510.1 and the breed-specific horn status did not segregate, e.g. as observed for 75 (98.7 %) males from the sex-dependent horned Rambouillet breed (Table 4).

In order to analyze LD between SNP OAR10_29511510.1 and the insertion polymorphism in the 3′-UTR of *RXFP2*, we genotyped 75 sheep from six completely polled and six completely horned breeds (n = 32) and from four breeds with variable horn status (n = 43) using available capture sequencing data from this study or Sanger sequencing, and the multiplex PCR developed in this study (details on the breeds included and genotyping data not shown). Although the resulting estimated LD (R package *GENETICS*) was relatively low over all 75 sheep ($r^2 = 0.369$), it was 3.27 times higher in the 32 sheep from completely polled and completely horned breeds ($r^2 = 0.635$) than in the 43 sheep from

breeds with variable horn status ($r^2 = 0.194$), which suggests a strong association of this SNP in mapping designs that include completely polled and completely horned breeds and a weak association in designs that include sheep from breeds with variable horn status.

Possible reasons for breed differences in the segregation of variants within *RXFP2* with horn status

Wiedemar and Drögemüller [24] postulated that the insertion in the 3′-UTR of the *RXFP2* gene had an effect on the processing and translation of this gene and therefore inhibited normal horn growth. Interestingly, the insertion was predicted to be a complete protein coding gene (*LOC101110773*) by automated computational analysis using Gnomon (http://www.ncbi.nlm.nih.gov/nuccore/NC_019467.2?report=genbank&from=29433047&to=29434884), which agrees with the results of Pickering et al. [20] that indicated a retrotransposition of a complete functional mRNA during the insertion event. The *LOC101110773* gene is predicted to encode an elongation factor 1-alpha 1-like protein, which may modify the tissue-specific transcription of *RXFP2* or other loci. Quite different from the prediction based on the reference sheep genome Oar_v3.1, only a short 3′UTR (without an insertion) is predicted for *RXFP2* based on Oar_v4.0, and thus, the complete 1.78-kb insertion is assigned to the next neighboring gene, i.e. *LOC101110773*. Taking this fact into account, there are different important questions that need to be answered: (1) are the 3′UTR of the *RXFP2* mRNA in polled and horned sheep with *der/der* and *anc/anc* genotypes different?, (2) assuming that there are no differences in the mRNA, can differential expression of the *LOC101110773* gene be involved?, and (3) could SNP rs421908034, which is positioned 214 bp away from *LOC101110773*, influence its expression and thus indirectly modify the transcription of *RXFP2*? The design used in our study is not suitable for answering these and other similar questions but provides information for an improved future mapping design, differential expression and gene editing studies. However, according to our observations, to explain an effect of the 1.78-kb insertion, it would be necessary to involve complementary causality of one or more other, currently unknown, variants in the sheep genome. This complementary causality of more alleles or genes has been speculated since the 1940s (see e.g. [5]). Nonetheless, all recent designs for genome-wide association analyses pointed towards a very strong effect of a single locus that is near or within *RXFP2* [14]. Our results (Fig. 2) also strongly suggest a signature of positive selection in the completely polled breeds of Central and Western European origin. This selection signature is most prominent in an interval of ~40 kb (between

29,436,000 and 29,476,000 bp). Therefore, it is reasonable to expect some causal relationships between polledness and the observed genetic variation within this region. However, although considerable efforts were made in this study to ensure comprehensive information and results, we still cannot exclude the presence of some other variants or combinations of variants within *RXFP2* that might trigger alternative splicing or similar mechanisms thus contributing to the polled phenotype. Therefore, there is still the possibility that the 1.78-kb insertion polymorphism has no causal effect on horn status and is only in strong LD with one or more unknown causal variants. In this context, the differences in segregation of the 1.78-kb insertion with horn status in various breeds could be due to allelic heterogeneity and to the presence of some currently unknown, breed-specific causal alleles with different levels of LD with the 1.78-kb insertion polymorphism and other polymorphisms near or within *RXFP2*. It is very likely that at least one of the complementary causal variants is in interaction with or located on the sex chromosomes. In this context, it should be mentioned that *RXFP2* codes for the receptor of the insulin-like peptide 3 (INSL3) that is a major secretory product of the testicular Leydig cells of male mammals [27] and seems, among others, to play an important role in the fetal abdominal *descensus testis* and in maintaining spermatogenesis [28].

Conclusions

Based on our findings, we conclude that the insertion polymorphism in the 3'-UTR of *RXFP2* and the SNPs in the 3'-UTR, exon 14 and intron 11 of this gene (some of these previously shown to be associated with horn status in several sheep breeds) are, at least in breeds with a variable and/or sex-dependent horn status, not in population-wide LD with the horn status and thus cannot be used as universal markers to define the genetic horn status in the global sheep population. Moreover, the 1.78-kb insertion can be excluded as the only cause of polledness in sheep. Our results suggest that future studies aimed at solving this question should focus on breeds with variable and sex-dependent horn status. To exploit historical recombinations, emphasis should be placed on collecting samples and precise phenotypes from breeds with larger spatial and genetic distance to breeds that originate from Northern and Central Europe. Moreover, the procedures should be able to fulfill the requirements for mapping under the hypothesis of a heterogeneous and/or polygenic mode of inheritance. Similarly, in order to increase the probability of success, the design of future differential expression, alternative splicing and gene editing studies should consider various combinations of neighboring variants such as the SNPs and the insertion in the 3'-UTR of *RXFP2*.

Additional files

Additional file 1: Table S1. Primers used for Sanger sequencing and genotyping the insertion in the 3'-UTR of *RXFP2*.

Additional file 2: Figure S1. Multiple sequence alignment of the region surrounding the insertion in the 3'-UTR of *RXFP2*. The 506-bp sequence amplified in sheep from horned breeds with primers F1/R1 was searched in the whole genome sequence of seven *Bovidae* using NCBI's blast server (https://blast.ncbi.nlm.nih.gov/Blast.cgi. Accessed 11 April 2016) with default settings (Megablast). The results are schematically shown by plotting the obtained alignments as grey rectangles and by highlighting SNPs and small indels with white vertical bars. The 1.78-kb insertion is only found in the sheep reference genome (that originates from a polled Texel sheep) but not in any of the reference genomes of horned *Bovidae*. The sequence is highly conserved in bovids (93 to 99 % sequence identity), in contrast to other mammals that only show a maximum of 78 % sequence identity (not shown).

Additional file 3: Table S2. Genotypes at the insertion polymorphism in the 3'-UTR of *RXFP2*, at SNP rs421908034 in the 3'-UTR and SNP rs414104606 in exon 14 of *RXFP2* for 61 sheep from three breeds with variable and sex-dependent horn status.

Authors' contributions

GL designed the experiment, contributed samples, contributed to establish sequencing and genotyping methods, contributed to data analysis, and wrote the main text of the manuscript. SK designed and conducted targeted sequencing, contributed to data analysis, contributed to establish genotyping methods, and was involved in drafting the manuscript. SR contributed to data analysis and was involved in drafting the manuscript. JK extracted and analyzed 50 k SNP chip data on international sheep breeds. BM contributed samples, breeding records and phenotype data. IR contributed samples and to targeted sequencing. IM designed the experiment, contributed samples, to establish sequencing and genotyping methods, to data analysis and wrote significant parts of the manuscript. All authors read and approved the final manuscript.

Author details

[1] Department of Animal Breeding and Genetics, Justus Liebig University of Gießen, Ludwigstrasse 21a, 35390 Giessen, Germany. [2] Laboratory for Functional Genome Analysis (LAFUGA), Gene Center Munich, LMU Munich, Feodor-Lynen-Strasse 25, 81377 Munich, Germany. [3] Chair of Animal Genetics and Husbandry, LMU Munich, Veterinaerstrasse 13, 80539 Munich, Germany. [4] Department of Animal Science and Technology, Faculty of Agriculture, University of Zagreb, Svetošimunska cesta 25, 10000 Zagreb, Croatia. [5] Tierzuchtforschung e.V. München, Senator-Gerauer-Strasse 23, 85586 Poing, Germany.

Acknowledgements

The authors are grateful to the sheep breeders for providing samples and to the International Sheep Genomics Consortium (ISGC) for providing 50 k SNP chip genotyping data. We thank Carina Crispens, Renate Damian and Martin Dinkel for extraction and genotyping of DNA samples and Elisabeth Kunz for revising the manuscript.

Competing interests

The authors declare that they have no competing interests.

References

1. Castle WE. Genetics of horns in sheep. J Hered. 1940;31:486–7.
2. Zeuner FE. Geschichte der Haustiere. München: Bayerischer Landwirtschaftsverlag; 1963.
3. Mason IL, Porter V. Mason's World dictionary of livestock breeds, types and varieties. 5th ed. Wallingford: CABI Publishing; 2002.

4. Ćurković M, Ramljak J, Ivanković S, Mioč B, Ivanković A, Pavić V, et al. The genetic diversity and structure of 18 sheep breeds exposed to isolation and selection. J Anim Breed Genet. 2015;133:71–80.

5. Warwick BL, Dunkle PB. Inheritance of horns in sheep. J Hered. 1939;30:325–9.

6. Wood TB. Note on the inheritance of horns and face color in sheep. J Agric Sci. 1905;1:364–5.

7. Wood TB. The inheritance of horns and face-colour in sheep. J Agric Sci. 1909;3:145–54.

8. Arkell TR. Some data on the inheritance of horns in sheep. Durham: Bulletin New Hampshire Agricultural Experiment Station, New Hampshire College of Agriculture and Mechanic Arts; 1912.

9. Dolling CHS. Hornedness and polledness in sheep. I. The inheritance of polledness in the Merino. Austr J Agric Res. 1960;11:427–38.

10. Dolling CHS. Hornedness and polledness in sheep. 1968. The inheritance of polledness in the Border Leicester. Austr J Agric Res. 1968;19:649–55.

11. Dolling CHS. Hornedness and polledness in sheep. II. The inheritance of horns in Merino ewes. Austr J Agric Res. 1960;11:618–27.

12. Dolling CHS. Hornedness and polledness in sheep. III. The inheritance of horns in Dorset Horn ewes. Austr J Agric Res. 1960;11:845–50.

13. Clutton-Brock T, Pemberton J. Soay sheep dynamics and selection in an island population. Cambridge: Cambridge University Press; 2004.

14. Johnston SE, McEwan JC, Pickering NK, Kijas JW, Beraldi D, Pilkington JG, et al. Genome-wide association mapping identifies the genetic basis of discrete and quantitative variation in sexual weaponry in a wild sheep population. Mol Ecol. 2011;20:2555–66.

15. Arkell TR, Davenport CB. Horns in sheep as a typical sex-limited character. Science. 1912;35:375–7.

16. Dolling CHS. Hornedness and polledness in sheep. IV. Tripple alleles affecting horn growth in the Merino. Austr J Agric Res. 1961;12:438–97.

17. Montgomery GW, Henry HM, Dodds KG, Beattie AE, Wuliji T, Crawford AM. Mapping the horns (Ho) locus in sheep: a further locus controlling horn development in domestic animals. J Hered. 1996;87:358–63.

18. Beraldi D, McRae AF, Gratten J, Slate J, Visscher PM, Pemberton JM. Development of a linkage map and mapping of phenotypic polymorphisms in a free-living population of Soay sheep (Ovis aries). Genetics. 2006;173:1521–37.

19. Johnston SE, Beraldi D, McRae AF, Pemberton JM, Slate J. Horn type and horn length genes map to the same chromosomal region in Soay sheep. Heredity (Edinb). 2010;104:196–205.

20. Pickering NK, Johnson PL, Auvray B, Dodds KG, McEwan JC. Mapping the horns locus in sheep. Proc Assoc Adv Anim Breed Genet. 2009;18:88–91.

21. Dominik S, Henshall JM, Hayes BJ. A single nucleotide polymorphism on chromosome 10 is highly predictive for the polled phenotype in Australian Merino sheep. Anim Genet. 2012;43:468–70.

22. Kijas JW, Lenstra JA, Hayes B, Boitard S, Porto Neto LR, San Cristobal M, et al. Genome-wide analysis of the world's sheep breeds reveals high levels of historic mixture and strong recent selection. PLoS Biol. 2012;10:e1001258.

23. Wang XL, Zhou GX, Li Q, Zhao DF, Chen YL. Discovery of SNPs in RXFP2 related to horn types in sheep. Small Ruminant Res. 2014;116:133–6.

24. Wiedemar N, Drögemüller C. A, 1.8-kb insertion in the 3'-UTR of RXFP2 is associated with polledness in sheep. Anim Genet. 2015;46:457–61.

25. Montgomery GW, Sise JA. Extraction of DNA from sheep white blood-cells. New Zeal J Agr Res. 1990;33:437–41.

26. Li H, Durbin R. Fast and accurate short read alignment with burrows-wheeler transform. Bioinformatics. 2009;25:1754–60.

27. Adham IM, Burkhardt E, Benahmed M, Engel W. Cloning of a cDNA for a novel insulin-like peptide of the testicular Leydig cells. J Biol Chem. 1993;268:26668–72.

28. Ivell R, Anand-Ivell R. Biological role and clinical significance of insulin-like peptide 3. Curr Opin Endocrinol Diabetes Obes. 2011;18:210–6.

12

The potential of shifting recombination hotspots to increase genetic gain in livestock breeding

Serap Gonen[1], Mara Battagin[1], Susan E. Johnston[2], Gregor Gorjanc[1] and John M. Hickey[1*]

Abstract

Background: This study uses simulation to explore and quantify the potential effect of shifting recombination hotspots on genetic gain in livestock breeding programs.

Methods: We simulated three scenarios that differed in the locations of quantitative trait nucleotides (QTN) and recombination hotspots in the genome. In scenario 1, QTN were randomly distributed along the chromosomes and recombination was restricted to occur within specific genomic regions (i.e. recombination hotspots). In the other two scenarios, both QTN and recombination hotspots were located in specific regions, but differed in whether the QTN occurred outside of (scenario 2) or inside (scenario 3) recombination hotspots. We split each chromosome into 250, 500 or 1000 regions per chromosome of which 10% were recombination hotspots and/or contained QTN. The breeding program was run for 21 generations of selection, after which recombination hotspot regions were kept the same or were shifted to adjacent regions for a further 80 generations of selection. We evaluated the effect of shifting recombination hotspots on genetic gain, genetic variance and genic variance.

Results: Our results show that shifting recombination hotspots reduced the decline of genetic and genic variance by releasing standing allelic variation in the form of new allele combinations. This in turn resulted in larger increases in genetic gain. However, the benefit of shifting recombination hotspots for increased genetic gain was only observed when QTN were initially outside recombination hotspots. If QTN were initially inside recombination hotspots then shifting them decreased genetic gain.

Discussion: Shifting recombination hotspots to regions of the genome where recombination had not occurred for 21 generations of selection (i.e. recombination deserts) released more of the standing allelic variation available in each generation and thus increased genetic gain. However, whether and how much increase in genetic gain was achieved by shifting recombination hotspots depended on the distribution of QTN in the genome, the number of recombination hotspots and whether QTN were initially inside or outside recombination hotspots.

Conclusions: Our findings show future scope for targeted modification of recombination hotspots e.g. through changes in zinc-finger motifs of the PRDM9 protein to increase genetic gain in production species.

Background

This study uses simulation to explore the potential of shifting recombination hotspots in the genome to increase genetic gain in livestock breeding. Genetic gain is influenced by four factors: (1) the accuracy of selection;

(2) the generation interval; (3) the intensity of selection; and (4) the additive genetic standard deviation. Advances in reproductive technologies, genotyping, sequencing and genomic selection in the last few decades have enabled the manipulation of the first three factors to deliver higher rates of genetic gain in many closed livestock breeding programs (e.g. [1–3]). The implementation of these new technologies has required substantial investment, and without continued investment and

*Correspondence: john.hickey@roslin.ed.ac.uk
[1] The Roslin Institute and Royal (Dick) School of Veterinary Studies, The University of Edinburgh, Easter Bush, Midlothian, Scotland, UK
Full list of author information is available at the end of the article

advancements in technology, the rate of genetic gain may decline in the future if only the first three of the above factors are addressed. Another possibility is to target the genetic variation that is available for selection in each generation. Large genetic variance enables large response to selection in the short-term, whereas careful maintenance and exploitation of genetic variance across generations enables large response to selection in the long-term.

While the ultimate origin of genetic variation is mutation, recombination through crossing-over can create new combinations of existing alleles, i.e. by releasing standing allelic (genic) variation, which in turn determines genotypic (genetic) variation. Recombination is advantageous if it uncouples favourable alleles that are tightly linked to unfavourable alleles and, which provides more opportunities for selection. Recombination is disadvantageous if it breaks favourable allele combinations [4, 5]. The amount of variation released by recombination depends on the rate of recombination and the locations of recombination events (i.e. crossovers) relative to the causal variants that underlie the trait(s) under selection.

A recent simulation study showed that increasing the rate of recombination could increase genetic gain [6], but achieving a twofold increase in genetic gain required a 20-fold increase in the rate of recombination. In livestock species, average genome lengths are generally constrained to between 20 and 40 Morgan (M) (i.e. on average one to two recombinations per chromosome per meiosis) [7–11]. In most species, recombinations are unevenly distributed along the genome and tend to be clustered in narrow (1 to 2 kb) regions of the genome known as "recombination hotspots" (e.g. [12, 13]). A strategy that changes the locations of these recombination hotspots rather than the rate of recombination within hotspots might be a more effective and feasible way of increasing genetic gain through the manipulation of recombination.

The mechanisms that underlie the locations of recombination events have and are being investigated in several model and non-model organisms. In most eukaryotic species, hotspots are temporally stable and occur primarily at transcription start sites and promoter regions, where the chromatin is more open [14–16]. In contrast, hotspot positions in most mammals (including humans, apes, mice and cattle [17–20]) evolve rapidly, and are determined by a DNA-binding zinc-finger domain in the protein PRDM9. The protein product of the PRDM9 gene has three functional domains: an N-terminal KRAB domain involved in protein–protein binding and interactions, a PR/SET domain involved in histone methylation, and a zinc finger domain involved in DNA sequence recognition and binding. The PR/SET and zinc finger domains are the primary determinants for the specification and initiation of recombination events. Upon binding to a zinc finger DNA recognition site, the PR/SET domain trimethylates lysine 4 of histone H3. This initiates chromatin remodelling to create active chromatin and the formation of a double-stranded DNA break, where the process of repair could involve a recombination event [17, 21–30]. The number of zinc finger domains influences the locations of recombination hotspots and the rate of recombination within a hotspot, and mutations in the zinc-finger domain can change the DNA sequence motifs to which it binds [26, 30]. The number of zinc finger domains is highly variable within and across species, therefore the locations of recombination hotspots are rarely conserved even between closely-related species such as humans and chimpanzees that otherwise share ~99% identity at the sequence level [28, 31]. In some species (including livestock species such as cattle), multiple paralogs of the PRDM9 gene have been identified, which further increases the variability in the location and number of recombination hotspots [28].

In livestock breeding programs that have been on-going for many generations, small changes in the recombination landscape could have occurred [32]. However, the number of generations in the majority of livestock species is unlikely to be large enough to see drastic changes in the distribution of recombination hotspots along the genome. Selection over many generations with a largely constant recombination landscape could have resulted in the accumulation of a large amount of standing allelic variation in recombination deserts, which has been largely inaccessible to selection due to quantitative trait nucleotides (QTN) alleles being linked in repulsion. This rich resource of available standing allelic variation could be released and used if the locations of recombination hotspots could be changed. For example, this may become possible by modification of the PRDM9 gene using new technologies such as genome editing. This has already been demonstrated in mice [33], and the benefit of such an approach in livestock could be estimated by simulation.

The increase in genetic gain that may be achieved by shifting recombination hotspots would depend on the distribution of causal QTN in relation to each other and to existing recombination hotspots. Currently, the distribution of QTN for traits of interest in livestock is largely unknown. QTN may be randomly distributed or clustered, and may be located inside or outside recombination hotspots. If QTN are partially or fully located in regions where very few recombination events occur, shifting recombination hotspots could yield large increases in genetic gain. The aim of this study was to quantify the potential of shifting recombination hotspots to increase genetic gain for quantitative traits in livestock breeding. Our results show that shifting recombination hotspots could release greater amounts of standing allelic variation and, through this, increase genetic gain.

Methods

Simulation was used to evaluate the potential of shifting recombination hotspots to increase genetic gain for quantitative traits in livestock breeding. We tested a number of scenarios using different strategies for shifting recombination hotspots and different distributions of QTN and recombination hotspots across the genome. All scenarios followed a common overall structure, where the simulation scheme was divided into historical and future components. The historical component was split into two parts: (1) evolution under the assumption that livestock populations have been evolving neutrally for tens of thousands of years prior to domestication, and (2) 21 recent generations of modern animal breeding with selection based on true breeding values (TBV). In the historical component, recombination events were constrained to recombination hotspots. The future component consisted of a further 80 generations of modern animal breeding with selection based on TBV with the option to shift recombination hotspots to adjacent regions. In the rest of the paper, historical generations are denoted −20 to 0 and future generations are denoted 1 to 80.

The simulations were designed to: (1) generate whole-genome sequence data; (2) generate QTN that affect phenotypes; (3) generate pedigree structures for a typical livestock population; and (4) perform selection. For each scenario, genetic gain, genetic variance $\left(\sigma_A^2\right)$ and genic variance $\left(\sigma_\alpha^2\right)$ were evaluated. Results are presented as the mean of ten replicates for each scenario on a per generation and/or cumulative basis (information on the standardised values for the replicate mean and between replicate variation is in Additional file 1).

Whole-genome sequence data and historical evolution

Sequence data was generated using the Markovian Coalescent Simulator (MaCS) [34] and AlphaSim [35, 36] for 1000 base haplotypes for each of 10 chromosomes. Chromosomes each comprised 10^8 bp and were simulated using a per site mutation rate of 2.5×10^{-8}. All chromosomes were assumed 1 M long, i.e., with an expectation of one recombination per meiosis. We constrained recombination to defined hotspots. The effective population size (N_e) varied over time in accordance with estimates for the Holstein cattle population. N_e was set to 500 in the final generation of the coalescent simulation, 1256 individuals 1000 years ago, 4350 individuals 10,000 years ago and 43,500 individuals 100,000 years ago, with linear changes in between these time-points. The resulting sequence had approximately 3,000,000 bi-allelic segregating sites in total.

Quantitative trait variants

A quantitative trait influenced by 10,000 QTN was simulated by sampling QTN from the segregating sequence sites in the base population, with the restriction that 1000 QTN were sampled from each of the 10 chromosomes. We simulated different locations of QTN in the genome depending on the scenario (Fig. 1). In scenario 1, QTN were randomly distributed along the genome (Fig. 1a). In scenarios 2 and 3, QTN were clustered into defined chromosome regions (Fig. 1b, c). QTN had their allele substitution effects (α) sampled from a normal distribution with a mean of zero and standard deviation of 0.01 (1.0 divided by the square root of the number of QTN). QTN additive effects were used to compute the TBV of an individual.

Pedigree structure, gamete inheritance and selection strategies

A pedigree of 101 generations of 1000 individuals in equal sex ratio in each generation was simulated. In generation −20, individuals had their chromosomes sampled from the 1000 base chromosomes. In each subsequent generation (−19 to 80), the chromosomes of each individual were sampled from parental chromosomes with recombination. A recombination rate of 1 M per chromosome was assumed, resulting in a 10 M genome. Recombination locations were simulated by ignoring interference and within defined hotspots. In each generation, 25 males were selected to be sires of the next generation using truncation selection on TBV. No selection was performed on females, and all 500 individuals were used as dams. Mating was at random.

Chromosome regions

To investigate the effect of co-located QTN and recombination hotspots, each chromosome was split into 250, 500 or 1000 regions per chromosome, each of equal length (Table 1). In each case, 90% of the regions were not QTN clusters (if QTN clusters were simulated) or recombination hotspots (i.e. recombination never occurred in these regions). The remaining 10% of the regions were either QTN clusters, recombination hotspots (i.e. recombination could occur in these regions), or both QTN clusters and recombination hotspots. QTN clusters and recombination hotspots were evenly spaced across the genome.

Recombination hotspots

Recombination events were simulated to occur within defined regions (see "Chromosome regions" section and Table 1). Each region had an equal probability for a recombination event to occur and probabilities remained constant across all generations. The probability for a

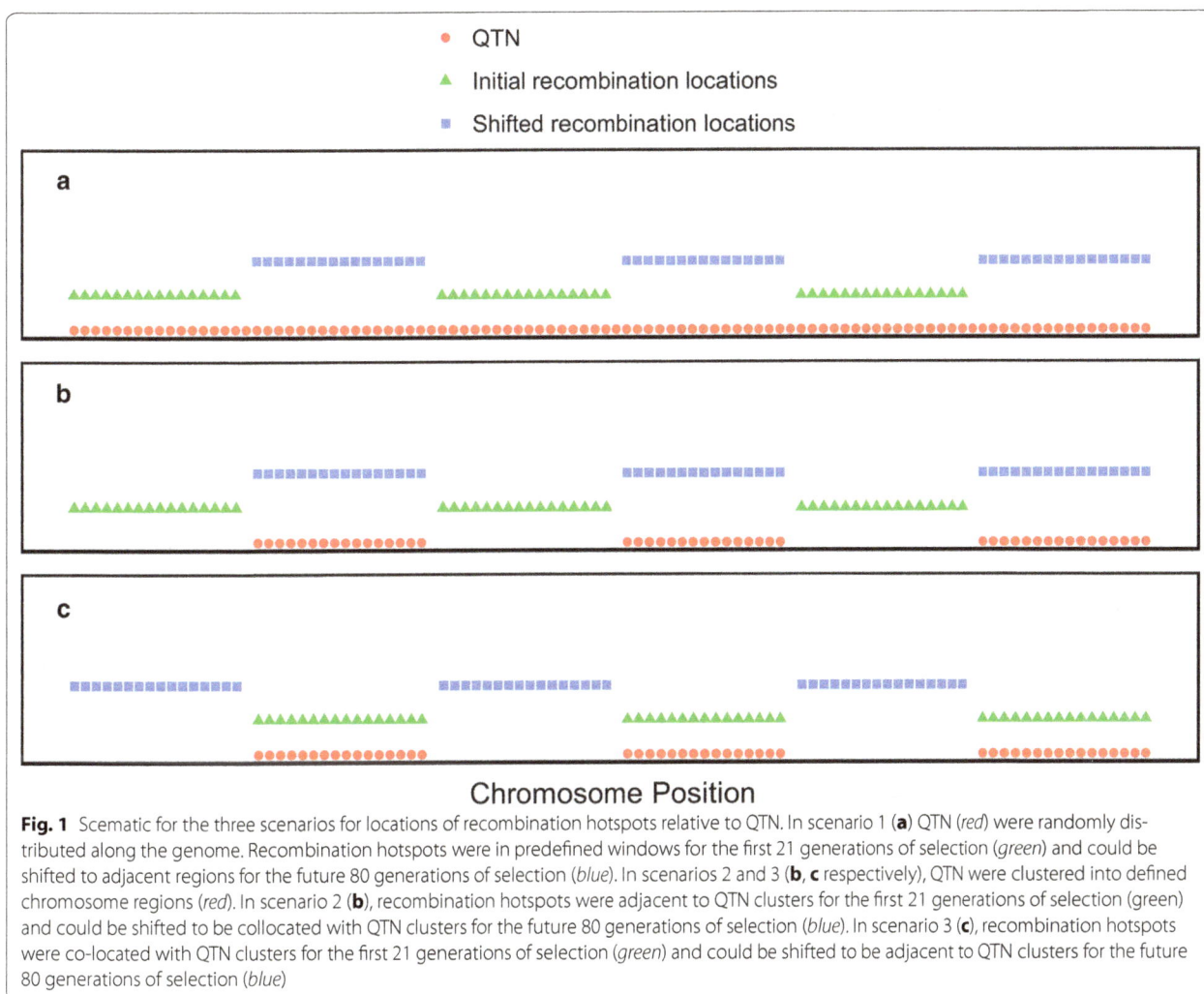

Fig. 1 Scematic for the three scenarios for locations of recombination hotspots relative to QTN. In scenario 1 (**a**) QTN (*red*) were randomly distributed along the genome. Recombination hotspots were in predefined windows for the first 21 generations of selection (*green*) and could be shifted to adjacent regions for the future 80 generations of selection (*blue*). In scenarios 2 and 3 (**b**, **c** respectively), QTN were clustered into defined chromosome regions (*red*). In scenario 2 (**b**), recombination hotspots were adjacent to QTN clusters for the first 21 generations of selection (green) and could be shifted to be collocated with QTN clusters for the future 80 generations of selection (*blue*). In scenario 3 (**c**), recombination hotspots were co-located with QTN clusters for the first 21 generations of selection (*green*) and could be shifted to be adjacent to QTN clusters for the future 80 generations of selection (*blue*)

Table 1 Chromosome regions and locations of QTN and recombination hotspots

Number of chromosome regions	Number of active regions[a]	Number of inactive regions[b]	Probability for recombination or QTN per region
250	25	225	0.04
500	50	450	0.02
1000	100	900	0.01

[a] Active region—is a recombination hotspot and/or a QTN cluster

[b] Inactive region—is never a recombination hotspot or a QTN cluster

recombination event to occur within a region depended on the number of regions simulated. For example, with 250 regions per chromosome, 25 were recombination hotspots and each had a probability of 0.04 for the occurrence of a recombination (i.e. 1/25 assuming a 1 M chromosome) in any given individual (Table 1).

Genetic gain

Genetic gain was calculated in units of the standard deviation of TBV in the base generation (generation 1) as $(\overline{TBV_{curr}} - \overline{TBV_{base}})/\sigma_{TBV_{base}}$, where $\overline{TBV_{curr}}$ is the mean TBV of the current generation and $\overline{TBV_{base}}$ and $\sigma_{TBV_{base}}$ are the mean and standard deviation of TBV in the base generation, respectively. The base generation represents the start of the breeding program whereas the current generation represents the number of generations since the breeding program started. These would be equal when the current generation is the base generation. The genetic variance (i.e. realised additive variance) in each generation was calculated as: $\sigma_A^2 = a'a/(n-1)$, where a is a zero mean vector of TBV of the n individuals in that generation. The genic variance (i.e. expected additive variance if all QTN were independent and in Hardy–Weinberg equilibrium) was calculated as: $\sigma_\alpha^2 = 2\sum_{i=1}^{n_{QTV}} p_i q_i \alpha_i^2$, where p_i and q_i are the allele frequencies in the current generation and α_i is the allele substitution effect of QTN i.

Scenarios

For each of the three numbers of regions (i.e. 250, 500 or 1000), three scenarios were simulated. In scenario 1 (Fig. 1a), QTN were randomly distributed across each chromosome (red) and recombination hotspots were in equally spaced regions for generations −20 to 0 (green). In generation 0 (i.e., the start of future breeding), there was an option to shift recombination hotspots to adjacent regions (blue) for the future 80 generations of selection.

In scenarios 2 and 3, the structure for choosing and shifting the recombination hotspot regions was as described above for scenario 1, but QTN were clustered. In scenario 2 (Fig. 1b), QTN (red) were outside recombination hotspots (green) in generations −20 to 0. In generation 0, there was an option to shift recombination hotspots so that QTN were inside recombination hotspots for the future 80 generations of selection (blue). In scenario 3 (Fig. 1c), QTN (red) were inside recombination hotspots (green). In generation 0, there was an option to shift recombination hotspots so that QTN were outside the recombination hotspots for the future 80 generations of selection (blue).

Results

Our results show that shifting recombination hotspots could release more of the standing allelic variation in each generation and, through this, increase genetic gain. However, the benefit of shifting recombination hotspots was only observed when QTN were initially outside recombination hotspots, and genetic gain decreased if QTN were initially inside recombination hotspots.

The default scenario for the results is 500 regions per chromosome with randomly distributed QTN. Within each simulation replicate, each scenario had the same first 21 generations (i.e. generations −20 to 0), thus these initial generations are omitted from the figures included in this paper. All results are standardised to generation 0 and are presented for generations 0 to 80 only. All figures represent the average of the 10 replicates of each scenario (information on the standardised values for the replicate mean and between replicate variation is in Additional file 1). In all the figures, the red lines indicate results for when recombination hotspots were kept constant and the blue lines indicate results for when recombination hotspots were shifted in generation 0. The results are split into four sections: (1) effect of shifting recombination hotspots; (2) effect of the distribution of QTN; (3) effect of collocated QTN and recombination hotspots; and (4) effect of the number of regions per chromosome. Within each of these sections, we evaluated the genetic gain achieved and the change in genetic and genic variance.

Effect of shifting recombination hotspots

Shifting recombination hotspots reduced the rate of decline in the genetic and genic variance. This in turn resulted in an increase in genetic gain compared to when recombination hotspots were not shifted. This is shown in Fig. 2, which plots the standardised (a) genetic variance, (b) genic variance and (c) genetic gain against time when QTN were randomly distributed (i.e. scenario 1). Figure 2 shows that the benefit was most apparent in the long term and that very little extra genetic gain was achieved with shifting in the short term.

Effect of the distribution of QTN

Figure 3 is a comparison of the effect of shifting recombination hotspots on (a) genetic variance, (b) genic variance and (c) genetic gain when QTN were randomly distributed (solid lines, scenario 1) versus when QTN were clustered (dashed lines, scenario 2). Figure 3 shows that shifting recombination hotspots reduced the decline in genetic and genic variance more when QTN were clustered than when they were randomly distributed. This in turn meant that shifting recombination hotspots increased genetic gain more when QTN were clustered than when they were randomly distributed. Figure 3 also shows that shifting recombination hotspots has a smaller effect on genetic variance, genic variance and genetic gain when QTN were randomly distributed compared to when QTN were clustered. This is due to the higher chance of recombination (with or without shifting) between a pair of randomly distributed QTN than between a pair of clustered QTN. This result also suggests that even in the absence of shifting, more genetic gain is likely to be achieved for traits that are influenced by randomly distributed QTN compared to traits influenced by clustered QTN.

Effect of co-located QTN and recombination hotspots

Figure 4 shows the comparison of the effect of shifting recombination hotspots on (a) genetic variance, (b) genic variance and (c) genetic gain when QTN were clustered and were initially outside recombination hotspots (solid lines, scenario 2) or were initially inside recombination hotspots (dashed lines, scenario 3). Figure 4 shows that the decline in genetic and genic variance was greater when recombination hotspots were shifted out of QTN clusters (scenario 3), which was reflected as a decrease in genetic gain.

Effect of the number of regions per chromosome

Figures 5 and 6 demonstrate the effect of the number of recombination hotspots (25, 50 or 100) on genetic variance (panels a, b, c), genic variance (panels d, e, f) and genetic gain (Fig. 6, panels a, b, c) when QTN were randomly distributed (scenario 1). Figures 5 and 6 show that shifting recombination hotspots reduced the decline of

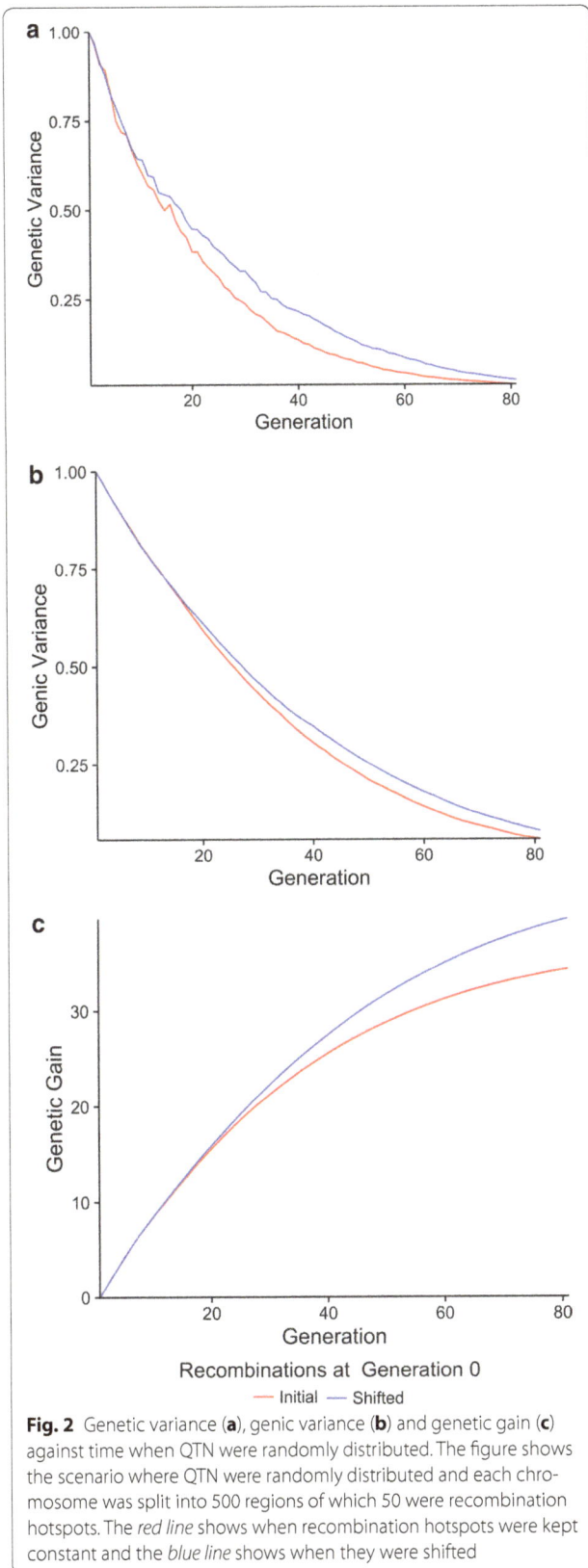

Fig. 2 Genetic variance (**a**), genic variance (**b**) and genetic gain (**c**) against time when QTN were randomly distributed. The figure shows the scenario where QTN were randomly distributed and each chromosome was split into 500 regions of which 50 were recombination hotspots. The *red line* shows when recombination hotspots were kept constant and the *blue line* shows when they were shifted

genetic and genic variance more and provided greater genetic gain when the number of recombination hotspots was small compared to when it was large. The benefit of shifting recombination hotspots was also observed much more quickly when the number of recombination hotspots was small. In summary, these results suggest that shifting recombination hotspots would have a larger and faster effect when there are long regions of the genome without recombination and where the number of recombination events per chromosome is small.

Discussion

We have split the discussion into two parts. The first part addresses the possible advantages and disadvantages of shifting recombination hotspots in livestock breeding. The second part addresses the assumptions of our analyses and the feasibility of shifting recombination hotspots in livestock.

Possible advantages and disadvantages of shifting recombination hotspots in livestock breeding

Our results show that shifting recombination hotspots to regions that have not recombined for 21 generations of selection released more standing allelic variation and increased genetic gain. However, the benefit of shifting recombination hotspots depended on the distribution of QTN and the number and location of recombination hotspots. We observed the largest benefit when QTN were clustered and the number of recombination hotspots was small. When the number of recombination hotspots was large, QTN alleles in different genomic regions recombined more often. This meant that a larger amount of variance was already available for selection without shifting recombination hotspots and so the benefit of shifting recombination hotspots for increasing genetic gain was smaller. When QTN were initially outside recombination hotspots, shifting recombination hotspots increased genetic gain. However, when QTN were initially in recombination hotspots, shifting decreased genetic gain compared to what would be achieved if recombination hotspot locations were kept constant. Although this result is not unexpected, it highlights the crucial point that care should be taken in selecting the genomic locations where recombination hotspots should be added or removed. Therefore, for a recombination hotspot shifting strategy to be effective in practice, knowledge on the locations of QTN and recombination hotspots along the genome would be useful. The ability to discover QTN underlying traits of interest, to map recombination hotspots, and the feasibility of manipulating recombination hotspot locations in the genome are further discussed below.

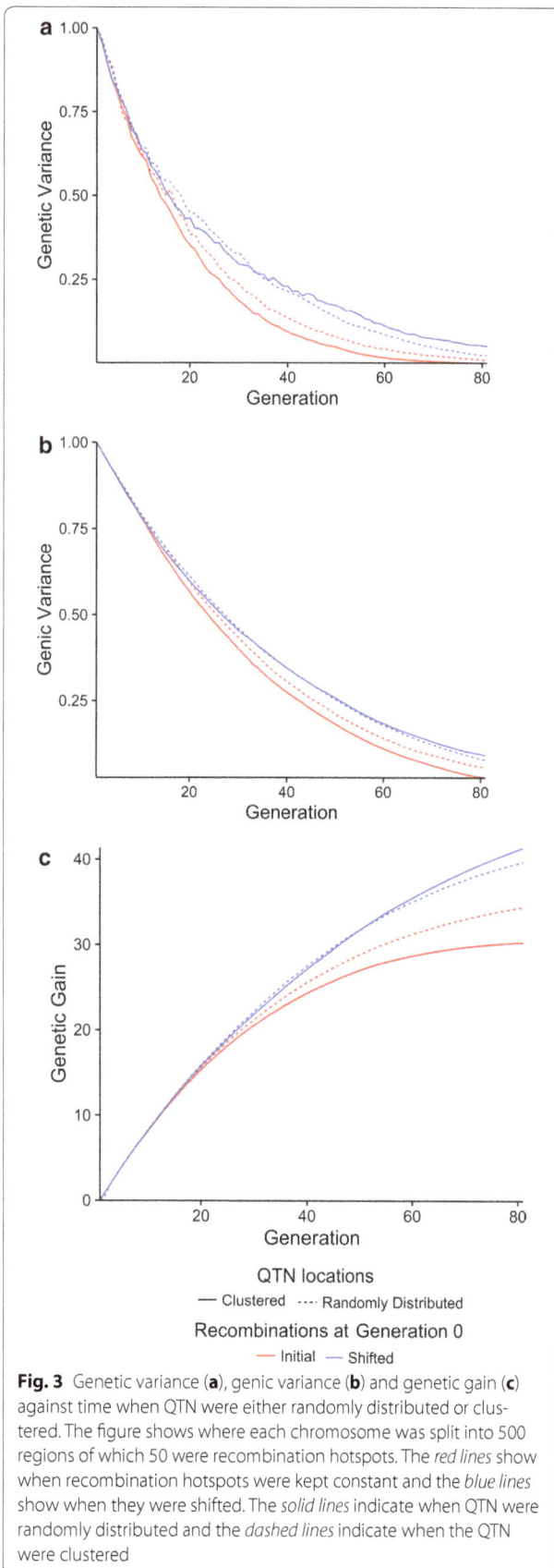

Fig. 3 Genetic variance (**a**), genic variance (**b**) and genetic gain (**c**) against time when QTN were either randomly distributed or clustered. The figure shows where each chromosome was split into 500 regions of which 50 were recombination hotspots. The *red lines* show when recombination hotspots were kept constant and the *blue lines* show when they were shifted. The *solid lines* indicate when QTN were randomly distributed and the *dashed lines* indicate when the QTN were clustered

Assumptions in our analyses and feasibility of shifting recombination hotspots in livestock

The benefit of shifting recombination hotspots to increase genetic gain that was observed in this study must be interpreted in the context that some of the assumptions made are patently oversimplified and currently not all technologically possible. We group these assumptions into the following two categories and expand on each assumption below: (1) genetic architecture of the trait, and (2) current state of technologies and the feasibility of shifting recombination hotspots in livestock.

Genetic architecture of the trait

We assumed a quantitative trait influenced by 10,000 QTN with known genomic locations, effect sizes and allele frequencies. When evaluating the value of shifting recombination hotspots in the various scenarios, we considered only additive effects of QTN (i.e. no epistasis and no dominance) and assumed independence between QTN. We also only evaluated a subset of all possible scenarios for the distribution of QTN and recombination hotspots in the genome. Specifically, we assumed that QTN were either clustered or randomly distributed and did not evaluate an intermediate scenario whereby some QTN would be clustered and some would be randomly distributed. We also assumed that recombinations always occurred within hotspots and never outside hotspots. We made these assumptions in order to minimise noise in the simulation and to help in the elucidation of the underlying mechanisms and effects of shifting recombination hotspots in different scenarios. We address the validity of these assumptions below and provide some discussion around the pitfalls should these assumptions not be fully met within real breeding programs.

We assumed that all QTN locations underlying the trait of interest were known. At present, knowledge of this information is sparse but it would be helpful for the practical implementation in order to know the genomic regions to where recombination hotspots should be shifted. Without this information, extra care would be required to prevent the introduction of recombination hotspots in regions where QTN alleles are in coupling phase (i.e. are favourably linked) or where QTN exist in permutations that have positive epistatic interactions. Although information of QTN at the nucleotide level is largely unknown, cruder measures derived from classical quantitative trait locus (QTL) mapping, regional heritability mapping (e.g. [37–42]) or functional genome annotation [43] are available and could be used to crudely identify regions of the genome that may be suitable for introducing recombination hotspots. We believe that much of the benefit of shifting recombination hotspots would likely be obtained with crude knowledge of

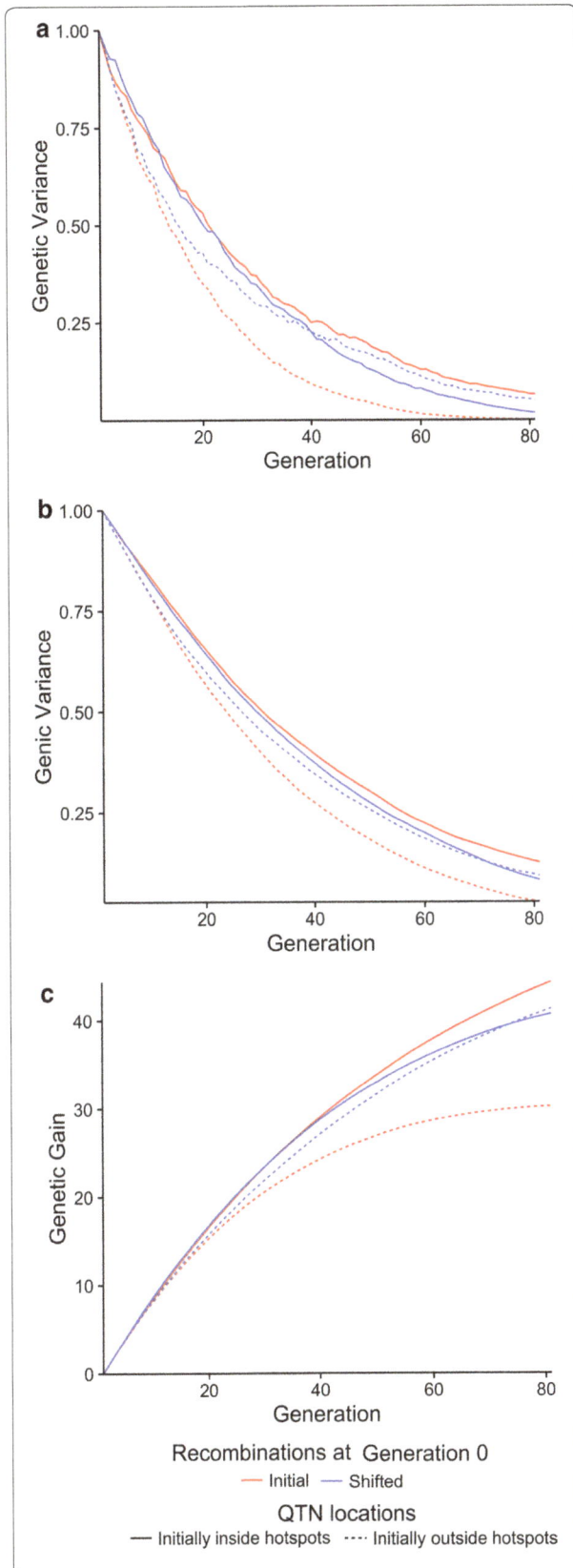

Fig. 4 Genetic variance (**a**), genic variance (**b**) and genetic gain (**c**) against time for generations 0 to 80. The figure shows where each chromosome was split into 500 regions of which 50 were recombination hotspots. The *red lines* indicate when recombination hotspots were kept constant and the *blue lines* show when they were shifted. The *solid lines* indicate when QTN were outside recombination hotspots in generations −20 to 0 and the *dashed lines* indicate when QTN were in recombination hotspots in generations −20 to 0

regions of the genome that harbour QTN rather than very refined knowledge of the precise location and effect of each QTN. That said, with the advances in genome science that have been and are likely to be made in the next few decades and the shift in livestock breeding programs towards the routine collection of sequence data, knowledge of the precise location and effect of QTN that underlie quantitative traits is likely to increase.

We assumed that the inheritance of the simulated quantitative trait was fully additive and did not simulate the effects of dominance or epistasis. In our view, dominance would just scale the benefits of shifting recombination up or down. It would not alter the general trends that were observed from a purely additive model because dominance, as with additivity, acts at each QTN independently of actions at other QTN. However, epistasis could greatly alter the general trends. If large epistatic effects exist, they could particularly affect the scenarios where QTN are clustered by function. For example, clustering of QTN could be caused by selection for specific combinations of favourable alleles or could be due to sharing of common regulatory elements, and introducing a recombination hotspot to within these clusters would break up these favourable allele combinations. This would reduce genetic gain but could also have fitness consequences. However, the properties of epistasis are largely unknown and thus difficult to simulate, the impact (if any) of epistasis is not well understood, and the data and theory suggest that epistasis has a minor contribution to the total variation for quantitative traits in livestock populations [44].

We assumed that QTN were either randomly distributed or were clustered in specific regions in the genome. There is some evidence that QTN may be distributed in clusters along the genome. For example, Wood et al. [45] found 697 significant hits from genome-wide association studies (GWAS) that together explained one-fifth of the heritability for human height in a large dataset. These 697 hits were distributed along the human genome in 423 distinct clusters that were enriched for genes. Regional heritability mapping suggests that other traits in other species are similarly distributed [46]. Such clustering may well be common in livestock populations and

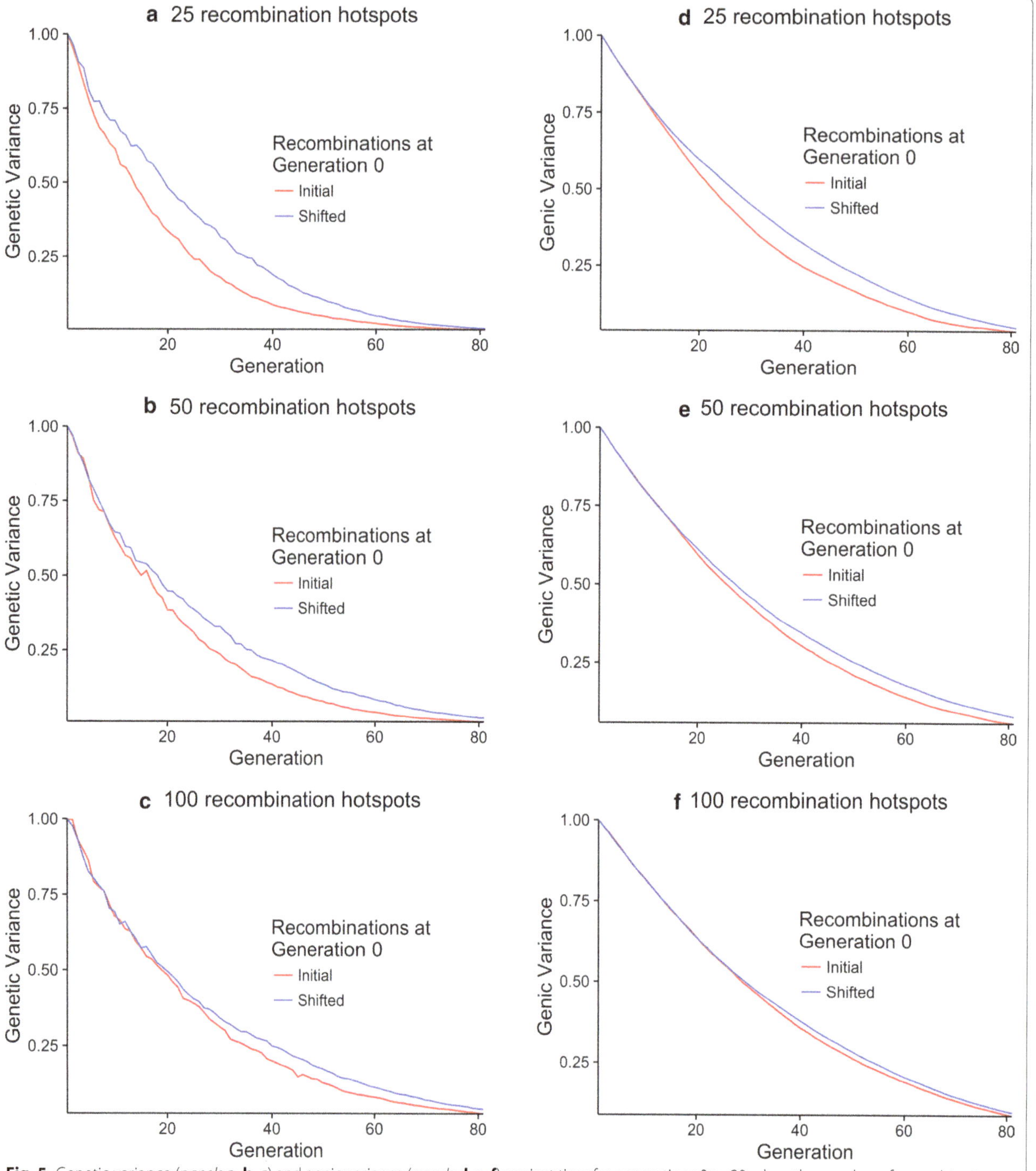

Fig. 5 Genetic variance (*panels* **a**, **b**, **c**) and genic variance (*panels* **d**, **e**, **f**) against time for generations 0 to 80 when the number of recombination hotspots was 25 (*panels* **a** and **d**), 50 (*panels* **b** and **e**) and 100 (*panels* **c** and **f**). The *red lines* show when recombination hotspots were kept constant and the *blue lines* show when they were shifted

knowledge of this clustering, combined with knowledge of the distribution of recombination hotspots, could be used to determine the potential extra genetic gain that could be achieved by shifting recombination hotspots.

In the present study, we chose two extremes of clustering (clustered or randomly distributed) for the purposes of simplicity and to demonstrate the effect of shifting recombination hotspots in these scenarios. Any benefit

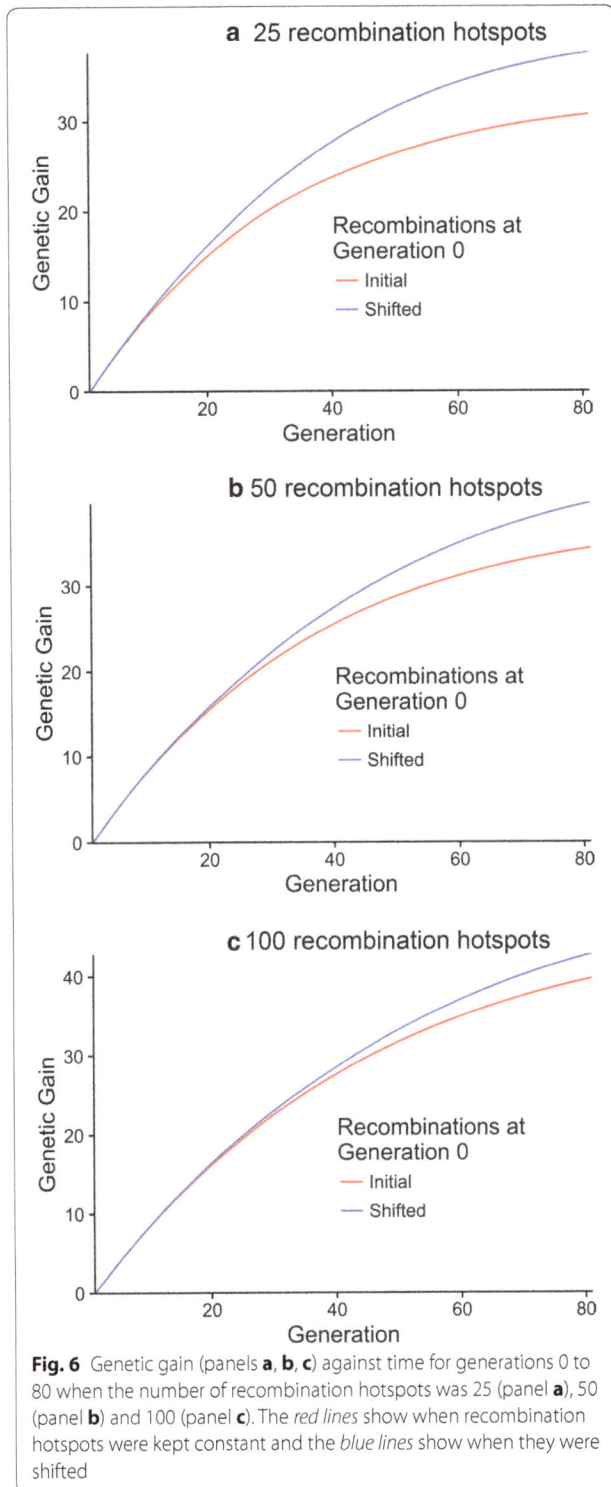

Fig. 6 Genetic gain (panels **a**, **b**, **c**) against time for generations 0 to 80 when the number of recombination hotspots was 25 (panel **a**), 50 (panel **b**) and 100 (panel **c**). The *red lines* show when recombination hotspots were kept constant and the *blue lines* show when they were shifted

We assumed that recombination events only occurred inside and never outside hotspots. This extreme scenario was chosen again with the aim of minimising potential sources of noise that would confound the effects due to shifting recombination hotspots alone. Furthermore, empirical studies across a number of species that aim to map recombination events have shown that most, if not all, recombination tends to occur in hotspots [47–49]. Further evidence from empirical studies in humans [12, 50] and livestock species such as cattle [19], pigs [9] and chicken [10] have shown that, in general, recombination tends to occur within defined regions of the genome. If recombination events outside of hotspots were more common and were more randomly distributed across the chromosome then shifting recombination hotspots in livestock would have small benefit.

Feasibility of shifting recombination hotspots

The biggest assumption in our study is that shifting recombination hotspots in livestock breeding programs is biologically, technologically and economically feasible. To effectively shift recombination hotspots, the biological mechanisms controling their exact locations and that initiate and complete a recombination need to be well characterised. As described above, this has been extensively done in model unicellular organisms including many bacterial species and yeast [47–49].

In many species including many livestock species, the mechanisms that underlie recombination are only partially understood. The major gene that determines the positions of recombination events in most mammal species is *PRDM9* [7, 28, 51]. The number of zinc finger domains in *PRDM9* is highly variable within and across species and even between individuals in the same population, which results in a high diversity in the regions of the genome where the PRDM9 protein will bind and thus the exact locations of recombination events in the genome [7, 28, 31, 52]. Using such knowledge, shifting recombination hotspots in mammalian livestock species could be achieved by (1) introducing a new *PRDM9* paralog, (2) changing the number of zinc finger domains in a single *PRDM9* gene, (3) changing the number of zinc finger recognition sites in a region of the genome where PRDM9 already binds, and/or (4) adding or removing new PRDM9 zinc finger recognition sites in the genome in mammals. All of these could potentially be achieved using genome editing technologies such as CRISPR-Cas9 [33], provided that these technologies are approved for use and could be shown to cost-effectively increase genetic gain in livestock.

Conclusions

Recombination is an important biological process for the release of standing allelic variation and could enable a longer sustained response to selection in breeding

accrued from shifting recombination hotspots in real breeding programs where QTN are both clustered and randomly distributed will be between these two extremes.

programs. In this study, we used simulation to show that shifting recombination hotspots to regions of the genome where recombination had not occurred for 21 generations of selection increased genetic gain. However, the benefit of shifting these depended on the locations of QTN and recombination hotspots in the genome. The greatest increase in genetic gain was achieved when QTN were clustered, the number of recombination hotspots was small, and QTN were initially located outside of recombination hotspots. If QTN were initially inside recombination hotspots, shifting them decreased genetic gain. Although currently not technologically possible, advances in genomic technologies such as large-scale sequencing and genome editing over the next decades could make the shifting of recombination hotspots feasible, beneficial and cost-effective for increasing genetic gain in breeding programs.

Authors' contributions
JMH conceived the study. SG and JMH designed the study. SG and GG modified the simulation program. SG performed the study and wrote the first draft of the manuscript. JMH, GG, MB and SEJ helped to interpret the results and refine the manuscript. All authors read and approved the final manuscript.

Author details
[1] The Roslin Institute and Royal (Dick) School of Veterinary Studies, The University of Edinburgh, Easter Bush, Midlothian, Scotland, UK. [2] Institute of Evolutionary Biology, The University of Edinburgh, Charlotte Auerbach Road, Edinburgh EH9 3FL, UK.

Acknowledgements
The authors acknowledge the financial support from the BBSRC ISPG to The Roslin Institute "BB/J004235/1", from Genus PLC and from Grant Numbers "BB/M009254/1", "BB/L020726/1", "BB/N004736/1", "BB/N004728/1", "BB/L020467/1", BB/N006178/1 and Medical Research Council (MRC) Grant Numbers "MR/M000370/1". This work has made use of the resources provided by the Edinburgh Compute and Data Facility (ECDF) (http://www.ecdf.ed.ac.uk). The authors thank Dr Andrew Derrington (Scotland, UK) for assistance in refining the manuscript.

Competing interests
The authors declare that they have no competing interests.

References
1. Daetwyler HD, Capitan A, Pausch H, Stothard P, van Binsbergen R, Brøndum RF, et al. Whole-genome sequencing of 234 bulls facilitates mapping of monogenic and complex traits in cattle. Nat Genet. 2014;46:858–65.
2. Harris BL, Johnson DL. Genomic predictions for New Zealand dairy bulls and integration with national genetic evaluation. J Dairy Sci. 2010;93:1243–52.
3. Pryce JE, Hayes BJ, Goddard ME. Genotyping dairy females can improve the reliability of genomic selection for young bulls and heifers and provide farmers with new management tools. In: Proceedings of the 38th international committee for animal recording meeting: 28 May–1 June 2012, Cork; 2012. p. 28.
4. Otto SP, Barton NH. Selection for recombination in small populations. Evolution. 2001;55:1921–31.
5. Hill WG, Robertson A. The effect of linkage on limits to artificial selection. Genet Res. 2007;89:311–36.
6. Battagin M, Gorjanc G, Faux AM, Johnston SE, Hickey JM. Effect of manipulating recombination rates on response to selection in livestock breeding programs. Genet Sel Evol. 2016;48:44.
7. Sandor C, Li W, Coppieters W, Druet T, Charlier C, Georges M. Genetic variants in REC8, RNF212, and PRDM9 influence male recombination in cattle. PLoS Genet. 2012;8:e1002854.
8. Weng ZQ, Saatchi M, Schnabel RD, Taylor JF, Garrick DJ. Recombination locations and rates in beef cattle assessed from parent-offspring pairs. Genet Sel Evol. 2014;46:34.
9. Tortereau F, Servin B, Frantz L, Megens HJ, Milan D, Rohrer G, et al. A high-density recombination map of the pig reveals a correlation between sex-specific recombination and GC content. BMC Genomics. 2012;13:586.
10. Groenen MAM, Wahlberg P, Foglio M, Cheng HH, Megens HJ, Crooijmans RPMA, et al. A high-density SNP-based linkage map of the chicken genome reveals sequence features correlated with recombination rate. Genome Res. 2009;19:510–9.
11. Maddox JF, Davies KP, Crawford AM, Hulme DJ, Vaiman D, Cribiu EP, et al. An enhanced linkage map of the sheep genome comprising more than 1000 loci. Genome Res. 2001;11:1275–89.
12. Mackiewicz D, de Oliveira PMC, Moss de Oliveira S, Cebrat S. Distribution of recombination hotspots in the human genome—a comparison of computer simulations with real data. PLoS One. 2013;8:e65272.
13. Paigen K, Petkov P. Mammalian recombination hot spots: properties, control and evolution. Nat Rev Genet. 2010;11:221–33.
14. Choi K, Henderson IR. Meiotic recombination hotspots—a comparative view. Plant J. 2015;83:52–61.
15. Singhal S, Leffler EM, Sannareddy K, Turner I, Venn O, Hooper DM, et al. Stable recombination hotspots in birds. Science. 2015;350:928–32.
16. Kaur T, Rockman MV. Crossover heterogeneity in the absence of hotspots in Caenorhabditis elegans. Genetics. 2014;196:137–48.
17. Baudat F, Imai Y, de Massy B. Meiotic recombination in mammals: localization and regulation. Nat Rev Genet. 2013;14:794–806.
18. Stevison LS, Woerner AE, Kidd JM, Kelley JL, Veeramah KR, McManus KF, et al. The time scale of recombination rate evolution in Great Apes. Mol Biol Evol. 2016;33:928–45.
19. Ma L, O'Connell JR, VanRaden PM, Shen B, Padhi A, Sun C, et al. Cattle sex-specific recombination and genetic control from a large pedigree analysis. PLoS Genet. 2015;11:e1005387.
20. Baker Z, Schumer M, Haba Y, Bashkirova L, Holland C, Rosenthal GG, et al. Repeated losses of PRDM9-directed recombination despite the conservation of PRDM9 across vertebrates. Elife. 2017;6:e24133 (pii).
21. Buard J, Barthes P, Grey C, de Massy B. Distinct histone modifications define initiation and repair of meiotic recombination in the mouse. EMBO J. 2009;28:2616–24.
22. Buard J, Rivals E, Dunoyer de Segonzac D, Garres C, Caminade P, de Massy B, et al. Diversity of Prdm9 zinc finger array in wild mice unravels new facets of the evolutionary turnover of this coding minisatellite. PLoS One. 2014;9:e85021.
23. Oliver PL, Goodstadt L, Bayes JJ, Birtle Z, Roach KC, Phadnis N, et al. Accelerated evolution of the Prdm9 speciation gene across diverse metazoan taxa. PLoS Genet. 2009;5:e1000753.
24. Groeneveld LF, Atencia R, Garriga RM, Vigilant L. High diversity at PRDM9 in chimpanzees and bonobos. PLoS One. 2012;7:e39064.
25. Thomas JH, Emerson RO, Shendure J. Extraordinary molecular evolution in the PRDM9 fertility gene. PLoS One. 2009;4:e8505.
26. Ségurel L, Leffler EM, Przeworski M. The case of the fickle fingers: how the PRDM9 zinc finger protein specifies meiotic recombination hotspots in humans. PLoS Biol. 2011;9:e1001211.
27. Neale MJ. PRDM9 points the zinc finger at meiotic recombination hotspots. Genome Biol. 2010;11:104.
28. Ahlawat S, Sharma P, Sharma R, Arora R, De S. Zinc Finger domain of the PRDM9 gene on chromosome 1 exhibits high diversity in ruminants but its paralog PRDM7 contains multiple disruptive mutations. PLoS One. 2016;11:e0156159.

29. Parvanov ED, Petkov PM, Paigen K. *Prdm9* controls activation of mammalian recombination hotspots. Science. 2009;327:835.

30. Billings T, Parvanov ED, Baker CL, Walker M, Paigen K, Petkov PM. DNA binding specificities of the long zinc-finger recombination protein PRDM9. Genome Biol. 2013;14:R35.

31. Winckler W, Myers SR, Richter DJ, Onofrio RC, McDonald GJ, Bontrop RE, et al. Comparison of fine-scale recombination rates in humans and chimpanzees. Science. 2005;308:107–11.

32. Úbeda F, Wilkins JF. The Red Queen theory of recombination hotspots. J Evol Biol. 2011;24:541–53.

33. Davies B, Hatton E, Altemose N, Hussin JG, Pratto F, Zhang G, et al. Re-engineering the zinc fingers of PRDM9 reverses hybrid sterility in mice. Nature. 2016;530:171–6.

34. Chen GK, Marjoram P, Wall JD. Fast and flexible simulation of DNA sequence data. Genome Res. 2009;19:136–42.

35. Hickey JM, Gorjanc G. Simulated data for genomic selection and genome-wide association studies using a combination of coalescent and gene drop methods. G3 (Bethesda). 2012;2:425–7.

36. Faux A-M, Gorjanc G, Gaynor RC, Battagin M, Edwards SM, Wilson DL, et al. AlphaSim: software for breeding program simulation. Plant Genome. 2016. doi:10.3835/plantgenome2016.02.0013.

37. Wang Y, Cao X, Gu X, Hu X. P5011 Fine mapping the QTL for growth traits in outbred chicken advanced intercross lines by improved ddGBS. J Anim Sci. 2016;94:120–1.

38. González-Prendes R, Quintanilla R, Cánovas A, Manunza A, Figueiredo Cardoso T, Jordana J, et al. Joint QTL mapping and gene expression analysis identify positional candidate genes influencing pork quality traits. Sci Rep. 2017;7:39830.

39. Müller MP, Rothammer S, Seichter D, Russ I, Hinrichs D, Tetens J, et al. Genome-wide mapping of 10 calving and fertility traits in Holstein dairy cattle with special regard to chromosome 18. J Dairy Sci. 2017;100:1987–2006.

40. Atlija M, Arranz JJ, Martinez-Valladares M, Gutiérrez-Gil B. Detection and replication of QTL underlying resistance to gastrointestinal nematodes in adult sheep using the ovine 50 K SNP array. Genet Sel Evol. 2016;48:4.

41. Lipkin E, Strillacci MG, Eitam H, Yishay M, Schiavini F, Soller M, et al. The use of Kosher phenotyping for mapping QTL affecting susceptibility to bovine respiratory disease. PLoS One. 2016;11:e0153423.

42. Fallahsharoudi A, de Kock N, Johnsson M, Bektic L, Ubhayasekera SJKA, Bergquist J, et al. QTL mapping of stress related gene expression in a cross between domesticated chickens and ancestral red junglefowl. Mol Cell Endocrinol. 2017;446:52–8.

43. Andersson L, Archibald AL, Bottema CD, Brauning R, Burgess SC, Burt DW, et al. Coordinated international action to accelerate genome-to-phenome with FAANG, the functional annotation of animal genomes project. Genome Biol. 2015;16:57.

44. Hill WG, Goddard ME, Visscher PM. Data and theory point to mainly additive genetic variance for complex traits. PLoS Genet. 2008;4:e1000008.

45. Wood AR, Esko T, Yang J, Vedantam S, Pers TH, Gustafsson S, et al. Defining the role of common variation in the genomic and biological architecture of adult human height. Nat Genet. 2014;46:1173–86.

46. Riggio V, Pong-Wong R. Regional heritability mapping to identify loci underlying genetic variation of complex traits. BMC Proc. 2014;8:S3.

47. Didelot X, Maiden MC. Impact of recombination on bacterial evolution. Trends Microbiol. 2010;18:315–22.

48. Ponticelli AS, Sena EP, Smith GR. Genetic and physical analysis of the M26 recombination hotspot of *Schizosaccharomyces pombe*. Genetics. 1988;119:491–7.

49. Lam I, Keeney S. Mechanism and regulation of meiotic recombination initiation. Cold Spring Harb Perspect Biol. 2015;7(1):a016634.

50. Jeffreys AJ, Neumann R, Panayi M, Myers S, Donnelly P. Human recombination hot spots hidden in regions of strong marker association. Nat Genet. 2005;37:601–6.

51. Berglund J, Quilez J, Arndt PF, Webster MT. Germline methylation patterns determine the distribution of recombination events in the dog genome. Genome Biol Evol. 2015;7:522–30.

52. Myers S, Bowden R, Tumian A, Bontrop RE, Freeman C, MacFie TS, et al. Drive against hotspot motifs in primates implicates the *PRDM9* gene in meiotic recombination. Science. 2010;327:876–9.

13

Construction of a large collection of small genome variations in French dairy and beef breeds using whole-genome sequences

Mekki Boussaha[1*], Pauline Michot[1,2], Rabia Letaief[1], Chris Hozé[1,2], Sébastien Fritz[1,2], Cécile Grohs[1], Diane Esquerré[4], Amandine Duchesne[1], Romain Philippe[3], Véronique Blanquet[3], Florence Phocas[1], Sandrine Floriot[1], Dominique Rocha[1], Christophe Klopp[5], Aurélien Capitan[1,2] and Didier Boichard[1]

Abstract

Background: In recent years, several bovine genome sequencing projects were carried out with the aim of developing genomic tools to improve dairy and beef production efficiency and sustainability.

Results: In this study, we describe the first French cattle genome variation dataset obtained by sequencing 274 whole genomes representing several major dairy and beef breeds. This dataset contains over 28 million single nucleotide polymorphisms (SNPs) and small insertions and deletions. Comparisons between sequencing results and SNP array genotypes revealed a very high genotype concordance rate, which indicates the good quality of our data.

Conclusions: To our knowledge, this is the first large-scale catalog of small genomic variations in French dairy and beef cattle. This resource will contribute to the study of gene functions and population structure and also help to improve traits through genotype-guided selection.

Background

In recent years, advances in high-throughput sequencing technologies have offered the opportunity to partially or completely re-sequence genomes, in a relatively cost-effective manner. The availability of whole-genome sequence (WGS) data for an increasing number of individuals offers new opportunities to study genetic variations at the genomic level with unprecedented accuracy.

In the past few years, several whole-genome sequencing studies have been carried out in different dairy and beef cattle breeds and identified a huge number of single nucleotide polymorphisms (SNPs) and small insertions and deletions (InDels) [1–5]. To date, the Ensembl (http://www.ensembl.org) short variation database contains over 99 million SNPs and InDels identified in

several cattle breeds. During the first phase of the 1000 bull genomes project, the genomes of 234 bulls were sequenced, which has enabled the identification of over 28 million reliable SNPs and InDels [5]. Only 13 French bulls were included in this phase.

In this work, we performed a large-scale study to investigate both SNPs and small InDels in whole-genome sequencing data for 274 animals from several major French dairy and beef breeds. The collection of genome variations reported in this study will be useful to study their potential links with the genetic variability of economically important traits.

Methods

Animal ethics

No animal experimentation was used in this study, since no new tissue samples were collected. All whole-genome sequence data used in this study were already

---

*Correspondence: mekki.boussaha@inra.fr
[1] GABI, INRA, AgroParisTech, Université Paris-Saclay, 78350 Jouy-en-Josas, France
Full list of author information is available at the end of the article

available in our laboratory and were produced as previously described [1].

Whole-genome sequencing and sequence alignment to the reference

The whole genome of 274 animals corresponding to both French dairy and beef breeds (Table 1) were used for 2 × 100 bp paired-end sequencing on an Illumina HiSeq 2000 with a TruSeq SBS v3-HS Kit (Illumina).

Sequence alignments were carried out using the Burrows-Wheeler Alignment tool (BWA-v0.6.1-r104) [6] with the aln option with default parameters for mapping reads to the UMD3.1 bovine reference genome [7]. Potential PCR duplicates, which can adversely affect the variant calls, were removed using the MarkDuplicates tools from the Picard package version 1.4.0 [8]. Only properly paired reads with a mapping quality of at least 30 (−q = 30) were retained. The resulting BAM files were then used for all subsequent analyses.

Identification of small insertions and deletions

Small genomic variations were detected using the Genome Analysis Tool Kit 2.4–9 (GATK) version and GATK-UnifiedGenotyper as SNP caller [9]. Prior to variant discovery, reads were subjected to local realignment, coordinate sorting, quality recalibration, and removal of PCR duplicates. In the GATK analysis, we used a minimum confidence score threshold of Q30 with default parameters. We also used multi-sample variant calling in order to distinguish between a homozygous reference genotype and a missing genotype in the analyzed samples.

Table 1 Number of animals used per breed

Breed	Number of animals
Abondance	1
Aubrac	8
Brown Swiss	3
Salers	3
Tarentaise	1
Limousine	20
Simmental	1
Charolaise	34
Rouge des Prés	5
Montbéliarde	59
Normande	43
Vosgienne	4
Holstein	63
Parthenaise	2
Blonde d'Aquitaine	26
Cross-breed	1

This table lists the distribution in each breed of the 274 sequenced animals

Variant annotation

All variants were annotated with the Ensembl variant effect predictor (VEP) pipeline v81 [10] based on the Ensembl version 81 transcript set and using dbSNP build 143. The effect of the amino acid changes was predicted using SIFT [11, 12], a sequence homology-based tool that can determine whether an amino acid substitution in a protein is deleterious or tolerant.

Functional characterization of protein-coding genes with LoF variants

A set of 8337 gene products was used for gene ontology (GO) enrichment and functional analyses, using the GO [13] and the KEGG (Kyoto Encyclopedia of Genes and Genomes) [14] database resources. The Cytoscape [15] ClueGO plugin [16] was used to identify the biological functions to which genes contribute. The enrichment of biological terms and groups were set as follows. First, we used the enrichment tests based on the hyper-geometric distribution. Second, we set the statistical significance to 0.05 (p ≤ 0.05), and we used the Benjamini-Hochberg adjustment to correct the p value for the terms and the groups created by ClueGO. Third, we used fusion criteria to reduce the redundancy of related terms that have similar associated genes. Finally, we set the Kappa-statistics score threshold to 0.6.

Gene Ontology (GO) enrichment was also performed using the MouseMine analysis tools available at the MGI international database resources (http://www.mousemine.org/mousemine/begin.do).

Validation of LoF variants by high-throughput genotyping

The efficiency of our calling approach and the relevance of the resulting variants were assessed by genotyping a selected panel containing 304 heterozygous deleterious missense and loss-of-function SNPs for which no homozygous individual for the alternative allele was observed in our population. Genotyping was performed using the already available Illumina BovineLD custom BeadChip [17] and a panel of 172,416 beef and dairy cattle animals (Table 2).

Results and discussion

Whole-genome sequencing and read mapping

Two hundred and seventy four animals corresponding to both French dairy and beef breeds were selected for whole-genome sequencing (Table 1), of which 62 whole-genome sequences were already published [1]; the Illumina short reads are available at the European Nucleotide Archive (ENA) with study accession number PRJEB9343 (http://www.ebi.ac.uk/ena/data/view/PRJEB9343). Overall, 103 billion raw paired-end reads 100-bases long were generated, which resulted in over

Table 2 Total number of animals genotyped using the Illumina Bovine low density BeadChip

Breed	Number of animals
Abondance	39
Brown Swiss	627
Tarentaise	49
Limousine	2084
Simmental	2
Montbéliarde	55,382
Normande	20,697
Holstein	90,970
Blonde d'Aquitaine	2566

This table summarizes the number and the distribution in each breed of the animals genotyped using the Illumina bovine low density BeadChip

ten thousand gigabases of data. On average, 95% (from 56 to 99%) of the paired-end reads were properly aligned on the UMD3.1 bovine reference genome (see Additional file 1), which is in agreement with previous studies [1, 18]. The average genome-wide sequence coverage from the mapped reads was $13.8\times$ and ranged from almost $5\times$ to around $36\times$ across the different genomes, with 236 samples sequenced at least at 10-fold average coverage (see Additional file 1).

Identification of SNPs and small InDels

A search for small genome variations with the GATK-UnifiedGenotyper software resulted in the identification of 28,164,518 variants, of which 25,210,883 were SNPs, 1769,413 small deletions and 1184,222 small insertions. Almost 87% of the deletions and 93% of the insertions identified in our study were 1 to 3 bp long (see Additional file 2). The largest deletions and insertions identified were respectively 58 and 29 bp long (see Additional file 2). Overall, 73% of the identified variants (20,647,361) were known in the Ensembl variation 83 database (build 143). The remaining 27% were considered as novel variants and should contribute to better highlight the genetic variability in cattle.

A total of 146,944 genome variants were identified as bi-allelic in our dataset but contained more than two alleles in the Ensembl variation 83 database. Of these 146,944 genome variants, only 95 positions that displayed a single variant type in our dataset overlapped with multiple variant types in the Ensembl variation 83 database. For the remaining 146,849 positions, a single variant type was observed in both databases, of which 129,356 (88.1%) SNPs and 17,493 (11.9%) InDels were identified in our dataset. Among the 129,356 discrepant SNPs, 99.3% (128,407) were reported to be tri-allelic SNPs and only 0.7% (949) corresponded to InDels in the Ensembl variation 83 database. Of the 17,493 discrepant InDels,

67.3% (11,770) corresponded to tri-allelic SNPs and 32.7% (5723) were also InDels but with multiple alleles in the Ensembl variation 83 database. In addition, we identified 88,289 positions that displayed one type of variant (i.e. SNP or InDel but not both) in our dataset but which overlapped with multiple variant types in the Ensembl variation 83 database. We also identified 517,417 variants for which the alleles differed between our dataset and the Ensembl 83 variation database. These inconsistencies could be partly explained by the use of different variant calling algorithms. Indeed, a previous study in Danish Holstein dairy cattle also reported similar inconsistencies [3]. In that study, genotype accuracy was assessed for 15 variants for which samtools-derived genotypes differed from those predicted by GATK. Their results revealed that GATK provided more accurate genotype calls than samtools.

Evaluation of sequencing genotypes

To evaluate the quality of our sequencing data-derived genotypes, we performed three different analyses. First, we used the ratio of transitions over transversions (Ts/Tv) as a diagnostic measure to assess the quality of our sequencing data. The average Ts/Tv ratio observed in our whole-genome sequencing data was 2.12 and ranged from 2.05 on BTA6 to 2.35 on BTA25 (Fig. 1). This average rate is within the same range as those observed in other species. For example, in human whole-genome sequence data, the genome-wide Ts/Tv ratio ranged from 2.0 to 2.2 [19, 20]. In mouse and pig, similar ratios were reported i.e. about 2.0 [21] and 2.04 ± 0.28 [22], respectively. DePristo et al. [19] indicated that the Ts/Tv ratio should be around 2.1 for whole-genome sequencing and that lower ratios may indicate that the sequencing data includes false positives caused by random sequencing errors. Therefore, the Ts/Tv ratio estimated in our study is indicative of good sequencing data quality.

Second, we measured the call rate by estimating the percentage of samples presenting a known genotype for each variant. On average, 95% of the variants were called in more than 90% of the samples with 13% (3,655,506) of the variants being genotyped in all 274 samples (Fig. 2).

Third, we compared our sequencing data-derived genotypes to SNP array-derived genotypes using the Illumina High-Density (HD) Bovine SNP BeadChip® which includes 777,962 SNPs [23]. Overall, both genotyping data sources were available for 152 samples. The average genotype concordance rate was around 99.1% and ranged from 91.7 to 99.8% (see Additional file 3). We also observed a dependency of chip genotype concordance on sequencing depth (see Additional file 3; Fig. 3). Lower accuracy rates were found for samples with a low depth of coverage (less than $10\times$). For 21 samples, the

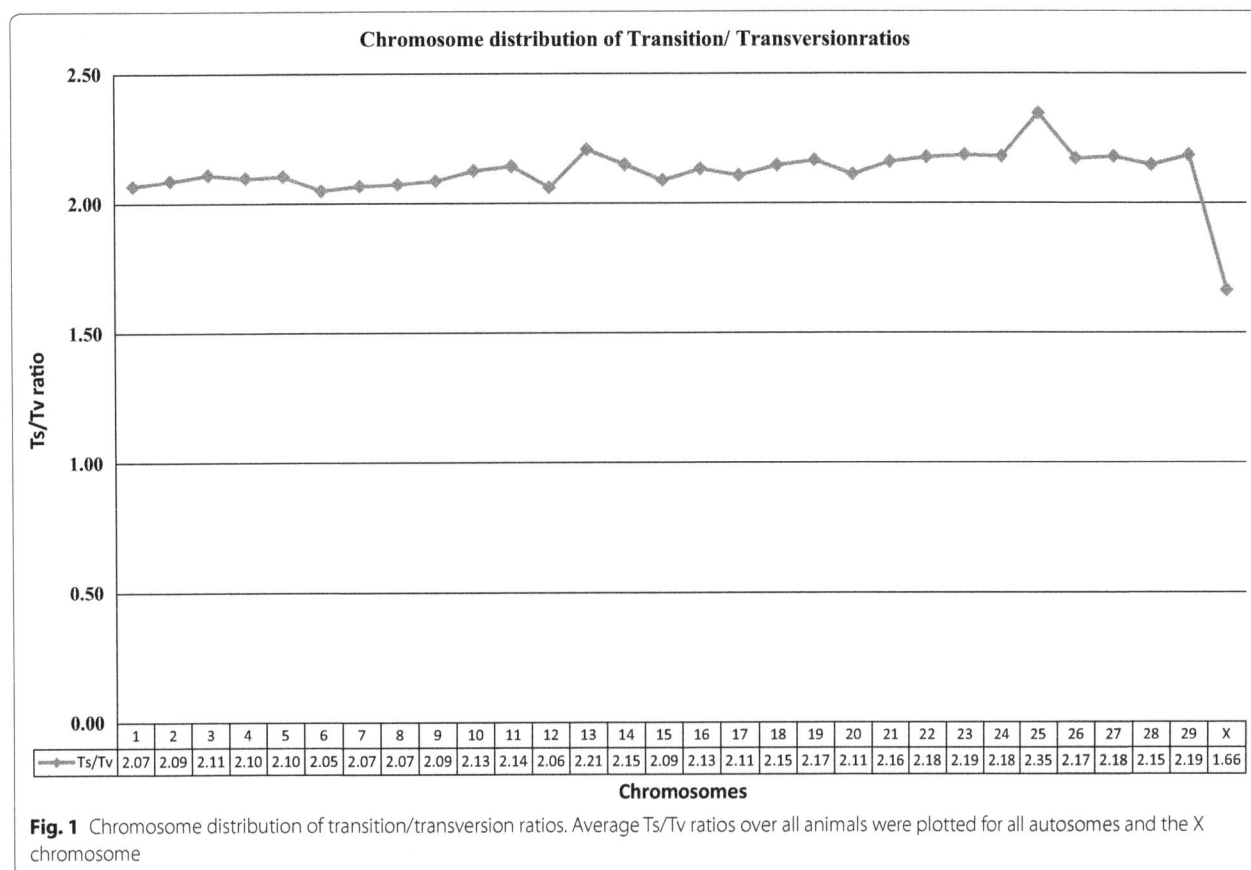

Fig. 1 Chromosome distribution of transition/transversion ratios. Average Ts/Tv ratios over all animals were plotted for all autosomes and the X chromosome

concordance rate was less than 98% but their depth of coverage was higher than 11×. Of these 21 samples, 20 had a concordance rate between 95 and 97% and were considered as acceptable. The observed lower concordance rates could be partly due to lower sequence data quality compared to the rest of our sample set.

The low missing rate and high concordance rate observed in our study can be explained by the good average genome-wide sequence coverage of the mapped reads in our data. Indeed, more than 86% of our samples were sequenced at least at an average 10-fold coverage. Another reason is the efficiency of our variation calling strategy.

Functional annotation of small genome variants

Functional annotation of the identified small genome variants was carried out using the Ensembl VEP annotation software [10]. Overall, 66% of the annotated variants were located in intergenic regions and almost 30% were identified within gene intronic sequences (Table 3). The remaining 4% were located within gene-coding, upstream and downstream regions. Of these, 85,038 variants were located within the 5′ or 3′ untranslated regions (UTR), 171 were located within genes coding for micro RNAs

(miRNAs), 96,711 missense mutations were identified within gene coding regions, 358 InDels were predicted to cause inframe insertions and 814 InDels were predicted to cause inframe deletions.

Overall, we identified 2120 variants that affected splice sites. These included 1471 splice donor and 649 splice acceptor site variants. In addition, 1159 variants were predicted to create a premature stop codon and 68 to disrupt a termination codon. Around 2287 InDels were predicted to cause a frameshift in coding sequences which were considered as loss-of-function (LOF) variants and may result in reduced or complete inactivation of protein functions by disrupting either the protein-coding gene itself or genetic regulatory elements. These LOF variant candidates are of particular interest since they might have effects on economically important traits.

Among the annotated deleterious missense and LOF variants, we identified several mutations that were previously reported to be associated with dairy and beef traits in cattle. For example, the amino acid change of phenylalanine to leucine at position 94 (F94L) of the myostatin (MSTN) protein was identified in 31 samples, among which six animals were heterozygous (three Charolaise, two Aubrac and one Rouge-des-Prés) and 25

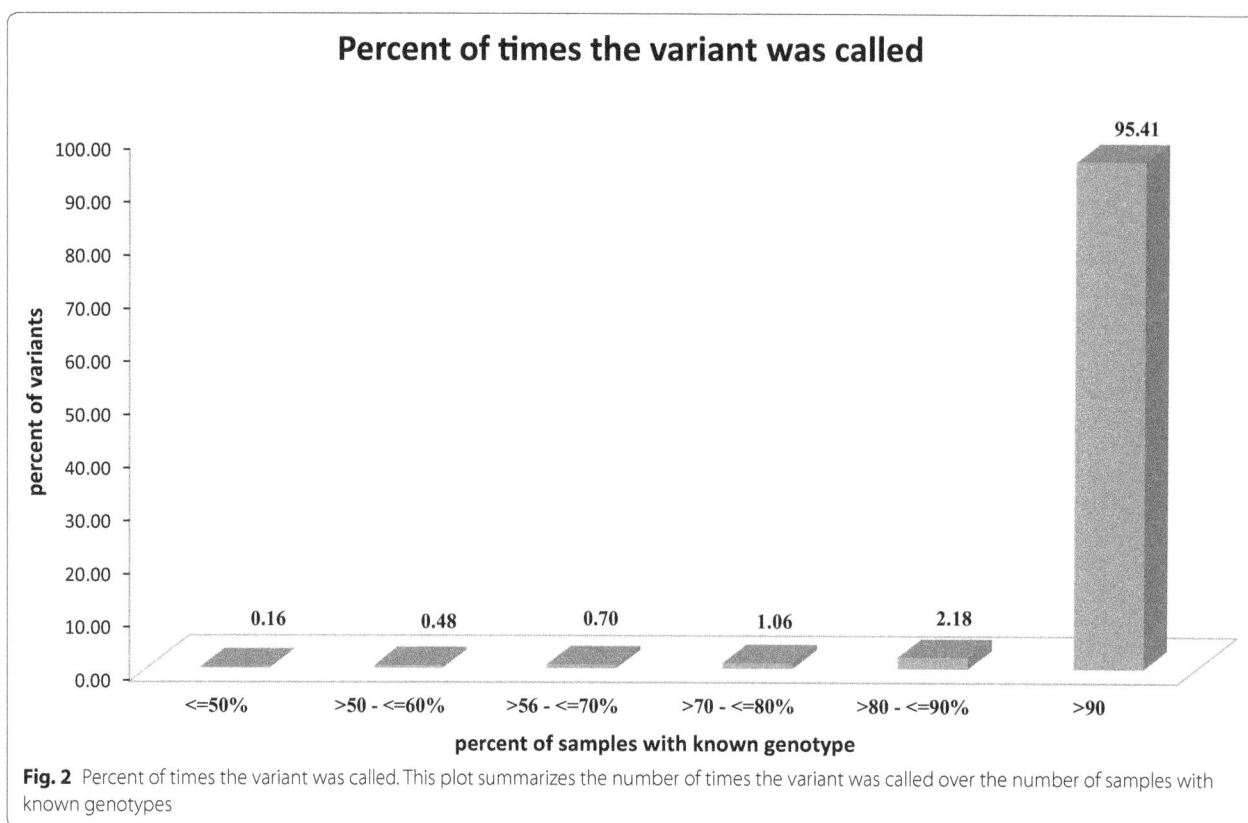

Fig. 2 Percent of times the variant was called. This plot summarizes the number of times the variant was called over the number of samples with known genotypes

were homozygous (19 Limousine and six Aubrac) for this locus. We also observed the *MSTN* pQ204* mutation in eight samples, all of which corresponded to the Charolaise breed and all animals were heterozygous. Both F94L and Q204* substitutions are associated with double muscling phenotypes in Limousine [24] and Charolaise [25] cattle, respectively.

The F279Y mutation within the *growth hormone receptor (GHR)* gene was observed in 35 samples corresponding to 29 dairy and six beef cattle animals (four Blonde d'Aquitaine, one Brown Swiss, one Charolaise, two Montbéliarde, five Normande and 22 Holstein) with the highest frequency observed in the Holstein breed (19 heterozygous and three homozygous individuals for the alternative allele). This SNP is located on BTA20 and has been shown to be associated with milk yield and composition [26, 27], feed intake, feed conversion efficiency and body energy traits [28].

Missense and LOF variants for which no homozygous individuals for the alternative allele are observed

Further analysis of the annotated variants revealed the presence of 14,469 missense and LOF variants with a significant biological impact based on SIFT predictions and for which no homozygous animal carrying the alternative

allele was observed among the 274 WGS (see Additional file 4). These were subsequently considered as our study panel in the rest of this paper.

This study panel contains 772 frameshift variants, 12,008 missense mutations with a deleterious effect predicted by SIFT with a score between 0 and 0.05, 67 start-lost variants, 583 stop variants (25 stop-lost and 558 stop-gained) and 1039 splice variants (264 splice-acceptor and 775 splice-donor variants).

The genotype distribution of our study panel revealed that seven frameshift variants were breed-specific (Table 4). Integrated Genome Viewer (IGV) visualization and inspections of BAM files for animals carrying these mutations revealed that four of the seven frameshift mutations were spurious variant calls (results not shown). The three remaining frameshift variants could be visualized and confirmed by IGV and were therefore considered as true variants. First, a five nucleotide insertion (-/CACGT) at position 66,552,044 on BTA1 was identified in two Blonde d'Aquitaine animals. This frameshift mutation was absent in both the Ensembl database and in the most recent 1000 bull genomes project dataset which contains small genomic variations for 1577 animals corresponding to 48 different breeds (Daetwyler HD, personal communication). This mutation affects the

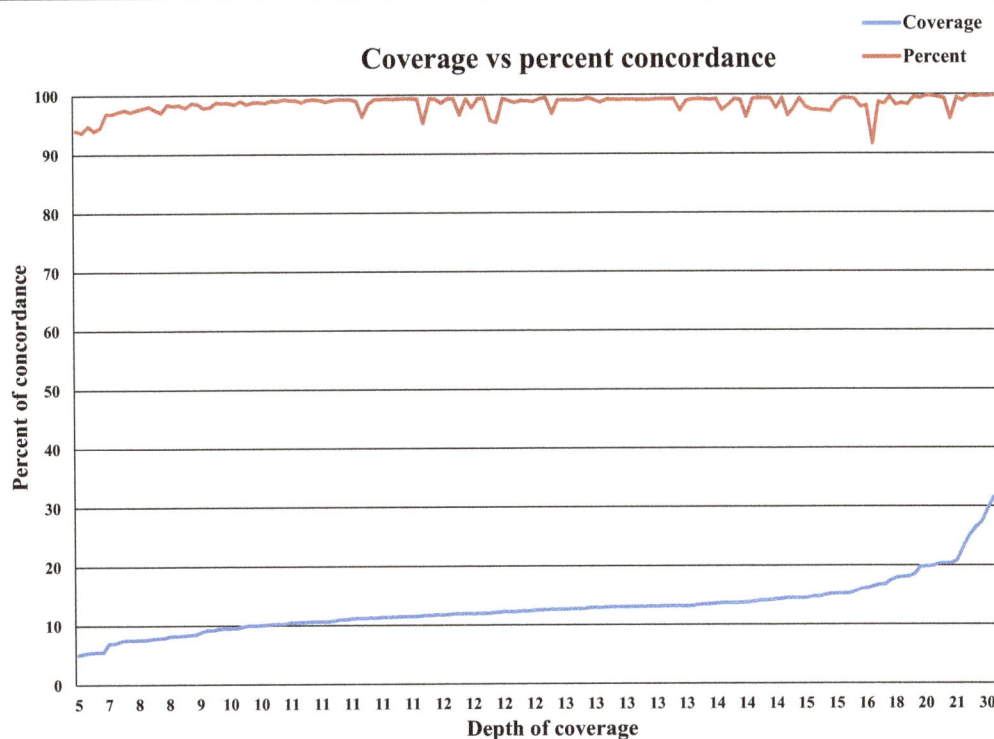

Fig. 3 Coverage versus percent concordance. This plot summarizes the sequence coverage versus the accuracy percent between sequences derived-genotypes and SNP-array ones

Table 3 Results of functional annotation by VEP

Functional class	All	SNP	InDels
3′ UTR variant	70,139	61,080	9059
5′ UTR variant	14,899	13,696	1203
Frameshift variant	2287	0	2287
Inframe deletion	814	0	814
Inframe insertion	358	0	358
Splice acceptor variant	649	510	139
Splice donor variant	1471	1378	93
Start lost	169	169	0
Stop gained	1159	1139	20
Stop lost	68	67	1
Mature miRNA variant	171	135	36
Intron variant	8,446,403	7,513,594	932,809
Downstream gene variant	1,335,987	1,179,502	156,485
Intergenic variant	18,568,837	16,664,272	1,904,565

This table summarizes the functional classification of all variants reported in this study

sequences of *A-family polymerase theta* (*POLQ*) gene by producing a frameshift insertion between amino acids number 2728 and 2729 just before the termination site. It induces a frameshift in the open reading frame which results in the addition of ten new amino acids followed by a new downstream termination site. The *POLQ* gene has been identified in several other species and was reported to play a major role in the DNA repair mechanism of double strand breaks (DSB) by alternative end-joining (alt-EJ; also called alternative non homologous end-joining (alt-NHEJ) or microhomology-mediated end joining)) [29–33]. Unlike the classical non homologous end-joining (c-NHEJ) mechanism, alt-EJ depends on resection of DNA ends to find microhomologies, which results in larger deletions and insertions [34, 35]. Inhibition of POLQ functions in mice were reported to be associated with chromosome instability phenotypes [36]. In mammalian cells, POLQ promotes the formation of chromosomal translocations and is essential for survival when the homology-directed repair (HDR) mechanism is impaired [31], which suggests that this mutation may cause embryonic lethality in cattle.

The two other frameshift mutations were identified in the Charolaise breed. The first one is a GACC insertion at position 149,472 on BTA19 and is located within an olfactory receptor gene coding sequence (*ENS-BTAG00000045560*). This variant was identified in three samples in our dataset and is also present in the Ensembl database. It leads to a frameshift mutation within the

Table 4 Distribution of LoF and deleterious variants

	Stop lost	Splice acceptor	Start lost	Frameshift	Missense deleterious	Stop gained	Splice donor
Holstein	1	15	10	0	1	5	18
Abondance	1	0	1	0	43	1	0
Cross-breed	0	0	0	1	10	1	0
Brown Swiss	0	6	1	0	4	3	3
Salers	0	1	0	0	47	1	2
Montbéliarde	1	1	3	0	705	34	18
Vosgienne	0	4	0	0	4	9	3
Normande	0	11	3	0	626	1	1
Simmental	0	2	0	0	33	3	1
Limousine	0	11	1	0	18	1	14
Charolaise	1	15	6	5	17	35	21
Parthenaise	0	0	0	0	1	9	2
Rouge des Prés	1	1	0	0	37	3	1
Tarentaise	0	1	0	0	54	0	0
Blonde d'Aquitaine	2	13	3	1	572	31	16
Aubrac	0	8	1	0	193	10	2

This table summarizes the distribution of LoF and deleterious variants in each breed and for each functional annotation class

7tm_1 (PF000001) pfam domain at amino acids 81 and 82 and creates 39 new amino acids followed by a termination site, thus producing a truncated protein, which contains only 26% (82 of 311 amino acids) of the wild type protein. The second frameshift mutation is a four nucleotide (-/AGTT) insertion identified at position 21,913,213 on BTA18. It was identified in two samples in our panel but it is absent in the Ensembl database. It is located within the *retinoblastoma-like 2* (*RBL2*) coding gene region and leads to a frameshift mutation within the RB_B box (PF01857 pfam domain) at amino acids 890–891, thus introducing 26 new amino acids before creating a premature termination site. Thus, a truncated protein representing only 78% (890/1140 amino acids) of the wild type protein is produced. RBL2, also called pRb2/p130, is a member of the retinoblastoma family of tumor suppressors [37] and its expression was reported to be altered in several cancer types [38–40]. RBL2 interacts with the E2F4 and E2F5 transcription factors and results in negative regulation of the cell cycle [41]. It is also involved in many other cellular processes, such as regulation of apoptosis and differentiation [37] and control of the length of telomeres [42].

Finally, we identified the p.Q579* mutation within the *APAF1* gene (HH1: Holstein Haplotype 1), the p.N290T deleterious missense mutation within the *GART* gene (HH4), the p.V180F deleterious missense mutation within the *SLC35A3* gene (CVM: complex vertebral malformation), the p.Q52* stop-gained variant within the *SHBG* gene (MH1: Montbéliarde Haplotype1) and the R12* stop-gained variant within the *SLC37A2* gene (MH2). All these substitutions are specific to the Holstein (HH1, HH4 and CVM) and Montbéliarde (MH1 and MH2) breeds, respectively and are considered to be strong candidate mutations for embryonic lethal defects [43].

Gene ontology and pathway analysis

In order to gain additional insight into the biological pathways and molecular functions that are affected by these variants, we performed a gene ontology (GO) enrichment and functional analysis using 8337 known Ensembl ID-associated genes retrieved from our variant annotation study (see Additional file 5). Several GO terms were significantly over-represented. For example, the six most enriched GO categories corresponding to biological processes were related to the regulation of GTPase-, Ras-, and Rho-mediated signal transduction. The three most enriched GO categories corresponding to cellular components were related to cytoskeleton and myosin complex and the five most enriched GO categories corresponding to molecular functions were related to ATP binding, adenine nucleotide binding, ATPase activity, motor activity and ribonucleotide binding.

Experimental validation of the study panel by high-throughput genotyping

Previous studies reported a significant rate of false positive calls among deleterious missense and loss-of-function variants [3, 44]. Thus, the efficiency of our calling approach and the relevance of the resulting variants were assessed by genotyping a selected panel containing 304 heterozygous deleterious missense and loss-of-function mutations for which no homozygous individual for the alternative allele was observed in our population. They were also selected based on their mapping quality (above 50) and their calling quality (above 30) scores. Genotyping was performed using the already available Illumina BovineLD custom BeadChip [17] and a panel of 172,416 animals corresponding to both beef and dairy cattle breeds (Table 2).

Overall, 276 (~91%) SNPs were polymorphic in all genotyped animals and were considered as true variants (see Additional file 6). Among these, 61 SNPs were private and were polymorphic only in one breed. Thus, they were considered as breed-specific variants i.e. two in Brown Swiss, three in Limousine, 12 in Montbéliarde, 27 in Normande, 16 in Holstein and one in Blonde d'Aquitaine. For 51 polymorphic SNPs, we observed only two genotypes. No homozygous individual for the alternative allele was observed among all genotyped samples. For these 51 variants, we determined the expected proportions of homozygous individuals for the alternative allele in each breed and then calculated the significance probability (p value) from the binomial distribution, with event probability equal to zero (which corresponded to the proportion of observed homozygous individuals for the alternative allele), and the number of observations was equal to the number of animals in each breed. For 41 of the 51 variants, there was no significant difference between the expected and the observed proportions (see Additional file 6). However, for the other 10 variants, the expected proportion was significantly different from the observed proportion in at least one breed (see Additional file 6). These corresponded to nine missense deleterious mutations and one LOF variant. This latter one corresponded to the p.Q579* mutation within the *APAF1* gene (HH1: Holstein Haplotype 1) which was previously reported as a strong candidate mutation for embryonic lethal defects [43]. As expected, significant differences between the observed and estimated proportions of homozygous individuals for the alternative allele were only observed in the Holstein breed. Two other deleterious missense mutations were also located within *CBX3* (*chromobox protein homolog 3*) and *RBBP5* (*RB binding protein 5, histone lysine methyltransferase complex subunit*) genes which are known to be associated with male germ cell survival and spermatogenesis [45] and sterility [46], respectively.

The 51 SNPs for which only two genotypes were observed were located within 42 known gene coding regions. Thus, these genes were used to carry out gene ontology (GO) and mammalian phenotype ontology (MPO) enrichment analyses using the MouseMine analysis tools (see Additional file 7). The most significant enriched MPO categories were related to abnormal nervous system morphology and phenotype, preweaning lethality, and abnormal brain development (see Additional file 7, sheet1). However, no significant GO category enrichment was obtained (see Additional file 7, sheet 2). It will be very interesting to study the effect of these variants on phenotypes of interest in cattle.

Conclusions

In this study, we performed a pan-genome assessment of small genome variations in cattle using whole-genome sequence data. Analysis of WGS data of 274 animals from both dairy and beef cattle breeds allowed the identification of over 28 millions small variations, among which we identified more than 25 million SNPs and around 3 million small insertions and deletions. To assess the quality of both our sequencing data and calling approach, we analyzed the transition to transversion ratio and the call rate, and we also compared the sequence-derived genotypes with array-derived ones. Results from all these analyses confirmed the efficiency of our sequencing data as well as the good quality of our variant calling procedure. Annotation of these variants revealed several deleterious missense and loss-of-function variants, among which we identified several mutations that were previously reported to be associated with either dairy or beef traits. Genotypic and allelic frequency distributions revealed the presence of more than 14,000 heterozygous candidate deleterious and LOF variants that segregated in the absence of individuals homozygous for the alternative allele in our population. Of these, we genotyped 172,416 animals from dairy and beef breeds with a panel of 304 SNPs, using the already available Illumina BovineLD custom BeadChip. Two hundred and seventy-six of these variants (~91%) were polymorphic in at least one breed and, thus, were considered as true variants. For 51 of the 276 polymorphic variants, we did not observe any homozygous individual for the alternative allele. These 51 variants will be useful to study their link with genetic variability of economically-important traits in cattle.

Additional files

Additional file 1. Summary of sequence coverage and sequence align-ment to the reference genome. Total and mapped reads, as well as total and percent of correctly mapped reads are indicated for each sample. The estimated sequence coverage was also obtained for each sample.

Additional file 2. Size distribution of small insertions and deletions iden-tified by GATK. Total number of small insertions and deletions and their distribution by size are indicated.

Additional file 3. Results of genotype accuracy analyses. Genotype accuracy analyses were performed by comparing sequence-derived genotypes and those obtained using the high-density SNP bead chip for 152 samples (39 Holstein, 49 Montbéliarde, 41 Normande and 23 Blonde d'Aquitaine).

Additional file 4. List of LoF and deleterious missense variants for which no homozygous individual for the alternative allele was observed in our samples. For each variant, overall and per breed genotype (0/0, 0/1 and 1/1) distribution as well as the frequency for the alternative allele are indicated. The functional annotation results and SIFT predictions are also indicated.

Additional file 5. GO enrichment analysis using the Cytoscape ClueGO plugin. The results of GO enrichment analyses were obtained using the overall list of LoF and deleterious missense variants for which no homozy-gous individual for the alternative allele was observed in our samples. The results for biological processes (BP), cellular components (CP) and molecular functions (MF) are in sheets 1, 2 and 3, respectively.

Additional file 6. Frequency of the alternative allele of SNPs in our vali-dation panel. (Sheet 1) Frequency of the alternative allele was calculated using the genotyping data obtained by genotyping all variants of our validation panel with the custom Illumina Bovine low-density BeadChip. Monomorphic variants (homozygous for the reference allele) in each breed are indicated by "0". Missing genotyping data are indicated by "ng". The numbers of observed types of genotype are in column 3 and coded as follows: (1) monomorphic variants; (2) and (3) correspond to the 276 polymorphic variants with (2) corresponding to the 51 variants for which no homozygous individual for the alternative allele was observed and (3) corresponding to variants for which we observed the three genotypes. (Sheet 2) Results of the exact binomial distribution test with event prob-ability being set to zero (which corresponds to the number of observed homozygous individuals for the alternative allele), and n was equal to the number of animals in each breed (indicated between parenthesis). For the Simmental breed, we had only two genotyped animals and therefore we did not do the binomial distribution test.

Additional file 7. Results of GO enrichment. GO enrichment analyses were performed using the MouseMine analysis tools and the results are summarized in sheets 1 and 2. Mammalian phenotype enrichment results are in sheet 1 and GO enrichment results in sheet 2.

Authors' contributions
MB and DB designed the study, and drafted the manuscript. PM, CH and AC analyzed the rare variants. MB, RL and CK carried out bioinformatics analysis. DE performed the whole-genome sequencing and sequence alignment to the reference. DB, DR, SFr, AC, RP, VB, FP, SFI, CG and AD conceived the whole-genome sequencing projects. All authors read and approved the final manuscript.

Author details
[1] GABI, INRA, AgroParisTech, Université Paris-Saclay, 78350 Jouy-en-Josas, France. [2] Allice, Maison Nationale des Eleveurs, 75012 Paris, France. [3] GMA, INRA, Université de Limoges, 87060 Limoges Cedex, France. [4] GenPhySE, INRA, INPT, ENVT, Université de Toulouse, Castanet Tolosan, France. [5] SIGENAE, UR 875, INRA, 31362 Castanet-Tolosan, France.

Acknowledgements
This work was funded by INRA (GenSSeq project from Selgen metaprogram), the Agence Nationale de la Recherche and Apis-Gene jointly (Cartoseq ANR-10-GENM-018, Gembal ANR-10-GENM-0014), the Limousin Région (BovSeq), Apis-Gene (ReproSeq, Akelos), Valogene (Valoseq) and the INRA Animal Genetics division (BOVATAX AAPGA 2012).

Competing interests
The authors declare that they have no competing interests.

References
1. Boussaha M, Esquerré D, Barbieri J, Djari A, Pinton A, Letaief R, et al. Genome-wide study of structural variants in bovine Holstein, Mont-béliarde and Normande dairy breeds. PLoS One. 2015;10:e0135931.
2. Stothard P, Liao X, Arantes AS, De Pauw M, Coros C, Plastow GS, et al. A large and diverse collection of bovine genome sequences from the Canadian cattle genome project. Gigascience. 2015;4:49.
3. Das A, Panitz F, Gregersen VR, Bendixen C, Holm LE. Deep sequencing of Danish Holstein dairy cattle for variant detection and insight into poten-tial loss-of-function variants in protein coding genes. BMC Genomics. 2015;16:1043.
4. Baes CF, Dolezal MA, Koltes JE, Bapst B, Fritz-Waters E, Jansen S, et al. Evaluation of variant identification methods for whole genome sequenc-ing data in dairy cattle. BMC Genomics. 2014;15:948.
5. Daetwyler HD, Capitan A, Pausch H, Stothard P, van Binsbergen R, Brøn-dum RF, et al. Whole-genome sequencing of 234 bulls facilitates mapping of monogenic and complex traits in cattle. Nat Genet. 2014;46:858–65.
6. Li H, Durbin R. Fast and accurate short read alignment with Burrows–Wheeler transform. Bioinformatics. 2009;25:1754–60.
7. Zimin AV, Delcher AL, Florea L, Kelley DR, Schatz MC, Puiu D, et al. A whole-genome assembly of the domestic cow, Bos taurus. Genome Biol. 2009;10:R42.
8. Picard Tools—by Broad Institute. http://broadinstitute.github.io/picard/.
9. McKenna A, Hanna M, Banks E, Sivachenko A, Cibulskis K, Kernytsky A, et al. The genome analysis toolkit: a MapReduce framework for analyzing next-generation DNA sequencing data. Genome Res. 2010;20:1297–303.
10. McLaren W, Pritchard B, Rios D, Chen Y, Flicek P, Cunningham F. Deriving the consequences of genomic variants with the Ensembl API and SNP Effect Predictor. Bioinformatics. 2010;26:2069–70.
11. Ng PC, Henikoff S. Predicting deleterious amino acid substitutions. Genome Res. 2001;11:863–74.
12. Kumar P, Henikoff S, Ng PC. Predicting the effects of coding non-synony-mous variants on protein function using the SIFT algorithm. Nat Protoc. 2009;4:1073–81.
13. Ashburner M, Ball CA, Blake JA, Botstein D, Butler H, Cherry JM, et al. Gene ontology: tool for the unification of biology. The Gene Ontology Consor-tium. Nat Genet. 2000;25:25–9.
14. Kanehisa M, Goto S, Kawashima S, Nakaya A. The KEGG databases at GenomeNet. Nucleic Acids Res. 2002;30:42–6.
15. Shannon P, Markiel A, Ozier O, Baliga NS, Wang JT, Ramage D, et al. Cytoscape: a software environment for integrated models of biomolecu-lar interaction networks. Genome Res. 2003;13:2498–504.
16. Bindea G, Mlecnik B, Hackl H, Charoentong P, Tosolini M, Kirilovsky A, et al. ClueGO: a Cytoscape plug-into decipher functionally grouped gene ontology and pathway annotation networks. Bioinformatics. 2009;25:1091–3.
17. Boichard D, Chung H, Dassonneville R, David X, Eggen A, Fritz S, et al. Design of a bovine low-density SNP array optimized for imputation. PLoS One. 2012;7:e34130.
18. Kawahara-Miki R, Tsuda K, Shiwa Y, Arai-Kichise Y, Matsumoto T, Kanesaki Y, et al. Whole-genome resequencing shows numerous genes with nonsynonymous SNPs in the Japanese native cattle Kuchinoshima-Ushi. BMC Genomics. 2011;12:103.
19. DePristo MA, Banks E, Poplin R, Garimella KV, Maguire JR, Hartl C, et al. A framework for variation discovery and genotyping using next-generation DNA sequencing data. Nat Genet. 2011;43:491–8.

20. Ebersberger I, Metzler D, Schwarz C, Pääbo S. Genomewide comparison of DNA sequences between humans and chimpanzees. Am J Hum Genet. 2002;70:1490–7.

21. Lindblad-Toh K, Winchester E, Daly MJ, Wang DG, Hirschhorn JN, Laviolette JP, et al. Large-scale discovery and genotyping of single-nucleotide polymorphisms in the mouse. Nat Genet. 2000;24:381–6.

22. Bianco E, Nevado B, Ramos-Onsins SE, Pérez-Enciso M. A deep catalog of autosomal single nucleotide variation in the pig. PLoS One. 2015;10:e0118867.

23. Matukumalli LK, Lawley CT, Schnabel RD, Taylor JF, Allan MF, Heaton MP, et al. Development and characterization of a high density SNP genotyping assay for cattle. PLoS One. 2009;4:e5350.

24. Sellick GS, Pitchford WS, Morris CA, Cullen NG, Crawford AM, Raadsma HW, et al. Effect of myostatin F94L on carcass yield in cattle. Anim Genet. 2007;38:440–6.

25. Grobet L, Poncelet D, Royo LJ, Brouwers B, Pirottin D, Michaux C, et al. Molecular definition of an allelic series of mutations disrupting the myostatin function and causing double-muscling in cattle. Mamm Genome. 1998;9:210–3.

26. Blott S, Kim JJ, Moisio S, Schmidt-Küntzel A, Cornet A, Berzi P, et al. Molecular dissection of a quantitative trait locus: a phenylalanine-to-tyrosine substitution in the transmembrane domain of the bovine growth hormone receptor is associated with a major effect on milk yield and composition. Genetics. 2003;163:253–66.

27. Viitala S, Szyda J, Blott S, Schulman N, Lidauer M, Mäki-Tanila A, et al. The role of the bovine growth hormone receptor and prolactin receptor genes in milk, fat and protein production in Finnish Ayrshire dairy cattle. Genetics. 2006;173:2151–64.

28. Banos G, Woolliams JA, Woodward BW, Forbes AB, Coffey MP. Impact of single nucleotide polymorphisms in leptin, leptin receptor, growth hormone receptor, and diacylglycerol acyltransferase (DGAT1) gene loci on milk production, feed, and body energy traits of UK dairy cows. J Dairy Sci. 2008;91:3190–200.

29. Ceccaldi R, Liu JC, Amunugama R, Hajdu I, Primack B, Petalcorin MIR, et al. Homologous-recombination-deficient tumours are dependent on Polθ-mediated repair. Nature. 2015;518:258–62.

30. Koole W, van Schendel R, Karambelas AE, van Heteren JT, Okihara KL, Tijsterman M. A polymerase theta-dependent repair pathway suppresses extensive genomic instability at endogenous G4 DNA sites. Nat Commun. 2014;5:3216.

31. Mateos-Gomez PA, Gong F, Nair N, Miller KM, Lazzerini-Denchi E, Sfeir A. Mammalian polymerase θ promotes alternative NHEJ and suppresses recombination. Nature. 2015;518:254–7.

32. Roerink SF, van Schendel R, Tijsterman M. Polymerase theta-mediated end joining of replication-associated DNA breaks in C. elegans. Genome Res. 2014;24:954–62.

33. Yousefzadeh MJ, Wyatt DW, Takata KI, Mu Y, Hensley SC, Tomida J, et al. Mechanism of suppression of chromosomal instability by DNA polymerase POLQ. PLoS Genet. 2014;10:e1004654.

34. Yu AM, McVey M. Synthesis-dependent microhomology-mediated end joining accounts for multiple types of repair junctions. Nucleic Acids Res. 2010;38:5706–17.

35. McVey M, Lee SE. MMEJ repair of double-strand breaks (director's cut): deleted sequences and alternative endings. Trends Genet. 2008;24:529–38.

36. Fernandez-Vidal A, Guitton-Sert L, Cadoret JC, Drac M, Schwob E, Baldacci G, et al. A role for DNA polymerase θ in the timing of DNA replication. Nat Commun. 2014;5:4285.

37. Indovina P, Marcelli E, Casini N, Rizzo V, Giordano A. Emerging roles of RB family: new defense mechanisms against tumor progression. J Cell Physiol. 2013;228:525–35.

38. Milde-Langosch K, Goemann C, Methner C, Rieck G, Bamberger AM, Löning T. Expression of Rb2/p130 in breast and endometrial cancer: correlations with hormone receptor status. Br J Cancer. 2001;85:546–51.

39. Li Q, Sakurai Y, Ryu T, Azuma K, Yoshimura K, Yamanouchi Y, et al. Expression of Rb2/p130 protein correlates with the degree of malignancy in gliomas. Brain Tumor Pathol. 2004;21:121–5.

40. D'Andrilli G, Masciullo V, Bagella L, Tonini T, Minimo C, Zannoni GF, et al. Frequent loss of pRb2/p130 in human ovarian carcinoma. Clin Cancer Res. 2004;10:3098–103.

41. Dyson N. The regulation of E2F by pRB-family proteins. Genes Dev. 1998;12:2245–62.

42. Kong LJ, Meloni AR, Nevins JR. The Rb-related p130 protein controls telomere lengthening through an interaction with a Rad50-interacting protein, RINT-1. Mol Cell. 2006;22:63–71.

43. Fritz S, Capitan A, Djari A, Rodriguez SC, Barbat A, Baur A, et al. Detection of haplotypes associated with prenatal death in dairy cattle and identification of deleterious mutations in GART, SHBG and SLC37A2. PLoS One. 2013;8:e65550.

44. Köks S, Reimann E, Lilleoja R, Lättekivi F, Salumets A, Reemann P, et al. Sequencing and annotated analysis of full genome of Holstein breed bull. Mamm Genome. 2014;25:363–73.

45. Brown JP, Bullwinkel J, Baron-Lühr B, Billur M, Schneider P, Winking H, Singh PB. HP1gamma function is required for male germ cell survival and spermatogenesis. Epigenet Chromatin. 2010;3(1):9.

46. Li T, Kelly WG. Li T, Kelly WG. A role for Set1/MLL-related components in epigenetic regulation of the Caenorhabditis elegans germ line. PLoS Genet. 2011;7:e1001349.

14

Sequence variants selected from a multi-breed GWAS can improve the reliability of genomic predictions in dairy cattle

Irene van den Berg[1,2]* (iD), Didier Boichard[2] and Mogens S. Lund[1]

Abstract

Background: Sequence data can potentially increase the reliability of genomic predictions, because such data include causative mutations instead of relying on linkage disequilibrium (LD) between causative mutations and prediction variants. However, the location of the causative mutations is not known, and the presence of many variants that are in low LD with the causative mutations may reduce prediction reliability. Our objective was to investigate whether the use of variants at quantitative trait loci (QTL) that are identified in a multi-breed genome-wide association study (GWAS) for milk, fat and protein yield would increase the reliability of within- and multi-breed genomic predictions in Holstein, Jersey and Danish Red cattle. A wide range of scenarios that test different strategies to select prediction markers, for both within-breed and multi-breed prediction, were compared.

Results: For all breeds and traits, the use of variants selected from a multi-breed GWAS resulted in substantial increases in prediction reliabilities compared to within-breed prediction using a 50 K SNP array. Reliabilities depended highly on the choice of the prediction markers, and the scenario that led to the highest reliability varied between breeds and traits. While genomic correlations across breeds were low for genome-wide sequence variants, the effects of the QTL variants that yielded the highest reliabilities were highly correlated across breeds.

Conclusions: Our results show that the use of sequence variants, which are located near peaks of QTL that are detected in a multi-breed GWAS, can increase reliability of genomic predictions.

Background

Accuracy of genomic predictions is highly influenced by the size of the reference population used [1–3]. In cattle, for breeds such as the Holstein breed, this is not a problem since large reference populations are available at both the national and international levels [4], but for breeds with a smaller reference populations, accuracies of genomic prediction may not be sufficiently high. Using a large multi-breed reference population could potentially increase the accuracy of genomic predictions, by allowing breeds that have a small reference population to use information from other breeds. However, in

practice, large increases in accuracy of genomic predictions are obtained only when the breeds included in the multi-breed reference population are closely related [5, 6]. When more distant breeds are combined together, increases in accuracies of genomic predictions are generally small or zero compared to within-breed predictions [7–11]. One reason for this could be that linkage disequilibrium (LD) is conserved over much shorter distances across breeds than within breeds [12]. With the availability of high-density single nucleotide polymorphism (SNP) chips, de Roos et al. [12] showed that the LD between single nucleotide polymorphisms (SNPs) on the high-density SNP chip across dairy cattle breeds is sufficiently high to make across-breed prediction feasible and it was then assumed that increasing marker density furthermore to the whole-genome sequence level would improve multi-breed prediction. However, reliabilities of genomic predictions that are obtained with the bovine

*Correspondence: irene.vandenberg@unimelb.edu.au
[1] Department of Molecular Biology and Genetics, Faculty of Science and Technology, Center for Quantitative Genetics and Genomics, Aarhus University, 8830 Tjele, Denmark
Full list of author information is available at the end of the article

high-density SNP chip (HD) are not much higher than those with the 50 K SNP chip [9, 10]. Increasing marker density to the HD or the sequence level adds a large number of genome-wide variants but only a few of these variants are close to the causative mutations. Unless only variants in perfect LD with the causative mutations are used, the variants in imperfect LD with the causative mutations will limit the reliability of genomic predictions [13]. While whole-genome sequence data contain causative mutations and variants in high LD with some causative mutations, most of the variants are in low LD with the causative mutations. Thus, it is not surprising, that the use of whole-genome sequence data for genomic prediction does not necessarily increase reliabilities of genomic predictions compared to the use of genome-wide SNPs [14, 15], especially if the models used do not allow for sufficiently different within-breed variances and across-breed covariances for different SNPs.

In a simulation study, Pérez-Enciso et al. [16] obtained very high reliabilities by including the causative mutations in the model, while either addition of non-causative variants or removal of some causative mutations decreased reliabilities. Studies in cattle [17, 18] and *Drosophila melanogaster* [19] showed that selecting prediction variants based on the results of genome-wide association studies (GWAS) can yield substantial increases in the reliability of genomic predictions.

Because LD is conserved over much shorter distances across than within breeds [12], increasing the distance between causative mutations and prediction variants had a stronger effect on across-breed prediction than on within-breed prediction. In a simulation study [20], reliability of genomic predictions decreased faster across breeds than within breeds as the distance between prediction variants and causative mutations increased. Therefore, in order to infer information across breeds, it is important to use variants that are in high LD with the causative mutations. Although the true causative mutations are unknown, with a few exceptions [21], a large number of quantitative trait loci (QTL) regions have been detected in dairy cattle [22–27], and this information could be used to select sequence variants for genomic prediction. However, variants that are linked to a QTL in one breed but not in another breed can introduce noise, and reduce accuracy of genomic prediction for the other breed. Thus, careful selection of QTL variants is likely to be relevant for multi-breed prediction. Because LD is conserved over shorter distances across breeds, fewer variants are associated with the same causative mutations across breeds. Consequently, multi-breed GWAS results in more precise QTL mapping for variants that are shared across breeds [11, 28, 29].

Another potential difficulty in multi-breed prediction is that variant effects differ across breeds, which can be due to dominance or epistasis. However, even for genes with additive effects, differences in effects could be due to allele frequencies differing among breeds, or simply to the LD between prediction variants and causative mutations differing among breeds [6]. Thus, considering that SNP effects can be correlated across breeds rather than assuming that they are the same in each breed may be important to take advantage of sequence data for genomic prediction.

When within-breed genomic predictions are used, they rely heavily on the structure of the relationships within the breed that create LD in relatively large regions. Such structures are disrupted when populations from different breeds are combined, which results in LD being persistent over shorter regions across breeds. In addition, SNP effects can be easily dominated by the SNP effects in the breed with the largest population, which may lead to the prediction of a non-existing effect in the other breeds. As a consequence, the SNP may lose its predictive ability for the other breeds or even introduce noise from the breed with the largest reference population. Thus, in order to allow for private genetic variation and efficient use of within-breed family relationships, it could be useful to include a genomic component that models the genomic covariances within a given breed in the model.

Our objective was to investigate whether the use of variants at QTL that are selected from a multi-breed GWAS for milk, fat and protein yields would increase the reliability of within- and multi-breed genomic predictions in three dairy cattle breeds that range from very related populations to unrelated breeds. We used a model with a 50 K SNP genomic component and a QTL genomic component that includes sequence variants. We assumed that reliability of genomic predictions would increase when QTL variants were included in the model compared to models using only 50 K SNPs and that if too many were included, this advantage would decrease. More precisely, we expected that:

1. single-trait models that assume equal variant effects across breeds would be efficient for closely related populations;
2. including a QTL component with sequence variants would increase the reliability of genomic predictions and increase the correlations of variant effects between breeds compared to the 50 K SNP component;
3. a restricted number of prediction markers per QTL interval would improve the reliability of genomic predictions, especially for distantly related breeds;

4. a multi-breed GWAS would select sequence variants more accurately than a within-breed GWAS, especially for multi-breed prediction.

We used different models to test these assumptions.

Methods
Data
All genotype and phenotype data used in this study were obtained from pre-existing routine genetic evaluation data for the dairy cattle populations and required no ethical approval. Data from 5852 French Holstein (HOLFR), 5411 Danish Holstein (HOLDK), 1203 Danish Jersey (JER) and 937 Danish Red (RDC) bulls were included in the analyses. Although the HOLFR and HOLDK populations belong to the same breed, they were considered as different breeds. Holstein and RDC breeds are weakly related, while the JER breed is much more distantly related from either the RDC or Holstein breeds [6]. For all the bulls, deregressed proofs (DRP) were available for milk, fat and protein yields. Since the French and Danish scales differ, it was necessary to standardize the DRP within each breed, so that they were comparable between countries. All individuals were genotyped with the 50 K SNP chip and a subset of the individuals was also genotyped with the HD SNP chip, or sequenced. Individuals that were genotyped with the 50 K SNP chip were first imputed to HD, and then to the whole-genome sequence level, so that full genome sequence information was available for all the individuals. Imputation of Danish bulls from 50 K to HD and imputation of both French and Danish bulls from HD to whole-genome sequence level were done by using IMPUTE2 [30], while imputation of French bulls from 50 K to HD was performed by using Beagle [31]. For the Danish bulls, imputation from HD to whole-genome sequence level was based on a multi-breed reference population that included 1228 individuals from the fourth run of the 1000 Bull Genomes project [32] and 80 bulls from other projects carried out at Aarhus University. The HOLFR bulls were imputed by using a joint multi-breed French-Danish reference population that included 122 Holstein, 27 Jersey, 28 Montbéliarde, 23 Normande and 45 Danish Red bulls. More details on the imputation of the Danish bulls are in Brøndum et al. [17] and for the imputation of the French bulls from 50 K to HD in Hozé et al. [33].

For each population, individuals were divided into a training and a validation population. The validation populations consisted of the youngest individuals of each breed, and their sires were excluded from the training population. The training populations included 4911 HOLDK, 5335 HOLFR, 957 JER and 745 RDC bulls, and

the validation populations consisted of 500 HOLDK, 517 HOLFR, 246 JER and 192 RDC bulls.

Selection of prediction markers included in the QTL component
Several scenarios with different sets of prediction markers and different models were investigated. All sets of prediction markers included only variants with a minor allele frequency (MAF) higher than 0.01 and an IMPUTE2 INFO score of at least 0.9, which resulted in the basic set (50 K) comprising 37,856 SNPs from the 50 K SNP chip. For the other sets, variants were selected based on their associations with milk, fat or protein yield that had been identified in previously performed GWAS.

The dataset used for the multi-breed GWAS included all the bulls of the four populations (HOLFR, HOLDK, JER and RDC) in the training populations, their sires, and an additional 1935 Montbéliarde and 1725 Normande bulls. First, a GWAS was performed within each of the six populations, using whole-genome sequence data. After filtering out variants with a MAF lower than 0.005 and an IMPUTE2 INFO score less than 0.60, 24,550,115 SNPs and indels remained in the dataset. A single-marker model was run for each of these polymorphisms, within each of the six populations:

$$y_{ik} = \mu + s_{ik} + \beta g_i + e_{ik},$$

where y_{ik} is the DRP of milk yield, fat yield or protein yield for individual i with sire k, s_{ik} the random effect of sire k, β the effect of the variant, g_i the allele dose (ranging from 0 to 2) for individual i and e_{ik} a random residual.

Subsequently, a multi-breed GWAS was performed combining all six populations. To reduce computing time, the multi-breed GWAS was only run for variants with a p value $<10^{-5}$ for the HOLDK or HOLFR bulls, or $<10^{-3}$ for one of the other breeds for at least one of the traits. A breed effect was added to the model to account for between-breed differences:

$$y_{ijk} = \mu + s_{ik} + b_{ij} + \beta g_{ijk} + e_{ijk},$$

where b_{ij} is the effect of breed j of individual i. A full description of the GWAS is in [29].

Within breeds, variants were selected based on their associations with milk, fat or protein yield, which had been identified in either the within-breed or multi-breed GWAS, while for multi-breed analyses, variants were selected based on their associations with milk, fat or protein yield, which were detected in the multi-breed GWAS. Thresholds for within-breed p values were equal to 10^{-t}, with t equal to 10, 12 or 14 for Holstein populations and 4, 6 or 8 for Jersey and Danish Red populations. For the multi-breed models, t was equal to 10, 14 or 20. Due to the large differences in number of individuals per breed, the power of the GWAS varied strongly between breeds. Therefore, different thresholds were used for each breed, i.e. the thresholds for the JER and RDC breeds were chosen so that the range of the number of selected variants included the number of variants used for the HOLDK and HOLFR populations. An overview of all scenarios can be found in Table 1. Within breeds (WB-50 K + QTLt scenario), all variants that passed these thresholds were selected. Subsequently, LD pruning was performed on the selected variants using PLINK [34], with a R^2 threshold of 0.95. Selection of variants was the same for the multi-breed and within-breed analyses in the MB-50 K + QTLt scenarios. In scenarios MB-50 K + QTLt-n/w, the number of variants per interval (n) was, after LD pruning, limited to the 1, 10 or 25 variants with the lowest p values, per window (w) of 1, 2 or 10 Mb. Intervals were defined starting from the highest peak, until there were no more variants with a p value below t. The number of QTL variants selected from the within- and multi-breed GWAS are in Tables 2 and 3, respectively. If a variant was included in the QTL component of one scenario, it was excluded from the 50 K component for that scenario.

Statistical models

Genomic estimated breeding values (GEBV) were estimated using a Bayesian SNP best linear unbiased prediction (BLUP) model as implemented in the Bayz software

Table 2 Different sets of QTL markers selected from within-breed GWAS

Set	Selection threshold	Number of selected variants		
		Milk yield	Fat yield	Protein yield
Danish Holstein				
WBQTL10	10^{-10}	2595	2523	1491
WBQTL12	10^{-12}	1868	1719	612
WBQTL14	10^{-14}	1511	1220	298
French Holstein				
WBQTL10	10^{-10}	2249	1924	921
WBQTL12	10^{-12}	1382	1108	330
WBQTL14	10^{-14}	958	782	168
Jersey				
WBQTL4	10^{-04}	14,101	6632	3219
WBQTL6	10^{-06}	2464	578	345
WBQTL8	10^{-08}	677	51	22
Danish Red				
WBQTL4	10^{-04}	9548	4925	5330
WBQTL6	10^{-06}	873	648	383
WBQTL8	10^{-08}	80	232	12

Table 1 Descriptions of the scenarios used in the paper

Scenario[a]	Model	QTL component[b]
WB-50 K	WB	–
WB-50 K + WBQTLt	WB	All variants with a p value below 10^{-t} in a within breed GWAS
WB-50 K + MBQTLt	WB	All variants with a p value below 10^{-t} in a multi breed GWAS
WB-50 K + MBQTLt-n/w	WB	Maximum n variants with a p value below 10^{-t} per interval of i Mb in a multi breed GWAS
MB-50 K	MB	–
MB-50 K + MBQTLt	MB	All variants with a p value below 10^{-t} in a multi breed GWAS
MB-50 K + MBQTLt-n/w	MB	Maximum n variants with a p value below 10^{-t} per interval of i Mb in a multi breed GWAS
MT-50 K	MT	–
MT-50 K + MBQTLt	MT	All variants with a p value below 10^{-t} in a multi breed GWAS
MT-50 K + MBQTLt-n/w	MT	Maximum n variants with a p value below 10^{-t} per interval of i Mb in a multi breed GWAS

WB within-breed, *MB* multi-breed, *MT* multi-trait model

[a] Acronym of the scenario

[b] Describes how the variants in the QTL component were selected

[35], using only the 50 K data or the 50 K data and a second marker component with QTL marker components. In the models using only the 50 K data, all SNP effects were assumed to come from a single normal distribution. In the models that included a QTL component, QTL marker effects were assumed to come from a second normal distribution. Both within- and multi-breed models were tested and in the multi-breed models, the same trait in different breeds was considered either as a single trait, using a fixed breed effect to account for differences between breeds, or as multiple correlated traits, using a multi-trait model. For all scenarios, the Markov chain Monte Carlo (MCMC) was run for 50,000 iterations, discarding the first 10,000 as burn-in.

Within-breed model with a 50 K component

In the basic model (WB-50 K), only the 50 K SNPs were used for within-breed prediction:

$$y_i = \mu + \sum_{m=1}^{M} z_{im} a_m + e_i,$$

where y_i is the deregressed proof (DRP) of individual i, μ the mean, M is the total number of 50 K SNPs, z_{im} the genotype of individual i for SNP m, a_m the allele substitution effect of SNP m and e_i a random residual for individual i. SNP effects and residuals were assumed to be drawn from normal distributions $\sim N(0, \sigma_a^2)$ and $\sim N(0, \sigma_e^2)$, respectively. Additive SNP variance σ_a^2 and

Table 3 Different sets of QTL markers selected from multi-breed GWAS

Set	Selection threshold	Window size (Mb)	n[a]	Number of selected variants		
				Milk yield	Fat yield	Protein yield
MBQTL10	10	–	–	8361	9615	6119
MBQTL10-1/1	10	1	1	375	448	522
MBQTL10-10/1	10	1	10	1954	2612	2773
MBQTL10-25/1	10	1	25	3130	4190	4096
MBQTL10-1/2	10	2	1	269	292	342
MBQTL10-10/2	10	2	10	1457	1856	2080
MBQTL10-25/2	10	2	25	2363	3189	3410
MBQTL10-1/10	10	10	1	111	109	107
MBQTL10-10/10	10	10	10	709	775	911
MBQTL10-25/10	10	10	25	1230	1454	1808
MBQTL14	14	–	–	3821	4077	1402
MBQTL14-1/1	14	1	1	102	155	134
MBQTL14-10/1	14	1	10	614	816	633
MBQTL14-25/1	14	1	25	1046	1341	894
MBQTL14-1/2	14	2	1	67	111	95
MBQTL14-10/2	14	2	10	416	635	534
MBQTL14-25/2	14	2	25	762	1065	801
MBQTL14-1/10	14	10	1	27	40	41
MBQTL14-10/10	14	10	10	194	279	295
MBQTL14-25/10	14	10	25	352	534	517
MBQTL20	20	–	–	2225	2252	299
MBQTL20-1/1	20	1	1	30	45	23
MBQTL20-10/1	20	1	10	203	251	130
MBQTL20-25/1	20	1	25	384	424	205
MBQTL20-1/2	20	2	1	18	35	19
MBQTL20-10/2	20	2	10	138	192	104
MBQTL20-25/2	20	2	25	257	314	162
MBQTL20-1/10	20	10	1	7	15	12
MBQTL20-10/10	20	10	10	48	94	57
MBQTL20-25/10	20	10	25	115	173	85

[a] Maximum number of variants per interval (n), and the number of selected variants for milk, fat and protein yields

residual variance σ_e^2 were assigned uniform non-informative priors.

Within-breed models with 50 K and QTL genomic components

In scenarios WB-50 K + WBQTL and WB-50 K + MBQTL, a second genetic component was added to the model, using WBQTLt, MBQTLt or MBQTLt-n/w variants:

$$y_i = \mu + \sum_{m=1}^{M} z_{im} a_m + \sum_{n}^{N} z_{in} q_n + e_i,$$

where N is the total number of QTL markers, z_{in} the genotype of individual i for marker n, and q_n the allele substitution effect of marker n. QTL marker effects were drawn

from a normal distribution $\sim N\left(0, \sigma_q^2\right)$, and additive QTL variance σ_q^2 was assigned a uniform non-informative prior.

Multi-breed models

MB-50 K was a single-trait multi-breed model that assumed that the same trait measured in different breeds was a single trait, with a breed effect to account for the difference in means between breeds:

$$y_i = \mu + b_{ij} + \sum_{m=1}^{M} z_{im} a_m + e_i,$$

where b_{ij} is the effect of breed j of individual i. A uniform non-informative prior was assigned to b_i.

Model MB-50 K + MBQTL was similar to the MB-50 K model, with the addition of one of the MBQTLt or MBQTLt-n/w sets as a multi-breed QTL component:

$$y_i = \mu + b_{ij} + \sum_{m=1}^{M} z_{im}a_m + \sum_{n}^{N} z_{in}q_n + e_i.$$

Multi-trait models

In the basic multi-trait model (MT-50 K), the same trait measured in different breeds was considered as multiple traits by assuming a correlation between allele substitution effects in the 50 K component across breeds:

$$y_{ij} = \mu_j + \sum_{m=1}^{M} z_{im}a_{jm} + e_{ij},$$

where y_{ij} is the DRP of individual i from breed j, μ_j the mean of breed j, and a_{jm} the allele substitution effect of marker m in breed j. Additive marker effects were assumed to be normally distributed $\sim N\left(0, \sigma_{aj}^2\right)$ with additive marker variance σ_{aj}^2. Uniform non-informative priors were assigned to σ_{aj}^2 and to covariance $\sigma_{aj,ak}$ between the additive marker effects on the DRP in breed j and on the DRP in breed k. Residual covariances between DRP for individuals of different traits were 0.

Model MT-50 K + MBQTL was similar to the MT-50 K model, except for the addition of one of the MBQTLt or MBQTLt-n/w sets as a multi-breed QTL component:

$$y_{ij} = \mu_j + \sum_{m=1}^{M} z_{im}a_{jm} + \sum_{n}^{N} z_{in}q_{jn} + e_{ij},$$

where q_{jn} is the additive QTL marker effect of marker n in breed j. Additive marker effects are assumed to be normally distributed $\sim N\left(0, \sigma_{qj}^2\right)$. Both σ_{qj}^2 and covariance $\sigma_{qj,qk}$ between the additive QTL marker effects on the DRP in breed j and on the DRP in breed k were assigned uniform, non-informative priors.

Genomic correlations between 50 K SNP effects on the DRP in different breeds were estimated with the MT-50 K and MT-50 K + MBQTL models, and genomic correlations between QTL marker effects on the DRP in different breeds were estimated with the MT-50 K + MBQTL model. Genomic correlations were considered significant if they were greater than twice the standard error.

Evaluation of scenarios

Reliabilities were estimated as the squared correlation between DRP and GEBV, divided by the mean reliability of DRP in the test population. Bias was assessed by regression of DRP on GEBV. In the WB-50 K + MBQTL14-10/2 scenario for milk yield, five MCMC chains were run to assess convergence. Correlations between GEBV

obtained by different runs were above 0.9999 for all breeds.

In the scenarios with a QTL component, the proportion of variants explained by the QTL component was estimated as:

$$h_{QTL}^2 = \frac{\sigma_{QTL}^2}{\sigma_{50K}^2 + \sigma_{QTL}^2 + \sigma_e^2},$$

where σ_{50K}^2 and σ_{QTL}^2 are the variances of the 50 K and QTL components, respectively. These variances were estimated using the Gbayz programme that is part of the Bayz software [35]. For each MCMC iteration, var(**Za**) and var(**Zq**) were estimated, where **Z** is a design matrix and **a** and **q** are vectors of the regression coefficients of 50 K and QTL marker effects, respectively. Subsequently, posterior estimates of σ_{50K}^2 and σ_{QTL}^2 were obtained by averaging var(**Za**) and var(**Zq**) over all MCMC cycles.

Results

Comparison between different scenarios and prediction models

Reliabilities of genomic predictions obtained by using the 50 K SNPs and the scenarios that led to the highest reliabilities for each breed and trait are in Table 4. The highest reliabilities of genomic predictions for the HOLDK and HOLFR populations and the JER and RDC populations were obtained in scenarios MB-50 K-MBQTLt-n/w and WB-50 K-MBQTLt-n/w, respectively. Averaged across traits, the increase in reliability of the best scenario compared to model WB-50 K was equal to 0.08, 0.08, 0.06 and 0.06 for HOLDK, HOLFR, JER and RDC, respectively. The set of QTL markers that resulted in the highest reliability and the number of QTL variants in that set varied between breeds and traits. Averaged across traits, the numbers of QTL markers that yielded the highest reliability were equal to 1359, 662, 265 and 561 for HOLDK, HOLFR, JER and RDC, respectively. The number of QTL variants that led to the highest reliability was much larger for milk yield than for fat and protein yields, with, averaged across breeds, 1080 variants for milk yield, 564 for fat yield and 490 for protein yield.

Increases in reliability of genomic predictions varied greatly depending on the set of QTL markers used compared with model WB-50 K. Figure 1 shows this variation among the scenarios investigated for milk yield, while the results for fat and protein yield are in Figure S1 (see Additional file 1: Figure S1).

Tables 5 and 6 compare reliabilities of genomic prediction between model WB-50 K and the other scenarios. Scenario WB-50 K-WBQTLt resulted in small increases up to 0.05 for the HOLDK and HOLFR populations, except for protein yield for HOLDK, which had

Table 4 Scenarios with best reliability (r^2) for each breed and trait

Breed	Trait	50 K[a]	Best scenario	n^b	r^2	Δ^c
HOLDK	Milk yield	0.44	MB-50 K + MBQTL10-25/1	3130	0.53	0.09
	Fat yield	0.48	MB-50 K + MBQTL20-25/1	424	0.58	0.10
	Protein yield	0.39	MB-50 K + MBQTL10-1/1	522	0.44	0.06
HOLFR	Milk yield	0.33	MB-50 K + MBQTL14-25/1	1046	0.41	0.08
	Fat yield	0.37	MB-50 K + MBQTL20-25/1	424	0.46	0.10
	Protein yield	0.37	MB-50 K + MBQTL14-25/10	517	0.44	0.06
JER	Milk yield	0.30	WB-50 K + MBQTL20-1/10	7	0.40	0.10
	Fat yield	0.16	MB-50 K + MBQTL10-10/10	775	0.20	0.04
	Protein yield	0.22	WB-50 K + MBQTL20-1/10	12	0.27	0.05
RDC	Milk yield	0.14	WB-50 K + MBQTL20-10/2	138	0.20	0.06
	Fat yield	0.11	WB-50 K + MBQTL14-10/2	635	0.19	0.07
	Protein yield	0.09	WB-50 K + MBQTL10-10/10	911	0.14	0.05

HOLDK Danish Holstein, *HOLFR* French Holstein, *JER* Jersey, *RDC* Danish Red

[a] Reliabilities when using 50 K SNPs

[b] Number of QTL variants included in the scenario

[c] Difference in reliability between the best scenario and that obtained with 50 K SNPs

a reliability that decreased by 0.01. For JER, reliabilities increased for milk and fat yield by 0.04 and 0.02, respectively, but no difference was found for protein yield. For RDC, only small differences were found, with decreases of 0.01 for fat and protein yield, and no difference for milk yield. The scenarios using QTL variants selected from the multi-breed GWAS showed increased reliabilities for all breeds. Larger increases were obtained when the number of variants per QTL region was limited. Averaged across breeds and traits, the differences in reliability between model WB-50 K and the other models using the QTL set that yielded the highest reliability (Δ_{max}) were equal to 0.02, 0.03 and 0.05 for scenarios WB-50 K + WBQTLt, WB-50 K + MBQTLt, and WB-50 K + MBQTLt-n/w, respectively.

Model MB-50 K led to substantial increases in reliability for HOLDK and HOLFR, only small differences for RDC, and a small decrease up to 0.03 for JER. For all breeds and traits, reliabilities were higher when variants selected from a multi-breed GWAS were used than when only 50 K SNPs were used. The best advantage was found when using the QTL variants for JER and RDC, while for HOLDK and HOLFR, the largest increases were obtained by combining the four populations. For most breeds and traits, reliabilities were higher when the number of QTL variants was limited than when all QTL variants were used. The largest difference between scenarios MB-50 K + MBQTLt and MB-50 K + MBQTLt-n/w was observed for JER for fat yield, with a Δ_{max} of −0.01 for the first and 0.04 for the latter model. Averaged across breeds and traits, Δ_{max} was equal to 0.03, 0.05 and 0.06 for the MB-50 K,

MB-50 K + MBQTLt and MB-50 K + MBQTLt-n/w models, respectively.

Bias

Regression coefficients of DRP on GEBV for all breeds and traits are in Table 7. In all scenarios, GEBV were overestimated, the bias being larger for JER and RDC than for HOLDK and HOLFR. Overall, using QTL variants for prediction had only a limited influence on the bias, with either an increase or a decrease in some scenarios compared to WB-50 K (see Table 7).

Influence of the number of QTL markers on reliability of genomic prediction and on the variance explained by QTL markers

The number of selected QTL markers varied markedly between scenarios. Figure 2 shows reliabilities of genomic predictions according to number of QTL markers used for the WB-50 K-MBQTLt-n/w scenarios for milk yield. Results for fat and protein yield are in Figure S2 (see Additional file 2: Figure S2). Although reliability of genomic prediction depended on the number of QTL variants used, there were no clear peaks, but overall reliabilities were highest when a relatively small number of QTL variants was used.

The proportion of variance explained by the QTL component (h_{QTL}^2) varied greatly between scenarios, as shown in Fig. 3 for milk yield, and Figure S3 (see Additional file 3: Figure S3) for fat and protein yield. The h_{QTL}^2 obtained in the WB-50 K + WBQTLt scenarios was much larger for the JER and RDC breeds than for the HOLDK and HOLFR populations. For HOLDK and

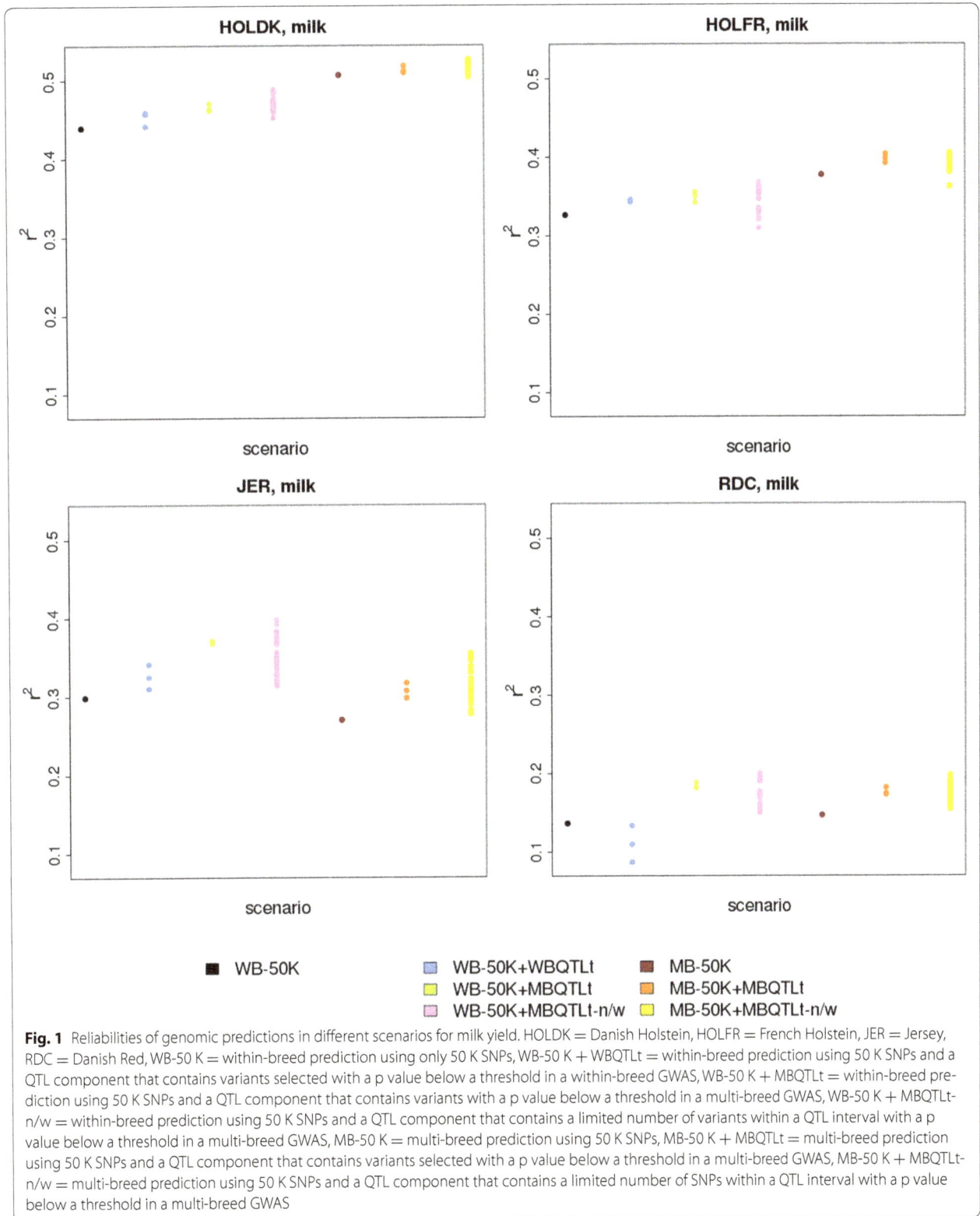

Fig. 1 Reliabilities of genomic predictions in different scenarios for milk yield. HOLDK = Danish Holstein, HOLFR = French Holstein, JER = Jersey, RDC = Danish Red, WB-50 K = within-breed prediction using only 50 K SNPs, WB-50 K + WBQTLt = within-breed prediction using 50 K SNPs and a QTL component that contains variants selected with a p value below a threshold in a within-breed GWAS, WB-50 K + MBQTLt = within-breed prediction using 50 K SNPs and a QTL component that contains variants with a p value below a threshold in a multi-breed GWAS, WB-50 K + MBQTLt-n/w = within-breed prediction using 50 K SNPs and a QTL component that contains a limited number of variants within a QTL interval with a p value below a threshold in a multi-breed GWAS, MB-50 K = multi-breed prediction using 50 K SNPs, MB-50 K + MBQTLt = multi-breed prediction using 50 K SNPs and a QTL component that contains variants selected with a p value below a threshold in a multi-breed GWAS, MB-50 K + MBQTLt-n/w = multi-breed prediction using 50 K SNPs and a QTL component that contains a limited number of SNPs within a QTL interval with a p value below a threshold in a multi-breed GWAS

Table 5 Differences in reliability between the WB-50 K model and other models (Δ_{max}) for within-breed prediction

Breed	Trait	WB-50 K + WBQTLt	WB-50 K + MBQTLt	WB-50 K + MBQTLt-n/w
HOLDK	Milk yield	0.02	0.03	0.05
	Fat yield	0.04	0.05	0.05
	Protein yield	−0.01	0.01	0.03
HOLFR	Milk yield	0.02	0.03	0.04
	Fat yield	0.05	0.05	0.05
	Protein yield	0.02	0.03	0.03
JER	Milk yield	0.04	0.07	0.10
	Fat yield	0.02	0.01	0.02
	Protein yield	0.00	0.02	0.05
RDC	Milk yield	0.00	0.05	0.06
	Fat yield	−0.01	0.03	0.07
	Protein yield	−0.01	0.01	0.05

The table provides only the Δ_{max}, i.e. the difference (Δ) obtained with the QTL set that resulted in the highest reliability

The QTL component consisted of all the variants selected with a p value less than a threshold for the within-breed GWAS (WB-50 K + WBQTLt) and the multi-breed GWAS (WB-50 K + MBQTLt), or of a limited number of SNPs per QTL interval that were selected with p value less than a threshold for the multi-breed GWAS (WB-50 K + MBQTLt-n/w)

HOLDK Danish Holstein, *HOLFR* French Holstein, *JER* Jersey, *RDC* Danish Red

Table 6 Differences in reliability between the WB-50 K model and other models for multi-breed prediction

Breed[a]	Trait	MB-50 K (Δ[b])	MB-50 K + MBQTLt (Δ^c_{max})	MB-50 K + MBQTLt-n/w (Δ^c_{max})
HOLDK	Milk	0.07	0.08	0.09
	Fat	0.06	0.09	0.10
	Protein	0.04	0.05	0.06
HOLFR	Milk	0.05	0.08	0.08
	Fat	0.06	0.09	0.10
	Protein	0.04	0.06	0.06
JER	Milk	−0.03	0.02	0.06
	Fat	0.01	−0.01	0.04
	Protein	−0.02	0.02	0.03
RDC	Milk	0.01	0.05	0.06
	Fat	0.02	0.06	0.05
	Protein	0.00	0.02	0.04

The model included one genetic component with all 50 K SNPs (MB-50 K), or an additional component that included either all SNPs selected with a p value less than a threshold for the multi-breed GWAS (MB-50 K + MBQTLt) or a limited number of variants per QTL interval selected with a p value less than a threshold for the multi-breed GWAS (MB-50 K + MBQTLt-n/w)

[a] HOLDK: Danish Holstein, HOLFR: French Holstein, JER: Jersey, RDC: Danish Red

[b] Differences in reliability between the reliability obtained with WB-50 K and the other models (Δ)

[c] For the models with a QTL component, the table provides only the Δ_{max}, i.e. the Δ obtained with the QTL set that resulted in the highest reliability

HOLFR, regardless of the scenarios using a QTL component selected from the multi-breed GWAS, h^2_{QTL} was either larger or smaller than that obtained in scenario WB-50 K + WBQTLt, depending on the criteria that were applied for QTL selection, while for JER and RDC, h^2_{QTL} was almost always substantially larger in scenarios WB-50 K + WBQTLt.

For all breeds, h^2_{QTL} was influenced by the number of QTL variants used in the QTL component, as shown in Fig. 4 for the WB-50 K + MBQTLt-n/w scenarios for milk

yield, and Figure S4 (see Additional file 4: Figure S4) for fat and protein yield. In scenarios WB-50 K + MBQTLt, the number of selected QTL variants depended solely on the threshold applied for QTL selection, and h^2_{QTL} increased approximately linearly with the number of QTL variants (results not shown). In the sets used for scenarios WB-50 K + MBQTLt-n/w, h^2_{QTL} was larger for the sets with a lower selection threshold and thus a larger number of QTL. For scenarios WB-50 K + MBQTLt-n/w in which the same threshold was applied, h^2_{QTL} fluctuated

Table 7 Regression coefficients of DRP on GEBV for milk, fat and protein yield

Scenario	HOLDK	HOLFR	JER	RDC
Milk yield				
WB-50 K	0.83	0.72	0.67	0.71
WB-50 K + WBQTLt	0.82–0.83	0.72–0.72	0.62–0.71	0.46–0.56
WB-50 K + MBQTLt	0.83–0.84	0.70–0.73	0.70–0.70	0.67–0.70
WB-50 K + MBQTLt-n/w	0.83–0.86	0.68–0.74	0.68–0.78	0.61–0.72
MB-50 K	0.89	0.73	0.69	0.53
MB-50 K-MBQTLt	0.88–0.89	0.74–0.75	0.67–0.69	0.55–0.57
MB-50 K-MBQTLt-n/w	0.87–0.90	0.72–0.75	0.69–0.80	0.51–0.59
Fat yield				
WB-50 K	0.83	0.78	0.55	0.58
WB-50 K + WBQTLt	0.82–0.82	0.79–0.81	0.49–0.58	0.45–0.49
WB-50 K + MBQTLt	0.82–0.82	0.80–0.80	0.59–0.59	0.52–0.54
WB-50 K + MBQTLt-n/w	0.79–0.82	0.78–0.82	0.50–0.61	0.49–0.62
MB-50 K	0.81	0.87	0.57	0.40
MB-50 K-MBQTLt	0.81–0.81	0.87–0.88	0.52–0.56	0.43–0.46
MB-50 K-MBQTLt-n/w	0.80–0.82	0.87–0.90	0.47–0.62	0.38–0.46
Protein yield				
WB-50 K	0.75	0.74	0.61	0.57
WB-50 K + WBQTLt	0.72–0.74	0.74–0.75	0.47–0.60	0.41–0.50
WB-50 K + MBQTLt	0.73–0.75	0.75–0.76	0.60–0.61	0.55–0.59
WB-50 K + MBQTLt-n/w	0.73–0.77	0.72–0.76	0.57–0.64	0.55–0.72
MB-50 K	0.80	0.77	0.60	0.43
MB-50 K-MBQTLt	0.78–0.80	0.78–0.78	0.63–0.64	0.44–0.49
MB-50 K-MBQTLt-n/w	0.79–0.82	0.76–0.79	0.62–0.68	0.36–0.54

HOLDK Danish Holstein, *HOLFR* French Holstein, *JER* Jersey, *RDC* Danish Red

a lot without necessarily increasing if a larger number of QTL variants was used. Sets used in scenarios MBQTLt-n/w and MBQTLt led to similar h^2_{QTL}, while MBQTLt-n/w included much fewer QTL variants than MBQTLt (results not shown).

Genomic correlations between breeds

For the multi-trait models, genomic correlations between the same traits in different breeds were estimated. Figure 5 shows the genomic correlations using the 50 K component in the MT-50 K, MT-50 K + MBQTLt and MT-50 K + MBQTLt-n/w models. Genomic correlations of the 50 K component ranged from 0.43 to 0.76 between HOLDK and HOLFR, from 0.03 to 0.28 between HOLDK or HOLFR and RDC, and from −0.12 to 0.05 between JER and any other breed.

Genomic correlations that were computed by using the QTL component and the MT-50 K + MBQTLt and MT-50 K + MBQTLt-n/w scenarios are in Fig. 6. All genomic correlations were larger when the QTL component was used than when the 50 K component was used. The largest correlations were obtained with the

MT-50 K + MBQTLt-n/w scenarios. Genomic correlations between HOLDK and HOLFR ranged from 0.73 to 0.86 for MT-50 K + MBQTt-n/w and from 0.79 to 0.97 for MT-50 K + MBQTLt-n/w. Between HOLDK or HOLFR and RDC, genomic correlations that were computed by using the QTL component ranged from 0.32 to 0.48 for MT-50 K + MBQTLt, and from 0.26 to 0.94 with the MT-50 K + MBQTLt-n/w and between JER and the other breeds, the lowest correlations were found for fat yield (ranging from −0.07 to 0.17 for MT-50 K + MBQTLt and from −0.13 to 0.50 for MT-50 K + MBQTLt-n/w), while for milk and protein yield, they were always positive (ranging from 0.19 to 0.46 for MT-50 K + MBQTLt and from 0.13 to 0.86 for MT-50 K + MBQTLt-n/w).

Posterior standard deviations of the genomic correlations are in Table S1 (see Additional file 5: Table S1); they ranged from 0.01 to 0.33 when the QTL components were used and from 0.03 to 0.17 when the 50 K component was used. Standard deviations were smallest between the two Holstein populations (on average 0.02 for the QTL components and 0.05 for the 50 K component), and larger for any other breed combination (on average 0.16 for the QTL components and 0.11 for the 50 K component).

Discussion

The advantage of whole-genome sequence data is that they include causative mutations. However, some causative mutations may be absent, for example because of partial variant calling that does not consider structural variants, and because some variants may be filtered out due to poor sequencing and imputation quality. Furthermore, the locations of the causative mutations present in the data are unknown. Thus, we attempted to identify variants that were in high LD with the causative mutations based on GWAS data. Using QTL variants that were selected from a multi-breed GWAS for within-breed prediction resulted in substantial increases in the reliability of genomic predictions for all breeds and traits compared to a 50 K within-breed model. While the reliability of multi-breed prediction increased when QTL markers were used rather than only 50 K SNPs, multi-breed reliabilities were very similar to within-breed reliabilities when markers in the QTL component were chosen based on multi-breed GWAS data.

Increases in reliabilities observed for the two Holstein populations when within-breed QTL variants were used were in the range of those reported by Brøndum et al. [17]. In RDC, inclusion of within-breed QTL variants decreased the reliability of genomic predictions. This can be explained by the large difference in population size

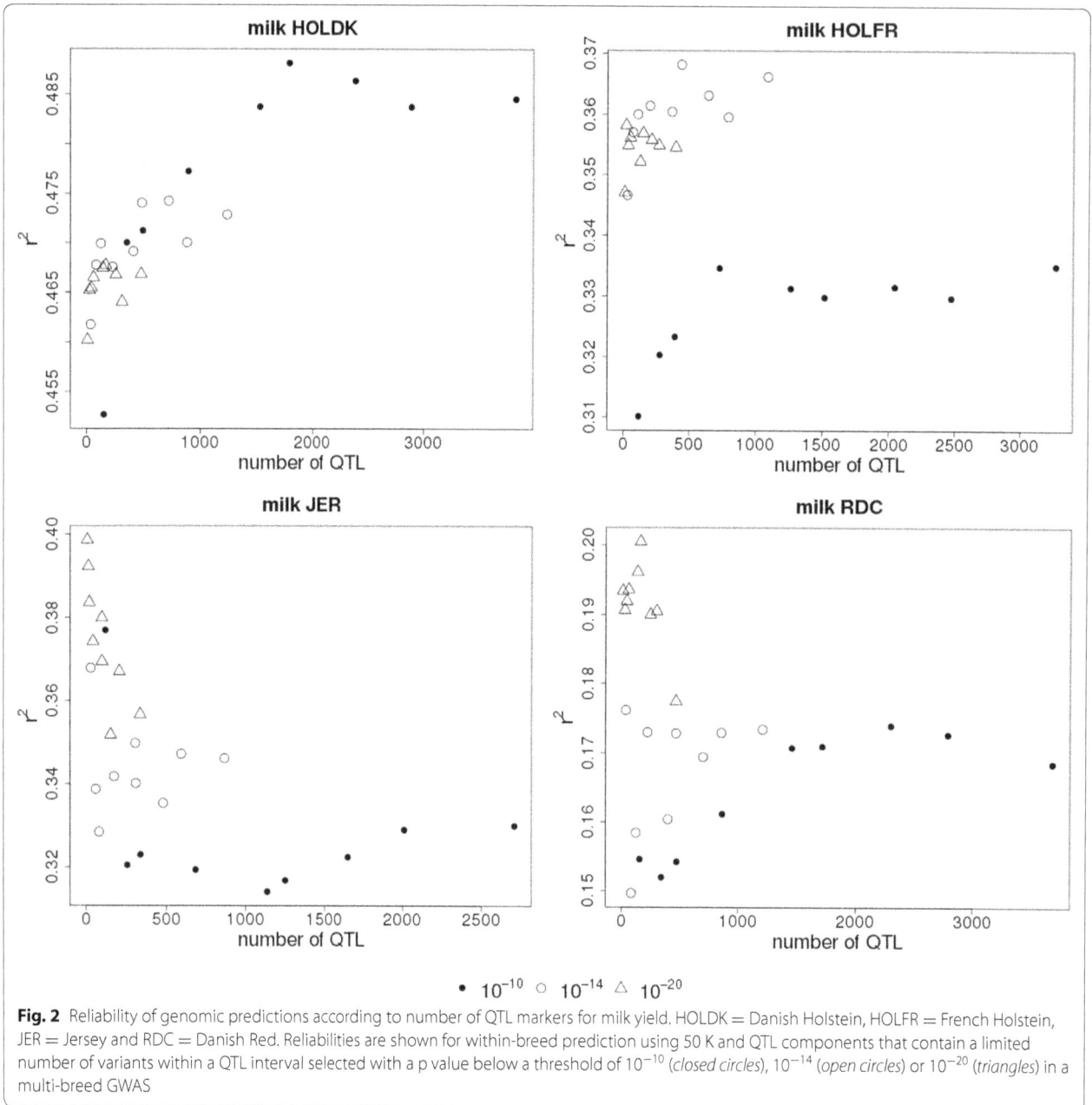

Fig. 2 Reliability of genomic predictions according to number of QTL markers for milk yield. HOLDK = Danish Holstein, HOLFR = French Holstein, JER = Jersey and RDC = Danish Red. Reliabilities are shown for within-breed prediction using 50 K and QTL components that contain a limited number of variants within a QTL interval selected with a p value below a threshold of 10^{-10} (*closed circles*), 10^{-14} (*open circles*) or 10^{-20} (*triangles*) in a multi-breed GWAS

used for the GWAS that was performed to select QTL variants. For the within-breed GWAS, each Holstein population included about 5000 individuals, while the JER and RDC populations each included less than 1000 individuals, which results in much more detection power for HOLFR and HOLDK than for JER and RDC. Thus, selection of variants for JER and RDC is likely less reliable, and they add noise rather than information on the causative mutations, which results in a reduced reliabilities of genomic predictions.

We expected that selected variants from the multi-breed GWAS would be beneficial mainly for multi-breed prediction, but not necessarily for within-breed prediction for JER and RDC, since the multi-breed GWAS was dominated by Holstein animals. Within-breed prediction using variants selected from the multi-breed GWAS did, however, increase reliabilities of genomic predictions for all breeds, including JER and RDC. Our findings confirm those from other studies [11, 28], which showed that a multi-breed GWAS results in more accurate QTL

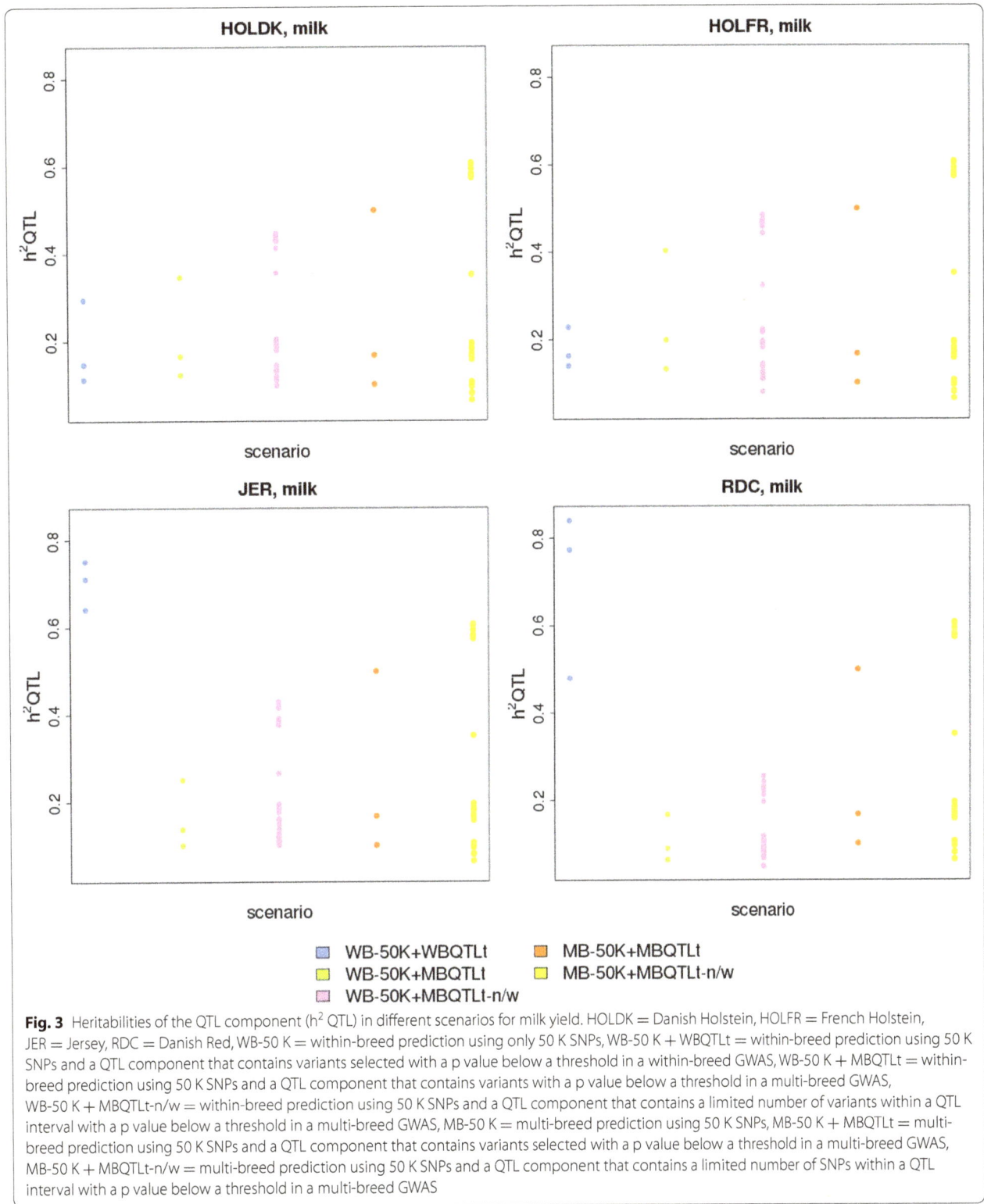

Fig. 3 Heritabilities of the QTL component (h^2 QTL) in different scenarios for milk yield. HOLDK = Danish Holstein, HOLFR = French Holstein, JER = Jersey, RDC = Danish Red, WB-50 K = within-breed prediction using only 50 K SNPs, WB-50 K + WBQTLt = within-breed prediction using 50 K SNPs and a QTL component that contains variants selected with a p value below a threshold in a within-breed GWAS, WB-50 K + MBQTLt = within-breed prediction using 50 K SNPs and a QTL component that contains variants with a p value below a threshold in a multi-breed GWAS, WB-50 K + MBQTLt-n/w = within-breed prediction using 50 K SNPs and a QTL component that contains a limited number of variants within a QTL interval with a p value below a threshold in a multi-breed GWAS, MB-50 K = multi-breed prediction using 50 K SNPs, MB-50 K + MBQTLt = multi-breed prediction using 50 K SNPs and a QTL component that contains variants selected with a p value below a threshold in a multi-breed GWAS, MB-50 K + MBQTLt-n/w = multi-breed prediction using 50 K SNPs and a QTL component that contains a limited number of SNPs within a QTL interval with a p value below a threshold in a multi-breed GWAS

mapping than a within-breed GWAS. While adding QTL variants selected from the multi-breed GWAS resulted in increased reliabilities for all breeds and traits, they were highly sensitive to the choice of the QTL markers. This is in line with results from a study by Ober et al. [19], who used variants that were selected from a GWAS for

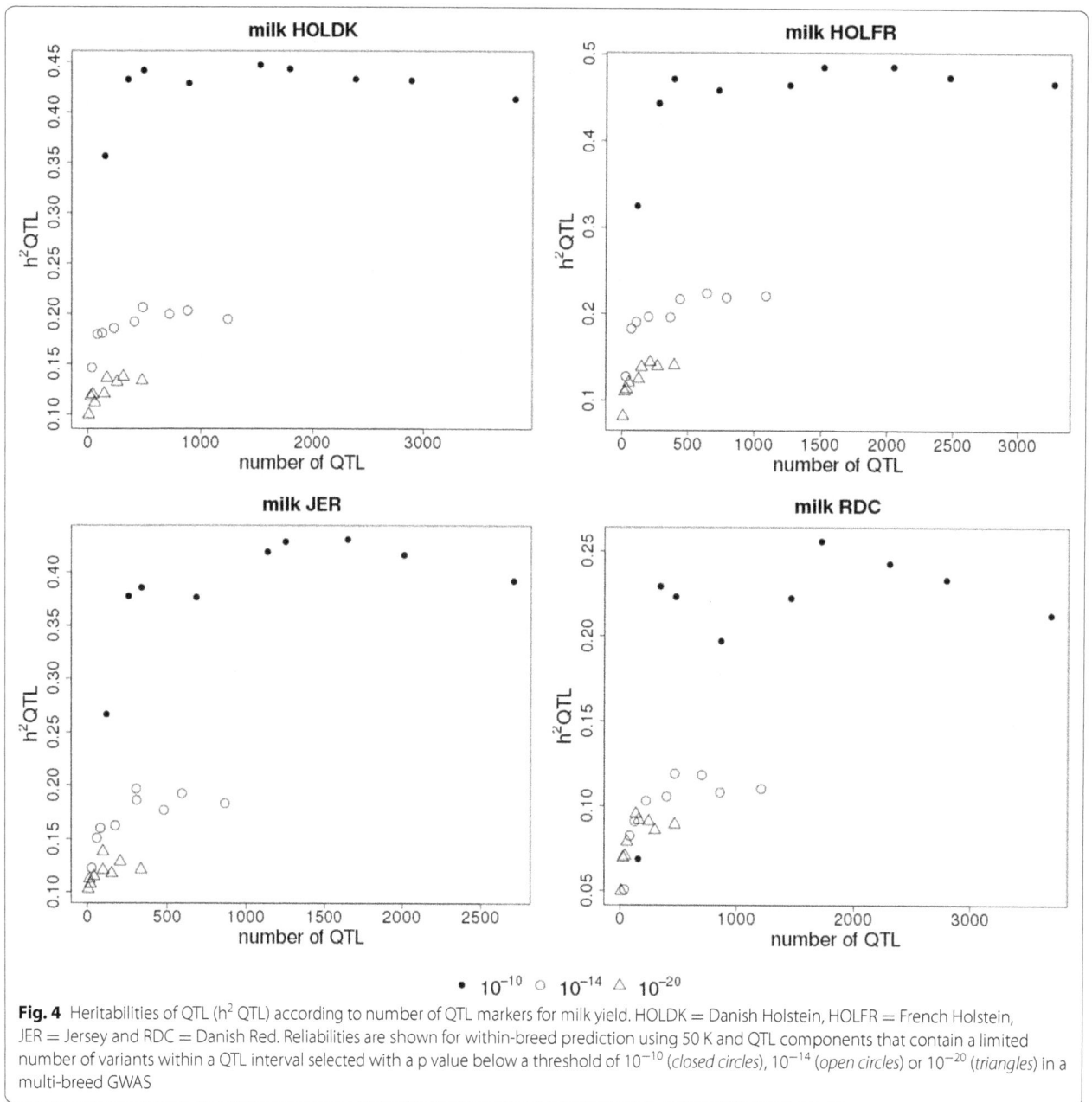

Fig. 4 Heritabilities of QTL (h^2 QTL) according to number of QTL markers for milk yield. HOLDK = Danish Holstein, HOLFR = French Holstein, JER = Jersey and RDC = Danish Red. Reliabilities are shown for within-breed prediction using 50 K and QTL components that contain a limited number of variants within a QTL interval selected with a p value below a threshold of 10^{-10} (*closed circles*), 10^{-14} (*open circles*) or 10^{-20} (*triangles*) in a multi-breed GWAS

genomic prediction of quantitative traits in *Drosophila*. They showed that accuracy of genomic prediction varied strongly with the threshold used to select prediction variants. In our study, the highest reliabilities were always obtained when the number of QTL variants per region was limited. This confirms our expectation that a restricted number of prediction markers per QTL interval leads to higher reliabilities than selecting a larger number of markers. Although the most significant variant in a GWAS is not necessarily the causative mutation, variants near the peak are more likely to be in high LD

with the causative mutation, while variants further away are likely to be in lower LD and therefore, introduce more noise in the prediction. Therefore, restricting the number of variants per QTL interval resulted in higher reliabilities than selecting all variants with p values below a threshold. The optimal filtering, regarding both p value and restriction of variants per region, depended on breed and trait. For the JER breed, reliabilities of genomic predictions were highest with much fewer variants than for the other breeds. Again, this can be explained by the short distances over which LD is conserved across

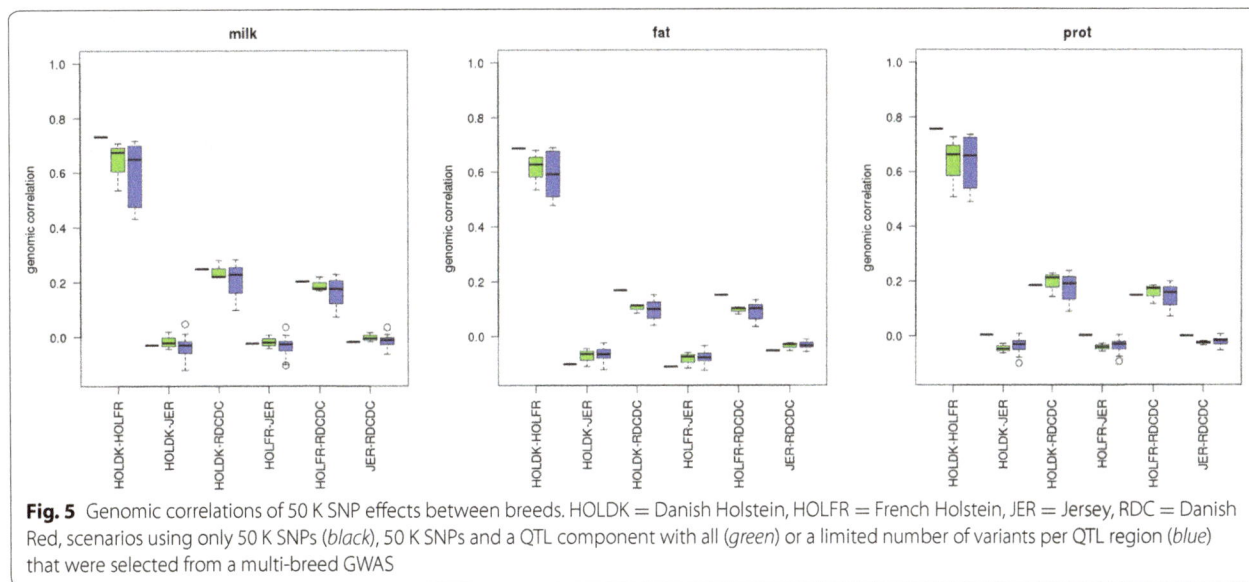

Fig. 5 Genomic correlations of 50 K SNP effects between breeds. HOLDK = Danish Holstein, HOLFR = French Holstein, JER = Jersey, RDC = Danish Red, scenarios using only 50 K SNPs (*black*), 50 K SNPs and a QTL component with all (*green*) or a limited number of variants per QTL region (*blue*) that were selected from a multi-breed GWAS

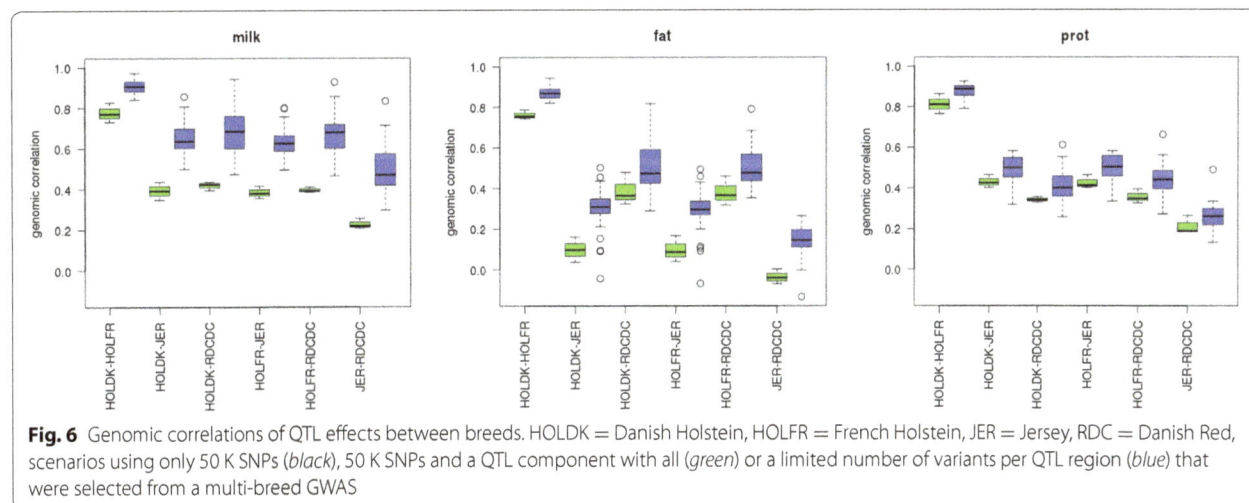

Fig. 6 Genomic correlations of QTL effects between breeds. HOLDK = Danish Holstein, HOLFR = French Holstein, JER = Jersey, RDC = Danish Red, scenarios using only 50 K SNPs (*black*), 50 K SNPs and a QTL component with all (*green*) or a limited number of variants per QTL region (*blue*) that were selected from a multi-breed GWAS

breeds. The multi-breed GWAS was dominated by Holstein animals, but also used the data from other breeds. Therefore, the variants that are in LD with the causative mutations in both Holstein populations and the other breeds, are likely to be among the variants with the most significant p values, while variants further away from the peak, may only be in LD with the causative mutation in the Holstein populations.

The variance explained by the QTL component varied strongly between breeds, traits and sets of prediction markers. Although for JER and RDC the WB-50 K + WBQTLt scenarios led to similar or lower reliabilities compared to model WB-50 K, the QTL markers used in those scenarios did explain a substantially larger part of the total genetic variance than the other

sets of prediction markers. In these scenarios, the QTL markers may estimate a polygenic effect rather than accurately estimate the effects of the largest QTL, but are actually less accurate in capturing the polygenic effect than WB-50 K, and thereby, reducing the reliability of genomic predictions. While the QTL markers used in the WB-50 K + MBQTLt-n/w scenarios only explained a small part of the total genetic variance, their use resulted in large increases in the reliability of genomic predictions for JER and RDC.

The advantage of having a second genetic component with QTL variants in the SNP BLUP model could be due to some variants having a larger effect rather than to the specific variants being included in the QTL component. If this is the case, the advantage of the

QTL component will be smaller with a mixture model. We tested the WB-50 K, WB-50 K + MBQTLt, and WB-50 K + MBQTLt-n/w scenarios also with a Bayesian mixture model that fitted two mixture distributions for the 50 K SNPs, and two different mixture distributions for the QTL markers. Reliabilities of genomic predictions obtained with the Bayesian mixture model and the SNP BLUP model were similar and increases obtained by adding a multi-breed QTL component were within the same range (results not shown).

Combining all populations in the multi-breed models led to higher reliabilities than within-breed prediction only for HOLDK and HOLFR, which is not surprising, since the Holstein reference population was approximately doubled by combining the two HOLDK and HOLFR populations. While the use of a multi-breed population and sequence information is valuable in pinpointing the location of variants that are in close LD with the causative variants, using these variants for multi-breed prediction is, however, straightforward. Variant effects can differ between breeds and multi-breed prediction models can carry noise from a large population to smaller populations. This confirms our expectation, that when combining data from multiple breeds, the single-trait models are suitable for closely related populations, but not for more distantly related breeds, because they assume equal variant effects across populations.

The multi-trait models allow the estimation of genomic correlations of marker effects across traits. The correlations obtained with the 50 K SNPs confirmed the relatedness between the different populations: while the Holstein populations are highly related and the RDC and Holstein populations are moderately related, genomic correlations between the JER breed and either of the other breeds are approximately 0. With such correlations, it is not surprising that with model MB-50 K, which assumes similar marker effects for all breeds, reliability of genomic predictions did not increase for RDC and decreased for JER. However, genomic correlations estimated for the multi-breed QTL components were moderate to high, even between JER and the other breeds, indicating that the multi-breed QTL components did contain variants that were associated with QTL segregating across breeds. The fact that higher genomic correlations were obtained in the MT-50 K + MBQTLt-n/w scenarios than in the MT-50 K + MBQTLt scenarios confirms that stricter selection criteria result in the selection of variants that are closely located to the causative mutations. However with such high across-breed correlations, it is surprising that the use of a multi-breed reference population yielded no advantage for JER and RDC.

This is probably due to the low across-breed correlations of the 50 K SNPs.

Although the multi-trait model allowed the 50 K SNPs and the QTL markers to have different genomic correlations, reliabilities of genomic predictions were similar to those obtained in the WB-50 K + MBQTLt and WB-50 K + MBQTLt-n/i scenarios (results not shown).

To take advantage of the highly correlated multi-breed QTL effects, without having to overcome the noise introduced by the 50 K SNPs, a model that includes a multi-breed QTL component but only within-breed 50 K components may result in increased reliabilities compared to within-breed prediction. Porto-Neto et al. [18] showed that to improve across-breed prediction, it is important to select variants that are highly correlated across breeds. In their study, variants were selected from a GWAS within Brahman and Tropical composite cattle. Variants with effects in the same direction in both breeds resulted in increased across-breed reliabilities of genomic predictions and high genomic correlations, while variants with opposite effects decreased reliabilities and resulted in negative genomic correlations. By fitting separate within- and multi-breed genomic relationship matrices, Khansefid et al. [36] reported increases in accuracy for some traits compared to a model using only within-breed relationships.

None of the sets of prediction markers used here yielded the highest reliability for all breeds. Although such a set would be ideal, it might not be realistic. Variants that play an important role in one breed, could actually introduce noise in another breed. Furthermore, QTL properties such as allele frequencies influence accuracy [37], and can differ between breeds and traits. Rather than testing a large number of prediction sets to find the optimal set for each breed and trait, as was done in this study, a multi-trait Bayesian variable selection model as described by Janss [38] could potentially select the most adequate variants for each breed.

Several studies have shown that, using full sequence data directly for genomic prediction, rather than pre-selecting variants, does not improve prediction reliability [14, 15]. Our results show that both prediction reliability and genomic correlations across populations and breeds are highly sensitive to the choice of the prediction markers. Full sequence data is likely to result in similar genomic relationships and correlations as genome-wide SNPs, and is therefore unlikely to improve prediction reliability. Bayesian variable selection models allow for heterogeneous variances and could potentially exploit the presence of causative mutations in the sequence data by assigning non-zero effects to variants that are close to the causative mutations, and zero effects to all other variants. However, in practice Van Binsbergen et al. [15] found no increase in prediction reliability using full sequence data compared to SNPs, even with a Bayesian variable selection model. A potential explanation for the lack of improvement in prediction reliability could be that the number of SNPs is much larger than the number of individuals. The number of SNPs can be significantly reduced by preselecting SNPs based on their functional annotations, for example by using only SNPs located within genes. By doing this, Hayes et al. [39] reported a 2% increase in prediction accuracy in Holstein cattle, averaged over production traits. Erbe et al. [9] showed that the use of variants from the transcribed regions of the genome resulted in higher accuracy for across-breed prediction compared to prediction based on 50 K genotypes. Selection of variants based on functional annotations could also be used to refine the selection of variants per interval by giving preference to variants located in genes rather than only selecting variants based on their statistical association detected in a GWAS.

Selecting prediction variants based on their association with a trait could result in prediction bias. While there was bias in all our results, the inclusion of QTL markers did not consistently increase the bias, i.e. it increased or decreased depending on the set of QTL markers used. Regression coefficients were always less than 1, which indicates that the GEBV were overestimated for the test animals. This may be due to inflated GEBV and strong selection of individuals in the test population for the traits in the analyses. Furthermore, this effect of selection was increased by the fact that the sires used for prediction were removed from the reference population.

While some sets of QTL markers resulted in substantial increases in prediction reliability for the populations that were tested in our study, this may not be true for other populations. The optimal set of prediction markers differed between populations, and the sets that we identified are not necessarily the best sets for other populations. Furthermore, we studied milk traits for which few QTL are known to have large effects. Increasing prediction reliability by adding sequence variants is likely to be more challenging for more polygenic traits. Brøndum et al. [17] found smaller increases in prediction reliability for mastitis and fertility than for production traits. Our results do not provide a list of markers that increase prediction, but they do demonstrate that sequence variants can potentially increase prediction reliability. In our analyses, we tested a large number of prediction sets, which is not practical for routine genomic evaluation. An alternative could be to make a less stringent selection of prediction markers, but subsequently use a more sophisticated prediction model, that allows marker effects to differ between breeds and traits. Further research is required to develop a more practical way to exploit sequence data for genomic prediction.

Conclusions

Prediction reliability increased substantially for all breeds and traits when sequence variants selected from a GWAS were used for genomic prediction. Even for within-breed prediction, a multi-breed GWAS was more efficient in identifying variants that increase prediction reliability than within-breed GWAS. Prediction reliabilities were highly sensitive to the choice of prediction markers, and limiting the number of variants per QTL region led to higher prediction reliabilities than selecting them on the basis of a p value threshold. While the highest prediction reliabilities were obtained within breed, multi-breed prediction reliabilities were higher than multi-breed prediction reliabilities when using only 50 K SNPs, and across breed genomic correlations of QTL variants were much higher than those obtained at 50 K SNPs. Our results show that sequence data can potentially increase reliabilities of genomic predictions, if the proper variants are used, which is more likely if they are selected from a multi-breed GWAS.

Additional files

Additional file 1: Figure S1. Reliabilities of genomic predictions in different scenarios for fat and protein yield. HOLDK = Danish Holstein, HOLFR = French Holstein, JER = Jersey, RDC = Danish Red, WB-50 K = within-breed prediction using only 50 K SNPs, WB-50 K + WBQTLt = within-breed prediction using 50 K SNPs and a QTL component that contains variants selected with a p value below a threshold in a within-breed GWAS, WB-50 K + MBQTLt = within-breed prediction using 50 K SNPs and a QTL component that contains variants with a p value below a threshold in a multi-breed GWAS, WB-50 K + MBQTLt-n/w = within-breed prediction using 50 K SNPs and a QTL component that contains a limited number of variants within a QTL interval with a p value below a threshold in a multi-breed GWAS, MB-50 K = multi-breed prediction using 50 K SNPs, MB-50 K + MBQTLt = multi-breed prediction using 50 K SNPs and a QTL component that contains variants selected with a p value below a threshold in a multi-breed GWAS, MB-50 K + MBQTLt-n/w = multi-breed prediction using 50 K SNPs and a QTL component that contains a limited number of SNPs within a QTL interval with a p value below a threshold in a multi-breed GWAS.

Additional file 2: Figure S2. Reliabilities of genomic predictions according to number of QTL markers for fat and protein yield. HOLDK = Danish Holstein, HOLFR = French Holstein, JER = Jersey and RDC = Danish Red. Reliabilities are shown for within breed prediction using 50 K and QTL components containing a restricted number of markers in a QTL interval with a p value below a threshold of 10^{-10} (closed circles), 10^{-14} (open circles) or 10^{-20} (triangles) in a multi breed GWAS.

Additional file 3: Figure S3. Heritabilities of the QTL component (h^2 QTL) in different scenarios for fat and protein yield. HOLDK = Danish Holstein, HOLFR = French Holstein, JER = Jersey, RDC = Danish Red, WB-50 K = within-breed prediction using only 50 K markers, WB-50 K + WBQTLt = within breed prediction using 50 K markers and a QTL component containing markers with a p value below a threshold in a within breed GWAS, WB-50 K + MBQTLt = within breed prediction using 50 K markers and a QTL component containing markers with a p value below a threshold in a multi breed GWAS, WB-50 K + MBQTLt-n/w = within breed prediction using 50 K markers and a QTL component containing a restricted number of markers in a QTL interval with a p value below a threshold in a multi breed GWAS, MB-50 K = multi breed prediction using 50 K markers, MB-50 K + MBQTLt = multi breed prediction using 50 K markers and a QTL component containing markers with a p value below a threshold in a multi breed GWAS, MB-50 K + MBQTLt-n/w = multi breed prediction using 50 K markers and a QTL component containing a restricted number of markers in a QTL interval with a p value below a threshold in a multi breed GWAS.

Additional file 4: Figure S4. Heritabilities of QTL (h^2 QTL) according to number of QTL markers for fat and protein traits. HOLDK = Danish Holstein, HOLFR = French Holstein, JER = Jersey and RDC = Danish Red. Reliabilities are shown for within-breed prediction using 50 K and QTL components containing a restricted number of markers in a QTL interval with a p value below a threshold of 10^{-10} (closed circles), 10^{-14} (open circles) or 10^{-20} (triangles) in a multi breed GWAS.

Additional file 5: Table S1. Posterior standard deviations of genomic correlations between breeds. Average standard deviation across scenarios followed by minimum and maximum standard deviation, for the 50 K component (σ_{50K}), QTL component where QTL markers are selected based on their p value (σ_{MBQTLt}) and QTL component where the maximum number of markers per QTL window is restricted ($\sigma_{MBQTLt-n/w}$), HOLDK = Danish Holstein, HOLFR = French Holstein, JER = Jersey and RDC = Danish Red.

Authors' contributions

All authors conceived the study and contributed to the manuscript. IB performed the analysis and wrote the paper. All authors read and approved the final manuscript.

Author details

[1] Department of Molecular Biology and Genetics, Faculty of Science and Technology, Center for Quantitative Genetics and Genomics, Aarhus University, 8830 Tjele, Denmark. [2] GABI, INRA, AgroParisTech, Université Paris Saclay, 78350 Jouy-en-Josas, France.

Acknowledgements

IB benefited from an Erasmus-Mundus fellowship and a Grant by Apisgene, within the framework of the European Graduate School in Animal Breeding and Genetics. French sequence data originated from the Cartoseq project funded by ANR and Apisgene (ANR10-GENM-0018). This research was supported by the center for Genomic Selection in Animals and Plants (GenSAP) funded by The Danish Council for Strategic Research. We acknowledge the 1000 bull genomes project (http://www.1000bullgenomes.com) for sharing data to impute to whole-genome sequence level.

Competing interests

The authors declare that they have no competing interests.

References

1. Goddard ME, Hayes BJ. Mapping genes for complex traits in domestic animals and their use in breeding programmes. Nat Rev Genet. 2009;10:381–91.
2. Habier D, Tetens J, Seefried FR, Lichtner P, Thaller G. The impact of genetic relationship information on genomic breeding values in German Holstein cattle. Genet Sel Evol. 2010;42:5.
3. Liu Z, Seefried FR, Reinhardt F, Rensing S, Thaller G, Reents R. Impacts of both reference population size and inclusion of a residual polygenic effect on the accuracy of genomic prediction. Genet Sel Evol. 2011;43:19.
4. Lund MS, de Roos APW, de Vries AG, Druet T, Ducrocq V, Fritz S, et al. A common reference population from four European Holstein populations increases reliability of genomic predictions. Genet Sel Evol. 2011;43:43.
5. Brøndum RF, Rius-Vilarrasa E, Strandén I, Su G, Guldbrandtsen B, Fikse WF, et al. Reliabilities of genomic prediction using combined reference data of the Nordic Red dairy cattle populations. J Dairy Sci. 2011;94:4700–7.
6. Lund MS, Su G, Janss L, Guldbrandtsen B, Brøndum RF. Invited review: genomic evaluation of cattle in a multi-breed context. Livest Sci. 2014;166:101–10.
7. Hayes BJ, Bowman PJ, Chamberlain AC, Verbyla K, Goddard ME. Accuracy of genomic breeding values in multi-breed dairy cattle populations. Genet Sel Evol. 2009;41:51.
8. Pryce JE, Gredler B, Bolormaa S, Bowman PJ, Egger-Danner C, Fuerst C, et al. Short communication: genomic selection using a multi-breed, across-country reference population. J Dairy Sci. 2011;94:2625–30.
9. Erbe M, Hayes BJ, Matukumalli LK, Goswami S, Bowman PJ, Reich CM, et al. Improving accuracy of genomic predictions within and between dairy cattle breeds with imputed high-density single nucleotide polymorphism panels. J Dairy Sci. 2012;95:4114–29.
10. Hozé C, Fritz S, Phocas F, Boichard D, Ducrocq V, Croiseau P. Efficiency of multi-breed genomic selection for dairy cattle breeds with different sizes of reference population. J Dairy Sci. 2014;97:3918–29.
11. Kemper KE, Reich CM, Bowman PJ, Vander Jagt CJ, Chamberlain AJ, Mason BA, et al. Improved precision of QTL mapping using a nonlinear Bayesian method in a multi-breed population leads to greater accuracy of across-breed genomic predictions. Genet Sel Evol. 2015;47:29.
12. de Roos APW, Hayes BJ, Spelman RJ, Goddard ME. Linkage disequilibrium and persistence of phase in Holstein-Friesian, Jersey and Angus cattle. Genetics. 2008;179:1503–12.
13. de los Campos G, Vazquez AI, Fernando R, Klimentidis YC, Sorensen D. Prediction of complex human traits using the genomic best linear unbiased predictor. PLoS Genet. 2013;9:e1003608.
14. Ober U, Ayroles JF, Stone EA, Richards S, Zhu D, Gibbs RA, et al. Using whole-genome sequence data to predict quantitative trait phenotypes in *Drosophila melanogaster*. PLoS Genet. 2012;8:e1002685.
15. van Binsbergen R, Calus MPL, Bink CAM, van Eeuwijk FA, Schrooten C, Veerkamp RF. Genomic prediction using imputed whole-genome sequence data in Holstein Friesian cattle. Genet Sel Evol. 2015;47:71.
16. Pérez-Enciso M, Rincón JC, Legarra A. Sequence- vs. chip-assisted genomic selection: accurate biological information is advised. Genet Sel Evol. 2015;47:43.
17. Brøndum RF, Su G, Janss L, Sahana G, Guldbrandtsen B, Boichard D, et al. Quantitative trait loci markers derived from whole genome sequence data increases the reliability of genomic prediction. J Dairy Sci. 2015;98:4107–16.
18. Porto-Neto LR, Barendse W, Henshall JM, McWilliam SM, Lehnert SA, Reverter A. Genomic correlation: harnessing the benefit of combining two unrelated populations for genomic selection. Genet Sel Evol. 2015;47:84.
19. Ober U, Huang W, Magwire M, Schlather M, Simianer H, Mackay TFC. Accounting for genetic architecture improves sequence based genomic prediction for a *Drosophila* fitness trait. PLoS One. 2015;10:e0126880.
20. van den Berg I, Guldbrandtsen B, Hozé C, Brøndum RF, Boichard D, Lund MS. Across breed QTL detection and genomic prediction in French and Danish dairy cattle breeds. In: Proceedings of the 10th world congress of genetics applied to livestock production: 17–22 August 2014; Vancouver. 2014. https://asas.org/docs/default-source/wcgalp-posters/490_paper_9308_manuscript_573_0.pdf?sfvrsn=2.
21. Braunschweig MH. Mutations in the bovine *ABCG2* and the ovine *MSTN* gene added to the few quantitative trait nucleotides identified in farm animals: a mini-review. J Appl Genet. 2010;51:289–97.
22. Boichard D, Grohs C, Bourgeois F, Cerqueira F, Faugeras R, Neau A, et al. Detection of genes influencing economic traits in three French dairy cattle breeds. Genet Sel Evol. 2003;35:77–101.
23. Ashwell MS, Heyen DW, Sonstegard TS, Van Tassell CP, Da Y, VanRaden PM, et al. Detection of quantitative trait loci affecting milk production, health, and reproductive traits in Holstein cattle. J Dairy Sci. 2004;87:468–75.
24. Khatkar MS, Thomson PC, Tammen I, Raadsma HW. Quantitative trait loci mapping in dairy cattle: review and meta-analysis. Genet Sel Evol. 2004;36:163–90.

25. Daetwyler HD, Schenkel FS, Sargolzaei M, Robinson JAB. A genome scan to detect quantitative trait loci for economically important traits in Holstein cattle using two methods and a dense single nucleotide polymorphism map. J Dairy Sci. 2008;91:3225–36.

26. Druet T, Fritz S, Boussaha M, Ben-Jemaa S, Guillaume F, Derbala D, et al. Fine mapping of quantitative trait loci affecting female fertility in dairy cattle on BTA03 using a dense single-nucleotide polymorphism map. Genetics. 2008;178:2227–35.

27. Cole JB, Wiggans GR, Ma L, Sonstegard TS, Lawlor TJ, Crooker BA, et al. Genome-wide association analysis of thirty one production, health, reproduction and body conformation traits in contemporary U.S. Holstein cows. BMC Genomics. 2011;12:408.

28. Raven LA, Cocks BG, Hayes BJ. Multibreed genome wide association can improve precision of mapping causative variants underlying milk production in dairy cattle. BMC Genomics. 2014;15:62.

29. van den Berg I, Boichard D, Lund MS. Multi breed genome wide association study and meta-analyses of production traits using whole genome sequence data for five French and Danish dairy cattle breeds. J Dairy Sci. 2016;99:8932–45.

30. Howie B, Fuchsberger C, Stephens M, Marchini J, Abecasis GR. Fast and accurate genotype imputation in genome-wide association studies through pre-phasing. Nat Genet. 2012;44:955–9.

31. Browning BL, Browning SR. A unified approach to genotype imputation and haplotype-phase inference for large data sets of trios and unrelated individuals. Am J Hum Genet. 2009;84:210–23.

32. Daetwyler HD, Capitan A, Pausch H, Stothard P, van Binsbergen R, Brøndum RF, et al. Whole-genome sequencing of 234 bulls facilitates mapping of monogenic and complex traits in cattle. Nat Genet. 2014;46:858–65.

33. Hozé C, Fouilloux MN, Venot E, Guillaume F, Dassonneville R, Fritz S, et al. High-density marker imputation accuracy in sixteen French cattle breeds. Genet Sel Evol. 2013;45:33.

34. Purcell S, Neale B, Todd-Brown K, Thomas L, Ferreira MAR, Bender D, et al. PLINK: a tool set for whole-genome association and population-based linkage analyses. Am J Hum Genet. 2007;81:559–75.

35. Kapell DN, Sorensen D, Su G, Janss LL, Ashworth CJ, Roehe R. Efficiency of genomic selection using Bayesian multi-marker models for traits selected to reflect a wide range of heritabilities and frequencies of detected quantitative traits loci in mice. BMC Genet. 2012;13:42.

36. Khansefid M, Pryce JE, Bolormaa S, Miller SP, Wang Z, Li C, et al. Estimation of genomic breeding values for residual feed intake in a multibreed cattle population. J Anim Sci. 2014;92:3270–83.

37. Wientjes YC, Calus MP, Goddard ME, Hayes BJ. Impact of QTL properties on the accuracy of multi-breed genomic prediction. Genet Sel Evol. 2015;47:42.

38. Janss LL. Disentangling pleiotropy along the genome using sparse latent variable models. In: Proceedings of the 10th world congress of genetics applied to livestock production: 17–22 August 2014; Vancouver. 2014. https://asas.org/docs/default-source/wcgalp-proceedings-oral/214_paper_10343_manuscript_1330_0.pdf?sfvrsn=2.

39. Hayes BJ, MacLeod IM, Daetwyler HD, Bowman PJ, Chamberlain AJ, vander Jagt CJ, et al. Genomic prediction from whole genome sequence in livestock: the 1000 bull genomes project. In: Proceedings of the 10th world congress of genetics applied to livestock production: 17–22 August 2014; Vancouver. 2014. https://asas.org/docs/default-source/wcgalp-proceedings-oral/183_paper_10441_manuscript_1644_0.pdf?sfvrsn=2.

15

Prediction of the reliability of genomic breeding values for crossbred performance

Jérémie Vandenplas* ⓘ, Jack J. Windig and Mario P. L. Calus

Abstract

Background: In crossbreeding programs, various genomic prediction models have been proposed for using phenotypic records of crossbred animals to increase the selection response for crossbred performance in purebred animals. A possible model is a model that assumes identical single nucleotide polymorphism (SNP) effects for the crossbred performance trait across breeds (ASGM). Another model is a genomic model that assumes breed-specific effects of SNP alleles (BSAM) for crossbred performance. The aim of this study was to derive and validate equations for predicting the reliability of estimated genomic breeding values for crossbred performance in both these models. Prediction equations were derived for situations when all (phenotyping and) genotyping data have already been collected, i.e. based on the genetic evaluation model, and for situations when all genotyping data are not yet available, i.e. when designing breeding programs.

Results: When all genotyping data are available, prediction equations are based on selection index theory. Without availability of all genotyping data, prediction equations are based on population parameters (e.g., heritability of the traits involved, genetic correlation between purebred and crossbred performance, effective number of chromosome segments). Validation of the equations for predicting the reliability of genomic breeding values without all genotyping data was performed based on simulated data of a two-way crossbreeding program, using either two closely-related breeds, or two unrelated breeds, to produce crossbred animals. The proposed equations can be used for an easy comparison of the reliability of genomic estimated breeding values across many scenarios, especially if all genotyping data are available. We show that BSAM outperforms ASGM for a specific breed, if the effective number of chromosome segments that originate from this breed and are shared by selection candidates of this breed and crossbred reference animals is less than half the effective number of all chromosome segments that are independently segregating in the same animals.

Conclusions: The derived equations can be used to predict the reliability of genomic estimated breeding values for crossbred performance using ASGM or BSAM in many scenarios, and are thus useful to optimize the design of breeding programs. Scenarios can vary in terms of the genetic correlation between purebred and crossbred performances, heritabilities, number of reference animals, or distance between breeds.

Background

Several livestock production systems are based on crossbreeding schemes (e.g., [1–3]), and take advantage of the increased performance of crossbred animals compared to purebred animals, along with breed complementarity. For such production systems based on crossbreeding, the breeding goal for the purebred populations is to optimize the performance of crossbred descendants. However, the selection of purebred animals for crossbred performance has not been extensively implemented in livestock, partly due to the difficulty of routine collection of pedigree information on crossbred animals [4].

With the advent of genomic selection, various genomic prediction models have been proposed, which use phenotypic records of crossbred animals to increase the selection response for crossbred performance in purebred animals (e.g., [2, 4–6]). These approaches predict breeding values for crossbred performance of selection

*Correspondence: jeremie.vandenplas@wur.nl
Animal Breeding and Genomics Centre, Wageningen UR Livestock Research, P.O. Box 338, 6700 AH Wageningen, The Netherlands

candidates using the estimated allele substitution effects of many single nucleotide polymorphisms (SNPs). The SNP allele substitution effects are estimated from phenotypes of genotyped reference animals. In the context of crossbreeding, several breeds and their crosses are involved in genomic prediction, and purebred and crossbred performances are often considered to be different but correlated traits (e.g., [1, 3, 5, 7, 8]). Therefore, estimates of SNP allele substitution effects for purebred and crossbred performance traits may not be the same for purebred and crossbred populations, e.g., due to genotype by environment interactions. Assuming only additive gene action, one approach to accommodate this is to model differences between allele substitution SNP effects using a multivariate genomic model that assumes a correlation structure between the effects of SNPs across the purebred and crossbred populations, or equivalently, by assuming a genetic correlation structure across the trait measured in purebred and crossbred populations [9, 10]. These multivariate genomic models are referred to hereafter as across-breed SNP genotype models (ASGM), since the estimates of SNP allele substitution effects for the crossbred performance trait are also used to predict breeding values for crossbred performance of purebred selection candidates, regardless of their breed of origin [4, 6]. Thus, estimates of SNP effects for the crossbred performance trait using ASGM are not breed-specific. However, a number of factors may have an impact on the effect that can be measured for a SNP for the crossbred performance trait. First, the two parental alleles at a SNP in a crossbred animal may have different effects on the phenotype due to different levels of linkage disequilibrium (LD) with a quantitative trait locus (QTL) in the parental purebred populations. Second, different genetic backgrounds, such as dominance or epistatic interactions, can result in the effects of the same QTL to be different in purebred versus crossbred animals. And third, purebred and crossbred animals may be exposed to different environments, leading to genotype by environment interactions. Because of these reasons, estimated allele substitution effects at SNPs for the crossbred performance trait may be breed-specific. To accommodate all these differences, previously an approach was proposed [3–6] that estimates breed-specific allele substitution effects for the crossbred performance trait (BSAM), assuming that the breed origin of SNP alleles in crossbred animals is known. Results from simulations have shown that BSAM can result in greater accuracy of genomic estimated breeding values (EBV) of purebreds for crossbred performance than ASGM under some conditions [2, 4, 11].

In order to be able to evaluate many different breeding program designs that apply genomic prediction for crossbreeding performance, it would be useful to be able to predict the reliability of genomic EBV using, for example, different genomic models or different breeding schemes. Prediction of reliability should preferably consider the genotype data of all reference animals and selection candidates when available, although it is also desirable to be able to predict the reliability when genotype data of, e.g., selection candidates is not available, i.e. when designing breeding programs. Various equations have been proposed in the literature to predict the reliability or the accuracy (i.e., the square root of reliability) of genomic EBV for (groups of) animals. The investigated genomic predictions rely on single-population genomic models [12, 13], and on ASGM [10, 14]. When genotypes are available for both reference animals and selection candidates, prediction equations are derived using selection index (SI) theory, while before availability of all genotyping data, they are derived using population parameters (e.g., heritability, number of reference animals) [10, 13, 15]. However, to our knowledge, equations for predicting the reliability of genomic EBV for crossbreeding performance for (groups of) animals have not yet been reported.

The primary aim of this study was to derive equations for predicting the reliability of genomic EBV for crossbred performance based on ASGM or BSAM. Prediction equations were derived for situations when all genotyping data are available for both reference animals and selection candidates (referred to as "with availability of genotyping data"), and for situations when the genotyping data are not available (referred to as "without availability of genotyping data"). The second aim was to compare the predictions of the reliability of genomic EBV without availability of genotyping data to the predictions obtained from the equations with availability of genotyping data, because the former are an approximation of the latter. Both reliabilities have the same expectation, since they both rely on prediction error variances (PEV) and assume absence of selection. Finally, the equation for predicting reliability without availability of genotyping data was used to investigate the expected ranges of reliabilities of genomic EBV using BSAM for a pig breeding program.

Methods

The first part of this section describes equations for predicting the reliability of genomic EBV for crossbred performance using ASGM or BSAM. For the derivations of these equations, we assumed a crossbreeding program with two breeds, A and B, with their F1 being crossbred AB animals. In order to simplify the derivation of the equations, we assumed that phenotypes are corrected for all fixed and random effects, other than additive genetic effects. Furthermore, reference animals are defined as animals with genotypes and phenotypes, and selection candidates are defined as animals with genotypes but without their own phenotype. The assumption that all reference animals have genotypes is likely to be correct in the near future, as genotyping costs continue to decrease. The aim is to predict the reliability of genomic EBV for crossbred performance for selection candidates of breed A. For the reference population, three scenarios were investigated: (1) the reference population includes only breed A animals (PB–PB), i.e. purebred (PB) phenotypes are used to predict EBV for crossbred (CB) performance of PB selection candidates; (2) the reference population includes only crossbred AB animals (CB–PB), i.e. CB phenotypes are used to predict EBV for CB performance of PB selection candidates; and (iii) the reference population includes both crossbred AB and breed A animals (CB + PB–PB), i.e. CB and PB phenotypes are used to predict EBV for CB performance of PB selection candidates. These scenarios represent situations where crossbred animals are terminal animals in commercial herds of pigs and chickens. The second part of this section describes simulations of the three scenarios used to validate the prediction equations without availability of genotyping data. In the equations below, reference animals are indicated by uppercase letters, while selection candidates are indicated by lowercase letters.

Across-breed SNP genotype models

Equations for predicting the reliability of genomic EBV for crossbred performance using ASGM were developed for the three scenarios. As ASGM is assumed, breed A and crossbred AB animals can be considered as belonging to different populations, assuming the genetic correlation between the PB and CB performance traits (r_{PC}) to be the genetic correlation between these breed A and crossbred AB populations. Therefore, equations for predicting the reliability of genomic EBV for crossbred performance for the three scenarios using ASGM can be derived from previous studies by, for example, Daetwyler et al. [12] and Wientjes et al. [10], without availability of genotyping data, and by VanRaden [15] with availability of genotyping data.

PB–PB scenario

The PB–PB scenario considers breed A animals for both reference animals and selection candidates. Phenotypes are therefore associated with the purebred performance trait, while the trait of interest is the crossbred performance trait. Indeed, selection candidates must be selected to optimize crossbred performance of their crossbred descendants.

Wientjes et al. [10] developed equations for predicting the accuracy of across-population genomic EBV values

without and with availability of genotyping data. Assuming additive gene action, differences in allele substitution effects that underlie the population-specific trait of interest were modelled by the genetic correlation between traits, which implies a multivariate genomic model. Similarly, for the PB–PB scenario, differences in allele substitution effects that underlie the purebred and crossbred performance traits can be considered in terms of the genetic correlation between the purebred and crossbred performance traits (r_{PC}) [10, 16]. Therefore, following Wientjes et al. [10], with availability of genotyping data, the average predicted reliability of genomic EBV for crossbred performance across-breed A selection candidates using breed A reference animals can be computed as follows, based on SI theory:

$$
r_{P\_ASGM\_with}^2 = \frac{1}{N_a} \sum_i r_{P\_ASGM\_with_i}^2
$$

$$
= \frac{1}{N_a} \sum_i r_{PC}^2 \frac{\mathbf{G}_{a_i,A} \left(\mathbf{G}_{AA} + \mathbf{I} \frac{1-h_a^2}{h_a^2} \right)^{-1} \mathbf{G}_{A,a_i}}{\mathbf{G}_{a_i,a_i}},
$$

(1)

where N_a is the number of breed A selection candidates; $h_a^2 = \frac{\sigma_a^2}{\sigma_a^2 + \sigma_{e_A}^2}$ is the heritability of the purebred performance trait, with σ_a^2 being the genetic variance of the purebred performance trait, and $\sigma_{e_A}^2$ the residual variance of purebred performance trait; $r_{PC} = \frac{\sigma_{a,c}}{\sqrt{\sigma_a^2 \sigma_c^2}}$ with $\sigma_{a,c}$ being the genetic covariance between the purebred and crossbred performance trait and σ_c^2 the genetic variance of the crossbred performance trait; matrix $\mathbf{G}_{A,A}$ is the $N_A \times N_A$ genomic relationship matrix for the N_A reference animals of breed A; vector $\mathbf{G}_{a_i,A}$ is the row corresponding to the ith selection candidate of breed A of the $N_a \times N_A$ genomic relationship matrix $\mathbf{G}_{a,A}$ between selection candidates of breed A and reference animals of breed A; $\mathbf{G}_{a_i,a_i}$ is the diagonal element corresponding to the ith selection candidate of breed A of the $N_a \times N_a$ genomic relationship matrix $\mathbf{G}_{a,a}$ between selection candidates of breed A; and matrix $\mathbf{I}$ is the identity matrix.

Matrices $\mathbf{G}_{A,A}$, $\mathbf{G}_{a,a}$, and $\mathbf{G}_{a,A}$ are parts of the genomic relationship matrix among all reference animals and selection candidates of breed A, i.e. $\mathbf{G} = \begin{bmatrix} \mathbf{G}_{A,A} & \mathbf{G}_{A,a} \\ \mathbf{G}_{a,A} & \mathbf{G}_{a,a} \end{bmatrix}$. Without loss of generality, and similar to Wientjes et al. [10], matrix $\mathbf{G}$ is computed following the second method of VanRaden [15], i.e., $\mathbf{G} = \frac{\mathbf{Z}\mathbf{Z}'}{m}$ where m is the number of SNP genotypes, and matrix $\mathbf{Z}$ contains the standardized genotypes as $\mathbf{Z}_{lk} = \frac{\mathbf{M}_{lk} - 2p_k}{\sqrt{2p_k(1-p_k)}}$, with $\mathbf{M}_{lk}$ being the SNP genotype (coded as 0 for one homozygous genotype, 1 for the heterozygous genotype, or 2 for the alternate homozygous genotype) of the *l*th animal of breed A for the *k*th locus, and p_k is the allele frequency at the *k*th locus.

Without availability of genotyping data, the predicted reliability of genomic EBV for crossbred performance of breed A selection candidates and using breed A reference animals can be computed as [10]:

$$r^2_{P\_ASGM\_without} = r^2_{PC} \frac{N_A h_a^2}{N_A h_a^2 + Me_{a,A}},\qquad(2)$$

where $Me_{a,A}$ is the effective number of chromosome segments that are shared between selection candidates and reference animals of breed A. If the term r^2_{PC} is ignored, or equal to 1, Eq. (2) has the same form as the equation proposed by Daetwyler et al. [12]. One of the assumptions in the derivation of this equation was that the error variance was approximately equal to the phenotypic variance, because only one locus was taken into account at a time and each locus explains only a small part of the additive genetic variance [10, 12, 14]. However, as explained by Daetwyler et al. [12] in the Appendix of their paper, this approximation results in slight underestimation of the predicted reliabilities, because the error variance decreases when multiple loci are used. In Additional file 1 of the current study, we proposed a derivation of Eq. (2) based on the mixed model theory and ignoring the term r^2_{PC}. We assumed that a single population was used and that effects of all independent loci are estimated simultaneously. Our derivation leads to the same equation as proposed by Daetwyler et al. [12] in the Appendix of their paper, which corrects for the fact that the error variance decreases when multiple loci are used. This derivation using the mixed model theory can be extended for deriving prediction equations using ASGM, and it will be also the basis for deriving prediction equations using BSAM.

CB–PB scenario

The reference population for the CB–PB scenario includes genotyped crossbred AB animals that have phenotypes for the crossbred performance trait. The selection candidates are breed A animals that are related to the reference population and that must be selected to optimize crossbred performance of their crossbred AB descendants. Because the trait of interest is the crossbred performance trait and because allele substitution SNP effects are estimated from crossbred data, the average predicted reliability of genomic EBV for crossbred performance across breed A selection candidates using a crossbred AB reference population can be computed with availability of genotyping data as follows [10]:

$$r^2_{C\_ASGM\_with} = \frac{1}{N_a} \sum_i r^2_{C\_ASGM\_with_i}$$

$$= \frac{1}{N_a} \sum_i \frac{\mathbf{G}_{a_i,AB} \left(\mathbf{G}_{AB,AB} + \mathbf{I} \frac{1-h_c^2}{h_c^2} \right)^{-1} \mathbf{G}_{AB,a_i}}{\mathbf{G}_{a_i,a_i}}$$

$$(3)$$

where $h_c^2 = \frac{\sigma_c^2}{\sigma_c^2 + \sigma_{e_c}^2}$ is the heritability of the crossbred performance trait, with $\sigma_{e_c}^2$ being the residual variance; matrix $\mathbf{G}_{AB,AB}$ is the $N_{AB} \times N_{AB}$ genomic relationship matrix between the N_{AB} crossbred AB reference animals; and vector $\mathbf{G}_{a_i,AB}$ is the row corresponding to the *i*th selection candidate of breed A of the $N_a \times N_{AB}$ genomic relationship matrix $\mathbf{G}_{aAB}$ between breed A selection candidates and crossbred AB reference animals. Similarly to Wientjes et al. [10], the genomic relationship matrix between breed A selection candidates and crossbred AB reference animals, $\mathbf{G}$, is computed following the second method of VanRaden [15] but taking into account that the selection candidates and reference animals belong to two different populations. It then follows that $\mathbf{G} = \begin{bmatrix} \mathbf{G}_{AB,AB} & \mathbf{G}_{AB,a} \\ \mathbf{G}_{a,AB} & \mathbf{G}_{a,a} \end{bmatrix} = \frac{\mathbf{Z}\mathbf{Z}'}{m}$, where *m* is the number of SNPs and matrix $\mathbf{Z}$ contains the standardized genotypes as $\mathbf{Z}_{ljk} = \frac{\mathbf{M}_{ljk} - 2p_{jk}}{\sqrt{2p_{jk}(1-p_{jk})}}$, with $\mathbf{M}_{ljk}$ being the SNP genotype (coded as previously) of the *l*th individual from the *j*th population (i.e., purebred or crossbred) for the *k*th locus, and p_{jk} is the allele frequency of the *j*th population at the *k*th locus.

Without availability of data, an equation that predicts the reliability of genomic EBV for crossbred performance of breed A selection candidates using N_{AB} crossbred AB reference animals can be simply written as follows:

$$r^2_{C\_ASGM\_without} = \frac{N_{AB} h_c^2}{N_{AB} h_c^2 + Me_{a,AB}},\qquad(4)$$

where $Me_{a,AB}$ is the effective number of chromosome segments shared by breed A selection candidates and crossbred AB reference animals [10].

CB + PB–PB scenario

The reference population for the CB + PB–PB scenario includes animals of breed A with phenotypes for the purebred performance trait and crossbred AB animals with phenotypes for the crossbred performance trait. The selection candidates are animals of breed A that are related to the reference population. Since the crossbred performance trait is the trait of interest, the average predicted reliability of genomic EBV across selection candidates for crossbred performance of breed A selection candidates using breed A and crossbred AB reference animals can be computed with availability of genotyping data as follows [10]:

$$r^2_{C+P\_ASGM\_with} = \frac{1}{N_a} \sum_i r^2_{C+P\_ASGM\_with_i}$$

$$= \frac{1}{N_a} \sum_i \frac{1}{\mathbf{G}_{a_i,a_i}} \begin{bmatrix} r_{PC}\mathbf{G}_{a_i,A} & \mathbf{G}_{a_i,AB} \end{bmatrix} \tag{5}$$

$$\times \begin{bmatrix} \mathbf{G}_{A,A} + \mathbf{I}\frac{1-h_a^2}{h_a^2} & \mathbf{G}_{A,AB}r_{PC} \\ r_{PC}\mathbf{G}_{AB,A} & \mathbf{G}_{AB,AB} + \mathbf{I}\frac{1-h_c^2}{h_c^2} \end{bmatrix}^{-1} \begin{bmatrix} r_{PC}\mathbf{G}_{a_i,A} \\ \mathbf{G}_{a_i,AB} \end{bmatrix}.$$

Without availability of genotyping data, the prediction equation for the reliability of genomic EBV for crossbred performance of breed A selection candidates using breed A and crossbred AB reference animals can be written as follows [14]:

$$r^2_{C+P\_ASGM\_without} = \begin{bmatrix} r_{PC}\sqrt{\frac{h_a^2}{Me_{a,A}}} & \sqrt{\frac{h_c^2}{Me_{a,AB}}} \end{bmatrix}$$

$$\times \begin{bmatrix} \frac{h_a^2}{Me_{a,A}} + \frac{1}{N_A} & r_{PC}\sqrt{\frac{h_a^2 h_c^2}{Me_{a,A}Me_{a,AB}}} \\ r_{PC}\sqrt{\frac{h_a^2 h_c^2}{Me_{a,A}Me_{a,AB}}} & \frac{h_c^2}{Me_{a,AB}} + \frac{1}{N_{AB}} \end{bmatrix}^{-1} \tag{6}$$

$$\times \begin{bmatrix} r_{PC}\sqrt{\frac{h_a^2}{Me_{a,A}}} \\ \sqrt{\frac{h_c^2}{Me_{a,AB}}} \end{bmatrix}.$$

Breed-specific allele substitution models

In crossbred populations, SNP effects may be breed-specific due to a number of factors [4], including different extents of LD between SNP and QTL between breeds, which can be accommodated by using BSAM, which fits breed-specific allele substitution effects [3, 4]. In this section, it is assumed that the breed origin of SNP alleles is known, as required by BSAM. Moreover, only the CB–PB and CB + PB–PB scenarios are considered, since the PB–PB scenario involves data on only one breed. To our knowledge, equations for predicting the reliability of genomic EBV using BSAM have not previously been developed.

CB–PB scenario

For the CB–PB scenario, assuming that each individual has one phenotypic record corrected for all effects other than the additive genetic effects, BSAM for the crossbred performance trait can be written as follows [3, 4]:

$$\mathbf{y}_{AB} = \mathbf{Z}_{AB}^{(A)}\boldsymbol{\beta}_c^{(A)} + \mathbf{Z}_{AB}^{(B)}\boldsymbol{\beta}_c^{(B)} + \mathbf{e}_{AB},$$

where $\mathbf{y}_{AB}$ is the vector of corrected records of crossbred performance; $\mathbf{Z}_{AB}^{(A)}$ ($\mathbf{Z}_{AB}^{(B)}$) contains the standardized breed A (B) SNP alleles of each crossbred animal; $\boldsymbol{\beta}_c^{(A)}$ ($\boldsymbol{\beta}_c^{(B)}$)

is the vector of breed A (B)-specific allele substitution effects for all SNPs; and $\mathbf{e}_{AB}$ is the residual vector. Entries of matrix $\mathbf{Z}_{AB}^{(A)}$ are defined as $Z_{AB_{lk}}^{(A)} = \frac{\mathbf{M}_{lk}-p_{Ak}}{\sqrt{2p_{Ak}(1-p_{Ak})}}$, where element $\mathbf{M}_{lk}$ is set to 0 or 1 when the kth locus of the lth individual has breed A allele 1 or 2, respectively; and p_{Ak} is the frequency at the kth locus for breed A. Matrix $\mathbf{Z}_{AB}^{(B)}$ is defined similarly. Expectations and variances of $\boldsymbol{\beta}_c^{(A)}$ and $\boldsymbol{\beta}_c^{(B)}$ are assumed to be $E\begin{bmatrix} \boldsymbol{\beta}_c^{(A)} \\ \boldsymbol{\beta}_c^{(B)} \end{bmatrix} = \begin{bmatrix} \mathbf{0} \\ \mathbf{0} \end{bmatrix}$ and

$$Var\begin{bmatrix} \boldsymbol{\beta}_c^{(A)} \\ \boldsymbol{\beta}_c^{(B)} \end{bmatrix} = \begin{bmatrix} \mathbf{I}\sigma_{\beta_c^{(A)}}^2 & \mathbf{0} \\ \mathbf{0} & \mathbf{I}\sigma_{\beta_c^{(B)}}^2 \end{bmatrix} = \begin{bmatrix} \mathbf{I}\frac{\sigma_{c_A}^2}{m} & \mathbf{0} \\ \mathbf{0} & \mathbf{I}\frac{\sigma_{c_B}^2}{m} \end{bmatrix}$$

where $\sigma_{\beta_c^{(A)}}^2$ ($\sigma_{\beta_c^{(B)}}^2$) is the variance of the breed A (B)-specific allele substitution effect, and $\sigma_{c_A}^2$ ($\sigma_{c_B}^2$) is the additive genetic variance due to alleles from population A (B) in the crossbred population for the crossbred performance trait [3, 4].

Equivalently, BSAM for the crossbred performance trait can be written as [3]:

$$\mathbf{y}_{AB} = \mathbf{c}_{AB}^{(A)} + \mathbf{c}_{AB}^{(B)} + \mathbf{e}_{AB},$$

where $\mathbf{c}_{AB}^{(A)} = \mathbf{Z}_{AB}^{(A)}\boldsymbol{\beta}_c^{(A)}$ ($\mathbf{c}_{AB}^{(B)} = \mathbf{Z}_{AB}^{(B)}\boldsymbol{\beta}_c^{(B)}$) is the vector of breed A (B) of origin additive genetic effects for the crossbred performance trait. It then follows that expectations and variances of $\mathbf{c}_{AB}^{(A)}$ and $\mathbf{c}_{AB}^{(B)}$ are defined as $E\begin{bmatrix} \mathbf{c}_{AB}^{(A)} \\ \mathbf{c}_{AB}^{(B)} \end{bmatrix} = \begin{bmatrix} \mathbf{0} \\ \mathbf{0} \end{bmatrix}$ and

$$Var\begin{bmatrix} \mathbf{c}_{AB}^{(A)} \\ \mathbf{c}_{AB}^{(B)} \end{bmatrix} = \begin{bmatrix} \mathbf{Z}_{AB}^{(A)}\mathbf{Z}_{AB}^{(A)'}\frac{\sigma_{c_A}^2}{m} & \mathbf{0} \\ \mathbf{0} & \mathbf{Z}_{AB}^{(B)}\mathbf{Z}_{AB}^{(B)'}\mathbf{I}\frac{\sigma_{c_B}^2}{m} \end{bmatrix}$$

$$= \begin{bmatrix} \mathbf{G}_{AB,AB}^{(A)}\sigma_{c_A}^2 & \mathbf{0} \\ \mathbf{0} & \mathbf{G}_{AB,AB}^{(B)}\sigma_{c_B}^2 \end{bmatrix},$$

where $\mathbf{G}_{AB,AB}^{(A)}$ ($\mathbf{G}_{AB,AB}^{(B)}$) is the breed A (B) partial genomic relationship among the N_{AB} crossbred AB animals [3]. These assumptions imply that $\mathbf{c}_{AB}^{(A)}$ and $\mathbf{c}_{AB}^{(B)}$, as well as $\boldsymbol{\beta}_c^{(A)}$ and $\boldsymbol{\beta}_c^{(B)}$, are independent of each other.

Based on SI theory, genomic EBV for crossbred performance of selection candidates from breed A ($\mathbf{c}_a^{(A)}$) can be predicted from records of crossbred AB reference animals as follows:

$$\hat{\mathbf{c}}_a^{(A)} = Cov\left(\mathbf{c}_a^{(A)}, \mathbf{y}_{AB}\right)\left(Var\left(\mathbf{y}_{AB}\right)\right)^{-1}\mathbf{y}_{AB}.$$

Using the model description, it can then be shown that the variance of $\mathbf{y}_{AB}$ is equal to:

$$Var\left(\mathbf{y}_{AB}\right) = \mathbf{G}_{ABAB}^{(A)}\sigma_{c_A}^2 + \mathbf{G}_{ABAB}^{(B)}\sigma_{c_B}^2 + \mathbf{I}\sigma_{e_{AB}}^2,$$

and that the covariance between $\mathbf{c}_a^{(A)}$ and $\mathbf{y}_{AB}$ is equal to:

$$Cov\left(\mathbf{c}_a^{(A)}, \mathbf{y}_{AB}\right) = Cov\left(\mathbf{c}_a^{(A)}, \mathbf{c}_{AB}^{(A)} + \mathbf{c}_{AB}^{(B)} + \mathbf{e}_{AB}\right)$$

$$= Cov\left(\mathbf{c}_a^{(A)}, \mathbf{c}_{AB}^{(A)}\right) + Cov\left(\mathbf{c}_a^{(A)}, \mathbf{c}_{AB}^{(B)}\right)$$

$$+ Cov\left(\mathbf{c}_a^{(A)}, \mathbf{e}_{AB}\right)$$

$$= Cov\left(\mathbf{c}_a^{(A)}, \mathbf{c}_{AB}^{(A)}\right) = \mathbf{G}_{a,AB}^{(A)}\sigma_{c_A}^2,$$

with matrix $\mathbf{G}_{a,AB}^{(A)} = \frac{1}{m}\mathbf{Z}_a^{(A)}\mathbf{Z}_{AB}^{(A)'}$ being the breed A-specific partial genomic relationship matrix between the N_a selection candidates of breed A and the N_{AB} crossbred AB reference animals.

The reliability of $\hat{c}_{a_i}^{(A)}$ of the ith selection candidate of breed A is then equal to:

$$r_{C\_BSAM\_with_i}^2 = \frac{\left(Cov\left(\hat{c}_{a_i}^{(A)}, c_{a_i}^{(A)}\right)\right)^2}{Var\left(\hat{c}_{a_i}^{(A)}\right)Var\left(c_{a_i}^{(A)}\right)} = \frac{Var\left(\hat{c}_{a_i}^{(A)}\right)}{Var\left(c_{a_i}^{(A)}\right)}$$

$$= \frac{1}{\mathbf{G}_{a_i,a_i}^{(A)}}\mathbf{G}_{a_i,AB}^{(A)}\left(\mathbf{G}_{AB,AB}^{(A)} + \mathbf{G}_{AB,AB}^{(B)}\frac{\sigma_{c_B}^2}{\sigma_{c_A}^2} + \mathbf{I}\frac{\sigma_e^2}{\sigma_{c_A}^2}\right)^{-1}\mathbf{G}_{AB,a_i}^{(A)}.$$

With availability of genotyping data, the average predicted reliability of genomic EBV across all breed A selection candidates is equal to:

$$r_{C\_BSAM\_with}^2 = \frac{1}{N_a}\sum_i r_{C\_BSAM\_with_i}^2 = \frac{1}{N_a}\sum_i \frac{1}{\mathbf{G}_{a_i,a_i}^{(A)}}\mathbf{G}_{a_i,AB}^{(A)}$$

$$\times \left(\mathbf{G}_{AB,AB}^{(A)} + \mathbf{G}_{AB,AB}^{(B)}\frac{h_{c_B}^2}{h_{c_A}^2} + \mathbf{I}\frac{1 - \frac{1}{2}h_{c_A}^2 - \frac{1}{2}h_{c_B}^2}{h_{c_A}^2}\right)^{-1}\mathbf{G}_{AB,a_i}^{(A)},$$

(7)

where $h_{c_A}^2 = \frac{\sigma_{c_A}^2}{\frac{\sigma_{c_A}^2}{2} + \frac{\sigma_{c_B}^2}{2} + \sigma_{e_{AB}}^2}$ $\left(h_{c_B}^2 = \frac{\sigma_{c_B}^2}{\frac{\sigma_{c_A}^2}{2} + \frac{\sigma_{c_B}^2}{2} + \sigma_{e_{AB}}^2}\right)$ is the breed A (B)-specific heritability of crossbred performance.

Since no equation has previously been proposed to predict the reliability of genomic EBV for BSAM without availability of genotyping data, here, we put forward a derivation based on mixed model theory [17], assuming that allele substitution effects for breeds A and B are estimated simultaneously. Equivalence between the mixed model and SI theories has previously been shown under certain conditions, including the use of the same estimates of the fixed effects [15, 17, 18]. Our derivation of the equation for predicting the reliability of genomic EBV for BSAM without availability of genotyping data [i.e., Eq. (8) below] is detailed in Additional file 2, and the result is briefly described in the following.

Consider N_{AB} unrelated genotyped crossbred AB reference animals. For simplicity, it is assumed that the breed A-specific effect $\beta_{c_k}^{*(A)}$ of each kth independent locus explains an equal amount of the breed A-specific additive genetic variance $\sigma_{c_A}^2$, i.e., $\sigma_{c_A}^2 = Me_{a,AB}^{(A)}\sigma_{\beta_c^{*(A)}}^2$, with $Me_{a,AB}^{(A)}$ being the effective number of chromosome segments underlying the crossbred performance trait for breed A and segregating in both breed A selection candidates and crossbred AB reference animals. The same assumption is made for the breed B-specific effect $\beta_{c_k}^{*(B)}$. The genomic EBV ($c_{a_i}^{(A)}$) for the ith selection candidate of breed A can be predicted as follows:

$$\hat{c}_{a_i}^{(A)} = \mathbf{z}_{a_i}^{*(A)}\hat{\boldsymbol{\beta}}_c^{*(A)},$$

where $\mathbf{z}_{a_i}^{*(A)}$ is a vector of the standardized genotypes for the $Me_{a,AB}^{(A)}$ independent loci of the ith selection candidate of breed A and $\hat{\boldsymbol{\beta}}_c^{*(A)}$ is the vector of the predictions of $\boldsymbol{\beta}_c^{*(A)}$. Following mixed model theory [17, 19], the reliability of $\hat{c}_{a_i}^{(A)}$ can be computed from the prediction error variance, $Var\left(\hat{c}_{a_i}^{(A)} - c_{a_i}^{(A)}\right)$, and is equal to:

$$r_{C_{BSAM_{without_i}}}^2 = 1 - \frac{Var\left(\hat{c}_{a_i}^{(A)} - c_{a_i}^{(A)}\right)}{Var\left(c_{a_i}^{(A)}\right)}$$

$$= \frac{Var\left(\hat{c}_{a_i}^{(A)}\right)}{Var\left(c_{a_i}^{(A)}\right)} = \frac{Var\left(\mathbf{z}_{a_i}^{*(A)}\hat{\boldsymbol{\beta}}_c^{*(A)}\right)}{Var\left(\mathbf{z}_{a_i}^{*(A)}\boldsymbol{\beta}_c^{*(A)}\right)}$$

$$= \frac{\mathbf{z}_{a_i}^{*(A)}Var\left(\hat{\boldsymbol{\beta}}_c^{*(A)}\right)\mathbf{z}_{a_i}^{*(A)'}}{\mathbf{z}_{a_i}^{*(A)}Var\left(\boldsymbol{\beta}_c^{*(A)}\right)\mathbf{z}_{a_i}^{*(A)'}}.$$

Assuming that the allele substitution effect $\beta_{c_k}^{*(A)}$ of each kth independent locus explains an equal amount of the breed A-specific additive genetic variance $\sigma_{c_A}^2$ and that the reliability of the estimated effect, $r_{\beta_c^{*(A)}}^2$, is the same for each locus, it follows that:

$$r_{C\_BSAM\_without_i}^2 = \frac{Var\left(\hat{\beta}_{c_k}^{*(A)}\right)}{Var\left(\beta_{c_k}^{*(A)}\right)} = r_{\beta_c^{*(A)}}^2.$$

Reliability $r_{\beta_c^{*(A)}}^2$ can be approximated as follows. Let $\widehat{\mathbf{y}_{AB}^*}$ be the vector of phenotypes corrected for all other fixed effects for the breed A-specific allele substitution effects other than the kth effect, $\hat{\boldsymbol{\beta}}_{c_{\neq k}}^{*(A)}$, as well as for the breed B-specific allele substitution effects, $\hat{\boldsymbol{\beta}}_c^{*(B)}$. The prediction of $\beta_{c_k}^{*(A)}$ for the kth locus can then be performed using the following model:

$$\widehat{\mathbf{y}_{AB}^*} = \mathbf{z}_{AB_k}^{*(A)}\beta_{c_k}^{*(A)} + \boldsymbol{\varepsilon}_{AB_k},$$

where vector $\mathbf{z}_{AB_k}^{*(A)}$ contains the standardized breed A alleles of crossbred AB reference animals, and $\boldsymbol{\varepsilon}_{AB_k}$ is the residual vector.

The variance of $\widehat{\mathbf{y}_{AB}^*}$ is equal to:

$$Var\left(\widehat{\mathbf{y}_{AB}^*}\right) = Var\left(\mathbf{z}_{AB_k}^{*(A)}\beta_{c_k}^{*(A)}\right) + Var\left(\boldsymbol{\varepsilon}_{AB_k}\right)$$
$$= \mathbf{z}_{AB_k}^{*(A)}\mathbf{z}_{AB_k}^{*(A)'}\sigma_{\beta_c^{*(A)}}^2 + Var\left(\boldsymbol{\varepsilon}_{AB_k}\right),$$

from which it follows that, after some algebra and assuming unrelated genotyped animals (see Additional file 2 for details):

$$Var\left(\boldsymbol{\varepsilon}_{AB_k}\right) = Var\left(\widehat{\mathbf{y}_{AB}^*}\right) - \mathbf{z}_{AB_k}^{*(A)}\mathbf{z}_{AB_k}^{*(A)'}\sigma_{\beta_c^{*(A)}}^2$$
$$= \mathbf{I}\left(\frac{\sigma_{c_A}^2}{2}\left(1 - r_{\beta_c^{*(A)}}^2\right) + \frac{\sigma_{c_B}^2}{2}\left(1 - r_{\beta_c^{*(B)}}^2\right) + \sigma_{e_{AB}}^2\right)$$
$$\approx \mathbf{I}\left(\sigma_{P_{AB}}^2 - \frac{\sigma_{c_A}^2}{2}r_{\beta_c^{*(A)}}^2 - \frac{\sigma_{c_B}^2}{2}r_{\beta_c^{*(B)}}^2\right) = \mathbf{I}\sigma_{\varepsilon_{AB}}^2,$$

where $\sigma_{P_{AB}}^2\left(\sigma_{e_{AB}}^2\right)$ is the phenotypic (residual) variance of the crossbred performance trait.

Therefore, following mixed model theory [17, 19], the prediction of $\beta_{c_k}^{*(A)}, \hat{\beta}_{c_k}^{*(A)}$, is equal to:

$$\hat{\beta}_{c_k}^{*(A)} = \left(\mathbf{z}_{AB_k}^{*(A)'}\mathbf{z}_{AB_k}^{*(A)}\sigma_{\varepsilon_{AB}}^{-2} + \sigma_{\beta_c^{*(A)}}^{-2}\right)^{-1}\sigma_{\varepsilon_{AB}}^{-2}\mathbf{z}_{AB_k}^{*(A)'}\widehat{\mathbf{y}_{AB}^*},$$

and the reliability of $\hat{\beta}_{c_k}^{*(A)}$ is equal to (see Additional file 2 for more details):

$$r_{\beta_c^{*(A)}}^2 = \frac{Var\left(\beta_{c_k}^{*(A)}\right) - Var\left(\hat{\beta}_{c_k}^{*(A)} - \beta_{c_k}^{*(A)}\right)}{Var\left(\beta_{c_k}^{*(A)}\right)}$$
$$= \frac{N_{AB}\sigma_{\varepsilon_{AB}}^{-2}\sigma_{\beta_c^{*(A)}}^2}{N_{AB}\sigma_{\varepsilon_{AB}}^{-2}\sigma_{\beta_c^{*(A)}}^2 + 2}.$$

It then follows that, without availability of genotyping data, the predicted reliability of the genomic EBV for breed A selection candidates is equal to (see Additional file 2 for more details):

$$r_{C\_BSAM\_without}^2 = r_{\beta_c^{*(A)}}^2$$
$$= \frac{N_{AB}h_{c_A}^2}{N_{AB}h_{c_A}^2 + 2Me_{a,AB}^{(A)}\left(1 - \frac{1}{2}h_{c_A}^2r_{c_a^{(A)}}^2 - \frac{1}{2}h_{c_B}^2r_{c_a^{(B)}}^2\right)}.$$

By ignoring the term $\left(-\frac{1}{2}h_{c_A}^2r_{c_a^{(A)}}^2 - \frac{1}{2}h_{c_B}^2r_{c_a^{(B)}}^2\right)$ for low $h_{c_A}^2$ and $h_{c_B}^2$, the prediction equation simplifies to:

$$r_{C\_BSAM\_without}^2 = \frac{N_{AB}h_{c_A}^2}{N_{AB}h_{c_A}^2 + 2Me_{a,AB}^{(A)}}. \tag{8}$$

CB + PB–PB scenario

For the CB + PB–PB scenario, the reference population includes breed A and crossbred AB reference animals, each with their own phenotypes. The BSAM for the crossbred performance trait, assuming that each individual has one record corrected for all effects other than additive genetic effects, can be written as follows [3, 4]:

$$\begin{bmatrix} \mathbf{y}_A \\ \mathbf{y}_{AB} \end{bmatrix} = \begin{bmatrix} \mathbf{Z}_A^{(A)} & \mathbf{0} \\ \mathbf{0} & \mathbf{Z}_{AB}^{(A)} \end{bmatrix}\begin{bmatrix} \boldsymbol{\beta}_a^{(A)} \\ \boldsymbol{\beta}_c^{(A)} \end{bmatrix}$$
$$+ \begin{bmatrix} \mathbf{0} & \mathbf{0} \\ \mathbf{0} & \mathbf{Z}_{AB}^{(B)} \end{bmatrix}\begin{bmatrix} \mathbf{0} \\ \boldsymbol{\beta}_c^{(B)} \end{bmatrix} + \begin{bmatrix} \mathbf{e}_A \\ \mathbf{e}_{AB} \end{bmatrix},$$

where $\mathbf{y}_A$ is the vector of corrected records of purebred performance, $\mathbf{Z}_A^{(A)}$ contains the standardized SNP genotypes of breed A reference animals, and $\boldsymbol{\beta}_a^{(A)}$ is the vector of breed A allele substitution effects for all SNPs for purebred performance.

Equivalently, the previous BSAM can be written as [3]:

$$\begin{bmatrix} \mathbf{y}_A \\ \mathbf{y}_{AB} \end{bmatrix} = \begin{bmatrix} \mathbf{a}_A^{(A)} \\ \mathbf{c}_{AB}^{(A)} \end{bmatrix} + \begin{bmatrix} \mathbf{0} \\ \mathbf{c}_{AB}^{(B)} \end{bmatrix} + \begin{bmatrix} \mathbf{e}_A \\ \mathbf{e}_{AB} \end{bmatrix},$$

where $\mathbf{a}_A^{(A)} = \mathbf{Z}_A^{(A)}\boldsymbol{\beta}_a^{(A)}$ is the vector of additive genetic effects for the purebred performance trait.

Expectations and variances and covariances of $\mathbf{a}_A^{(A)}, \mathbf{c}_{AB}^{(A)}$ and $\mathbf{c}_{AB}^{(B)}$ are assumed to be $E\begin{bmatrix} \mathbf{a}_A^{(A)} \\ \mathbf{c}_{AB}^{(A)} \\ \mathbf{c}_{AB}^{(B)} \end{bmatrix} = \begin{bmatrix} \mathbf{0} \\ \mathbf{0} \\ \mathbf{0} \end{bmatrix}$ and

$$Var\begin{bmatrix} \mathbf{a}_A^{(A)} \\ \mathbf{c}_{AB}^{(A)} \\ \mathbf{c}_{AB}^{(B)} \end{bmatrix} = \begin{bmatrix} \mathbf{Z}_A^{(A)}\mathbf{Z}_A^{(A)'}\frac{\sigma_{a_A}^2}{m} & \mathbf{Z}_A^{(A)}\mathbf{Z}_{AB}^{(A)'}\frac{\sigma_{a_Ac_A}}{m} & \mathbf{0} \\ \mathbf{Z}_{AB}^{(A)}\mathbf{Z}_A^{(A)'}\frac{\sigma_{a_Ac_A}}{m} & \mathbf{Z}_{AB}^{(A)}\mathbf{Z}_{AB}^{(A)'}\frac{\sigma_{c_A}^2}{m} & \mathbf{0} \\ \mathbf{0} & \mathbf{0} & \mathbf{Z}_{AB}^{(B)}\mathbf{Z}_{AB}^{(B)'}\frac{\sigma_{c_B}^2}{m} \end{bmatrix}$$
$$= \begin{bmatrix} \mathbf{G}_{A,A}^{(A)}\sigma_{a_A}^2 & \mathbf{G}_{A,AB}^{(A)}\sigma_{a_Ac_A} & \mathbf{0} \\ \mathbf{G}_{AB,A}^{(A)}\sigma_{a_Ac_A} & \mathbf{G}_{AB,AB}^{(A)}\sigma_{c_A}^2 & \mathbf{0} \\ \mathbf{0} & \mathbf{0} & \mathbf{G}_{AB,AB}^{(B)}\sigma_{c_B}^2 \end{bmatrix},$$

where $\mathbf{G}_{A,A}^{(A)}$ is the breed A genomic relationship matrix between N_A reference animals of breed A and $\mathbf{G}_{A,AB}^{(A)}$ is the breed A-specific partial genomic relationship matrix between N_A selection candidates of breed A and N_{AB} crossbred AB reference animals [3].

Based on the SI theory, genomic EBV for the crossbred performance trait for breed A selection candidates ($\mathbf{c}_a^{(A)}$) can be predicted from records of breed A reference animals and of crossbred AB reference animals:

$$\hat{\mathbf{c}}_a^{(A)} = Cov\left(\mathbf{c}_a^{(A)}, \mathbf{y}\right)\left(Var(\mathbf{y})\right)^{-1}\mathbf{y},$$

with $\mathbf{y} = \begin{bmatrix} \mathbf{y}_A \\ \mathbf{y}_{AB} \end{bmatrix}$.

Based on the model description, the variance of $\mathbf{y}$ is equal to:

$$Var(\mathbf{y}) = Var\left(\begin{bmatrix} \mathbf{y}_A \\ \mathbf{y}_{AB} \end{bmatrix}\right)$$

$$= \begin{bmatrix} \mathbf{G}_{AA}^{(A)}\sigma_{c_A}^2 + \mathbf{I}\sigma_{e_A}^2 & \mathbf{G}_{A,AB}^{(A)}\sigma_{a_A c_A} \\ \mathbf{G}_{A,AB}^{(A)}\sigma_{a_A c_A} & \mathbf{G}_{ABAB}^{(A)}\sigma_{c_A}^2 + \mathbf{G}_{ABAB}^{(B)}\sigma_{c_B}^2 + \mathbf{I}\sigma_{e_{AB}}^2 \end{bmatrix},$$

since:

$$Cov(\mathbf{y}_A, \mathbf{y}_{AB}) = Cov\left(\mathbf{a}_A^{(A)} + \mathbf{e}_A, \mathbf{c}_{AB}^{(A)} + \mathbf{c}_{AB}^{(B)} + \mathbf{e}_{AB}\right)$$

$$= Cov\left(\mathbf{a}_A^{(A)}, \mathbf{c}_{AB}^{(A)} + \mathbf{c}_{AB}^{(B)}\right)$$

$$= Cov\left(\mathbf{a}_A^{(A)}, \mathbf{c}_{AB}^{(A)}\right) = \mathbf{G}_{A,AB}^{(A)}\sigma_{a_A c_A}.$$

Similarly, the covariance between $\mathbf{c}_a^{(A)}$ and $\mathbf{y}$ is equal to:

$$Cov\left(\mathbf{c}_a^{(A)}, \mathbf{y}\right) = \begin{bmatrix} \mathbf{G}_{a,A}^{(A)}\sigma_{a_A c_A} & \mathbf{G}_{a,AB}^{(A)}\sigma_{c_A}^2 \end{bmatrix}.$$

The reliability of $\hat{c}_{a_i}^{(A)}$ of the ith selection candidate of breed A is then equal to:

Without availability of genotyping data, the prediction equation for the reliability of genomic EBV based on BSAM, $r_{C+P\_BSAM\_without}^2$, can be derived similarly to the prediction equation for the CB–PB scenario, $r_{C\_BSAM\_with}^2$. The derivation is based on mixed model theory and assumes that independent allele substitution effects for breeds A and B for both purebred and crossbred performances were estimated simultaneously. The detailed derivation can be found in Additional file 3.

Consider N_A unrelated genotyped breed A reference animals and N_{AB} unrelated genotyped crossbred AB reference animals. Similar to the CB–PB scenario, the genomic EBV ($\mathbf{c}_{a_i}^{(A)}$) for the ith selection candidate of breed A can be predicted as $\hat{\mathbf{c}}_{a_i}^{(A)} = \mathbf{z}_{a_i}^{*(A)}\hat{\boldsymbol{\beta}}_c^{*(A)}$ and its reliability is equal to:

$$r_{C+P\_BSAM\_without_i}^2 = \frac{Var\left(\hat{\mathbf{c}}_{a_i}^{(A)}\right)}{Var\left(\mathbf{c}_{a_i}^{(A)}\right)} = \frac{Var\left(\hat{\boldsymbol{\beta}}_{c_k}^{*(A)}\right)}{Var\left(\boldsymbol{\beta}_{c_k}^{*(A)}\right)} = r_{\beta_c^{*(A)}}^2.$$

$$r_{C+P\_BSAM\_with_i}^2 = \frac{\left(Cov\left(\hat{c}_{a_i}^{(A)}, c_{a_i}^{(A)}\right)\right)^2}{Var\left(\hat{c}_{a_i}^{(A)}\right)Var\left(c_{a_i}^{(A)}\right)} = \frac{Var\left(\hat{c}_{a_i}^{(A)}\right)}{Var\left(c_{a_i}^{(A)}\right)}$$

$$= \frac{1}{Var\left(c_{a_i}^{(A)}\right)}\begin{bmatrix} \mathbf{G}_{a_i,A}^{(A)}\sigma_{a_A c_A} & \mathbf{G}_{a_i,AB}^{(A)}\sigma_{c_A}^2 \end{bmatrix}$$

$$\times \begin{bmatrix} \mathbf{G}_{A,A}^{(A)}\sigma_{a_A}^2 + \mathbf{I}\sigma_{e_A}^2 & \mathbf{G}_{A,AB}^{(A)}\sigma_{a_A c_A} \\ \mathbf{G}_{A,AB}^{(A)}\sigma_{a_A c_A} & \mathbf{G}_{AB,AB}^{(A)}\sigma_{c_A}^2 + \mathbf{G}_{AB,AB}^{(B)}\sigma_{c_B}^2 + \mathbf{I}\sigma_{e_{AB}}^2 \end{bmatrix}^{-1}\begin{bmatrix} \mathbf{G}_{A,a_i}^{(A)}\sigma_{a_A c_A} \\ \mathbf{G}_{AB,a_i}^{(A)}\sigma_{c_A}^2 \end{bmatrix}$$

$$= \frac{1}{\mathbf{G}_{a_i,a_i}^{(A)}}\begin{bmatrix} r_{PC}^{(A)}\mathbf{G}_{a_i,A}^{(A)} & \mathbf{G}_{a_iA,B}^{(A)} \end{bmatrix}$$

$$\times \begin{bmatrix} \mathbf{G}_{A,A}^{(A)} + \mathbf{I}\frac{1-h_{a_A}^2}{h_{a_A}^2} & r_{PC}^{(A)}\mathbf{G}_{A,AB}^{(A)} \\ r_{PC}^{(A)}\mathbf{G}_{AB,A}^{(A)} & \mathbf{G}_{AB,AB}^{(A)} + \mathbf{G}_{AB,AB}^{(B)}\frac{h_{c_B}^2}{h_{c_A}^2} + \mathbf{I}\frac{1-\frac{1}{2}h_{c_A}^2 - \frac{1}{2}h_{c_B}^2}{h_{c_A}^2} \end{bmatrix}^{-1}\begin{bmatrix} r_{PC}^{(A)}\mathbf{G}_{A,a_i}^{(A)} \\ \mathbf{G}_{AB,a_i}^{(A)} \end{bmatrix},$$

where the breed A-specific genetic correlation between purebred and crossbred performance traits $r_{PC}^{(A)}$ is equal to $r_{PC}^{(A)} = \frac{\sigma_{a_A c_A}}{\sqrt{\sigma_{a_A}^2\sigma_{c_A}^2}}$.

With availability of genotyping data, the average predicted reliability of genomic EBV across all breed A selection candidates is therefore equal to:

Reliability $r_{\beta_c^{*(A)}}^2$ can be approximated as follows. The prediction of $\beta_{c_k}^{*(A)}$ for the kth independent locus can be performed using the phenotypes of both purebred and crossbred performances, $\begin{bmatrix} \widetilde{\mathbf{y}_A^*} \\ \mathbf{y}_{AB}^* \end{bmatrix}$, corrected for all other

$$r_{C+P\_BSAM\_with}^2 = \frac{1}{N_a}\sum_i r_{C+P\_BSAM\_with_i}^2 = \frac{1}{\mathbf{G}_{a_i,a_i}^{(A)}}\begin{bmatrix} r_{PC}^{(A)}\mathbf{G}_{a_i,A}^{(A)} & \mathbf{G}_{a_iA,B}^{(A)} \end{bmatrix}$$

$$\times \frac{1}{N_a}\sum_i \begin{bmatrix} \mathbf{G}_{A,A}^{(A)} + \mathbf{I}\frac{1-h_{a_A}^2}{h_{a_A}^2} & r_{PC}^{(A)}\mathbf{G}_{A,AB}^{(A)} \\ r_{PC}^{(A)}\mathbf{G}_{AB,A}^{(A)} & \mathbf{G}_{AB,AB}^{(A)} + \mathbf{G}_{AB,AB}^{(B)}\frac{h_{c_B}^2}{h_{c_A}^2} + \mathbf{I}\frac{1-\frac{1}{2}h_{c_A}^2 - \frac{1}{2}h_{c_B}^2}{h_{c_A}^2} \end{bmatrix}^{-1} \tag{9}$$

$$\times \begin{bmatrix} r_{PC}^{(A)}\mathbf{G}_{A,a_i}^{(A)} \\ \mathbf{G}_{AB,a_i}^{(A)} \end{bmatrix}$$

fixed effects, as well as for the breed B-specific allele substitution effects and correlated effects, using the model:

$$\begin{bmatrix} \widehat{\mathbf{y}_A^*} \\ \mathbf{y}_{AB}^* \end{bmatrix} = \begin{bmatrix} \mathbf{z}_{A_k}^{*(A)} & \mathbf{0} \\ \mathbf{0} & \mathbf{z}_{AB_k}^{*(A)} \end{bmatrix} \begin{bmatrix} \hat{\beta}_{a_k}^{*(A)} \\ \hat{\beta}_{c_k}^{*(A)} \end{bmatrix} + \begin{bmatrix} \varepsilon_{A_k} \\ \varepsilon_{AB_k} \end{bmatrix},$$

where ε_{A_k} is the residual vector. For simplicity, we will assume that $Var\left(\begin{bmatrix} \varepsilon_{A_k} \\ \varepsilon_{AB_k} \end{bmatrix} \right) = \begin{bmatrix} \mathbf{I}\sigma_{\varepsilon_A}^2 & \mathbf{0} \\ \mathbf{0} & \mathbf{I}\sigma_{\varepsilon_{AB}}^2 \end{bmatrix}$, with $\sigma_{\varepsilon_A}^2$ being the residual variance associated with $\widehat{\mathbf{y}_A^*}$. Then, following mixed model theory [17, 19], the prediction of $\begin{bmatrix} \beta_{a_k}^{*(A)} \\ \beta_{c_k}^{*(A)} \end{bmatrix}$ is equal to:

$$\begin{bmatrix} \hat{\beta}_{a_k}^{*(A)} \\ \hat{\beta}_{c_k}^{*(A)} \end{bmatrix} = \left(\begin{bmatrix} \mathbf{z}_{A_k}^{*(A)'} & \mathbf{0} \\ \mathbf{0} & \mathbf{z}_{AB_k}^{*(A)'} \end{bmatrix} \begin{bmatrix} \mathbf{I}\sigma_{\varepsilon_A}^{-2} & \mathbf{0} \\ \mathbf{0} & \mathbf{I}\sigma_{\varepsilon_{AB}}^{-2} \end{bmatrix} \begin{bmatrix} \mathbf{z}_{A_k}^{*(A)} & \mathbf{0} \\ \mathbf{0} & \mathbf{z}_{AB_k}^{*(A)} \end{bmatrix} + \begin{bmatrix} \sigma_{\beta_a^{*(A)}}^2 & \sigma_{\beta_a^{*(A)}\beta_c^{*(A)}} \\ \sigma_{\beta_a^{*(A)}\beta_c^{*(A)}} & \sigma_{\beta_c^{*(A)}}^2 \end{bmatrix}^{-1} \right)^{-1}$$
$$\begin{bmatrix} \mathbf{z}_{A_k}^{*(A)'} & \mathbf{0} \\ \mathbf{0} & \mathbf{z}_{AB_k}^{*(A)'} \end{bmatrix} \begin{bmatrix} \mathbf{I}\sigma_{\varepsilon_A}^{-2} & \mathbf{0} \\ \mathbf{0} & \mathbf{I}\sigma_{\varepsilon_{AB}}^{-2} \end{bmatrix} \begin{bmatrix} \widehat{\mathbf{y}_A^*} \\ \mathbf{y}_{AB}^* \end{bmatrix},$$

and the reliability of $\hat{\beta}_{c_k}^{*(A)}$ is equal to:

$$r_{\beta_c^{*(A)}}^2 = \frac{Var\left(\beta_{c_k}^{*(A)}\right) - Var\left(\hat{\beta}_{c_k}^{*(A)} - \beta_{c_k}^{*(A)}\right)}{Var\left(\beta_{c_k}^{*(A)}\right)} = \frac{\sigma_{\beta_c^{*(A)}}^2 - PEV_{\hat{\beta}_c^{*(A)}}}{\sigma_{\beta_c^{*(A)}}^2},$$

where $PEV_{\hat{\beta}_{c_k}^{*(A)}}$ is the prediction error variance of $\hat{\beta}_{c_k}^{*(A)}$ and is equal to the diagonal element of the inverse of the left-hand side of the mixed model equations associated with the prediction of $\begin{bmatrix} \beta_{a_k}^{*(A)} \\ \beta_{c_k}^{*(A)} \end{bmatrix}$. After some algebra, which is detailed in Additional file 3, it follows that the predicted reliability of genomic EBV for breed A selection candidates without data is equal to:

$$r_{C+P\_BSAM\_without}^2 = r_{\beta_c^{*(A)}}^2 = \left[r_{PC}^{(A)}\sqrt{\frac{h_{a_A}^2}{Me_{a,A}}} \quad \sqrt{\frac{h_{c_A}^2}{2Me_{a,AB}^{(A)}}} \right]$$
$$\times \begin{bmatrix} \frac{h_{a_A}^2}{Me_{a,A}} + \frac{1}{N_A} & r_{PC}^{(A)}\sqrt{\frac{h_{c_A}^2}{2Me_{a,AB}^{(A)}} \frac{h_{a_A}^2}{Me_{a,A}}} \\ r_{PC}^{(A)}\sqrt{\frac{h_{c_A}^2}{2Me_{a,AB}^{(A)}} \frac{h_{a_A}^2}{Me_{a,A}}} & \frac{h_{c_A}^2}{2Me_{a,AB}^{(A)}} + \frac{1}{N_{AB}} \end{bmatrix}^{-1}$$
$$\times \begin{bmatrix} r_{PC}^{(A)}\sqrt{\frac{h_{a_A}^2}{Me_{a,A}}} \\ \sqrt{\frac{h_{c_A}^2}{2Me_{a,AB}^{(A)}}} \end{bmatrix} \tag{10}$$

Computation of the effective number of chromosome segments (Me)

The prediction equations without availability of genotyping data require the effective number of chromosome segments that are independently segregating in a population, including selection candidates (S) and reference animals (R) (i.e., that are shared between the two populations), $Me_{S,R}$. The value of $Me_{S,R}$ can be computed as proposed by Wientjes et al. [14]:

$$Me_{S,R} = \frac{1}{S_{\mathbf{G}_{S,R}-\mathbf{A}_{S,R}}^2},$$

where $\mathbf{G}_{S,R}$ is the genomic relationship matrix between selection candidates and reference animals, $\mathbf{A}_{S,R}$ is the pedigree relationship matrix, and $S_{\mathbf{G}_{S,R}-\mathbf{A}_{S,R}}^2$ is the empirical variance of the differences between corresponding elements of $\mathbf{G}_{S,R}$ and $\mathbf{A}_{S,R}$. In our study, $\mathbf{G}_{S,R}$ and $\mathbf{A}_{S,R}$ were scaled to the same base population by rescaling the inbreeding level in $\mathbf{G}_{S,R}$ to the inbreeding in $\mathbf{A}_{S,R}$ as follows [20]:

$$\mathbf{G}_{S,R}^* = (1 - \bar{F})\mathbf{G}_{S,R} + 2\bar{F}\mathbf{J},$$

where $\bar{F}$ is the average pedigree inbreeding level computed from the pedigree, and $\mathbf{J}$ is a matrix filled with 1s.

The proposed computation of Me requires genotypes for both selection candidates and reference animals, which may be inconsistent with its use in the computation of reliabilities without availability of genotyping data. However, it is reasonable to assume that genotypes are already available for a limited number of animals, for example at least 100, that have the right family structure that is representative of the evaluated scenario, such that an accurate approximation of Me can be computed [14].

The effective number of chromosome segments originating from a specific breed (b) and that are shared between purebred selection candidates (S) of this breed and crossbred reference animals (Rc), $Me_{S,Rc}^{(b)}$, is required for the prediction equations for BSAM. In this study, $Me_{a,AB}^{(A)}$, is required in Eqs. (8) and (10) and was assumed to be equal to $Me_{a,A}$, which is required in Eqs. (2) and (6). The equality $Me_{a,AB}^{(A)} = Me_{a,A}$ was assumed since the selection candidates were the same for Eqs. (2), (4), (6), (8), and (10) the number of reference animals R and Rc was large, and the parents of breed A and crossbred AB reference animals were sampled from the same finite pool.

Simulated data

Data were simulated to validate Eqs. (2), (4), (6), (8), and (10), which predict the reliability of genomic EBV for crossbred performance using ASGM or BSAM, without availability of genotyping data. Two extreme scenarios were considered, in which either two closely-related or two unrelated breeds were used to produce crossbred animals. The reliabilities predicted by Eqs. (2), (4), (6), (8), and (10) were validated against the reliabilities computed with the corresponding prediction equations with availability of genotyping data, that is, Eqs. (1), (3), (5), (7), and (9). The reliabilities predicted by equations with availability of genotyping data are equivalent to those computed from PEV associated with selection candidates of a genomic best linear unbiased prediction including both reference animals and selection candidates, based on phenotypes corrected with the best linear unbiased estimates of the fixed effects, and assuming the absence of selection [15].

Populations

First, historical and breed populations were simulated using the QMSim software [21]; second, a two-way crossbreeding program with five generations of random selection was simulated using a customized Fortran program. For the historical population, 1000 discrete random mating generations with a constant size of 10,000 individuals were simulated, which was followed by 1000 generations in which the population size was gradually decreased to 2000 individuals. In these 2000 historical generations, half of the simulated animals were males and the other half were females. Offspring were produced by the random union of gametes from the male and female gametic pools, and the number of offspring was equal to the number of animals required in the next generation. To simulate the two breed populations, A and B, two random samples were drawn from the last generation of the historical population (i.e., generation 2000), each including 500 males and 500 females. Subsequently, within each of the breeds, 10 or 100 generations of random mating were simulated before the two-way crossbreeding scheme was begun. These two scenarios (i.e., a common origin either 10 or 100 generations ago) will be referred to as related and unrelated breeds, respectively. For the 10 and 100 generations of random mating, a litter of four offspring (two males and two females) per female was simulated. From these offspring, 500 males were selected at random for the next generation. The number of females selected randomly for the next generation was gradually increased from 500 to 800 during the first four generations of the simulation of the breed populations, in order to enlarge the size of the population (Fig. 1).

In a second step, a two-way crossbreeding program with five generations of random selection was simulated. The animals of breeds A and B that were used to start the crossbreeding program were sampled from generation 2010 for the related breeds and from generation 2100 for

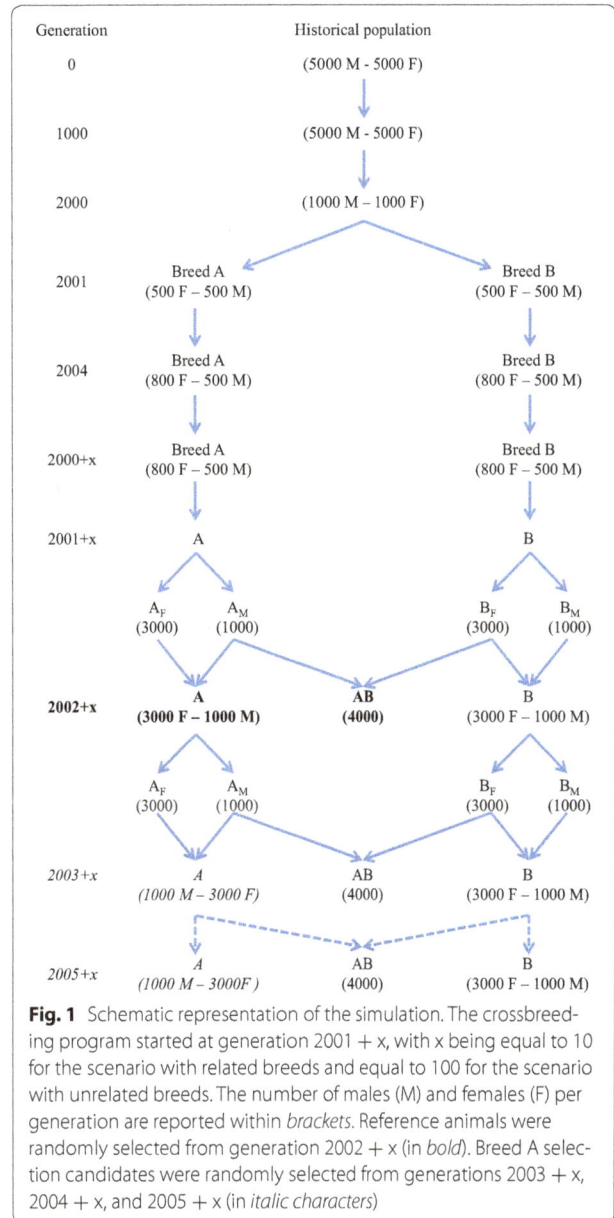

Fig. 1 Schematic representation of the simulation. The crossbreeding program started at generation 2001 + x, with x being equal to 10 for the scenario with related breeds and equal to 100 for the scenario with unrelated breeds. The number of males (M) and females (F) per generation are reported within *brackets*. Reference animals were randomly selected from generation 2002 + x (in *bold*). Breed A selection candidates were randomly selected from generations 2003 + x, 2004 + x, and 2005 + x (in *italic characters*)

the unrelated breeds. During the crossbreeding program, and for both breeds, animals of breeds A and B were randomly selected and mated to simulate the next generation of a constant size of 1000 males and 3000 females for each breed. From each of these five generations, animals of breeds A and B were randomly crossed to produce five generations of 4000 crossbred AB animals. Purebred animals used as parents of crossbred animals could also be parents of the next generation of purebred animals (Fig. 1).

Genotypes

The total length of the simulated genome was 10 Morgans (M) (10 chromosomes of 1 M and 4000 SNPs each). The positions of SNPs and of recombinations were randomized per chromosome and a recurrent mutation rate of 2.5×10^{-4} was assumed. All SNPs with a minor allele frequency (MAF) higher than or equal to 0.05 in the last historical generation (i.e., generation 2000) and were used to simulate the SNP genotypes of the purebred and crossbred animals. For subsequent analyses, 2000 SNPs were randomly selected from these SNPs for each chromosome. The breed origin of each allele for each crossbred animal was recorded. All scenarios (including the historical populations) were replicated 10 times.

Validation of prediction equations without availability of genotyping data

The validation required a set of known genotypes, as described previously, but no phenotype, since the reliabilities predicted without availability of genotyping data were validated against the reliabilities predicted with availability of genotyping data. However, estimates of heritabilities and genetic correlations between purebred and crossbred performance were required. Heritabilities of 0.20, 0.40, and 0.95 were used for both the purebred and crossbred performance traits. A high heritability, such as 0.95, and a single record per reference animal can be assumed when phenotypes of reference animals are derived from highly reliable EBV (e.g., deregressed EBV) [10]. Genetic correlations between purebred and crossbred performance traits were assumed to be equal to 0.30 or 0.70.

In the simulated data, two groups of reference animals and one group of selection candidates were defined for each scenario of related and unrelated breeds. For the scenarios with related and unrelated breeds, the two groups of reference animals were randomly selected from generations 2012 and 2102, respectively. For scenarios PB–PB and CB–PB, the two groups of reference animals included 2000 and 4000 animals that were randomly chosen from breed A and crossbred AB animals, respectively. For scenario CB + PB–PB, the first group included 4000 randomly chosen breed A animals and 2000 randomly chosen crossbred AB animals and the second group included 4000 breed A animals and 4000 crossbred AB animals. For the selection candidates for scenarios PB–PB, CB–PB and CB + PB–PB, 1000 breed A animals were randomly selected from each generation, starting from generation 2013 for the related breeds scenario and from generation 2103 for the unrelated breeds scenario, to create the groups of selection candidates. In the following, selection candidates from generations 2013 or 2103 are referred to as "G1" selection candidates. Similarly, selection candidates from generations 2014 and 2104 and from generations 2015 and 2105 are referred to as "G2" and "G3" selection candidates, respectively.

For each 'reference population-selection candidates' combination and for each scenario, reliabilities of the genomic EBV for crossbred performance were computed using Eqs. (1), (3), (5), (7), and (9) for the scenarios in which all data was available, and using Eqs. (2), (4), (6), (8), and (10) for scenarios without availability of genotyping data. The required genomic relationship matrices and values of Me were computed using our in-house software calc_grm [22]. The predicted reliabilities were averaged across the 10 replicates.

Application of a prediction equation

The proposed equations can be used to investigate the reliability of genomic EBV for crossbred performance in crossbreeding schemes. As an illustration, Eq. (10), which predicts the reliability of genomic EBV using both purebred and crossbred animals as reference animals by BSAM, was used to predict the reliability of genomic EBV for a pig production system for which 10,000 breed A animals were previously genotyped and phenotyped. The aim was to investigate the effect of the addition of crossbred AB animals to the reference population on the reliability of genomic EBV for crossbred performance. A heritability of 0.20 was assumed for both purebred and crossbred performance traits and the genetic correlation between purebred and crossbred performance traits for breed A, $(r_{PC}^{(A)})$, ranged from 0.0 to 1.0. Both values of Me required by Eq. (10) (i.e. $Me_{a,AB}^{(A)}$ and $Me_{a,A}$) were assumed to be equal to 476.6, based on the equation $Me = 2N_e L/(\ln(4N_e L))$ [23], with N_e being the effective population size and L being the total length of the genome in M. For N_e and L, we assumed values of 80 and 27 respectively, based on the study of Landrace pigs by Uimari and Tapio [24] and the study by Lin et al. [25]. The use of equal values of Me for the purebred and crossbred populations was based on the assumption that breed A parents of purebred and crossbred animals were sampled from the same pool.

Results

This section first presents the results of the validation of the equations for predicting reliability without availability of genotyping data. As defined previously, the reliabilities without availability of genotyping data were validated against the reliabilities computed with availability of genotyping data. The second part of this section describes the increase in reliabilities from the addition of crossbred animals to a purebred reference population in a pig breeding program.

PB–PB scenario

For the PB–PB scenario, the results show that reliabilities predicted without availability of genotyping data were of the same order of magnitude as reliabilities computed with availability of genotyping data (Figs. 2, 3). For the scenario with related breeds and $r_{PC} = 0.3$ (Fig. 2), the predicted reliabilities with availability of genotyping data were around 0.01 for $h_a^2 = 0.2$, in the range [0.02; 0.03] for $h_a^2 = 0.4$, and in the range [0.04; 0.05] for $h_a^2 = 0.95$, across all three groups of G1, G2 or G3 selection candidates and with 2000 reference animals from breed A. When $r_{PC} = 0.7$ (Fig. 3), the corresponding

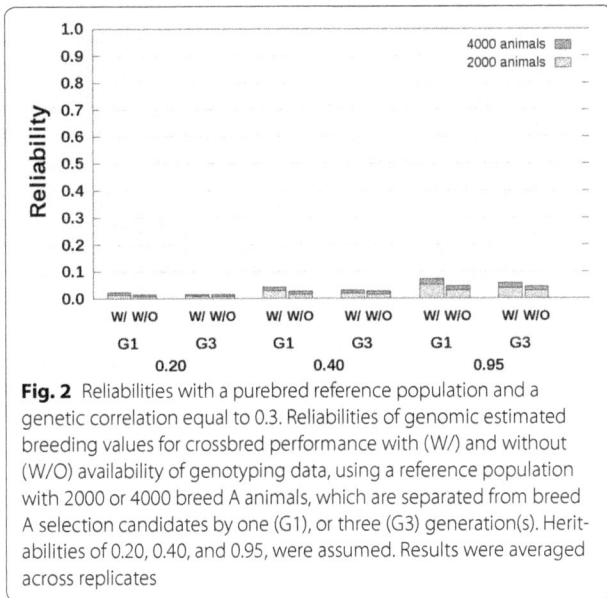

Fig. 2 Reliabilities with a purebred reference population and a genetic correlation equal to 0.3. Reliabilities of genomic estimated breeding values for crossbred performance with (W/) and without (W/O) availability of genotyping data, using a reference population with 2000 or 4000 breed A animals, which are separated from breed A selection candidates by one (G1), or three (G3) generation(s). Heritabilities of 0.20, 0.40, and 0.95, were assumed. Results were averaged across replicates

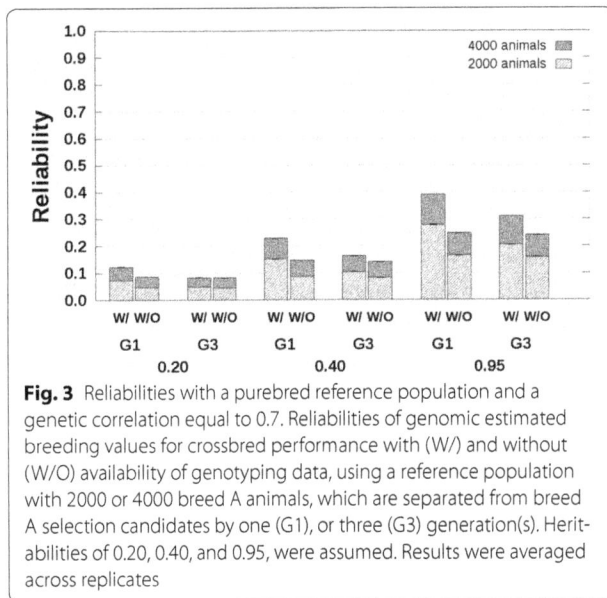

Fig. 3 Reliabilities with a purebred reference population and a genetic correlation equal to 0.7. Reliabilities of genomic estimated breeding values for crossbred performance with (W/) and without (W/O) availability of genotyping data, using a reference population with 2000 or 4000 breed A animals, which are separated from breed A selection candidates by one (G1), or three (G3) generation(s). Heritabilities of 0.20, 0.40, and 0.95, were assumed. Results were averaged across replicates

predicted reliabilities with availability of genotyping data were in the range [0.05; 0.07] for $h_a^2 = 0.2$, in the range [0.10; 0.15] for $h_a^2 = 0.4$, and in the range [0.20; 0.28] for $h_a^2 = 0.95$. For both scenarios with $r_{PC} = 0.3$ and $r_{PC} = 0.7$, the addition of 2000 breed A reference animals slightly increased the predicted reliabilities (Figs. 2, 3).

Reliabilities predicted without availability of genotyping data were always lower than those predicted with availability of genotyping data, which agrees with theory (see "PB–PB scenario" section in the "Methods" section). For the scenario with related breeds and $r_{PC} = 0.3$ (Fig. 2), the differences between reliabilities predicted without and with availability of genotyping data were around 0.00 for $h_a^2 = 0.2$, in the range [−0.02; 0.00] for $h_a^2 = 0.4$, and in the range [−0.02; −0.01] for $h_a^2 = 0.95$ across all three groups of G1, G2 or G3 selection candidates and with 2000 breed A reference animals. When $r_{PC} = 0.7$ (Fig. 3), the corresponding differences between reliabilities predicted without and with availability of genotyping data were in the range [−0.03; 0.00] for $h_a^2 = 0.2$, in the range [−0.06; −0.02] for $h_a^2 = 0.4$, and in the range [−0.11; −0.04] for $h_a^2 = 0.95$. The largest differences between reliabilities predicted without and with availability of genotyping data were always observed for the G1 selection candidates.

Similar results were obtained for the scenario with unrelated breeds (see Additional file 4: Tables S1, S2). Such similar results were expected since the distance between breeds is not taken into account by ASGM. The SD of the reliabilities across replicates were in the range [0.000; 0.001] (see Additional file 4: Tables S1, S2).

CB–PB scenario

Reliabilities with and without availability of genotyping data are presented in Fig. 4 for related breeds and in Fig. 5 for unrelated breeds. The CB–PB scenario included both ASGM and BSAM. For both models, the reliabilities predicted without availability of genotyping data underestimated the reliabilities predicted with availability of genotyping data. Underestimation was close to 0 when $h_c^2 = 0.20$, and increased up to 0.1 with increasing h_c^2 and number of crossbred AB reference animals. Similar to the PB–PB scenario, the underestimation of reliabilities predicted with availability of genotyping data by the reliabilities predicted without availability of genotyping data was largest for the G1 selection candidates.

For the G1 selection candidates, the reliabilities for ASGM with availability of genotyping data were around 0.09 with 2000 crossbred reference animals, independent of the relationship between the breeds, and around 0.16 with 4000 crossbred reference animals, using $h_c^2 = 0.20$ (Figs. 4, 5). Differences between the reliabilities predicted without and with availability of genotyping data were

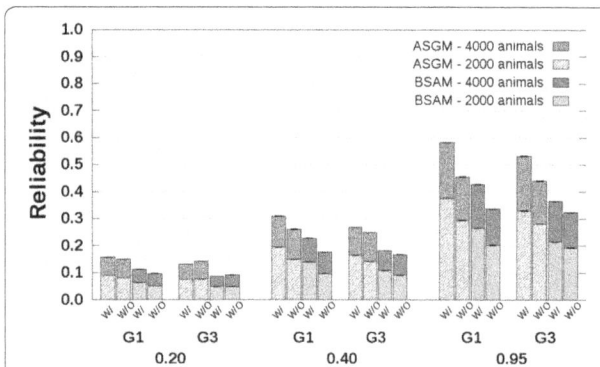

Fig. 4 Reliabilities with a crossbred reference population that originated from two related breeds. Reliabilities of genomic estimated breeding values for crossbred performance with (W/) and without (W/O) availability of genotyping data, based on an across-breed SNP genotype model (ASGM) or on a breed-specific allele substitution effects model (BSAM), and using a reference population with 2000 or 4000 crossbred AB animals. Reference animals were separated from breed A selection candidates by one (G1), or three (G3) generation(s). Heritabilities of 0.20, 0.40, and 0.95, were assumed. Results were averaged across replicates

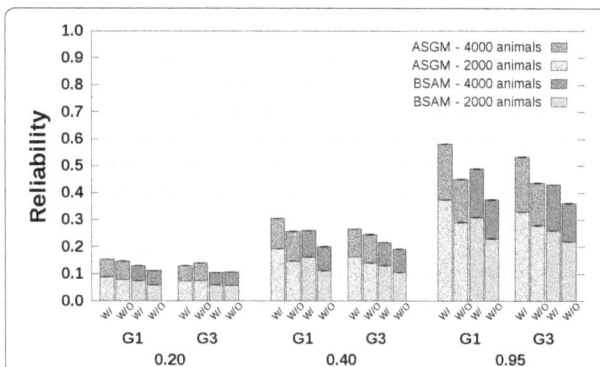

Fig. 5 Reliabilities with a crossbred reference population that originated from two unrelated breeds. Reliabilities of genomic estimated breeding values for crossbred performance with (W/) and without (W/O) availability of genotyping data, based on an across-breed SNP genotype model (ASGM) or on a breed-specific allele substitution effect model (BSAM), and using a reference population with 2000 or 4000 crossbred AB animals. Reference animals were separated from breed A selection candidates by one (G1), or three (G3) generation(s). Heritabilities of 0.20, 0.40, and 0.95, were assumed. Results were averaged across replicates

around −0.01 for both 2000 and 4000 crossbred reference animals. The corresponding reliabilities using $h_c^2 = 0.95$ were around 0.37 and 0.58 with 2000 and 4000 crossbred reference animals, respectively. The corresponding differences between reliabilities predicted without and with availability of genotyping data were in the range [−0.13; −0.08].

For G1 selection candidates with related breeds, the reliabilities for BSAM with availability of genotyping data

were around 0.06 and 0.11 with 2000 and 4000 crossbred reference animals, respectively, when using $h_c^2 = 0.20$ (Fig. 4). Differences between reliabilities predicted without and with availability of genotyping data were around −0.01 with both 2000 and 4000 crossbred reference animals. The corresponding reliabilities using $h_c^2 = 0.95$ were around 0.27 and 0.43 with 2000 and 4000 crossbred reference animals, respectively. Corresponding differences between reliabilities predicted without and with availability of genotyping data were in the range [−0.09; −0.06]. Similar differences were observed with unrelated breeds (Fig. 5). The SD of reliabilities across replicates were in the range [0.000; 0.002] (see Additional file 4: Tables S3, S4).

A comparison of reliabilities with availability of genotyping data between ASGM and BSAM showed that ASGM consistently performed better than BSAM. However, reliabilities for BSAM increased with increasing distance between breeds, while reliabilities for ASGM were only slightly affected (Figs. 4, 5). The increase in reliabilities with increasing distance between breeds, which compensates for the larger number of effects fitted in BSAM compared to ASGM, is in agreement with previous studies, e.g., Ibanez-Escriche et al. [4].

CB + PB–PB scenario
The CB + PB–PB scenario included both breed A and crossbred AB animals in the reference population. The number of breed A reference animals was always 4000. The number of crossbred AB animals was equal to 2000 or 4000. The CB + PB–PB scenario also included both ASGM and BSAM.

For related breeds, reliabilities without and with availability of genotyping data are presented in Fig. 6 for $r_{PC} = 0.3$ and in Fig. 7 for $r_{PC} = 0.7$. Reliabilities predicted without and with availability of genotyping data were of the same order of magnitude, for both ASGM and BSAM. Differences between the two predicted reliabilities were in the range [−0.09; 0.05]. Similar to previous results, these differences increased with heritability. Reliabilities for BSAM with availability of genotyping data were about 0.03 to 0.04 lower than the corresponding reliabilities for ASGM when heritabilities were assumed to be 0.20. This difference between reliabilities for BSAM and for ASGM increased with increasing heritability and r_{PC}, and with decreasing distance between breeds. For example, reliabilities for BSAM with availability of genotyping data were between 0.07 and 0.12 points lower than the corresponding reliabilities for ASGM when heritabilities were equal to 0.95. These lower reliabilities for BSAM can be attributed to the additional breed-specific effects fitted in the model for a given number of records. Similar

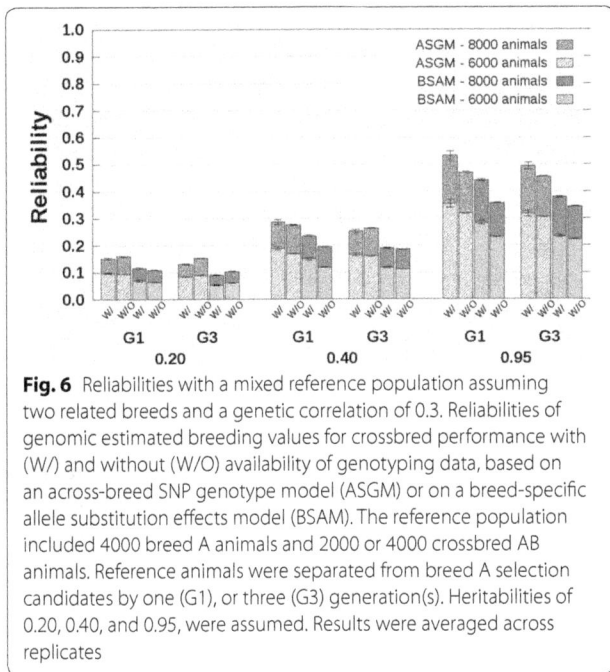

Fig. 6 Reliabilities with a mixed reference population assuming two related breeds and a genetic correlation of 0.3. Reliabilities of genomic estimated breeding values for crossbred performance with (W/) and without (W/O) availability of genotyping data, based on an across-breed SNP genotype model (ASGM) or on a breed-specific allele substitution effects model (BSAM). The reference population included 4000 breed A animals and 2000 or 4000 crossbred AB animals. Reference animals were separated from breed A selection candidates by one (G1), or three (G3) generation(s). Heritabilities of 0.20, 0.40, and 0.95, were assumed. Results were averaged across replicates

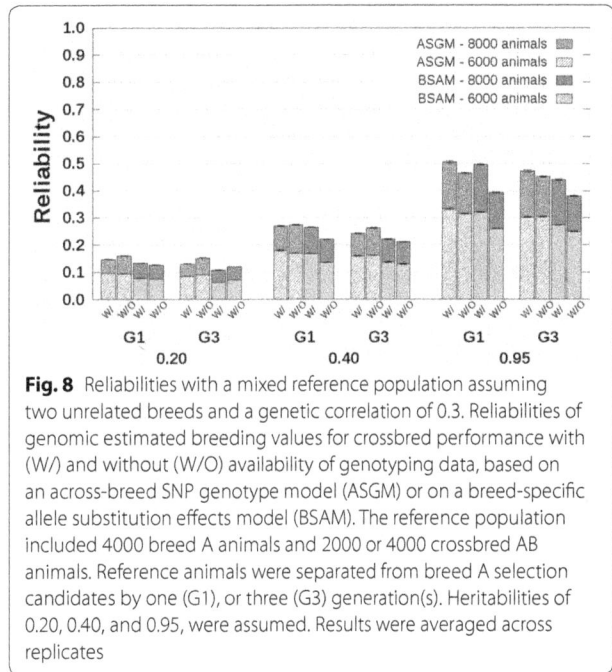

Fig. 8 Reliabilities with a mixed reference population assuming two unrelated breeds and a genetic correlation of 0.3. Reliabilities of genomic estimated breeding values for crossbred performance with (W/) and without (W/O) availability of genotyping data, based on an across-breed SNP genotype model (ASGM) or on a breed-specific allele substitution effects model (BSAM). The reference population included 4000 breed A animals and 2000 or 4000 crossbred AB animals. Reference animals were separated from breed A selection candidates by one (G1), or three (G3) generation(s). Heritabilities of 0.20, 0.40, and 0.95, were assumed. Results were averaged across replicates

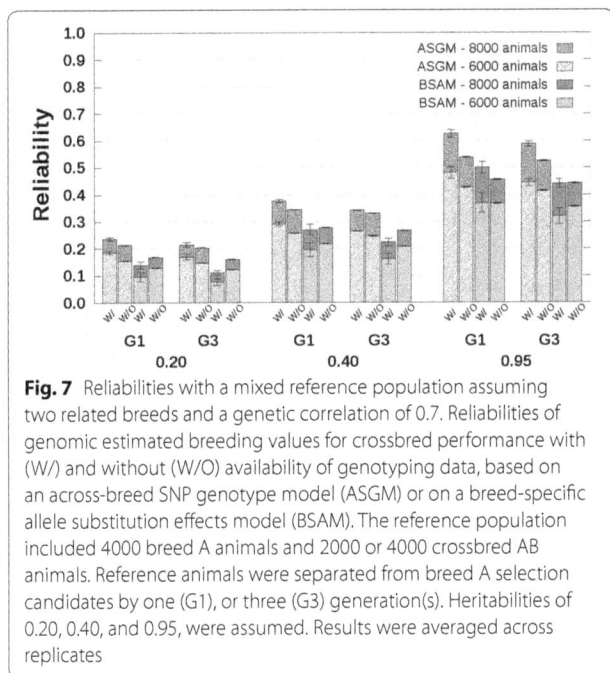

Fig. 7 Reliabilities with a mixed reference population assuming two related breeds and a genetic correlation of 0.7. Reliabilities of genomic estimated breeding values for crossbred performance with (W/) and without (W/O) availability of genotyping data, based on an across-breed SNP genotype model (ASGM) or on a breed-specific allele substitution effects model (BSAM). The reference population included 4000 breed A animals and 2000 or 4000 crossbred AB animals. Reference animals were separated from breed A selection candidates by one (G1), or three (G3) generation(s). Heritabilities of 0.20, 0.40, and 0.95, were assumed. Results were averaged across replicates

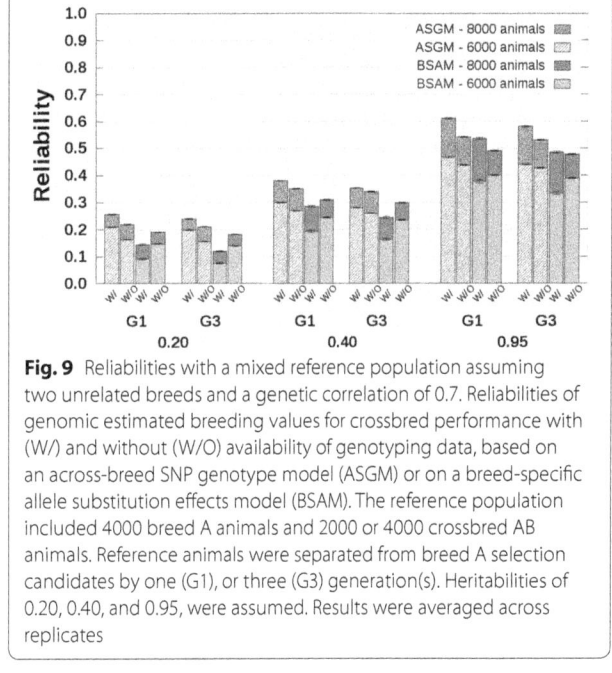

Fig. 9 Reliabilities with a mixed reference population assuming two unrelated breeds and a genetic correlation of 0.7. Reliabilities of genomic estimated breeding values for crossbred performance with (W/) and without (W/O) availability of genotyping data, based on an across-breed SNP genotype model (ASGM) or on a breed-specific allele substitution effects model (BSAM). The reference population included 4000 breed A animals and 2000 or 4000 crossbred AB animals. Reference animals were separated from breed A selection candidates by one (G1), or three (G3) generation(s). Heritabilities of 0.20, 0.40, and 0.95, were assumed. Results were averaged across replicates

trends were observed for reliabilities without availability of genotyping data.

With unrelated breeds, the reliabilities without and with availability of genotyping data, averaged across replicates, are presented in Fig. 8 for $r_{PC} = 0.3$ and in Fig. 9 for $r_{PC} = 0.7$. Differences between reliabilities predicted without and with availability of genotyping data were in the range $[-0.04; 0.07]$ for all scenarios with heritabilities equal to 0.20 and to 0.40, and in the range $[-0.10; 0.05]$ for all scenarios with heritabilities equal to 0.95. Reliabilities predicted with availability of genotyping data for BSAM with unrelated breeds were higher by about 0.01 to 0.06 than the reliabilities predicted with availability of genotyping data for BSAM with related breeds. Similar trends were observed for reliabilities predicted without availability of genotyping data, showing a reliable prediction of reliability computed with availability of genotyping data. The SD of reliabilities across replicates were in the range [0.000; 0.006] (see Additional file 4: Tables S5, S6, S7, S8).

Reliabilities in a pig-breeding program

Predicted reliabilities for BSAM when up to 10,000 crossbred AB animals were added to a reference population of 10,000 breed A animals are in Fig. 10, showing that predicted reliabilities increased when crossbred AB animals were added to the reference population. The increase in reliabilities decreased with increasing $r_{PC}^{(A)}$. The reliabilities obtained for ASGM with $r_{PC}^{(A)} = 0.92$ based on 10,000 breed A reference animals (and no crossbred AB reference animals) (0.68) was the same as that for BSAM based on only 10,000 crossbred AB reference animals (i.e., with $r_{PC}^{(A)} = 0.0$). Therefore, for $r_{PC}^{(A)} < 0.92$, BSAM with only crossbred reference animals can be at least as accurate as ASGM with a larger number of purebred reference animals.

Discussion

In this study, the term "reliability" refers to the precision of genomic EBV obtained by relating their PEV to the additive genetic variance of the base population, i.e., assuming absence of selection. Equations for predicting the reliability of genomic EBV for crossbred performance are proposed for reference populations that include purebred animals, crossbred animals, or both. Reliabilities were predicted for two models: ASGM and BSAM. For

the BSAM, we used the true breed-of-origin of all alleles for the crossbred animals, which would have to be estimated in practice, which may negatively impact the reliability obtained. However, we expect this to have only a very minor effect, since we showed in previous studies that it is possible to accurately derive breed-of-origin of alleles in three-breed crossbred pigs [26, 27].

Reliabilities of genomic EBV can be predicted when genotype data are already available, i.e., with availability of genotyping data, or without availability of genotyping data. For scenarios without availability of genotyping data, it is assumed that the required genetic parameters are computed using pedigree instead of genomic data, or that estimates are available from the literature. The results of this study showed that the reliabilities of genomic EBV for crossbred performance predicted without availability of genotyping data were of the same order of magnitude as those predicted with availability of genotyping data. Therefore, while prediction of reliability should preferably take the genotype data of selection candidates into account when available, both methods can predict the reliability of genomic EBV for crossbred performance for different reference populations, heritabilities, and r_{PC}. The derived equations can therefore be useful to optimize the design of breeding programs.

Reliabilities predicted without and with availability of genotyping data

The aim of this study was to predict the precision of genomic EBV based on PEV in the absence of selection. Thus, the derivation of our prediction equations without and with availability of genotyping data was based on the SI and mixed model theories and assumed that phenotypes were corrected for all fixed and random effects other than the considered genetic additive effects. The equivalence between SI and mixed model theories under certain conditions, such as the use of the same estimates for the fixed effects, has previously been shown by several studies (e.g., [15, 17, 28, 29]). Therefore, reliabilities predicted with availability of genotyping data would be expected to be close to reliabilities computed from PEV obtained from genomic best linear unbiased prediction, in the absence of selection. Equations for predicting the reliability of genomic EBV without availability of genotyping data were validated against the equations for predicting reliability with availability of genotyping data, and not against the reliability of selection, i.e., the squared correlation between estimated and true genomic breeding values, which is often obtained by cross-validation. Indeed, the reliability of genomic EBV is not equivalent to the reliability of selection for populations that are under selection, although they are equivalent for populations without selection [30–33]. Reliability of selection

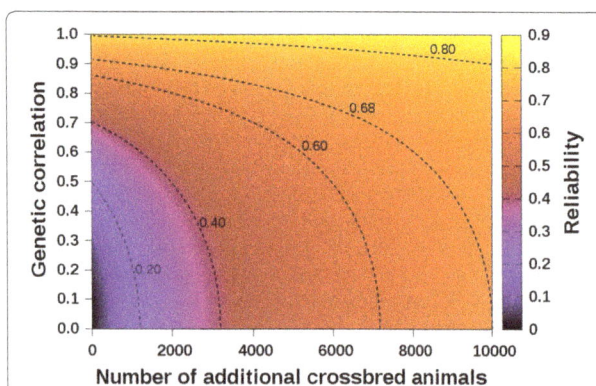

Fig. 10 Reliabilities with additional crossbred reference animals and different genetic correlations. Reliabilities predicted without availability of genotyping data for genomic estimated breeding values of crossbred performance using a breed-specific allele substitution effects model. The reference population included 10,000 purebred animals and a number of crossbred animals that varied from 0 to 10,000. A heritability of 0.20 was assumed for both purebred and crossbred performance traits, and the genetic correlation between purebred and crossbred performance traits varied from 0.0 to 1.0. All the required values of Me were assumed to be equal to 476.6

can be predicted from the reliability of genomic EBV by considering the intensity of selection using, e.g., the equations proposed by Dekkers [30] and Bijma [31].

Our prediction equations with availability of genotyping data can be extended to situations when phenotypes are already available for selection candidates and even to situations when some reference animals are not genotyped. First, in our study, selection candidates were defined as animals with genotypes but without phenotypes. In the context of poultry and pig breeding, this reflects for instance carcass, disease, or fertility traits. However, for growth-related traits, phenotypes are typically available for selection candidates at the time of selection. For these situations, prediction equations with availability of genotyping data can be used simply by including the selection candidates with genotypes and phenotypes as both reference animals and selection candidates, such that they are used both for computing the genomic relationship matrix between reference animals and selection candidates and the genomic relationship matrix between reference animals (e.g., for computing both $G_{a_i,A}$ and $G_{A,A}$ for the PB–PB scenario using ASGM). Second, situations in which some reference animals are not genotyped can also be modelled by our prediction equations with availability of genotyping data through the use of a combined pedigree-genomic relationship matrix H [34–36] instead of the genomic relationship matrix **G** used in this study. Third, our prediction equations with availability of genotyping data can be extended to situations in which reference animals have repeated phenotypes, or pseudo-phenotypes, such as deregressed proofs and associated weights. For instance, the predicted reliability for the PB–PB scenario using ASGM can be computed as:

$$r^2_{P\_ASGM\_with} = \frac{1}{N_a} \sum_i r^2_{PC} \frac{G_{a_i,A}W'\left(WG_{A,A}W' + R\frac{1-h_a^2}{h_a^2}\right)^{-1}WG_{A,a_i}}{G_{a_i,a_i}},$$

where **W** is the incidence matrix relating (pseudo-)phenotypes to reference animals and **R** is a diagonal matrix with elements equal to 1 for real phenotypes, or equal to the inverse of weights associated with pseudo-phenotypes [17, 28]. Unlike the prediction equations with availability of genotyping data, the extension of prediction equations without availability of genotyping data to more complex scenarios is not as straightforward.

We also assumed that all additive genetic variance was captured by the SNPs in the derivation of the prediction equations. When only a portion of the additive genetic variance is captured by the SNPs, the prediction equations need to take this into account, as proposed by Goddard et al. [13] and Wientjes et al. [14]. This proportion could be empirically estimated when the reference

population includes only one population by comparing predicted and realized (cross-validation) reliabilities [14].

For most scenarios, predicted reliabilities without availability of genotyping data underestimated the reliabilities predicted with availability of genotyping data (Figs. 2, 3, 4, 5, 6, 7, 8, 9). While this is in agreement with the theory, only a part of the underestimation is due to the fact that the decrease of the error variance when multiple loci are used was ignored (see "Methods" section; [12, 13]). This underestimation is greater when heritability and reliability increase to a value of 1 [13], as observed in our results. Most of the underestimation is, however, primarily due to an overestimation of Me, especially for the PB–PB and CB–PB scenarios with only one generation separating reference animals and selection candidates, for which the largest underestimations were observed. For instance, the fractional underestimation of $r^2_{P\_ASGM\_with}$ that can be attributed to not considering the reduction in error variance $\left(1 - h_a^2 r^2_{P\_ASGM\_without}\right)$ (see Additional file 1) is approximately equal to 0.03 (i.e., 3% error) for the PB–PB scenario with one generation separating 4000 reference animals and selection candidates, $h_a^2 = 0.95$, $r_{PC} = 1.0$, and $Me_{a,A} = 3730$ (obtained from one random replicate). This does, however, explains only part of the fractional underestimation of about 0.57 that is observed in Fig. 3. Thus, the underestimation appears to be mainly due to the overestimation of Me, particularly when only one generation separates the reference animals and selection candidates. Indeed, while estimates of Me increased with decreasing predicted reliabilities with availability of genotyping data, the results show that the reliabilities predicted without availability of genotyping data decreased at a lower rate than reliabilities predicted with availability of genotyping data when the relationships between the reference population and the selection candidates decreased. Further work to improve estimation of Me is needed, especially for scenarios in which reference animals and selection candidates are highly related.

While predicted reliabilities without availability of genotyping data were underestimated for most scenarios, overestimations were observed for some scenarios with reference populations that included both purebred and crossbred animals (Figs. 6, 7, 8, 9). These overestimations may be the result of estimation of Me and assumptions taken for the derivation of the equations without availability of genotyping data (e.g., a diagonal residual (co)variance matrix for corrected phenotypes associated with BSAM and using purebred and crossbred reference animals).

Potential use of the prediction equations

The equations derived in this study can be used to compare the effects of modifying the values of various factors (e.g., r_{PC}, numbers of reference animals, or relationships between the reference population and the selection candidates) on the reliability of genomic EBV for crossbred performance and for the optimization of the design of breeding programs. However, the effects of some factors should be compared carefully. For example, the results show that the prediction equations without availability of genotyping data should be used with care for the comparison of the effects of different relationships between the reference population and the selection candidates. The prediction equations without availability of genotyping data should also be used with care for the comparison of the reliabilities of the ASGM and BSAM models, especially when the reference population includes both purebred and crossbred animals (e.g., for the PB + CB– PB scenario with unrelated breeds and $r_{PC} = 0.7$; Fig. 9). Nevertheless, the prediction equations without availability of genotyping data can still provide some insight into the reliability of both models in different scenarios. For instance, the results (Figs. 4, 5, 6, 7, 8, 9) showed that reliabilities for BSAM tended to increase with increasing distance between breeds, while the reliabilities for ASGM were only slightly affected. The increase in reliabilities with increasing distance between breeds, which compensates for fitting more effects in BSAM in comparison to ASGM, is in agreement with previous studies, e.g., Ibanez-Escriche et al. [4]. For instance, assume that a reference population of a fixed number of crossbred AB animals is available, and that heritabilities of crossbred performance traits estimated for ASGM and BSAM for the breed A are equal. Therefore, from Eq. (4), $r^2_{C\_ASGM\_without} = \frac{N_{AB}h_c^2}{N_{AB}h_c^2 + Me_{a,AB}}$, and Eq. (8), $r^2_{C\_BSAM\_without} = \frac{N_{AB}h_{c_A}^2}{N_{AB}h_{c_A}^2 + 2Me_{a,AB}^{(A)}}$, it follows that the reliability of genomic EBV based on BSAM would be higher than the reliability based on ASGM if $Me_{a,AB} > 2Me_{a,AB}^{(A)}$. This will be the case if the LD patterns between breeds A and B are sufficiently different, which is more likely in the case when the breeds have diverged for many generations [37]. This is in agreement with our results (Figs. 4, 5) and previous studies based on simulated data (e.g., [4, 11]) which show that reliabilities for BSAM increase with increasing distance between breeds. The additional effects fitted in BSAM are taken into account in Eq. (8) by the factor of 2, which was also considered by van Grevenhof and van der Werf [38], who evaluated the benefit of including crossbred animals in the reference population of a crossbreeding program using genomic selection.

Computation of Me

The evaluation of different scenarios based on the prediction equations without availability of genotyping data requires accurate estimates of all parameters, and especially of Me (e.g., [10, 14, 39, 40]). Parameters such as heritabilities and correlations, if estimated inaccurately, would similarly bias reliabilities predicted without and with availability of genotyping data, since these parameters are used in both equations. However, Me, the effective number of segments that are shared and segregating in both selection candidates and reference animals, is only used when predicting reliability without availability of genotyping data, and has a large impact. In our study, the estimates of Me were computed from the differences between genomic and pedigree relationships between reference animals and selection candidates, as proposed by Wientjes et al. [14]. However, our results showed that these estimates of Me did not adequately consider the close relationships that can exist between reference animals and selection candidates. As already proposed by Daetwyler et al. [39] and Brard and Ricard [40], another approach would be to reverse the prediction equations without availability of genotyping data for computing Me. Required reliabilities and other parameters should be obtained from a reference population and different generations of selection candidates in which genomic prediction is already applied. However, estimates of Me obtained by the reversion of prediction equations would be underestimates, since this would include a correction for the fact that the error variance decreases when multiple loci are used, which is trait-dependent.

This study has introduced the concept of the effective number of chromosome segments originating from a specific breed (b), and shared by selection candidates (S) from this breed and crossbred reference animals (Rc), $Me_{S,Rc}^{(b)}$. This $Me_{S,Rc}^{(b)}$ is different from $Me_{S,Rc}$ as defined previously, since the latter does not take the breed origin of the chromosome segments of the crossbred animals into consideration. Indeed, each purebred population has its own value of Me, while the genome of crossbred animals combines segments from the different populations they originated from. Thus, the value of $Me_{S,Rc}$ includes both the effective number of chromosome segments segregating in breed b, and the effective number of chromosome segments segregating in the other breed(s) of origin for the crossbred animals, while $Me_{S,Rc}^{(b)}$ only involves the effective number of chromosome segments segregating in breed b. For this study, it was assumed that $Me_{S,Rc}^{(b)}$ (i.e., $Me_{a,AB}^{(A)}$) was equal to $Me_{S,R}$ (i.e., $Me_{a,A}$) for which the breed b selection candidates and reference animals (R) share the same parents as the crossbred Rc animals. This assumption was valid based on the results obtained. In practice, such an assumption would not be possible, since the purebred and crossbred reference animals may not share the same parents, or reference animals may belong to different generations. Further research on accurate estimation of Me is therefore required.

Conclusions

Several equations for predicting the reliability of genomic EBV for crossbred performance based on ASGM or on BSAM were derived for three different scenarios. These three scenarios involved a reference population that included only purebred animals, only crossbred animals, or both. The prediction equations were derived for application either without or with availability of genotyping data. Results showed that the reliabilities predicted without availability of genotyping data were of the same order of magnitude as the predictions of reliabilities predicted with availability of genotyping data. Thus, the proposed equations applied either without or with availability of genotyping data can be used to evaluate the effects of several parameters on the reliability of genomic EBV for crossbred performance (e.g., the genetic correlation between purebred and crossbred performances, heritabilities of the traits, number of reference animals, distance between breeds), and for the optimization of the design of breeding programs. Moreover, we showed that model BSAM can outperform model ASGM for a breed, if the effective number of chromosome segments originating from this breed and shared by selection candidates of this breed and crossbred reference animals is less than half the effective number of all chromosome segments that are independently segregating in these same animals, provided all other parameters remain equal. It is necessary to improve estimation of the effective number of chromosome segments to predict the reliability of genomic EBV without availability of genotyping data more accurately.

Additional files

Additional file 1. Derivation of the equation for predicting the reliability of genomic estimated breeding values without availability of data. Based on the mixed model theory, a derivation of the equation for predicting the reliability of genomic estimated breeding values is detailed in this document, assuming that effects of all independent loci are estimated simultaneously, and assuming a single population.

Additional file 2. Derivation of the prediction equation without availability of data, using a breed-specific allele substitution effects model and only crossbred reference animals. A derivation of the equation that predicts the reliability of genomic estimated breeding values for crossbred performance using a breed-specific allele substitution effects model and only crossbred animals without availability of genotyping data is detailed in this document.

Additional file 3. Derivation of the prediction equation without availability of data, using a breed-specific allele substitution effects model, and purebred and crossbred reference animals. A derivation of the equation for predicting the reliability of genomic estimated breeding values for crossbred performance using a breed-specific allele substitution effects model and a reference population including both purebred and crossbred animals, without availability of genotyping data, is detailed in this document.

Additional file 4: Table S1. Reliabilities for crossbred performance with a purebred reference population and a genetic correlation equal to 0.3.

Table S2. Reliabilities for crossbred performance with a purebred reference population and a genetic correlation equal to 0.7. **Table S3.** Reliabilities for crossbred performance with a crossbred reference population originated from two related breeds. **Table S4.** Reliabilities for crossbred performance with a crossbred reference population originated from two unrelated breeds. **Table S5.** Reliabilities for crossbred performance with a mixed reference population assuming two related breeds and a genetic correlation of 0.3. **Table S6.** Reliabilities for crossbred performance with a mixed reference population assuming two related breeds and a genetic correlation of 0.7. **Table S7.** Reliabilities for crossbred performance with a mixed reference population assuming two unrelated breeds and a genetic correlation of 0.3. **Table S8.** Reliabilities for crossbred performance with a mixed reference population assuming two unrelated breeds and a genetic correlation of 0.7.

Authors' contributions

JV derived the equations, performed the analyses, and drafted the manuscript. JJW wrote the simulation program. All authors discussed the design of the simulations. All authors provided valuable insights throughout the analysis and writing process. All authors read and approved the final manuscript.

Acknowledgements

Financial support from the Dutch Ministry of Economic Affairs, Agriculture, and Innovation (Public–private partnership "Breed4Food" Code BO-22.04-011-001-ASG-LR-3) is acknowledged. Discussions with Yvonne Wientjes and Piter Bijma, and useful comments of the two anonymous reviewers are acknowledged.

Competing interests

The authors declare that they have no competing interests.

References

1. Wei M, van der Werf JHJ. Maximizing genetic response in crossbreds using both purebred and crossbred information. Anim Sci. 1994;59:401–13.
2. Toosi A, Fernando RL, Dekkers JCM. Genomic selection in admixed and crossbred populations. J Anim Sci. 2010;88:32–46.
3. Christensen OF, Madsen P, Nielsen B, Su G. Genomic evaluation of both purebred and crossbred performances. Genet Sel Evol. 2014;46:23.
4. Ibáñez-Escriche N, Fernando RL, Toosi A, Dekkers JC. Genomic selection of purebreds for crossbred performance. Genet Sel Evol. 2009;41:12.
5. Dekkers JCM. Marker-assisted selection for commercial crossbred performance. J Anim Sci. 2007;85:2104–14.
6. Zeng J, Toosi A, Fernando RL, Dekkers JC, Garrick DJ. Genomic selection of purebred animals for crossbred performance in the presence of dominant gene action. Genet Sel Evol. 2013;45:11.
7. Lourenco DAL, Tsuruta S, Fragomeni BO, Chen CY, Herring WO, Misztal I. Crossbreed evaluations in single-step genomic best linear unbiased predictor using adjusted realized relationship matrices. J Anim Sci. 2016;94:909–19.
8. Hidalgo AM, Bastiaansen JWM, Lopes MS, Harlizius B, Groenen MAM, de Koning D-J. Accuracy of predicted genomic breeding values in purebred and crossbred pigs. G3 (Bethesda). 2015;5:1575–83.
9. Karoui S, Carabaño MJ, Díaz C, Legarra A. Joint genomic evaluation of French dairy cattle breeds using multiple-trait models. Genet Sel Evol. 2012;44:39.

10. Wientjes YC, Veerkamp RF, Bijma P, Bovenhuis H, Schrooten C, Calus MP. Empirical and deterministic accuracies of across-population genomic prediction. Genet Sel Evol. 2015;47:5.

11. Esfandyari H, Sørensen AC, Bijma P. A crossbred reference population can improve the response to genomic selection for crossbred performance. Genet Sel Evol. 2015;47:76.

12. Daetwyler HD, Villanueva B, Woolliams JA. Accuracy of predicting the genetic risk of disease using a genome-wide approach. PLoS One. 2008;3:e3395.

13. Goddard ME, Hayes BJ, Meuwissen THE. Using the genomic relationship matrix to predict the accuracy of genomic selection. J Anim Breed Genet. 2011;128:409–21.

14. Wientjes YCJ, Bijma P, Veerkamp RF, Calus MPL. An equation to predict the accuracy of genomic values by combining data from multiple traits, populations, or environments. Genetics. 2016;202:799–823.

15. VanRaden PM. Efficient methods to compute genomic predictions. J Dairy Sci. 2008;91:4414–23.

16. Vitezica ZG, Varona L, Elsen JM, Misztal I, Herring W, Legarra A. Genomic BLUP including additive and dominant variation in purebreds and F1 crossbreds, with an application in pigs. Genet Sel Evol. 2016;48:6.

17. Henderson CR. Applications of linear models in animal breeding. 2nd ed. Guelph: University of Guelph; 1984.

18. de los Campos G, Hickey JM, Pong-Wong R, Daetwyler HD, Calus MPL. Whole-genome regression and prediction methods applied to plant and animal breeding. Genetics. 2012;193:327–45.

19. Henderson CR. Best linear unbiased estimation and prediction under a selection model. Biometrics. 1975;31:423–47.

20. Powell JE, Visscher PM, Goddard ME. Reconciling the analysis of IBD and IBS in complex trait studies. Nat Rev Genet. 2010;11:800–5.

21. Sargolzaei M, Schenkel FS. QMSim: a large-scale genome simulator for livestock. Bioinformatics. 2009;25:680–1.

22. Calus MPL, Vandenplas J. Calc_grm—a program to compute pedigree, genomic, and combined relationship matrices. Wageningen: ABGC, Wageningen UR Livestock Research; 2016.

23. Goddard ME. Genomic selection: prediction of accuracy and maximisation of long term response. Genetica. 2009;136:245–57.

24. Uimari P, Tapio M. Extent of linkage disequilibrium and effective population size in Finnish Landrace and Finnish Yorkshire pig breeds. J Anim Sci. 2011;89:609–14.

25. Lin Z, Hayes BJ, Daetwyler HD. Genomic selection in crops, trees and forages: a review. Crop Pasture Sci. 2014;65:1177–91.

26. Sevillano CA, Vandenplas J, Bastiaansen JWM, Calus MPL. Empirical determination of breed-of-origin of alleles in three-breed cross pigs. Genet Sel Evol. 2016;48:55.

27. Vandenplas J, Calus MPL, Sevillano CA, Windig JJ, Bastiaansen JWM. Assigning breed origin to alleles in crossbred animals. Genet Sel Evol. 2016;48:61.

28. Mrode RA. Linear models for the prediction of animal breeding values. 2nd ed. Wallingford: CABI Publishing; 2005.

29. Strandén I, Garrick DJ. Technical note: derivation of equivalent computing algorithms for genomic predictions and reliabilities of animal merit. J Dairy Sci. 2009;92:2971–5.

30. Dekkers JCM. Asymptotic response to selection on best linear unbiased predictors of breeding values. Anim Sci. 1992;54:351–60.

31. Bijma P. Accuracies of estimated breeding values from ordinary genetic evaluations do not reflect the correlation between true and estimated breeding values in selected populations. J Anim Breed Genet. 2012;129:345–58.

32. Van Grevenhof EM, Van Arendonk JA, Bijma P. Response to genomic selection: the Bulmer effect and the potential of genomic selection when the number of phenotypic records is limiting. Genet Sel Evol. 2012;44:26.

33. Gorjanc G, Bijma P, Hickey JM. Reliability of pedigree-based and genomic evaluations in selected populations. Genet Sel Evol. 2015;47:65.

34. Aguilar I, Misztal I, Johnson DL, Legarra A, Tsuruta S, Lawlor TJ. Hot topic: a unified approach to utilize phenotypic, full pedigree, and genomic information for genetic evaluation of Holstein final score. J Dairy Sci. 2010;93:743–52.

35. Christensen OF, Lund MS. Genomic prediction when some animals are not genotyped. Genet Sel Evol. 2010;42:2.

36. Legarra A, Christensen OF, Aguilar I, Misztal I. Single step, a general approach for genomic selection. Livest Sci. 2014;166:54–65.

37. de Roos APW, Hayes BJ, Spelman RJ, Goddard ME. Linkage disequilibrium and persistence of phase in Holstein-Friesian, Jersey and Angus cattle. Genetics. 2008;179:1503–12.

38. van Grevenhof IE, van der Werf JH. Design of reference populations for genomic selection in crossbreeding programs. Genet Sel Evol. 2015;47:14.

39. Daetwyler HD, Pong-Wong R, Villanueva B, Woolliams JA. The impact of genetic architecture on genome-wide evaluation methods. Genetics. 2010;185:1021–31.

40. Brard S, Ricard A. Is the use of formulae a reliable way to predict the accuracy of genomic selection? J Anim Breed Genet. 2015;132:207–17.

16

Dimensionality of genomic information and performance of the Algorithm for Proven and Young for different livestock species

Ivan Pocrnic*[iD], Daniela A. L. Lourenco, Yutaka Masuda and Ignacy Misztal

Abstract

Background: A genomic relationship matrix (GRM) can be inverted efficiently with the Algorithm for Proven and Young (APY) through recursion on a small number of core animals. The number of core animals is theoretically linked to effective population size (N_e). In a simulation study, the optimal number of core animals was equal to the number of largest eigenvalues of GRM that explained 98% of its variation. The purpose of this study was to find the optimal number of core animals and estimate N_e for different species.

Methods: Datasets included phenotypes, pedigrees, and genotypes for populations of Holstein, Jersey, and Angus cattle, pigs, and broiler chickens. The number of genotyped animals varied from 15,000 for broiler chickens to 77,000 for Holsteins, and the number of single-nucleotide polymorphisms used for genomic prediction varied from 37,000 to 61,000. Eigenvalue decomposition of the GRM for each population determined numbers of largest eigenvalues corresponding to 90, 95, 98, and 99% of variation.

Results: The number of eigenvalues corresponding to 90% (98%) of variation was 4527 (14,026) for Holstein, 3325 (11,500) for Jersey, 3654 (10,605) for Angus, 1239 (4103) for pig, and 1655 (4171) for broiler chicken. Each trait in each species was analyzed using the APY inverse of the GRM with randomly selected core animals, and their number was equal to the number of largest eigenvalues. Realized accuracies peaked with the number of core animals corresponding to 98% of variation for Holstein and Jersey and closer to 99% for other breed/species. N_e was estimated based on comparisons of eigenvalue decomposition in a simulation study. Assuming a genome length of 30 Morgan, N_e was equal to 149 for Holsteins, 101 for Jerseys, 113 for Angus, 32 for pigs, and 44 for broilers.

Conclusions: Eigenvalue profiles of GRM for common species are similar to those in simulation studies although they are affected by number of genotyped animals and genotyping quality. For all investigated species, the APY required less than 15,000 core animals. Realized accuracies were equal or greater with the APY inverse than with regular inversion. Eigenvalue analysis of GRM can provide a realistic estimate of N_e.

Background

Genomic best linear unbiased prediction (GBLUP) methods [1] for genomic evaluation use single-nucleotide polymorphism (SNP) effects indirectly via the genomic relationship matrix (GRM). Therefore, GBLUP-based methods require a GRM inverse, which has a cubic cost and can be computed efficiently for perhaps up to 150,000 individuals. Because of widely available commercial genotyping tools, some populations such as the U.S. Holstein cattle have over one million genotyped animals, and computing a GRM inverse can be prohibitively expensive. In addition, a GRM often is not positive definite, and additional steps (e.g., blending with a numerator relationship matrix) are required to make the GRM positive definite [1]. Misztal et al. [2] suggested an efficient computation of the GRM inverse by using recursion on a small subset of animals. Initially, this subset of animals was labeled as high accuracy or "proven"; therefore, the method was named the Algorithm for Proven

*Correspondence: ipocrnic@uga.edu
Department of Animal and Dairy Science, University of Georgia, Athens, GA 30602, USA

and Young (APY). In this paper, we will refer to the GRM inverse calculated with this algorithm as the APY inverse and animals in the small subset as core animals. Compared with the regular GRM inverse, computing costs for the APY inverse are cubic only for the core subset and are linear for animals that are not in the subset. The estimated optimal subset size was approximately 8000 for Angus cattle [3] and 2000 to 6000 for commercial pigs [4]. Using U.S. Holstein data with 100,000 genotyped animals, Fragomeni et al. [5] found that any subset (including only bulls, only cows, and random animals) with at least 10,000 animals resulted in an accurate inverse. The APY inverse was successfully computed for about 570,000 genotyped Holsteins in less than 2 h of computing time on an average server with fewer than 20,000 core animals [6]. Using more than 10,000 animals as the core subset did not add any improvement in genetic prediction. For comparison, a regular inverse for 570,000 individuals would require several weeks of computing time and an amount of memory, which is available only in the largest computing clusters.

The theoretical framework of the APY inverse was proposed by Misztal [7]. For a population, the additive information is assumed to be in a limited number (n) of independent chromosome segments (M_e) or effective SNP markers (ESM). If M_e or ESM completely explain the additive variation, the breeding values of n animals are linear functions of M_e or ESM and contain nearly all the information in M_e or ESM. Defining any subset of n animals as core animals, a recursion on any n animals is sufficient. The magnitude of M_e is a function of effective population size (N_e), but the number of ESM could be computed as the number of eigenvalues explaining nearly all the variation in the GRM. Subsequently, the optimal number of core animals is a function of N_e and can be derived from eigenvalue analysis of the GRM.

The theory for APY inverse was tested by Pocrnic et al. [8] using six simulated populations with N_e ranging from 20 to 200. Each simulated population consisted of 10 non-overlapping generations under random mating and without selection, with 25,000 animals per generation and phenotypes available for generations 1 through 9. The last three generations (8 through 10) were completely genotyped, with 75,000 genotyped animals for each population. Their simulation assumed a total genome length of 30 Morgan and approximately 50,000 evenly allocated biallelic SNPs. They found that the number of largest eigenvalues that explain at least 90% of the variation in the GRM is almost a linear function of N_e. For the number of largest eigenvalues that explain from 95 to 99% of the variation, the curve was curvilinear, with departure from linearity attributed to a limited number of SNPs and a limited number of genotyped animals. True accuracies

were highest when the number of core animals corresponded to the number of eigenvalues explaining 98% of the variation, and they were slightly lower with the regular inverse or with half of the number of core animals.

The purpose of this study was to determine whether APY conclusions based on simulated data are valid with actual data across species. In particular, we wanted to find the optimal number of core animals per species, to investigate the changes in accuracy when recursions in APY are based on fractions of the optimal number of core animals, and to approximate the N_e for each species.

Methods
Data and models
Five previously collected datasets were used in this study. Analyses included the same models as those routinely used for national or commercial genetic evaluations of dairy (Holstein, Jersey) and beef (Angus) cattle, pigs, and broilers. The datasets and models were described in earlier studies [3, 6, 9–11]. Data for 11,626,576 Holstein final score records from 7093,380 cows were provided by Holstein Association USA, Inc. (Brattleboro, VT). Production data for Jerseys consisted of 4,168,048 records for 305-day milk, fat, and protein yields and were provided by the Animal Genomics and Improvement Laboratory, Agricultural Research Service, USDA (Beltsville, MD). For Angus cattle, more than 6 million records for birth weight and weaning weight and almost 3.4 million records for post-weaning gain were provided by the American Angus Association (St. Joseph, MO). More than 400,000 pig records for litter size and number of stillborn were provided by PIC (a Genus company, Hendersonville, TN). Finally, 196,613 records for body weight at grading, 51,774 records for residual feed intake, 9778 records for breast meat percentage, and 52,102 records for weight gain during feed conversion test were provided for broiler chickens by Cobb-Vantress Inc. (Siloam Springs, AR). The number of pedigrees used in the numerator relationship matrix (**A**) varied: 198,915 for broiler chickens, 2,429,392 for pigs, 2,468,914 for Jerseys, 8,236,425 for Angus, and 10,710,380 for Holsteins. The number of single-nucleotide polymorphisms used for genomic prediction and number of genotyped awwnimals also varied: 60,671 SNPs for Jerseys and Holsteins with 75,033 and 77,066 genotyped animals, respectively; 38,321 SNPS and 80,933 genotyped Angus; 36,551 SNPs and 22,575 genotyped pigs; and 39,102 SNPs and 15,720 genotyped broiler chickens.

Computations
Computations were similar to those described by Pocrnic et al. [8] except for the use of actual datasets and different validation strategies. The initial GRM ($\mathbf{G}_0$) was created

for each dataset by using the methodology of VanRaden [1] as $\mathbf{G}_0 = \mathbf{Z}\mathbf{Z}'/2\Sigma p_j(1 - p_j)$ where $\mathbf{Z}$ is a centered matrix of gene content adjusted for gene frequencies and p_j is allele frequency p for marker j. The observed allele frequencies were calculated directly from the SNP data of the genotyped population. The number of largest eigenvalues for $\mathbf{G}_0$ that explained 90, 95, 98, or 99% of variation was calculated using the DSYEV subroutine in LAPACK [12]. To obtain a positive definite GRM ($\mathbf{G}$), $\mathbf{A}$ was blended with $\mathbf{G}_0$ as $\mathbf{G} = w\mathbf{G}_0 + (1 - w)\mathbf{A}_{22}$, where w is a weight different for each breed/species ranging from 0.90 to 0.95, and $\mathbf{A}_{22}$ is the pedigree-based numerator relationship matrix for genotyped animals [1].

Single-step GBLUP was used for genomic evaluation, and analyses were performed with BLUP90IOD2 software [13] either with the regular (direct) inverse of the $\mathbf{G}$ matrix [14] or the APY inverse [2, 6]. If $\mathbf{G}$ was partitioned into blocks corresponding to core (c) and non-core (n) animals:

$$\mathbf{G} = \begin{bmatrix} \mathbf{G}_{cc} & \mathbf{G}_{cn} \\ \mathbf{G}_{nc} & \mathbf{G}_{nn} \end{bmatrix},$$

then the APY inverse [2, 7] was:

$$\mathbf{G}_{APY}^{-1} = \begin{bmatrix} \mathbf{G}_{cc}^{-1} & 0 \\ 0 & 0 \end{bmatrix} + \begin{bmatrix} -\mathbf{G}_{cc}^{-1}\mathbf{G}_{cn} \\ \mathbf{I} \end{bmatrix} \mathbf{M}_{nn}^{-1} \begin{bmatrix} -\mathbf{G}_{nc}\mathbf{G}_{cc}^{-1} & \mathbf{I} \end{bmatrix},$$

where $\mathbf{M}_{nn} = \text{diag}\{m_{nn,i}\} = \text{diag}\{g_{ii} - \mathbf{g}_{ic}\mathbf{G}_{cc}^{-1}\mathbf{g}_{ci}\}$, g_{ii} is the diagonal element of $\mathbf{G}_{nn}$ for non-core animal i, and $\mathbf{g}_{ic}$ is a vector of the genomic relationships of non-core animal i with all core animals. The number of core animals varied across datasets and corresponded to the number of largest eigenvalues in $\mathbf{G}_0$ that explained 90, 95, 98, or 99% of retained variation. The computational details for this algorithm were described by Masuda et al. [6].

Validation

The validation method depended on the amount of information available for the animals. For Holsteins and Jerseys, daughter deviations [15] were calculated in the complete dataset without genomic information and used as the dependent variable. Genomic estimated breeding values (GEBV) were calculated based on truncated data and used as the independent variable in a linear regression model. The truncation point was defined by the year when the phenotype was recorded: 2009 for Holsteins and 2010 for Jerseys. Coefficient of determination (R^2) for validation animals was used as a measure of reliability. For Holsteins, we defined the validation population as young genotyped bulls that had no daughters recorded in the truncated data, but had at least 30 daughters recorded in the complete dataset. For Jerseys, we defined the validation population as young genotyped bulls that had no daughters recorded in the truncated data, but

had estimated breeding values (EBV) with at least 75% reliability in the complete data. The Holstein and Jersey validation populations included 2948 and 449 bulls, respectively.

For the other datasets, validation was done by predictive ability [16] based on correlations between GEBV and phenotypes adjusted for fixed effects. The Angus validation population consisted of 27,528 genotyped animals born in 2013 that had their phenotypes excluded from the truncated data. Among those 27,528 animals, 18,204 had phenotypes for body weight, 18,524 for weaning weight, and 10,471 for post-weaning gain. For pigs, the validation population consisted of 881 genotyped animals born in 2014 with repeated records for litter size and number of stillborn (1166 and 1229, respectively); their phenotypes were excluded from the truncated data. The broiler validation population consisted of 2975 genotyped birds from the last generation that had their phenotypes excluded from the truncated data. Among the validation birds, 2975 had records for body weight at grading, 1954 for residual feed intake, 215 for breast meat percentage, and 1964 for weight gain during feed conversion test.

Validation parameters (reliability or predictive ability) were computed for genomic evaluations that used the APY inverse with the corresponding number of randomly chosen core animals based on eigenvalues that explained 90–99% of original variation. Validation parameters were computed similarly for genomic evaluations that used the regular inverse of $\mathbf{G}$.

Results and discussion

Numbers of largest eigenvalues that explain 90, 95, 98, and 99% of variation in $\mathbf{G}_0$ are in Table 1 by breed/species. Number of eigenvalues that accounted for 90% of the original variation ranged from 1239 for pigs to 4527 for Holsteins, and those that accounted for 99% ranged from 5570 for broiler chickens to 19,397 for Holsteins. For each population, the total number of positive eigenvalues in $\mathbf{G}_0$ is limited by the number of SNPs and the number of genotyped animals.

The distributions of eigenvalues that we obtained here for Holstein, Jersey, and Angus cattle, broiler chicken, and pig datasets were compared with those reported by Pocrnic et al. [8] for populations with an N_e of 20, 40, 80, 120, and 160 from a simulation study. In both cases, when the number of eigenvalues was plotted on a logarithmic scale, the curves were nearly linear. The distribution of eigenvalues observed for the Holstein dataset was nearly identical to that reported for a simulated population with an N_e of 160. The distribution of eigenvalues for the Angus and Jersey datasets were quite similar and intermediate to those found for simulated populations with an N_e of 80

Table 1 Numbers of largest eigenvalues that explain a given percentage of variation and estimated effective population size (N_e)

Population	Number of genotyped animals	Number of SNPs	90%	95%	98%	99%	N_e
Broiler chicken	15,720	39,102	1655	2606	4171	5570	44[a]
Pig	22,575	36,551	1239	2183	4103	6083	32[a] (48)[b]
Angus cattle	80,993	38,321	3654	6166	10,605	14,555	113[a]
Jersey cattle	75,053	60,671	3325	6074	11,500	16,645	101[a]
Holstein cattle	77,066	60,671	4527	7981	14,026	19,379	149[a]

[a] Based on chromosome length of 30 Morgan

[b] Based on chromosome length of 20 Morgan

and 120. For the pig dataset, the distribution of eigenvalues was intermediate to those found for simulated populations with an N_e of 20 and 40. Finally, for the broiler chicken dataset, the number of eigenvalues that explain 90% of the variation was close to that observed for a simulated population with an N_e of 40. As the proportion of explained variation increased, the number of eigenvalues for the broiler chicken decreased relative to those found for a simulated population with an N_e of 40. In general, the rank of the GRM was equal to or less than the number of genotyped animals and the number of SNPs. Smaller numbers of eigenvalues for the higher percentages of explained variation for the pig and broiler chicken datasets were likely the result of fewer genotyped animals (22,575 pigs and 15,720 broiler chickens) compared with the simulated population (75,000), since the rank of the GRM cannot exceed, and is likely smaller than, the number of genotyped animals. Another possible explanation is that fewer SNPs were used (36,000 for pigs and 39,000 for broiler chickens) compared with the 50,000 SNPs used in the simulation. MacLeod et al. [17] reported that the identification of 90% of the ancestral junctions between chromosome segments required 12 times as many SNPs as the number of junctions. Therefore, the number of chromosome segments that is determined by eigenvalue analysis will be underestimated if the number of SNPs (and genotyped animals) is too small. This may be generalized into a simple rule: the number of largest eigenvalues explaining a given percentage of variation is noticeably smaller than expected unless the corresponding number of SNPs (and perhaps genotyped animals) is at least 12 times larger. This condition was fulfilled when 90% of the variation was explained for all breeds/species but not when this percentage was higher.

Assuming that the number of eigenvalues for 90% of explained variation was the least affected by the limited number of genotyped individuals and SNPs, N_e can be estimated by interpolation of real to simulated data (Fig. 1) at 90% of explained variation. Thus, estimated

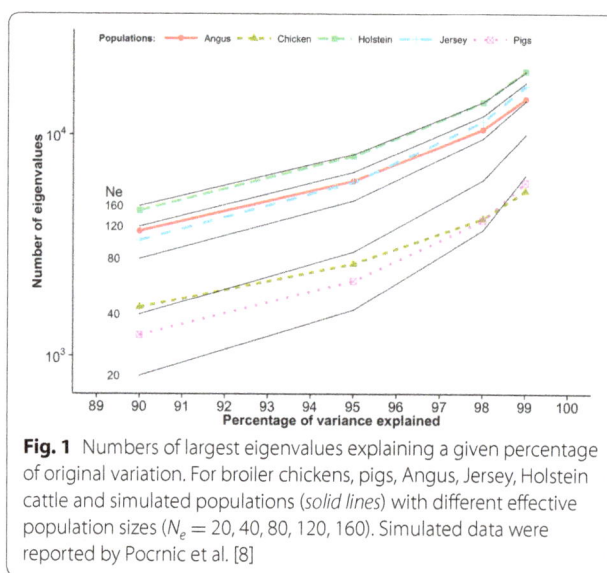

Fig. 1 Numbers of largest eigenvalues explaining a given percentage of original variation. For broiler chickens, pigs, Angus, Jersey, Holstein cattle and simulated populations (*solid lines*) with different effective population sizes (N_e = 20, 40, 80, 120, 160). Simulated data were reported by Pocrnic et al. [8]

N_e were 149 for the Holstein, 113 for the Angus, 101 for the Jersey, 44 for the broiler chicken, and 32 for the pig populations (Table 1). Estimates of N_e based on genotypic information can be influenced by several factors. First, the estimates can be affected by genotype imputation because most of the animals are genotyped with lower density chips and their genotypes are then imputed to higher density (sometimes with multiple imputations). The final number of SNPs used for evaluation, the quality control of genomic data, and the length of the genome can vary by breed and species. The simulation study reported by Pocrnic et al. [8] assumed a genome length of 30 Morgan, which is appropriate for many species including cattle and broiler chickens [18–21]. Estimates of the genome length for pigs are consistently lower and range from 18 to 23 Morgan [22–25]. Assuming a genome length of 20 Morgan for pigs, the N_e would be 50% larger than that estimated from the simulated population since $N_e \sim 1/L$ at a constant M_e, where L is genome length in Morgan. Therefore, assuming a genome length of 20 Morgan, estimated N_e for pigs in our study would be 48. Many other factors including different recombination rates, different genome lengths for each sex and different genotyping patterns for each sex can influence the estimated N_e. The assumptions in the simulations reported in [8] were idealistic in terms of population genetics (non-overlapping generations, random mating, no selection, and no migration), and differences in N_e resulted only from variation in sex ratios.

In the literature, estimates of N_e vary widely, and several approaches to calculate N_e have been reported (e.g., [26–28]). Leroy et al. [29] demonstrated variation in N_e estimates using different approaches. For Holsteins, N_e estimates range from 50 [30] to 150 [31], with many intermediate estimates in between [32–35]. Estimates for Jerseys range from 73 [34] to 135 [33]. For Angus, the N_e estimates vary from 26 [36] to 207 [37]. For various breeds of pigs, estimates can be as small as 55 [38] to as large as 113 [39]. Although N_e estimates for Holsteins and Jerseys are likely to be similar worldwide because of international breeding that is partially facilitated by the availability of Interbull evaluations, N_e estimates for pigs and broilers can vary because of the specific breeding structure used by individual companies. However, if different breeding companies use similar breeding plans, their individual populations may have a similar N_e. Eitan and Soller [40] found that broiler companies that led

breeding programs independently experienced similar problems (e.g., skeletal problems, metabolic disorders, hatchability problems, etc.) at the same time, indicating similar breeding plans.

Figures 2, 3 and 4 show correlations between GEBV based on regular and APY inverses of **G** for Angus cattle, pig, and broiler chicken populations, respectively. These correlations are for validation animals that were obtained from the analysis with different numbers of core animals. For all species and traits, correlations were 0.99 when the number of core animals was equal to the number of largest eigenvalues of $\mathbf{G}_0$ that explained either 98 or 99% of the original variation. The linearity of the curves suggests that correlations between regular and APY GEBV are nearly a linear function of percentage of explained variation. Somewhat different slopes for different traits and breeds/species could be explained by the fact that GEBV for young animals are a weighted sum of parent average and direct genomic value with additional variation that depends on whether genotyped animals have genotyped parents [1, 11]. A smaller slope is usually observed for traits with a lower heritability because the weight on parent average is larger, does not depend on direct genomic value, and subsequently does not depend on the number of core animals.

Figures 5, 6, 7, 8 and 9 show measures of accuracies as a function of the number of core animals: R^2 for Holstein and Jersey cattle and predictive ability for Angus cattle,

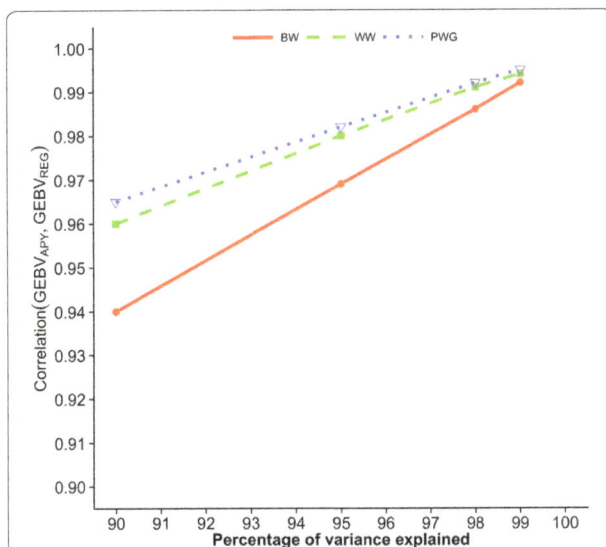

Fig. 2 Correlations between GEBV$_{REG}$ and GEBV$_{APY}$ of validation animals for Angus cattle. Genomic estimated breeding values (GEBV) are based on the regular inverse (GEBV$_{REG}$) and the Algorithm of Proven and Young inverse (GEBV$_{APY}$) of the genomic relationship matrix. Traits are birth weight (BW), weaning weight (WW), and post-weaning gain (PWG). The number of core animals is defined as the number of eigenvalues that explain 90, 95, 98, and 99% of the original variation

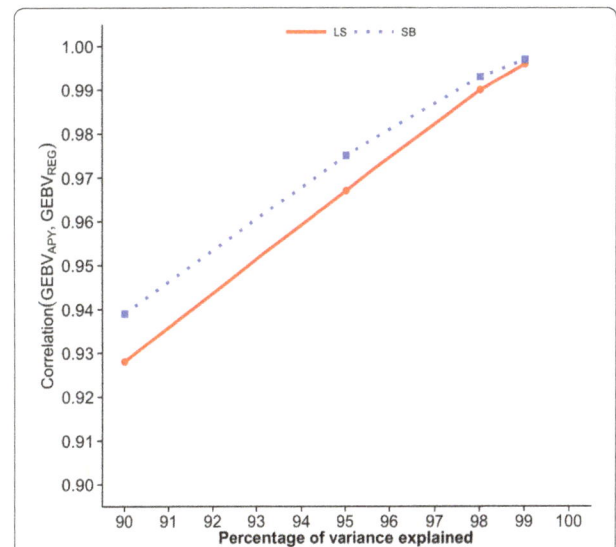

Fig. 3 Correlations between GEBV$_{REG}$ and GEBV$_{APY}$ of validation animals for pigs. Genomic estimated breeding values (GEBV) are based on the regular inverse (GEBV$_{REG}$) and the Algorithm of Proven and Young inverse (GEBV$_{APY}$) of the genomic relationship matrix. Traits are litter size (LS) and number of stillborn (SB). The number of core animals is defined as the number of eigenvalues that explain 90, 95, 98, and 99% of the original variation

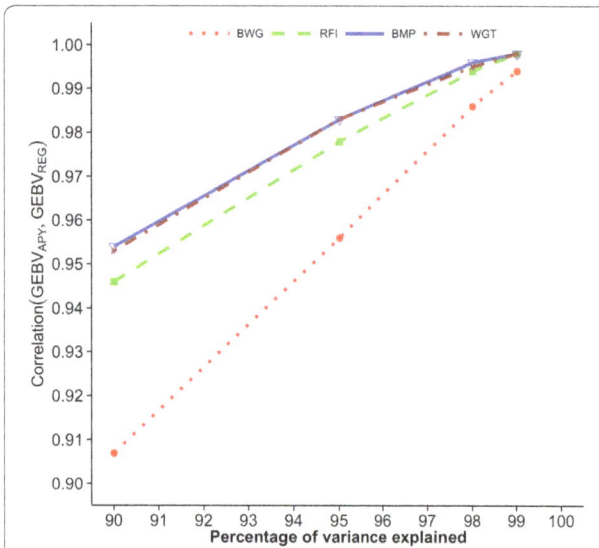

Fig. 4 Correlations between $GEBV_{REG}$ and $GEBV_{APY}$ of validation animals for broiler chickens. Genomic estimated breeding values (GEBV) are based on the regular inverse ($GEBV_{REG}$) and the Algorithm of Proven and Young inverse ($GEBV_{APY}$) of the genomic relationship matrix. Traits are body weight at grading (BWG), residual feed intake (RFI), breast meat percentage (BMP), and weight gain during feed conversion test (WGT). The number of core animals is defined as the number of eigenvalues that explain 90, 95, 98, and 99% of the original variation

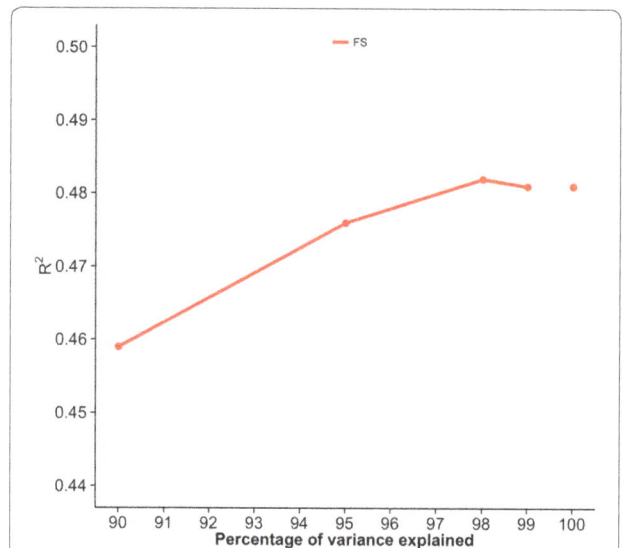

Fig. 5 Coefficients of determination (R^2) for final score (FS) of Holstein cattle. Value for 100% corresponds to the regular inverse of the genomic relationship matrix

pigs, and broiler chickens. Realized accuracies (or reliabilities) were plotted as a function of the number of eigenvalues that explain a given percentage of variation, and values for 100% correspond to the regular inverse of the GRM. The highest accuracy for Holsteins and Jerseys (Figs. 5, 6, respectively) corresponded to 98% of explained variation as in the simulation study of Pocrnic et al. [8]. However, the curves for the remaining breed/species, which are based on predictive ability, were different. For Angus (Fig. 7), accuracy increased only slightly from 90 to 99% of explained variation. For pigs (Fig. 8), accuracy increases were again small, with almost no increase for litter size. For broilers (Fig. 9), the trend also was for small increases for all traits except breast meat percentage, which had an unexpected decrease at 95% of explained variation.

All flat trends occurred when accuracy was calculated based on predictive ability. Such accuracies are affected by model quality, especially the inclusion of less than optimal parameters in multiple-trait models. The flat trends and especially the anomalies can also be attributed to imputation issues as companies usually work with low- and medium-density SNP chips, which, in addition, are modified over the years.

An important question with the APY is whether the random choice of core animals as used in this study is optimal. In a Holstein study [9], the use of about 10,000 proven bulls plus their dams as core animals provided an increase in reliability of 0.01 over random choices. In a pig study [4], correlations of GEBV based on full and APY inverses were higher than 0.98 with a random sample of about 2000 core animals (10% sample) and higher than 0.99 with about 6000 core animals (20% sample); correlations were lower than 0.95 when using only the youngest or only the oldest generations as core animals. Breeding values of n animals in the core group are assumed to contain all the additive information about the population in terms of ESM or M_e [7]. For most complete information with as few animals as possible, the subset of animals should be representative of the population and (almost) linearly independent. These conditions seem to be fulfilled if choice is at random and clones are avoided. Ostersen et al. [4] reported marginally higher correlations of GEBV obtained with APY than with the regular inverse although higher correlations do not necessarily mean higher accuracy; the highest accuracy in a simulation [8] and partially in this study was obtained when these correlations were about 0.98–0.99.

Another question with the APY is whether the number and selection of core animals should change over time. In general, realized accuracy (reliability) was maximized when the number of randomly selected core animals was about 100 N_e or about 3 N_eL. That number is not critical since the accuracy (or reliability) decreased less than 0.01 when the number of core animals increased or was reduced by 50%. If breeding practices do not cause fast changes in N_e over generations, the same number of core

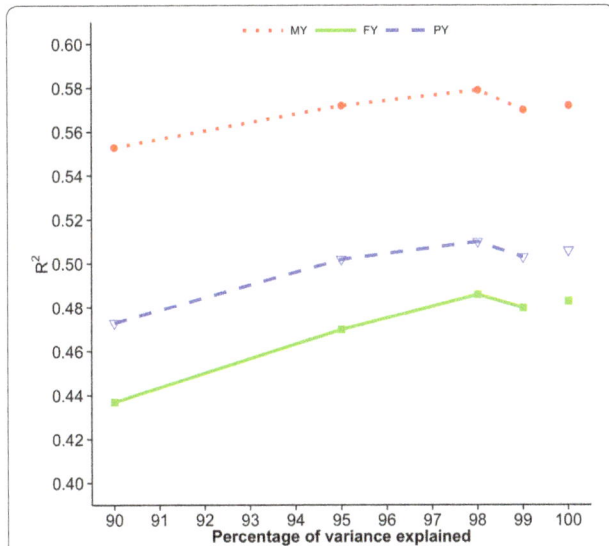

Fig. 6 Coefficient of determination (R^2) for 305-day milk yield (MY), fat yield (FY) and protein yield (PY) of Jersey cattle. Values for 100% correspond to the regular inverse of the genomic relationship matrix

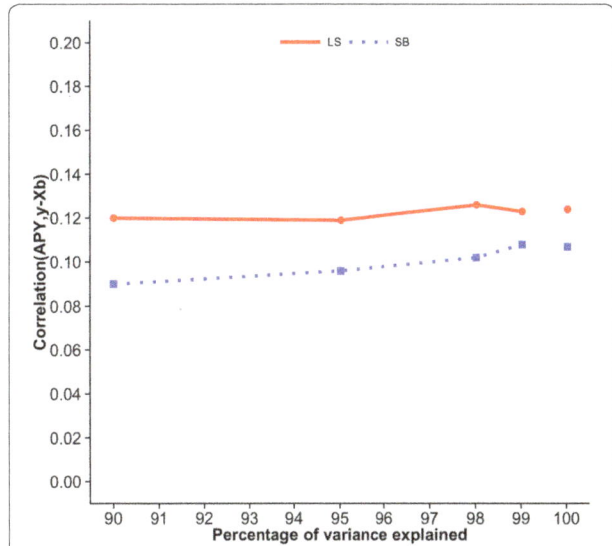

Fig. 8 Predictive ability for litter size (LS) and number of stillborn (SB) of pigs. Predictive ability is the correlation between genomic estimated breeding values based on the Algorithm of Proven and Young inverse of the genomic relationship matrix and phenotypes adjusted for fixed effects. The number of core animals is defined as the number of eigenvalues that explain 90, 95, 98, and 99% of the original variation; values for 100% correspond to the regular inverse of the genomic relationship matrix

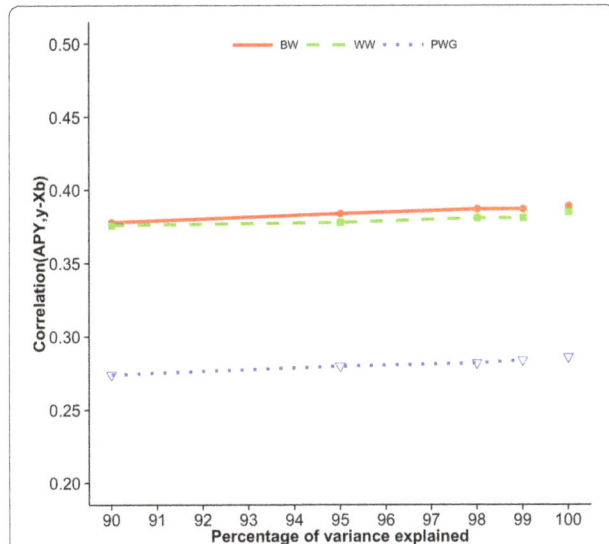

Fig. 7 Predictive ability for birth weight (BW), weaning weight (WW), and post-weaning gain (PWG) of Angus cattle. Predictive ability is the correlation between genomic estimated breeding values based on the Algorithm of Proven and Young inverse of the genomic relationship matrix and phenotypes adjusted for fixed effects. The number of core animals is defined as the number of eigenvalues that explain 90, 95, 98, and 99% of the original variation; values for 100% correspond to the regular inverse of the genomic relationship matrix

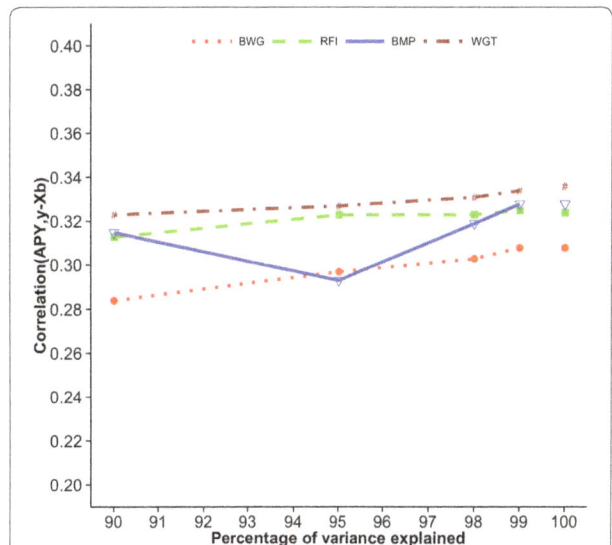

Fig. 9 Predictive ability for body weight at grading (BWG), residual feed intake (RFI), breast meat percentage (BMP), and weight gain during feed conversion test (WGT) of broiler chickens. Predictive ability is the correlation between genomic estimated breeding values based on the Algorithm of Proven and Young inverse of the genomic relationship matrix and phenotypes adjusted for fixed effects. The number of core animals is defined as the number of eigenvalues that explain 90, 95, 98, and 99% of the original variation; values for 100% correspond to the regular inverse of the genomic relationship matrix

animals selected randomly are likely to result in close to optimal evaluation accuracy. An exception could arise when the number of genotyped generations is large; under selection, older generations have little predictive power for selection candidates [41]. Further studies will determine whether the optimal approach in such a case is to choose core animals from younger generations or to remove old generations.

In this study, eigenvalue computations were done on an explicitly constructed $\mathbf{G}_0$, which actually shares the same eigenvalue distribution as the SNP BLUP matrix $\mathbf{Z'Z}$. When large datasets are used, singular value decomposition of matrix $\mathbf{Z}$ can be applied instead, since it is equivalent to eigenvalue decomposition of $\mathbf{Z'Z}$ and $\mathbf{ZZ'}$ and to the eigenvalues of $\mathbf{G}_0$ multiplied by a constant. Therefore, the number of largest eigenvalues for $\mathbf{G}_0$ is identical between two quantities. Let the singular value decomposition of matrix $\mathbf{Z}$ be $\mathbf{Z} = \mathbf{UDV'}$, where $\mathbf{D}$ is a diagonal matrix of singular values that correspond to the square root of the non-zero eigenvalues of $\mathbf{Z'Z}$ and $\mathbf{ZZ'}$. The columns of $\mathbf{U}$ are left singular vectors ($\mathbf{U'U} = \mathbf{UU'} = \mathbf{I}$), and the columns of $\mathbf{V}$ are right singular vectors ($\mathbf{V'V} = \mathbf{VV'} = \mathbf{I}$). They correspond to eigenvectors of $\mathbf{ZZ'}$ and $\mathbf{Z'Z}$, respectively. Then, $\mathbf{Z'Z} = \mathbf{VD'U'UDV'} = \mathbf{VD^2V'}$, and $(\mathbf{Z'Z})\mathbf{V} = \mathbf{VD^2}$, where $\mathbf{D^2}$ is a diagonal matrix of eigenvalues of $\mathbf{Z'Z}$ (squares of singular values of matrix $\mathbf{Z}$) and the columns of $\mathbf{V}$ are eigenvectors of $\mathbf{Z'Z}$. Similarly, $\mathbf{ZZ'} = \mathbf{UD^2U'}$. The singular value decomposition of $\mathbf{Z}$ can be computed using subroutine DGESVD in LAPACK [12], and computation cost will be quadratic for the number of markers but only linear for the number of individuals.

Conclusions

The optimal number of core animals for efficient inversion of GRM by APY is about 14,000 for Holstein and Angus cattle, 12,000 for Jersey cattle, and 6000 for pigs and broiler chickens, which corresponds approximately to $3\,N_eL$. These numbers are not critical since reduction in GEBV accuracy is minimal if using half the optimal numbers. Approximate N_e with a genome length of 30 Morgan is 149 for Holsteins, 101 for Jerseys, 113 for Angus, and 44 for broiler chickens; for pigs and a genome length of 20 Morgan, approximate N_e is 48.

Authors' contributions

IP analyzed the data and drafted the manuscript; DALL helped with important instructions on the experimental design, computations, and structure of the manuscript; YM analyzed the dairy data; IM supervised the project and provided crucial ideas. All authors read and approved the final manuscript.

Acknowledgements

We would like to thank Paul M. VanRaden, Melvin E. Tooker, and George R. Wiggans (retired) of the Animal Genomics and Improvement Laboratory, Agricultural Research Service, USDA, for providing computing environment and data for Jerseys. Reference dairy genotypes were provided by the Council on Dairy Cattle Breeding. Editing by Suzanne M. Hubbard and Heather L. Bradford is gratefully acknowledged. This research was primarily supported by grants from Holstein Association USA, American Angus Association, Zoetis, Cobb-Vantress, Smithfield Premium Genetics, Pig Improvement Company, and USDA's National Institute of Food and Agriculture (Agriculture and Food Research Initiative competitive Grant 2015-67015-22936).

Competing interests

The authors declare that they have no competing interests.

References

1. VanRaden PM. Efficient methods to compute genomic predictions. J Dairy Sci. 2008;91:4414–23.
2. Misztal I, Legarra A, Aguilar I. Using recursion to compute the inverse of the genomic relationship matrix. J Dairy Sci. 2014;97:3943–52.
3. Lourenco DAL, Tsuruta S, Fragomeni BO, Masuda Y, Aguilar I, Legarra A, et al. Genetic evaluation using single-step genomic best linear unbiased predictor in American Angus. J Anim Sci. 2015;93:2653–62.
4. Ostersen T, Christensen OF, Madsen P, Henryon M. Sparse single-step method for genomic evaluation in pigs. Genet Sel Evol. 2016;48:48.
5. Fragomeni BO, Lourenco DAL, Tsuruta S, Masuda Y, Aguilar I, Legarra A, et al. Hot topic: use of genomic recursions in single-step genomic best linear unbiased predictor (BLUP) with a large number of genotypes. J Dairy Sci. 2015;98:4090–4.
6. Masuda Y, Misztal I, Tsuruta S, Legarra A, Aguilar I, Lourenco DAL, et al. Implementation of genomic recursions in single-step genomic best linear unbiased predictor for US Holsteins with a large number of genotyped animals. J Dairy Sci. 2016;99:1968–74.
7. Misztal I. Inexpensive computation of the inverse of the genomic relationship matrix in populations with small effective population size. Genetics. 2016;202:401–9.
8. Pocrnic I, Lourenco DAL, Masuda Y, Legarra A, Misztal I. The dimensionality of genomic information and its effect on genomic prediction. Genetics. 2016;203:573–81.
9. Masuda Y, Misztal I, Tsuruta S, Lourenco DAL, Fragomeni BO, Legarra A, et al. Single-step genomic evaluations with 570 K genotyped animals in US holsteins. Interbull Bull. 2015;49:85–9.
10. Lourenco DAL, Tsuruta S, Fragomeni BO, Chen CY, Herring WO, Misztal I. Crossbreed evaluations in single-step genomic best linear unbiased predictor using adjusted realized relationship matrices. J Anim Sci. 2016;94:909–19.
11. Lourenco DAL, Fragomeni BO, Tsuruta S, Aguilar I, Zumbach B, Hawken RJ, et al. Accuracy of estimated breeding values with genomic information on males, females, or both: an example on broiler chicken. Genet Sel Evol. 2015;47:56.
12. Anderson E, Bai Z, Bischof C, Blackford S, Demmel J, Dongarra J, et al. LAPACK users' guide. 3rd ed. Philadelphia: Society for Industrial and Applied Mathematics; 1999.
13. Tsuruta S, Misztal I, Stranden I. Use of the preconditioned conjugate gradient algorithm as a generic solver. J Anim Sci. 2001;79:1166–72.
14. Aguilar I, Misztal I, Legarra A, Tsuruta S. Efficient computation of the genomic relationship matrix and other matrices used in single-step evaluation. J Anim Breed Genet. 2011;128:422–8.
15. VanRaden PM, Wiggans GR. Derivation, calculation, and use of national animal model information. J Dairy Sci. 1991;74:2737–46.
16. Legarra A, Robert-Granié C, Manfredi E, Elsen JM. Performance of genomic selection in mice. Genetics. 2008;180:611–8.
17. MacLeod AK, Haley CS, Woolliams JA, Stam P. Marker densities and the mapping of ancestral junctions. Genet Res. 2005;85:69–79.
18. Kappes SM, Keele JW, Stone RT, McGraw RA, Sonstegard TS, Smith TP, et al. A second-generation linkage map of the bovine genome. Genome Res. 1997;7:235–49.
19. Burt DW, Cheng HH. The Chicken gene map. ILAR J. 1998;39:229–36.
20. Arias JA, Keehan M, Fisher P, Coppieters W, Spelman R. A high density linkage map of the bovine genome. BMC Genet. 2009;10:18.
21. Groenen MAM, Wahlberg P, Foglio M, Cheng HH, Megens HJ, Crooijmans RPMA, et al. A high-density SNP-based linkage map of the chicken genome reveals sequence features correlated with recombination rate. Genome Res. 2009;19:510–9.
22. Rohrer GA, Alexander LJ, Keele JW, Smith TP, Beattie CW. A microsatellite linkage map of the porcine genome. Genetics. 1994;136:231–45.
23. Archibald AL, Haley CS, Brown JF, Couperwhite S, McQueen HA, Nicholson D, et al. The PiGMaP consortium linkage map of the pig (Sus scrofa). Mamm Genome. 1995;6:157–75.

24. Marklund L, Johansson Moller M, Hoyheim B, Davies W, Fredholm M, Juneja RK, et al. A comprehensive linkage map of the pig based on a wild pig-Large White intercross. Anim Genet. 1996;27:255–69.

25. Tortereau F, Servin B, Frantz L, Megens HJ, Milan D, Rohrer G, et al. A high density recombination map of the pig reveals a correlation between sex-specific recombination and GC content. BMC Genomics. 2012;13:586.

26. Caballero A. Developments in the prediction of effective population size. Heredity (Edinb). 1994;73:657–79.

27. Charlesworth B. Fundamental concepts in genetics: effective population size and patterns of molecular evolution and variation. Nat Rev Genet. 2009;10:195–205.

28. Luikart G, Ryman N, Tallmon DA, Schwartz MK, Allendorf FW. Estimation of census and effective population sizes: the increasing usefulness of DNA-based approaches. Conserv Genet. 2010;11:355–73.

29. Leroy G, Mary-Huard T, Verrier E, Danvy S, Charvolin E, Danchin-Burge C. Methods to estimate effective population size using pedigree data: examples in dog, sheep, cattle and horse. Genet Sel Evol. 2013;45:1.

30. Brotherstone S, Goddard M. Artificial selection and maintenance of genetic variance in the global dairy cow population. Philos Trans R Soc Lond B Biol Sci. 2005;360:1479–88.

31. Hayes BJ, Visscher PM, McPartlan HC, Goddard ME. Novel multilocus measure of linkage disequilibrium to estimate past effective population size. Genome Res. 2003;13:635–43.

32. Sargolzaei M, Schenkel FS, Jansen GB, Schaeffer LR. Estimating effective population size in North American Holstein cattle based on genome-wide linkage disequilibrium. In: Proceedings of the Dairy Cattle Breeding and Genetics Committee Meeting: Guelph; 2007.

33. de Roos AP, Hayes BJ, Spelman RJ, Goddard ME. Linkage disequilibrium and persistence of phase in Holstein-Friesian, Jersey and Angus cattle. Genetics. 2008;179:1503–12.

34. Bovine HapMap Consortium, Gibbs RA, Taylor JF, Van Tassell CP, Barendse W, Eversole KA, et al. Genome-wide survey of SNP variation uncovers the genetic structure of cattle breeds. Science. 2009;324:528–32.

35. Rodriguez-Ramilo ST, Fernandez J, Toro MA, Hernandez D, Villanueva B. Genome-wide estimates of coancestry, inbreeding and effective population size in the Spanish Holstein population. PLoS One. 2015;10:e0124157.

36. Falleiro VB, Malhado CHM, Malhado ACM, Carneiro PLS, Carrillo JA, Song J. Population structure and genetic variability of Angus and Nellore herds. J Agric Sci. 2014;6:276–85.

37. Lu D, Sargolzaei M, Kelly M, Li C, Vander Voort G, Wang Z, et al. Linkage disequilibrium in Angus, Charolais, and Crossbred beef cattle. Front Genet. 2012;3:152.

38. Uimari P, Tapio M. Extent of linkage disequilibrium and effective population size in Finnish Landrace and Finnish Yorkshire pig breeds. J Anim Sci. 2011;89:609–14.

39. Welsh CS, Blacburn HD, Schwab C. Population status of major U.S. swine breeds. In: Proceedings of the American Society of Animal Science Western Section: 16-18 June 2009; Fort Collins. 2009.

40. Eitan Y, Soller M. Poultry breeding: the broiler chicken as a harbinger of the future. In: Meyers RA, editor. Encyclopedia of Sustainability Science and Technology. New York: Springer; 2012. p. 8307–28.

41. Muir WM. Comparison of genomic and traditional BLUP-estimated breeding value accuracy and selection response under alternative trait and genomic parameters. J Anim Breed Genet. 2007;124:342–55.

17

Transcriptional profile of breast muscle in heat stressed layers is similar to that of broiler chickens at control temperature

Imran Zahoor[1,2], Dirk-Jan de Koning[1,3] and Paul M. Hocking[1*] (ID)

Abstract

Background: In recent years, the commercial importance of changes in muscle function of broiler chickens and of the corresponding effects on meat quality has increased. Furthermore, broilers are more sensitive to heat stress during transport and at high ambient temperatures than smaller egg-laying chickens. We hypothesised that heat stress would amplify muscle damage and expression of genes that are involved in such changes and, thus, lead to the identification of pathways and networks associated with broiler muscle and meat quality traits. Broiler and layer chickens were exposed to control or high ambient temperatures to characterise differences in gene expression between the two genotypes and the two environments.

Results: Whole-genome expression studies in breast muscles of broiler and layer chickens were conducted before and after heat stress; 2213 differentially-expressed genes were detected based on a significant ($P < 0.05$) genotype × treatment interaction. This gene set was analysed with the BioLayout Express^{3D} and Ingenuity Pathway Analysis software and relevant biological pathways and networks were identified. Genes involved in functions related to inflammatory reactions, cell death, oxidative stress and tissue damage were upregulated in control broilers compared with control and heat-stressed layers. Expression of these genes was further increased in heat-stressed broilers.

Conclusions: Differences in gene expression between broiler and layer chickens under control and heat stress conditions suggest that damage of breast muscles in broilers at normal ambient temperatures is similar to that in heat-stressed layers and is amplified when broilers are exposed to heat stress. The patterns of gene expression of the two genotypes under heat stress were almost the polar opposite of each other, which is consistent with the conclusion that broiler chickens were not able to cope with heat stress by dissipating their body heat. The differentially expressed gene networks and pathways were consistent with the pathological changes that are observed in the breast muscle of heat-stressed broilers.

Background

Modern broiler chickens are characterised by relatively fast growth rate, greater muscle mass and better feed conversion ratio compared with layer and traditional chicken breeds [1, 2]. The carcasses of some broiler chickens show changes in the appearance of breast meat, such as a pale colour with reduced water holding capacity, or dark, firm and dry muscle with different functional properties [3]. More recently, white striping, which is characterised by white parallel striations in the direction of the muscle fibres and "wooden breast" muscles, have been reported [4, 5]. Elevated activity of creatine kinase and histopathological changes in affected muscles are suggestive of a degenerative myopathy [4, 6]. These changes have implications for meat quality and, potentially, have a significant economic cost. Several factors affect the proportion of affected carcasses, including different genetic background, growth rate, season, heat and transport stress, and abattoir practices [7–9].

*Correspondence: paul.hocking@roslin.ed.ac.uk
[1] Division of Genetics and Genomics, Roslin Institute and R(D)SVS, University of Edinburgh, Easter Bush, Midlothian EH25 9RG, UK
Full list of author information is available at the end of the article

Genetic variation in muscle and meat quality traits has been quantified [2, 10] but these traits usually involve measuring slaughtered sibs. Recent technological innovations have opened the way for genomic selection (GS) based on DNA markers (single nucleotide polymorphisms, SNPs) [11, 12]. Therefore, our objective was to identify genetic networks and pathways that might be useful for the detection of causal genetic factors that are involved in breast muscle and meat quality disorders of broiler chickens. It is also likely that the identified genetic factors would be helpful in updating the existing SNP chips to enable scientists to perform genomic selection for better muscle and meat quality in broilers.

Through the use of high-throughput microarray technology, it is possible to identify differentially-expressed genes as a result of a specific treatment [13]. In this study, we used microarray analysis to identify candidate genes that may contribute to differences in muscle damage between broilers and layers. Spontaneous and stress-induced myopathies in broiler skeletal muscles are exacerbated by heat stress [14, 15] and, thus, we compared gene expression profiles in the breast muscles of broiler and layer genotypes that were subjected to control or heat stress conditions. Our experimental strategy was based on the hypothesis that the expression of genes that are differentially expressed in broilers and layers under normal conditions is increased and therefore more easily detected after heat stress. However, it is often difficult to assign biological significance to the large number of genes that are detected in a microarray experiment. This problem can be solved when the differentially-expressed genes are organised via hierarchical clustering methods [16] and, for this purpose, we used BioLayout Express[3D] [17, 18] and Ingenuity Pathway Analysis (IPA) (http://www.ingenuity.com/). In addition, we compared the results from these analyses with those obtained with the DAVID [19, 20] (https://david.ncifcrf.gov/) and Reactome [21, 22] (http://reactome.org/) software using more recent databases.

Methods
Animals and husbandry
We used 40 male broiler chicks of a male line (Ross 308, Aviagen, Newbridge, UK) from a commercial hatchery and 74 layer chicks (White Leghorn) from a line maintained at the Roslin Institute. For the first 2 weeks, birds were reared in groups of 20 individuals until the layers had been sexed by a DNA method [23]. At 2 weeks of age, the birds were distributed to eight pens by sex and genotype, with each pen containing 12 male layers and nine or ten broilers, in a completely randomised design. The birds were provided with feed (a commercial layer starter diet) and water ad libitum and the daily photoperiod was 16 h light and 8 h darkness.

The birds were subjected to experimental treatments over four days from 42 to 46 days of age. On each day, we randomly selected two pens for each breed and the birds were transferred into four controlled environment chambers. On each day, we randomly selected four chambers, i.e. two for the heat treatment (32 °C, 75% relative humidity or RH) and two as controls (21 °C, 50% RH). Each chamber contained two crates with two male broilers or two male layers, with pens and crates confounded. The crates were placed on a wooden pallet and the order of the pairs (crates) in each room was randomised. Sixty-four birds were used in the experiment.

About 30 min before the birds were transferred to the chambers, the relevant chamber was turned on, such that it could reach the required temperature and humidity before birds were placed into the chamber for the following 2 h. Birds were introduced in each chamber at intervals of 45 min to allow for sampling of the birds.

After completing the 2-h treatment, birds were removed from the crate and rectal temperatures were measured using a thermistor probe (Model 612-849; RS Components Ltd., Corby, Northants, UK). Then, they were euthanized by an intravenous injection of sodium pentobarbitone into the wing vein and two tissue samples of 100–120 mg were taken from the left pectoral muscle and snap frozen in liquid nitrogen for subsequent RNA extraction.

RNA extraction and microarray experiment
Samples of breast muscle from male chickens were randomised prior to extraction of RNA using Trizol (Life Technologies, Paisley, UK) following the manufacturer's recommended protocol. Briefly, the frozen tissue was homogenised in 1 ml of Trizol using the FastPrep® system with Lysing matrix D (MP Biochemicals). The phases were separated by addition of 200 µl of 2-bromo-chloropropane (Sigma Aldrich) and centrifuged for 15 min. A 500-µl sample of the clear upper aqueous layer was transferred to a fresh tube and 500 µl of isopropanol was added. The samples were centrifuged for 30 min to pellet the RNA, which was washed twice with 70% ethanol before air-drying. The RNA was resuspended in 100 µl of RNAse-free water prior to quantification and quality assessment. All RNA samples had a RNA integrity number (RIN) value higher than 8.0, as determined by the Agilent Bioanalyser RNA 6000 Nano Chip. Samples were diluted to 50 ng/µl with deionised and RNAse-free water. Aliquots of 20 µl from each sample were used for pooling the two samples from each crate to obtain eight replicates for each breed × treatment combination.

Microarray hybridisation was completed in the Ark-Genomics laboratory at the Roslin Institute (http://genomics.ed.ac.uk). Total RNA was prepared for hybridisation to the Affymetrix chicken GeneChip array using the Affymetrix IVT express kit according to the manufacturer's protocol. The generated cRNA was hybridised overnight to the cartridge arrays according to Affymetrix's protocols. The cartridges were washed and stained in the Affymetrix fluidic station using the hybridisation, wash and stain kit from Affymetrix. After staining, the arrays were scanned with the Affymetrix GeneChip system 3000 scanner. The resultant CEL files were reviewed using the Expression Console software from Affymetrix.

Thrity-two Affymetrix chicken array chips (38.5K; each GeneChip included 38,535 probes) were used in the microarray experiment. After scanning, the CEL files were analysed in four batches of eight slides to obtain expression values in GenStat (www.vsni.co.uk/software/genstat). Each batch contained slides from birds treated on the same day. The Robust Multichip Average (RMA) algorithm [24] was used to extract the gene expression data.

Statistical analysis

The experiment was a 2 × 2 factorial design (breed × treatment), with day/chambers/crates as blocking factors. Standard analysis of variance methods was used to analyse body temperature and body weight using GenStat v13 (https://www.vsni.co.uk/software/genstat/). Transformation to natural logarithms was necessary to achieve normally distributed residuals of body weight.

For the analysis of differentially-expressed genes, we used a model with fixed effect terms for breed and treatment and their interaction. The normalised data were analysed by using Microarray One-Channel ANOVA in GenStat, with a model that included breed × treatment as treatment structure and the hierarchical structure of day/chamber/breed as blocking factor. Genes that showed a significant breed × treatment interaction ($P < 0.05$) were used for subsequent investigation because they were expected to be most relevant for genetic differences between broilers and layers in response to heat stress. Based on these ANOVA results, the false discovery rate (FDR) was calculated for three probability values ($P < 0.05$, <0.01 and <0.001) for the effects of treatment, breed, and their interaction. FDR was calculated using the Mixture Model of GenStat and the maximum number of iteration cycles was set to 300.

Cluster analysis in BioLayout Express³D

Gene annotations were downloaded from the NetAffx analysis centre of Affymetrix (http://www.affymetrix.com/analysis/index.affx; downloaded 15 December

2016). Expression values for the selected subset of genes/probes were unlogged, entered into BioLayout Express³D (BLE, http://www.biolayout.org/) and analysed using a Pearson correlation threshold of 0.80. Clusters were viewed in the Class Viewer, after running the Markov Clustering Algorithm (MCL). For cluster size, a minimum threshold of four genes/probes per cluster was selected to limit the size of the smallest clusters [25]. Selected clusters were identified on the basis of a clear difference in expression pattern of the genes between treatments (control vs. heat treatment) and breeds. For functional analyses, clusters were combined into 'categories' on the basis of similarity in mean expression pattern across breeds and treatments.

Analysis of pathways and networks in IPA

The gene expression data for each of the six selected categories were combined into a single Excel sheet for analysis in Ingenuity Pathway Analysis (IPA, http://www.ingenuity.com/products/ipa) of the four breed × treatment combinations (broiler control, BC; broiler heat stress, BH; layer control, LC and layer heat stress, LH). The lists of genes for each category were analysed in IPA by using Fisher's exact test to identify biological functions and pathways that were enriched in the dataset using the 'Core Analysis' function of the IPA program. Genes were mapped against the 'Tissues and Cell Lines' available in the Ingenuity Pathway Analysis Knowledge Base (IPAKB). Because information in the IPA originates mainly from mammals (human, mouse and rat), the submitted lists of genes were mapped against all available species and changes to avian terminology, e.g. neutrophil to heterophil, were made. For network generation, we set a threshold of 35 molecules per network and 25 networks per analysis. Both direct and indirect relationships of molecules were considered.

Additional analyses of pathways and networks

To reconfirm the initial results, we repeated the analyses on pathways and networks with more recent databases. We used two software programs, i.e. DAVID (https://david.ncifcrf.gov/) and Reactome (http://reactome.org/),

Table 1 Number of significant genes for treatment, breed, and breed × treatment interaction at different levels of significance

Significance (P<)	Treatment (heat-stress vs. control)	Breed (broiler vs. layer)	Breed × treatment interaction
0.001	107	5208	93
0.01	617	8182	635
0.05	1922	10,733	2213

both accessed on 2nd July 2017. Further information is in Additional file 1: Table S1.

Results

Differentially-expressed genes

The Affymetrix Genechips were filtered for expression levels higher than 1, which reduced the number of probes from 38,535 to 19,038. The results of the ANOVA for the filtered set of genes are in Table 1. The false discovery rate (FDR) for statistically significant genes ($P < 0.05$) was less than 31.5% for the treatment × breed interaction, 44% for treatment, and 3% for breed. A total of 2213 genes were differentially expressed among the four treatment comparisons. The numbers of differentially-regulated genes that overlapped between the two treatments are in Fig. 1. We found 1361 upregulated genes in the comparison between BH and BC, of which 1316 (97%) were shared with downregulated genes in the comparison between LH and LC. Similarly, we found 852 downregulated genes in the comparison between BH and BC, of which 753 (88%) were shared with upregulated genes in the comparison between LH and LC.

Categorisation of candidate genes on the basis of their biological functions

Based on their biological function, genes that were differentially expressed for the breed × treatment interaction were divided into 12 categories (Table 2). More than 43% (959) of the genes had no gene ontology (GO) term for a biological process or function. These genes fell in two major groups: 424 genes had no known function and 534 genes were not involved in a known biological process.

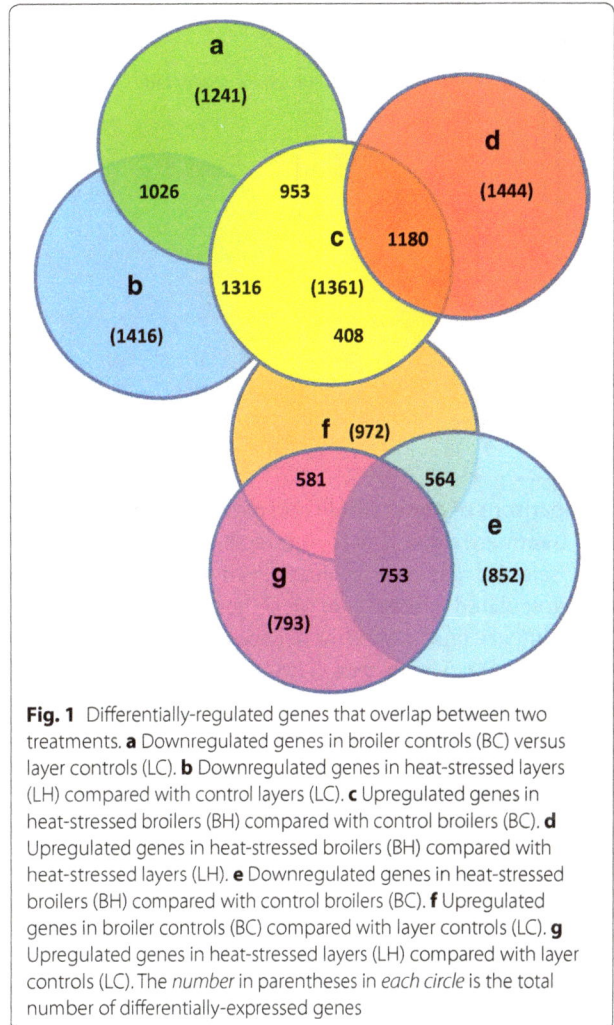

Fig. 1 Differentially-regulated genes that overlap between two treatments. **a** Downregulated genes in broiler controls (BC) versus layer controls (LC). **b** Downregulated genes in heat-stressed layers (LH) compared with control layers (LC). **c** Upregulated genes in heat-stressed broilers (BH) compared with control broilers (BC). **d** Upregulated genes in heat-stressed broilers (BH) compared with heat-stressed layers (LH). **e** Downregulated genes in heat-stressed broilers (BH) compared with control broilers (BC). **f** Upregulated genes in broiler controls (BC) compared with layer controls (LC). **g** Upregulated genes in heat-stressed layers (LH) compared with layer controls (LC). The *number* in parentheses in *each circle* is the total number of differentially-expressed genes

Table 2 Significant differentially-expressed genes for breed × treatment interaction (P < 0.05) grouped by function

Group	Biological functions	Number of genes
1	Transcripts with no known gene name	424
2	Genes with no GO terms for biological functions	534
3	Signal transduction	130
4	Stress-related response, inflammatory, angiogenesis, apoptotic, and proteolytic functions	334
5	Metabolic, and catabolic processes	190
6	Inter and intracellular transport of proteins, ions, and muscle contraction	162
7	Cellular proliferation, and organ development	142
8	Transcription and translation	138
9	Protein phosphorylation, dephosphorylation, modification, and folding	95
10	Signal transduction	92
11	DNA damage, repair, metabolism, and catabolic processes	60
12	Cytoskeleton organization and polymerization of filaments	42

Table 3 Numbers of genes, pathways and networks associated with different categories of genes based on function (see Fig. 2)

Category	Genes[a]	Pathways	Networks	Selected Pathways	Selected Networks	Functions
I	180	35	23	9	5	Stress response, cellular damage, connective tissue and muscle disorders
II	74	40	7	12	4	Cellular development, anti-apoptotic, anti-inflammatory and anti-stress functions
III	55	3	9	2	0	Anti-apoptotic, anti-oxidant, anti-inflammatory, energy production
IV	13	9	0	10	0	Stress, inflammatory, tissue damage, anti-oxidative, wound healing
V	7	9	4	5	0	Inflammation, immune functions, oxidative stress, phospholipid degradation
VI	16	7	4	3	0	Cell death, inflammatory and immune response, dellular development, haematopoiesis

[a] Genes mapped to corresponding identifiers

Comparisons of genes within and between breed and treatment significant for interaction

The selected genes were further divided into up- and downregulated patterns of gene expression for different comparisons within- and between-breed and treatment. Of the 54 clusters, 21 were selected for further analysis on the basis of their clear expression pattern, which included 509 genes that were grouped into six distinct categories (Table 3) according to the nature of their expression patterns corresponding to the (statistically significant) interactions of heat stress and genotype (Fig. 2). The expression values of the genes in category I were higher for broilers than for layers. Heat-stress resulted in a further increase in expression levels for broiler but a decrease for layers, compared to their respective controls. In the case of category II, the expression level of genes was higher for broilers than layers under control temperatures (as for category I). However, after heat-stress expression levels were lower in broilers compared with control broilers and conversely, higher in layers compared with LC. Expression values of category III genes were substantially higher for LC than BC whereas heat-stress resulted in further increases in gene expression in layers and decreases in broilers. Category IV genes were upregulated in BH compared with BC, whereas they were upregulated in LC compared with LH.

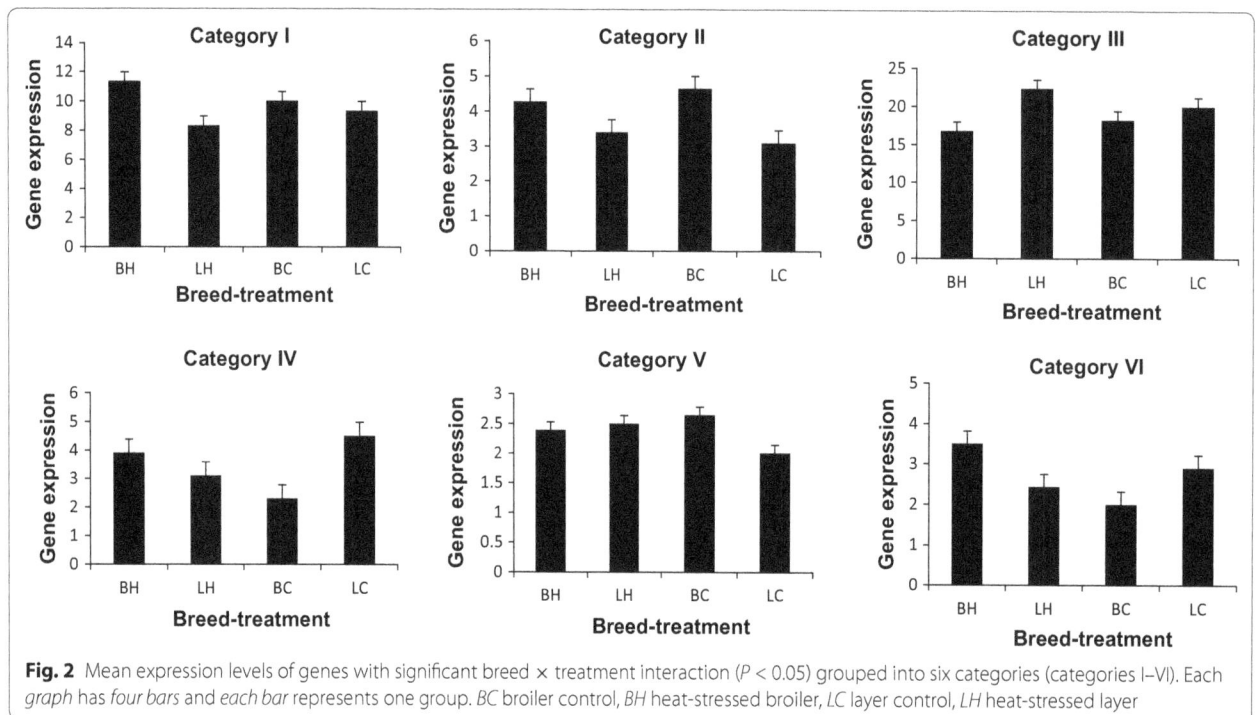

Fig. 2 Mean expression levels of genes with significant breed × treatment interaction ($P < 0.05$) grouped into six categories (categories I–VI). Each *graph* has *four bars* and *each bar* represents one group. *BC* broiler control, *BH* heat-stressed broiler, *LC* layer control, *LH* heat-stressed layer

Expression of category V genes was low in LC compared with all other groups and genes were upregulated in LH compared with LC, whereas they were downregulated in BH compared with BC. In the case of category VI genes, expression values for control layers were higher than for the respective broilers. After treatment, the expression of these genes increased in broilers but decreased in layers. Each of the six patterns of gene expression were analysed separately in IPA and significant ($P < 0.05$) pathways and networks were identified (see Additional file 1: Tables S2, S3, S4, S5, S6, S7). The set of genes which were filtered out by BioLayout Express (i.e. genes with a correlation coefficient of less than 0.80) was analysed in IPA separately, using the same procedure, to determine significant pathways and networks for this gene set, as shown in Additional file 1: Table S8.

Body weight and rectal temperature

Mean rectal temperatures for the control and heat-stressed conditions were 41.0 and 43.7 °C, respectively, in broilers, and 42.0 and 42.3 °C, respectively, in layers [standard error of difference (SED) 0.15 between breed and 0.14 between treatments]. The increase in rectal temperature in the heat-stressed birds was significantly larger in broilers (2.6 °C) than in layers (0.3 °C) which resulted in a significant breed × treatment interaction ($P < 0.001$). Average body weights (back-transformed) of broiler and layer males were 8.38 (4384 g) and 6.54 (693 g), respectively (SED 0.017, $P < 0.001$).

Discussion

Phenotypic responses validate experimental treatments

The large increase in rectal body temperature for broilers compared with layers is consistent with early reports in the literature [15, 26]. The results confirm the difficulty that broiler chickens have in coping with high ambient temperatures and other stressors, such as shackling, that may ultimately lead to detrimental consequences for both muscle function and meat quality [27, 28]. The results confirm that the heat treatment had the expected effect on the metabolism of broiler chickens and that the response in broilers was greater than in layers.

Microarray analysis

The microarray results showed large differences between broilers and layers. Nevertheless, comparatively few significant genes (107, 617 and 1922 at $P < 0.001$, $P < 0.01$ and $P < 0.05$, respectively) were differentially-expressed in the comparison between treatments, which indicated that the differences in gene expression between heat-stressed and control birds were not as large as those between breeds. The number of upregulated genes in BH compared with BC (97%) that were shared with downregulated genes in the LH and LC comparison, and the number of downregulated genes in the former (BH vs. BC) compared with the latter (LH vs. LC) (88%), suggest that changes in gene expression in response to heat-stress are opposite in broilers compared to layers, which is consistent with the conclusion that broiler chickens do not manage heat stress appropriately. Furthermore, differential gene expression in breast muscles of BC and LH compared with LC, separately, involved a similar set of genes, which suggests that, in terms of gene expression, control broilers are similar to heat-stressed layers. We found that 1026 downregulated genes overlapped in the comparison of LH vs. LC (71%) and BC versus LC (83%) and likewise 753 genes were common/overlapped in the set of upregulated genes in the LH vs. LC (95%) comparison and in the list of downregulated genes in BH versus BC (88%) comparison. Taken together, these results are consistent with the physiological changes and muscle disorders that were reported for broiler chickens reared at conventional temperatures [1, 4].

The 2213 genes that were differentially-expressed for the breed × treatment interaction term were classified into categories according to their function and the biological processes in which they are involved. For 424 transcripts (19.2% of the total), we found no gene symbol and no gene name, which indicates that many genes involved in heat-stress induced responses in chicken skeletal muscle are not characterised to date. Similarly, the second largest group of genes, representing 13.1% of the significant genes, had no GO term for a biological function at the time the GO terms for this gene set were retrieved from the NetAffx Analysis Centre of Affymetrix (http://www.affymetrix.com/estore/analysis/index.affx, re-accessed 15th December 2016).

About 15% of the 2213 genes, which were significant ($P < 0.05$) for breed × treatment interaction, are directly involved in stress-related response, inflammatory, angiogenesis, apoptotic, and proteolytic functions, which is consistent with the physiological changes in broiler muscle caused by heat-stress [15, 29]. Similarly, 4% of the genes are involved in signal transduction and are associated with various biological processes, including oxidative stress, inflammation, muscle contraction, glycogen metabolism, and the concentrations of intracellular ions [30–34], and 7.3% are involved in inter and intracellular transport of proteins associated with muscle contraction and muscle damage-related functions [35–37]. Other smaller categories of genes are involved in cellular proliferation, development and DNA damage repair.

Stress is known to accelerate metabolic rate, mainly through carbohydrate metabolism to produce larger amounts of energy and facilitate "fight or flight" responses [38–40]; about 6% of all 2213 genes were involved in metabolic and catabolic functions. The cytoskeleton is required for cell shape and motility and is involved in cell division [41, 42]. It has been suggested that the genes in Group 11 (Table 2) have a role in the movement and division of leukocytes, such as heterophils and macrophages, as secondary mediators of the genes in Groups 3, 4 and 5 to shape the stress and inflammatory response to heat stress. Of all the significant genes for breed × treatment interaction, 49 encode proteins located in the mitochondria and about 200 affect the cell membrane directly. These results suggest that damage to mitochondria and cell membrane are potentially important components of heat-stress induced pathogenesis in chicken breast muscles.

Taken together, these results suggest a picture of stress responses, inflammation, oxidative stress, and tissue damage, which is consistent with histological and physiological changes in broiler breast muscle [1]. Confirmatory evidence was also reported in a recent IPA analysis of differentially-expressed genes in "wooden breast" and control broiler muscles [43].

IPA analysis

Heat stress in broilers led to further increases in the expression of category I genes of the α-adrenergic signalling network (see Additional file 1: Table S2), which are involved in glycogenolysis under stressful conditions to provide energy for muscle contraction. However, stress hormones are also known to alter the activities of immune cells and lead to the production of various pro-inflammatory cytokines and chemokines [44, 45]. In agreement with these findings, genes of several chemokine pathways were also present in this category, which are involved in cytokine signalling, tissue damage and related functions (see Additional file 1: Table S2). Upregulation of these pathways in control broilers indicates that breast muscles in broilers at conventional ambient temperatures show physiological and functional changes that are further exacerbated by exposure to heat stress. However, the upregulated vascular endothelial growth factor (VEGF) signalling pathway is also a significant mediator of hypoxia-induced angiogenesis and is usually upregulated in hypoxia-like situations. Upregulation of this pathway in control broilers compared with layers suggests that broiler muscle cells were under hypoxic-stress even under control conditions. The reason for this may lie in the larger size of muscle fibres in broilers and an inadequate capillary supply, which are, in turn, considered to induce metabolic stress due to the larger diffusion distances for nutrients, metabolites and waste products [1]. This is consistent with reports that thermal stress leads to oxidative stress and muscle damage, as indicated by higher plasma creatine kinase activity

[26, 46–48]. Upregulation of the *nuclear factor erythroid 2-related factor 2* (*NRF2*)-mediated oxidative stress response pathway may be a protective measure to minimise the damaging effects of heat stress on anti-oxidant functions [49–51].

Expression of category II genes was highest in BC and decreased after heat stress. Upregulation of protein synthesis and angiogenic pathways in BC is logical, in the sense, that broilers have substantially higher growth rates and larger body mass than layers [52, 53]. Exposure to heat stress resulted in downregulation of these pathways in broilers, which is consistent with the negative effects of heat stress on growth-related traits [54]. Conversely, inflammatory and anti-inflammatory pathways were upregulated in layers after heat stress, possibly as a mechanism to protect the body from tissue damage. However, these results are in agreement with the physiological data (body temperature) from the current study that show that the increase in body temperature was much smaller in layers than in broilers. Consistent with this, Sandercock et al. [14] reported that the effects of heat stress on body temperature and plasma creatine kinase activity were much smaller in layers than in broilers. Similarly, the extent of heat stress induced oxidative stress in skeletal muscles was much smaller in layers than in broilers [55].

Hypoxia is known to decrease the efficiency of oxidative phosphorylation [56] and, thus, the downregulation of this pathway in broilers (Additional file 1: Table S4) could be due to hypoxia-like conditions in skeletal muscles. In contrast to our finding, Toyomizu et al. [57] reported that oxidative phosphorylation in skeletal muscles was much more efficient in broilers than in layers at 14–28 days of age when body weights were about 1.0 and 0.2 kg for broilers and layers, respectively. This greater efficiency of oxidative phosphorylation in broilers at that age is a logical outcome of selection for rapid growth. In the present study, broilers at 6 weeks of age were over 3.5 kg heavier than 28-day layers and the occurrence of an hypoxia-like situation in their muscles is consistent with a higher muscle to capillary ratio and larger diffusion distances for nutrients and metabolic wastes [1]. Consistent with this conclusion, some angiogenic pathways in category I that are involved in hypoxia-induced angiogenesis, such as the *VEGF* signalling pathway, were upregulated in broilers.

Category VI genes, such as *Janus kinase 1* (*JAK1*), *Janus kinase 2* (*JAK2*) and *tyrosine kinase 2* (*TYK2*) were up-regulated in BH compared with BC and have a role in wound healing and tissue regeneration [58], in agreement with categories I and IV genes, which indicates that muscle damage is much more important in heat-stressed broilers than BC, LC and LH. However, these pathways were downregulated in LH compared with LC, possibly because upregulation of survival-related pathways reduced damaging effects in LH [59].

Conclusions

The experimental paradigm of combining genetic and environmental differences was successful in identifying a limited number of pathways and networks that underlie muscle function and meat quality. Our findings provide new insights into the genetics and pathogenesis of muscle damage induced by heat stress through the identification of previously unknown pathways and networks. Importantly, our study also showed that the gene expression pattern for breast muscle of broiler chickens that were raised under a conventional (control) temperature was similar to that of heat-stressed layers and that the expression of these genes was further enhanced in heat-stressed broilers. These results provide a resource for the identification of candidate genes for muscle function and meat quality, which we will use in an accompanying paper to determine statistically significant associations of SNPs with muscle and meat quality traits in chicken.

Additional file

> **Additional file 1: Table S1.** Comparison of pathways analyses of microarray (gene expression) data on IPA, DAVID, and REACTOME. **Table S2.** Pathways selected from category I cluster analysis. **Table S3.** Pathways selected from category II cluster analysis. **Table S4.** Pathways selected from category III cluster analysis. **Table S5.** Pathways selected from category IV cluster analysis. **Table S6.** Pathways selected from category V cluster analysis. **Table S7.** Pathways selected from category VI cluster analysis. **Table S8.** Selected pathways from IPA analysis of Biolayout filtered genes.

Authors' contributions

IZ assisted with the experiment, collated the data, carried out the RNA extraction, analysed the microarray data, conducted the bioinformatics analyses, evaluated the pathways and networks and wrote the draft manuscript. DJK participated in data and bioinformatics analyses and interpretation and helped to draft the manuscript. PMH devised the experiment, assisted with data collection, supervised and coordinated the research and helped to draft the manuscript. All authors read and approved the final manuscript.

Author details
[1] Division of Genetics and Genomics, Roslin Institute and R(D)SVS, University of Edinburgh, Easter Bush, Midlothian EH25 9RG, UK. [2] Department of Animal Breeding and Genetics, University of Veterinary and Animal Sciences, Lahore 54000, Pakistan. [3] Department of Animal Breeding and Genetics, Swedish University of Agricultural Sciences, 750 07 Uppsala, Sweden.

Acknowledgements
The authors are grateful for assistance from the staff of the poultry unit at the Roslin Institute, to Caroline McCorquodale for assistance in the analysis of the microarray data and to Alison Downing of Edinburgh Genomics, The University of Edinburgh, for conducting the microarray hybridisations.

Competing interests
The authors declare that they have no competing interests.

Funding
Edinburgh Genomics is partly supported through core Grants from NERC (R8/H10/56), MRC (MR/K001744/1) and BBSRC (BB/J004243/1). The broiler chicks were kindly donated by Aviagen Ltd (Newbridge, UK). Funding for IZ was provided by the Higher Education Commission, Government of Pakistan through a PhD scholarship to IZ. The Roslin Institute is supported by a BBSRC Institute Strategic Programme Grant and this research was conducted under programme grant ISP1 (BB/J004316/1).

References

1. MacRae VE, Mahon M, Gilpin S, Sandercock DA, Mitchell MA. Skeletal muscle fibre growth and growth associated myopathy in the domestic chicken (Gallus domesticus). Br Poult Sci. 2006;47:264–72.
2. Hocking PM, Sandercock DA, Wilson S, Fleming RH. Quantifying genetic (co)variation and effects of genetic selection on tibial bone morphology and quality in 37 lines of broiler, layer and traditional chickens. Br Poult Sci. 2009;50:443–50.
3. Sheard PR, Hughes SI, Jaspal MH. Colour, pH and weight changes of PSE, normal and DFD breast fillets from British broilers treated with a phosphate-free, low salt marinade. Br Poult Sci. 2012;53:57–65.
4. Kuttappan VA, Shivaprasad HL, Shaw DP, Valentine BA, Hargis BM, Clark FD, McKee SR, Owens CM. Pathological changes associated with white striping in broiler breast muscles. Poult Sci. 2013;92:331–8.
5. Petracci M, Mudalal S, Bonfiglio A, Cavani C. Occurrence of white striping under commercial conditions and its impact on breast meat quality in broiler chickens. Poult Sci. 2013;92:1670–5.
6. Kuttappan VA, Huff GR, Huff WE, Hargis BM, Apple JK, Coon C, Owens CM. Comparison of hematologic and serologic profiles of broiler birds with normal and severe degrees of white striping in breast fillets. Poult Sci. 2013;92:339–45.
7. Petracci M, Bianchi M, Cavani C. The European perspective on pale, soft, exudative conditions in poultry. Poult Sci. 2009;88:1518–23.
8. Bailey RA, Watson KA, Bilgili SF, Avendano S. The genetic basis of pectoralis major myopathies in modern broiler chicken lines. Poult Sci. 2015;94:2870–9.
9. Trocino A, Piccirillo A, Birolo M, Radaelli G, Bertotto D, Filiou E, Petracci M, Xiccato G. Effect of genotype, gender and feed restriction on growth, meat quality and the occurrence of white striping and wooden breast in broiler chickens. Poult Sci. 2015;94:2996–3004.
10. Le Bihan-Duval E, Berri C, Baeza E, Millet N, Beaumont C. Estimation of the genetic parameters of meat characteristics and of their genetic correlations with growth and body composition in an experimental broiler line. Poult Sci. 2001;80:839–43.
11. Meuwissen THE, Hayes BJ, Goddard ME. Prediction of total genetic value using genome-wide dense marker maps. Genetics. 2001;157:1819–29.
12. Kranis A, Gheyas AA, Boschiero C, Turner F, Yu L, Smith S, Talbot R, Pirani A, Brew F, Kaiser P, Hocking PM, Fife M, Salmon N, Fulton J, Strom TM, Haberer G, Weigend S, Preisinger R, Gholami M, Qanbari S, Simianer H, Watson KA, Woolliams JA, Burt DW. Development of a high density 600K SNP genotyping array for chicken. BMC Genomics. 2013; 14.
13. Whitworth G. An introduction to microarray data analysis and visualization. Methods Mol Biol. 2010;470:19–50.
14. Sandercock DA, Hunter RR, Mitchell MA, Hocking PM. Thermoregulatory capacity and muscle membrane integrity are compromised in broilers compared with layers at the same age or body weight. Br Poult Sci. 2006;47:322–9.
15. Sandercock DA, Hunter RR, Nute GR, Mitchell MA, Hocking PM. Acute heat stress-induced alterations in blood acid-base status and skeletal muscle membrane integrity in broiler chickens at two ages: Implications for meat quality. Poult Sci. 2001;80:418–25.
16. Lim WK, Wang K, Lefebvre C, Califano A. Comparative analysis of microarray normalization procedures: effects on reverse engineering gene networks. Bioinformatics (Oxford, England). 2007;23:i282–8.
17. Freeman TC, Goldovsky L, Brosch M, van Dongen S, Mazière P, Grocock RJ, Freilich S, Thornton J, Enright AJ. Construction, visualisation, and clustering of transcription networks from microarray expression data. PLoS Comput Biol. 2007;3:2032–42.

18. Theocharidis A, van Dongen S, Enright AJ, Freeman TC. Network visualization and analysis of gene expression data using BioLayout Express(3D). Nat Protoc. 2009;4:1535–50.

19. Huang DW, Sherman BT, Lempicki RA. Systematic and integrative analysis of large gene lists using DAVID bioinformatics resources. Nat Protoc. 2009;4:44–57.

20. Huang DW, Sherman BT, Lempicki RA. Bioinformatics enrichment tools: paths toward the comprehensive functional analysis of large gene lists. Nucleic Acids Res. 2009;37:1–13.

21. Fabregat A, Sidiropoulos K, Garapati P, Gillespie M, Hausmann K, Haw R, Jassal B, Jupe S, Korninger F, McKay S, Matthews L, May B, Milacic M, Rothfels K, Shamovsky V, Webber M, Weiser J, Williams M, Wu GM, Stein L, Hermjakob H, D'Eustachio P. The reactome pathway knowledgebase. Nucleic Acids Res. 2016;44:D481–7.

22. Croft D, Mundo AF, Haw R, Milacic M, Weiser J, Wu GM, Caudy M, Garapati P, Gillespie M, Kamdar MR, Jassal B, Jupe S, Matthews L, May B, Palatnik S, Rothfels K, Shamovsky V, Song H, Williams M, Birney E, Hermjakob H, Stein L, D'Eustachio P. The reactome pathway knowledgebase. Nucleic Acids Res. 2014;42:D472–7.

23. Clinton M, Haines L, Belloir B, McBride D. Sexing chick embryos: a rapid and simple protocol. Br Poult Sci. 2001;42:134–8.

24. Irizarry RA, Hobbs B, Collin F, Beazer-Barclay YD, Antonellis KJ, Scherf U, Speed TP. Exploration, normalization, and summaries of high density oligonucleotide array probe level data. Biostatistics. 2003;4:249–64.

25. Theocharidis A, Dongen SV, Enright AJ, Freeman TC. Network visualization and analysis of gene expression data using BioLayout Express 3D. Nat Protoc. 2009;4:1535–50.

26. Mitchell MA, Sandercock DA. Increased hyperthermia induced skeletal muscle damage in fast growing broiler chickens? Poult Sci. 1995;74:74.

27. Debut M, Berri C, Arnould C, Guemene D, Sante-Lhoutellier V, Sellier N, Baeza E, Jehl N, Jego Y, Beaumont C, Le Bihan-Duval E. Behavioural and physiological responses of three chicken breeds to pre-slaughter shackling and acute heat stress. Br Poult Sci. 2005;46:527–35.

28. Petracci M, Bianchi M, Cavani C. Pre-slaughter handling and slaughtering factors influencing poultry product quality. Worlds Poult Sci J. 2010;66:17–26.

29. Sandercock DA, Mitchell MA. Myopathy in broiler chickens: a role for Ca^{2+}-activated phospholipase A(2)? Poult Sci. 2003;82:1307–12.

30. Allen SJ, Crown SE, Handel TM. Chemokine: receptor structure, interactions, and antagonism. Ann Rev Immunol. 2007; 787–820.

31. Amir R, Argoff CE, Bennett GJ, Cummins TR. The role of sodium channels in chronic inflammatory and neuropathic pain. J Pain. 2006;7:S1–29.

32. Ashton AW, Ware GM, Kaul DK, Ware JA. Inhibition of tumor necrosis factor alpha-mediated NF kappa B activation and leukocyte adhesion, with enhanced endothelial apoptosis, by G protein-linked receptor (TP) ligands. J Biol Chem. 2003;278:11858–66.

33. Baker OJ, Camden JM, Rome DE, Seye CI, Weisman GA. P2Y(2) nucleotide receptor activation up-regulates vascular cell adhesion molecular-1 expression and enhances lymphocyte adherence to a human submandibular gland cell line. Mol Immunol. 2008;45:65–75.

34. Lattin J, Zidar DA, Schroder K, Kellie S, Hume DA, Sweet MJ. G-protein-coupled receptor expression, function, and signaling in macrophages. J Leukoc Biol. 2007;82:16–32.

35. Buraei Z, Yang J. The beta subunit of voltage-gated Ca^{2+} channels. Physiol Rev. 2010;90:1461–506.

36. Allen DG. Skeletal muscle function: role of ionic changes in fatigue, damage and disease. Clin Exp Pharmacol Physiol. 2004;31:485–93.

37. Allen DG, Gervasio OL, Yeung EW, Whitehead NP. Calcium and the damage pathways in muscular dystrophy. Can J Physiol Pharmacol. 2010;88:83–91.

38. Shini S, Kaiser P. Effects of stress, mimicked by administration of corticosterone in drinking water, on the expression of chicken cytokine and chemokine genes in lymphocytes. Stress. 2009;12:388–99.

39. Shini S, Shini A, Huff GR. Effects of chronic and repeated corticosterone administration in rearing chickens on physiology, the onset of lay and egg production of hens. Physiol Behav. 2009;98:73–7.

40. Kaiser P, Zhiguang W, Lisa R, Mark F, Gibson M, Poh T-Y, Agim S, Wayne B, Shaniko S. Prospects for understanding immune-endocrine interactions in the chicken. Gen Comp Endocrinol. 2009;163:83–91.

41. Krupp M, Weinmann A, Galle PR, Teufel A. Actin binding LIM protein 3 (abLIM3). Int J Mol Med. 2006;17:129–33.

42. Maekawa M, Ishizaki T, Boku S, Watanabe N, Fujita A, Iwamatsu A, Obinata T, Ohashi K, Mizuno K, Narumiya S. Signaling from Rho to the actin cytoskeleton through protein kinases ROCK and LIM-kinase. Science. 1999;285:895–8.

43. Mutryn MF, Brannick EM, Fu W, Lee WR, Abasht B. Characterization of a novel chicken muscle disorder through differential gene expression and pathway analysis using RNA-sequencing. BMC Genomics. 2015;16.

44. Shini S, Huff GR, Shini A, Kaiser P. Understanding stress-induced immunosuppression: exploration of cytokine and chemokine gene profiles in chicken peripheral leukocytes. Poult Sci. 2010;89:841–51.

45. Shini S, Shini A, Kaiser P. Cytokine and chemokine gene expression profiles in heterophils from chickens treated with corticosterone. Stress. 2010;13:185–94.

46. Mujahid A, Yoshiki Y, Akiba Y, Toyomizu M. Superoxide radical production in chicken skeletal muscle induced by acute heat stress. Poult Sci. 2005;84:307–14.

47. Mujahid A, Sato K, Akiba Y, Toyomizu M. Acute heat stress stimulates mitochondrial superoxide production in broiler skeletal muscle, possibly via downregulation of uncoupling protein content. Poult Sci. 2006;85:1259–65.

48. Mujahid A, Pumford NR, Bottje W, Nakagawa K, Miyazawa T, Akiba Y, Toyomizu M. Mitochondrial oxidative damage in chicken skeletal muscle induced by acute heat stress. Journal of Poultry Science. 2007;44:439–45.

49. Alamdari N, Smith IJ, Aversa Z, Hasselgren PO. Sepsis and glucocorticoids upregulate p300 and downregulate HDAC6 expression and activity in skeletal muscle. Am J Physiol Regul Integr Comp Physiol. 2010;299:R509–20.

50. Frost RA, Lang CH. Skeletal muscle cytokines: regulation by pathogen-associated molecules and catabolic hormones. Curr Opin Clin Nutr Metab Care. 2005;8:255–63.

51. Vaarmann A, Fortin D, Veksler V, Momken I, Ventura-Clapier R, Garnier A. Mitochondrial biogenesis in fast skeletal muscle of CK deficient mice. Biochim Biophys Acta-Bioenerg. 2008;1777:39–47.

52. Sandercock DA, Nute GR, Hocking PM. Quantifying the effects of genetic selection and genetic variation for body size, carcass composition, and meat quality in the domestic fowl (Gallus domesticus). Poult Sci. 2009;88:923–31.

53. Griffin HD, Goddard C. Rapidly growing broiler (meat-type) chickens: their origin and use for comparative studies of the regulation of growth. Int J Biochem. 1994;26:19–26.

54. Khan RU, Naz S, Nikousefat Z, Tufarelli V, Javdani M, Rana N, Laudadio V. Effect of vitamin E in heat-stressed poultry. Worlds Poult Sci J. 2011;67:469–77.

55. Mujahid A, Akiba Y, Toyomizu M. Acute heat stress induces oxidative stress and decreases adaptation in young white leghorn cockerels by downregulation of avian uncoupling protein. Poult Sci. 2007;86:364–71.

56. Cerretelli P, Gelfi C. Energy metabolism in hypoxia: reinterpreting some features of muscle physiology on molecular grounds. Eur J Appl Physiol. 2011;111:421–32.

57. Toyomizu M, Kikusato M, Kawabata Y, Azad MAK, Inui E, Amo T. Meat-type chickens have a higher efficiency of mitochondrial oxidative phosphorylation than laying-type chickens. Comp Biochem Physiol A Mol Integr Physiol. 2011;159:75–81.

58. Flammer JR, Dobrovolna J, Kennedy MA, Chinenov Y, Glass CK, Ivashkiv LB, Rogatsky I. The type I interferon signaling pathway is a target for glucocorticoid inhibition. Mol Cell Biol. 2010;30:4564–74.

59. Dogra C, Changotra H, Mohan S, Kumar A. Tumor necrosis factor-like weak inducer of apoptosis inhibits skeletal myogenesis through sustained activation of nuclear factor-kappa B and degradation of MyoD protein. J Biol Chem. 2006;281:10327–36.

60. Caveggion E, Continolo S, Pixley FJ, Stanley ER, Bowtell DDL, Lowell CA, Berton G. Expression and tyrosine phosphorylation of Cbl regulates macrophage chemokinetic and chemotactic movement. J Cell Physiol. 2003;195:276–89.

61. Kedzierska K, Ellery P, Mak J, Lewin SR, Crowe SM, Jaworowski A. HIV-1 down-modulates gamma signaling chain of Fc gamma R in human macrophages: a possible mechanism for inhibition of phagocytosis. J Immunol. 2002;168:2895–903.

62. Lee WL, Cosio G, Ireton K, Grinstein S. Role of CrkII in Fcgamma receptor-mediated phagocytosis. J Biol Chem. 2007;282:11135–43.

63. Lehmann DM, Seneviratne AM, Smrcka AV. Small molecule disruption of G protein beta gamma subunit signaling inhibits neutrophil chemotaxis and inflammation. Mol Pharmacol. 2008;73:410–8.

64. Omori K, Ohira T, Uchida Y, Ayilavarapu S, Batista EL, Yagi M, Iwata T, Liu H, Hasturk H, Kantarci A, Van Dyke TE. Priming of neutrophil oxidative burst in diabetes requires preassembly of the NADPH oxidase. J Leukoc Biol. 2008;84:292–301.

65. Appleman LJ, van Puijenbroek AA, Shu KM, Nadler LM, Boussiotis VA. CD28 costimulation mediates down-regulation of p27kip1 and cell cycle progression by activation of the PI3K/PKB signaling pathway in primary human T cells. J Immunol. 2002;168:2729–36.

66. Borchers MT, Ansay T, DeSalle R, Daugherty BL, Shen H, Metzger M, Lee NA, Lee JJ. In vitro assessment of chemokine receptor-ligand interactions mediating mouse eosinophil migration. J Leukoc Biol. 2002;71:1033–41.

67. Pease JE. Asthma, allergy and chemokines. Curr Drug Targets. 2006;7:3–12.

68. Stojkov NJ, Janjic MM, Bjelic MM, Mihajlovic AI, Kostic TS, Andric SA. Repeated immobilization stress disturbed steroidogenic machinery and stimulated the expression of cAMP signaling elements and adrenergic receptors in Leydig cells. Am J Physiol Endocrinol Metab. 2012;302:E1239–51.

69. Dhakshinamoorthy S, Jain AK, Bloom DA, Jaiswal AK. Bach1 competes with Nrf2 leading to negative regulation of the antioxidant response element (ARE)-mediated NAD(P)H: quinone oxidoreductase 1 gene expression and induction in response to antioxidants. J Biol Chem. 2005;280:16891–900.

70. Karkkainen MJ, Petrova TV. Vascular endothelial growth factor receptors in the regulation of angiogenesis and lymphangiogenesis. Oncogene. 2000;19:5598–605.

71. Lu X, Masic A, Li Y, Shin Y, Liu Q, Zhou Y. The PI3K/Akt pathway inhibits influenza A virus-induced Bax-mediated apoptosis by negatively regulating the JNK pathway via ASK1. J Gen Virol. 2010.

72. Lee C, Liu QH, Tomkowicz B, Yi Y, Freedman BD, Collman RG. Macrophage activation through CCR5- and CXCR4-mediated gp120-elicited signaling pathways. J Leukoc Biol. 2003;74:676–82.

73. Oh JE, So KS, Lim SJ, Kim MY. Induction of apoptotic cell death by a ceramide analog in PC-3 prostate cancer cells. Arch Pharm Res. 2006;29:1140–6.

74. Woodcock J. Sphingosine and ceramide signalling in apoptosis. IUBMB Life. 2006;58:462–6.

75. Cao Q, Kim JH, Richter JD. CDK1 and calcineurin regulate Maskin association with eIF4E and translational control of cell cycle progression. Nat Struct Mol Biol. 2006;13:1128–34.

76. Karlsson HK, Nilsson PA, Nilsson J, Chibalin AV, Zierath JR, Blomstrand E. Branched-chain amino acids increase p70S6k phosphorylation in human skeletal muscle after resistance exercise. Am J Physiol Endocrinol Metab. 2004;287:E1–7.

77. Jivotovskaya AV, Valásek L, Hinnebusch AG, Nielsen KH. Eukaryotic translation initiation factor 3 (eIF3) and eIF2 can promote mRNA binding to 40S subunits independently of eIF4G in yeast. Mol Cell Biol. 2006;26:1355–72.

78. Park JS, Gamboni-Robertson F, He Q, Svetkauskaite D, Kim JY, Strassheim D, Sohn JW, Yamada S, Maruyama I, Banerjee A, Ishizaka A, Abraham E. High mobility group box 1 protein interacts with multiple Toll-like receptors. Am J Physiol Cell Physiol. 2006;290:C917–24.

79. Holmlund U, Wähämaa H, Bachmayer N, Bremme K, Sverremark-Ekström E, Palmblad K. The novel inflammatory cytokine high mobility group box protein 1 (HMGB1) is expressed by human term placenta. Immunology. 2007;122:430–7.

80. Yang H, Wang H, Czura CJ, Tracey KJ. The cytokine activity of HMGB1. J Leukoc Biol. 2005;78:1–8.

81. Chalfant CE, Spiegel S. Sphingosine 1-phosphate and ceramide 1-phosphate: expanding roles in cell signaling. J Cell Sci. 2005;118:4605–12.

82. Alemany R, van Koppen CJ, Danneberg K, Braak TM, Heringdorf MD. Regulation and functional roles of sphingosine kinases. Naunyn Schmiedebergs Arch Pharmacol. 2007;374:413–28.

83. Kono M, Mi Y, Liu Y, Sasaki T, Allende ML, Wu YP, Yamashita T, Proia RL. The sphingosine-1-phosphate receptors S1P1, S1P2, and S1P3 function coordinately during embryonic angiogenesis. J Biol Chem. 2004;279:29367–73.

84. Liu X, Zhang Q-H, Yi G-H. Regulation of metabolism and transport of sphingosine-1-phosphate in mammalian cells. Mol Cell Biochem. 2012;363:21–33.

85. Garcia-Echeverria C, Sellers WR. Drug discovery approaches targeting the PI3K/Akt pathway in cancer. Oncogene. 2008;27:5511–26.

86. Hossain MA, Rosengren KJ, Haugaard-Jonsson LM, Zhang S, Layfield S, Ferraro T, Daly NL, Tregear GW, Wade JD, Bathgate RA. The A-chain of human relaxin family peptides has distinct roles in the binding and activation of the different relaxin family peptide receptors. J Biol Chem. 2008;283:17287–97.

87. Frost RA, Lang CH. Regulation of muscle growth by pathogen-associated molecules. J Anim Sci. 2008;86:E84–93.

88. Grant S. Cotargeting survival signaling pathways in cancer. J Clin Invest. 2008;118:3003–6.

89. Yokogami K, Wakisaka S, Avruch J, Reeves SA. Serine phosphorylation and maximal activation of STAT3 during CNTF signaling is mediated by the rapamycin target mTOR. Curr Biol. 2000;10:47–50.

90. Hamal KR, Wideman RF, Anthony NB, Erf GF. Differential gene expression of proinflammatory chemokines and cytokines in lungs of ascites-resistant and -susceptible broiler chickens following intravenous cellulose microparticle injection. Vet Immunol Immunopathol. 2010;133:250–5.

91. Jaeschke H, Hasegawa T. Role of neutrophils in acute inflammatory liver injury. Liver Int. 2006;26:912–9.

92. Zhou HY, Shin EM, Guo LY, Youn UJ, Bae K, Kang SS, Zou LB, Kim YS. Anti-inflammatory activity of 4-methoxyhonokiol is a function of the inhibition of iNOS and COX-2 expression in RAW 264.7 macrophages via NF-kappaB, JNK and p38 MAPK inactivation. Eur J Pharmacol. 2008;586:340–9.

93. Vila-del Sol V, Díaz-Muñoz MD, Fresno M. Requirement of tumor necrosis factor alpha and nuclear factor-kappaB in the induction by IFN-gamma of inducible nitric oxide synthase in macrophages. J Leukoc Biol. 2007;81:272–83.

94. Willoughby DS, Wilborn CD. Estradiol in females may negate skeletal muscle myostatin mRNA expression and serum myostatin propeptide levels after eccentric muscle contractions. J Sports Sci Med. 2006;5:672–81.

95. Enns DL, Tiidus PM. Estrogen influences satellite cell activation and proliferation following downhill running in rats. J Appl Physiol. 2008;104:347–53.

96. Chen X-N, Zhu H, Meng Q-Y, Zhou J-N. Estrogen receptor-alpha and -beta regulate the human corticotropin-releasing hormone gene through similar pathways. Brain Res. 2008;1223:1–10.

97. Szendroedi J, Phielix E, Roden M. The role of mitochondria in insulin resistance and type 2 diabetes mellitus. Nat Rev Endocrinol. 2012;8:92–103.

98. Masana MI, Dubocovich ML. Melatonin receptor signaling: finding the path through the dark. Sci STKE. 2001;2001: PE39.

99. Ivanina T, Blumenstein Y, Shistik E, Barzilai R, Dascal N. Modulation of L-type Ca²⁺ channels by gbeta gamma and calmodulin via interactions with N and C termini of alpha 1C. J Biol Chem. 2000;275:39846–54.

100. Qin S, Stadtman ER, Chock PB. Regulation of oxidative stress-induced calcium release by phosphatidylinositol 3-kinase and Bruton's tyrosine kinase in B cells. Proc Natl Acad Sci USA. 2000;97:7118–23.

101. Mujahid A, Furuse M. Central administration of corticotropin-releasing factor induces tissue specific oxidative damage in chicks. Comp Biochem Physiol A Mol Integr Physiol. 2008;151:664–9.

102. Thaxton JP, Stayer P, Ewing M, Rice J. Corticosterone in commercial broilers. J Appl Poul Res. 2005;14:745–9.

103. Post J, Rebel JMJ, ter Huurne A. Physiological effects of elevated plasma corticosterone concentrations in broiler chickens. An alternative means by which to assess the physiological effects of stress. Poul Sci. 2003;82:1313–8.

104. Sak K, Boeynaems JM, Everaus H. Involvement of P2Y receptors in the differentiation of haematopoietic cells. J Leukoc Biol. 2003;73:442–7.

105. Burnstock G. Purinergic signaling and vascular cell proliferation and death. Arterioscler Thromb Vasc Biol. 2002;22:364–73.

106. Datta SR, Brunet A, Greenberg ME. Cellular survival: a play in three Akts. Genes Dev. 1999;13:2905–27.

107. Seino S, Shibasaki T. PKA-dependent and PKA-independent pathways for cAMP-regulated exocytosis. Physiol Rev. 2005;85:1303–42.

108. Wehrens XH, Lehnart SE, Reiken S, Vest JA, Wronska A, Marks AR. Inaugural article: ryanodine receptor/calcium release channel PKA phosphorylation: a critical mediator of heart failure progression. Proc Natl Acad Sci USA. 2006;103:511–8.

109. Gardai S, Whitlock BB, Helgason C, Ambruso D, Fadok V, Bratton D, Henson PM. Activation of SHIP by NADPH oxidase-stimulated Lyn leads to enhanced apoptosis in neutrophils. J Biol Chem. 2002;277:5236–46.

110. Yamasaki S, Saito T. Progress in allergy signal research on mast cells: signal regulation of multiple mast cell responses through FcepsilonRI. J Pharmacol Sci. 2008;106:336–40.

111. Balistreri CR, Caruso C, Grimaldi MP, Listi F, Vasto S, Orlando V, Campagna AM, Lio D, Candore G. CCR5 receptor—biologic and genetic implications in age-related diseases. In: Rattan SISAS (ed) Biogerontology: Mechanisms and Interventions, 2007; 162–72.

112. Viejo-Borbolla A, Martinez-Martin N, Nel HJ, Rueda P, Martin R, Blanco S, Arenzana-Seisdedos F, Thelen M, Fallon PG, Alcami A. Enhancement of chemokine function as an immunomodulatory strategy employed by human herpesviruses. Plos Pathogens. 2012;8.

113. Sun L, Ye RD. Role of G protein-coupled receptors in inflammation. Acta Pharmacol Sin. 2012;33:342–50.

114. Schwartz EA, Reaven PD. Lipolysis of triglyceride-rich lipoproteins, vascular inflammation, and atherosclerosis. Biochim Biophys Acta Mol Cell Biol Lipids. 1821;2012:858–66.

115. Ren GH, Takano T, Papillon J, Cybulsky AV. Cytosolic phospholipase A(2)-alpha enhances induction of endoplasmic reticulum stress. Biochim Biophys Acta Mol Cell Res. 1803;2010:468–81.

116. Calandra T, Roger T. Macrophage migration inhibitory factor: a regulator of innate immunity. Nat Rev Immunol. 2003;3:791–800.

117. Roger T, David J, Glauser MP, Calandra T. MIF regulates innate immune responses through modulation of Toll-like receptor 4. Nature. 2001;414:920–4.

118. Daun JM, Cannon JG. Macrophage migration inhibitory factor antagonizes hydrocortisone-induced increases in cytosolic IkappaBalpha. Am J Physiol Regul Integr Comp Physiol. 2000;279:R1043–9.

119. Santos LL, Dacumos A, Yamana J, Sharma L, Morand EF. Reduced arthritis in MIF deficient mice is associated with reduced T cell activation: down-regulation of ERK MAP kinase phosphorylation. Clin Exp Immunol. 2008;152:372–80.

120. Ohkawara T, Takeda H, Miyashita K, Nishiwaki M, Nakayama T, Taniguchi M, Yoshiki T, Tanaka J, Takana J, Imamura M, Sugiyama T, Asaka M, Nishihira J. Regulation of Toll-like receptor 4 expression in mouse colon by macrophage migration inhibitory factor. Histochem Cell Biol. 2006;125:575–82.

121. Spik I, Brénuchon C, Angéli V, Staumont D, Fleury S, Capron M, Trottein F, Dombrowicz D. Activation of the prostaglandin D2 receptor DP2/CRTH2 increases allergic inflammation in mouse. J Immunol. 2005;174:3703–8.

122. Nagata M, Saito K, Tsuchiya K, Sakamoto Y. Leukotriene D4 upregulates eosinophil adhesion via the cysteinyl leukotriene 1 receptor. J Allergy Clin Immunol. 2002;109:676–80.

123. Yang CH, Murti A, Valentine WJ, Du Z, Pfeffer LM. Interferon{alpha} Activates NF-{kappa}B in JAK1-deficient Cells through a TYK2-dependent Pathway. J Biol Chem. 2005;280:25849–53.

124. Simoncic PD, Lee-Loy A, Barber DL, Tremblay ML, McGlade CJ. The T cell protein tyrosine phosphatase is a negative regulator of janus family kinases 1 and 3. Curr Biol. 2002;12:446–53.

125. Ning Y, Riggins RB, Mulla JE, Chung H, Zwart A, Clarke R. IFNgamma restores breast cancer sensitivity to fulvestrant by regulating STAT1, IFN regulatory factor 1, NF-kappaB, BCL2 family members, and signaling to caspase-dependent apoptosis. Mol Cancer Ther. 2010;9:1274–85.

126. Varfolomeev EE, Ashkenazi A. Tumor necrosis factor: an apoptosis JuNKie? Cell. 2004;116:491–7.

127. Jiang S, Zu Y, Fu Y, Zhang Y, Efferth T. Activation of the mitochondria-driven pathway of apoptosis in human PC-3 prostate cancer cells by a novel hydrophilic paclitaxel derivative, 7-xylosyl-10-deacetylpaclitaxel. Int J Oncol. 2008;33:103–11.

128. Bots M, Medema JP. Granzymes at a glance. J Cell Sci. 2006;119:5011–4.

129. Ding S-Q, Li Y, Zhou Z-G, Wang C, Zhan L, Zhou B. Toll-like receptor 4-mediated apoptosis of pancreatic cells in cerulein-induced acute pancreatitis in mice. Hepatobiliary Pancreat Dis Int. 2010;9:645–50.

130. Barton GM, Medzhitov R. Toll-like receptor signaling pathways. Science. 2003;300:1524–5.

131. Rolo AP, Palmeira CM. Diabetes and mitochondrial function: Role of hyperglycemia and oxidative stress. Toxicol Appl Pharmacol. 2006;212:167–78.

132. Pinkoski MJ, Waterhouse NJ, Heibein JA, Wolf BB, Kuwana T, Goldstein JC, Newmeyer DD, Bleackley RC, Green DR. Granzyme B-mediated apoptosis proceeds predominantly through a Bcl-2-inhibitable mitochondrial pathway. J Biol Chem. 2001;276:12060–7.

133. Kim R, Emi M, Tanabe K. Caspase-dependent and -independent cell death pathways after DNA damage (Review). Oncol Rep. 2005;14:595–9.

134. Kuwano K, Yoshimi M, Maeyama T, Hamada N, Yamada M, Nakanishi Y. Apoptosis signaling pathways in lung diseases. Med Chem. 2005;1:49–56.

135. Kuai J, Nickbarg E, Wooters J, Qiu Y, Wang J, Lin LL. Endogenous association of TRAF2, TRAF3, cIAP1, and Smac with lymphotoxin beta receptor reveals a novel mechanism of apoptosis. J Biol Chem. 2003;278:14363–9.

136. You RI, Chen MC, Wang HW, Chou YC, Lin CH, Hsieh SL. Inhibition of lymphotoxin-beta receptor-mediated cell death by survivin-DeltaEx3. Cancer Res. 2006;66:3051–61.

137. Shi C-S, Shenderov K, Huang N-N, Kabat J, Abu-Asab M, Fitzgerald KA, Sher A, Kehrl JH. Activation of autophagy by inflammatory signals limits IL-1 beta production by targeting ubiquitinated inflammasomes for destruction. Nat Immunol. 2012;13:U255–74.

138. Sigala I, Zacharatos P, Toumpanakis D, Michailidou T, Noussia O, Theocharis S, Roussos C, Papapetropoulos A, Vassilakopoulos T. MAPKs and NF-kappa B differentially regulate cytokine expression in the diaphragm in response to resistive breathing: the role of oxidative stress. Am J Physiol Regul Integr Comp Physiol. 2011;300:R1152–62.

139. Yang CM, Luo SF, Hsieh HL, Chi PL, Lin CC, Wu CC, Hsiao LD. Interleukin-1 beta induces ICAM-1 expression enhancing leukocyte adhesion in human rheumatoid arthritis synovial fibroblasts: involvement of ERK, JNK, AP-1, and NF-kappa B. J Cell Physiol. 2010;224:516–26.

Permissions

All chapters in this book were first published in GSE, by BioMed Central; hereby published with permission under the Creative Commons Attribution License or equivalent. Every chapter published in this book has been scrutinized by our experts. Their significance has been extensively debated. The topics covered herein carry significant findings which will fuel the growth of the discipline. They may even be implemented as practical applications or may be referred to as a beginning point for another development.

The contributors of this book come from diverse backgrounds, making this book a truly international effort. This book will bring forth new frontiers with its revolutionizing research information and detailed analysis of the nascent developments around the world.

We would like to thank all the contributing authors for lending their expertise to make the book truly unique. They have played a crucial role in the development of this book. Without their invaluable contributions this book wouldn't have been possible. They have made vital efforts to compile up to date information on the varied aspects of this subject to make this book a valuable addition to the collection of many professionals and students.

This book was conceptualized with the vision of imparting up-to-date information and advanced data in this field. To ensure the same, a matchless editorial board was set up. Every individual on the board went through rigorous rounds of assessment to prove their worth. After which they invested a large part of their time researching and compiling the most relevant data for our readers.

The editorial board has been involved in producing this book since its inception. They have spent rigorous hours researching and exploring the diverse topics which have resulted in the successful publishing of this book. They have passed on their knowledge of decades through this book. To expedite this challenging task, the publisher supported the team at every step. A small team of assistant editors was also appointed to further simplify the editing procedure and attain best results for the readers.

Apart from the editorial board, the designing team has also invested a significant amount of their time in understanding the subject and creating the most relevant covers. They scrutinized every image to scout for the most suitable representation of the subject and create an appropriate cover for the book.

The publishing team has been an ardent support to the editorial, designing and production team. Their endless efforts to recruit the best for this project, has resulted in the accomplishment of this book. They are a veteran in the field of academics and their pool of knowledge is as vast as their experience in printing. Their expertise and guidance has proved useful at every step. Their uncompromising quality standards have made this book an exceptional effort. Their encouragement from time to time has been an inspiration for everyone.

The publisher and the editorial board hope that this book will prove to be a valuable piece of knowledge for researchers, students, practitioners and scholars across the globe.

List of Contributors

Laurence Puillet and Nicolas C. Friggens
UMR Modélisation Systémique Appliquée aux Ruminants, INRA, AgroParis- Tech, Université Paris-Saclay, 75005 Paris, France

Denis Réale
Département des Sciences Biologiques, Université du Québec à Montréal, Montréal, QC H3C 3P8, Canada

Goutam Sahana, Peipei Ma, Guosheng Su, Mogens Sandø Lund and Peter Sørensen
Department of Molecular Biology and Genetics, Center for Quantitative Genetics and Genomics, Aarhus University, 8830 Tjele, Denmark

Lingzhao Fang
Department of Molecular Biology and Genetics, Center for Quantitative Genetics and Genomics, Aarhus University, 8830 Tjele, Denmark
Key Laboratory of Animal Genetics, Breeding and Reproduction, Ministry of Agriculture and National Engineering Laboratory for Animal Breeding, College of Animal Science and Technology, China Agricultural University, Beijing 100193, China

Ying Yu and Shengli Zhang
Key Laboratory of Animal Genetics, Breeding and Reproduction, Ministry of Agriculture and National Engineering Laboratory for Animal Breeding, College of Animal Science and Technology, China Agricultural University, Beijing 100193, China

Dalinne C. C. Santos, Mariana M. Moraes, Andresa E. M. Araújo, José A. G. Bergmann, Eduardo M. Turra and Fabio L. B. Toral
Departamento de Zootecnia, Escola de Veterinária, Universidade Federal de Minas Gerais, Belo Horizonte, MG 31270-901, Brazil

Fernanda S. S. Raidan
Departamento de Zootecnia, Escola de Veterinária, Universidade Federal de Minas Gerais, Belo Horizonte, MG 31270-901, Brazil
School of Chemistry and Molecular Biosciences, The University of Queensland, 4072 Brisbane, QLD, Australia

Henrique T. Ventura
Associação Brasileira dos Criadores de Zebu, Uberaba, MG 38022-330, Brazil

Ole Fredslund Christensen
Department of Molecular Biology and Genetics, Center for Quantitative Genetics and Genomics, Aarhus University, 8830 Tjele, Denmark

Tao Xiang
Department of Molecular Biology and Genetics, Center for Quantitative Genetics and Genomics, Aarhus University, 8830 Tjele, Denmark
UR1388 GenPhySE, INRA, CS-52627, 31326 Castanet-Tolosan, France

Andres Legarra
UR1388 GenPhySE, INRA, CS-52627, 31326 Castanet-Tolosan, France

Zulma Gladis Vitezica
INP, ENSAT, GenPhySE, Université de Toulouse, 31326 Castanet-Tolosan, France

Jiangli Ren, Zhuolin Huang, Ran Zhang, Ning Li and Xiaoxiang Hu
State Key Laboratory for Agrobiotechnology, China Agricultural University, Beijing 100193, China

Cheng Tan
State Key Laboratory for Agrobiotechnology, China Agricultural University, Beijing 100193, China
Department of Animal Science, University of Minnesota, Saint Paul, MN 55108, USA

Dzianis Prakapenka and Yang Da
Department of Animal Science, University of Minnesota, Saint Paul, MN 55108, USA

Zhenfang Wu, Dewu Liu and Xiaoyan He
National Engineering Research Center for Breeding Swine Industry, South China Agricultural University, Guangdong 510642, China

Aldemar González-Rodríguez and Elena F. Mouresan
Departamento de Anatomía, Embriología y Genética, Universidad de Zaragoza, 50013 Saragossa, Spain

Sebastián Munilla
Departamento de Anatomía, Embriología y Genética, Universidad de Zaragoza, 50013 Saragossa, Spain Departamento de Producción Animal, Facultad de Agronomía, Universidad de Buenos Aires, 1417 Buenos Aires, Argentina

Juan Altarriba and Luis Varona
Departamento de Anatomía, Embriología y Genética, Universidad de Zaragoza, 50013 Saragossa, Spain Instituto Agroalimentario de Aragón (IA2), 50013 Saragossa, Spain

Jhon J. Cañas-Álvarez and Jesús Piedrafita
Grup de Recerca en Remugants, Departament de Ciència Animal I dels Aliments, Universitat Autònoma de Barcelona, 08193 Bellaterra, Barcelona, Spain

Clara Díaz
Departamento de Mejora Genética Animal, INIA, 28040 Madrid, Spain

Jesús Á. Baro
Departamento de Ciencias Agroforestales, Universidad de Valladolid, 34004 Palencia, Spain

Antonio Molina
MERAGEM, Universidad de Córdoba, 14071 Córdoba, Spain

Yu Wang, Jörn Bennewitz and Robin Wellmann
Institute of Animal Science, University of Hohenheim, 70593 Stuttgart, Germany

Panya Sae-Lim and Marie Lillehammer
Nofima Ås, Osloveien 1, P.O. Box 210, 1431 Ås, Norway

Antti Kause
Biometrical Genetics, Natural Resources Institute Finland, 31600 Jokioinen, Finland

Han A. Mulder
Animal Breeding and Genomics Centre, Wageningen University and Research, P.O. Box 338, 6700 AH Wageningen, The Netherlands

Agustín Blasco and Marina Martínez-Álvaro
Institute for Animal Science and Technology, Universitat Politècnica de València, Valencia, Spain

Maria-Luz García and María-José Argente
Departamento de Tecnología Agroalimentaria, Universidad Miguel Hernández de Elche, Orihuela, Spain

Noelia Ibáñez-Escriche
Genètica i Millora Animal, Institut de Recerca i Tecnologia Agroalimentàries, Caldes de Montbui, Spain

Sithembile O. Makina and Mahlako L. Makgahlela
Agricultural Research Council-Animal Production Institute, Private Bag X 2, Irene 0062, South Africa

Michiel M. Scholtz and Azwihangwisi Maiwashe
Agricultural Research Council-Animal Production Institute, Private Bag X 2, Irene 0062, South Africa Department of Animal, Wildlife and Grassland Sciences, University of Free State, Bloemfontein 9300, South Africa

Michael D. MacNeil
Agricultural Research Council-Animal Production Institute, Private Bag X 2, Irene 0062, South Africa Department of Animal, Wildlife and Grassland Sciences, University of Free State, Bloemfontein 9300, South Africa Delta G, Miles City, MT 59301, USA

Lindsey K. Whitacre, Jared E. Decker and Jeremy F. Taylor
Division of Animal Sciences, University of Missouri, Columbia, MO 65211, USA

Este van Marle-Köster
Department of Animal and Wildlife Sciences, University of Pretoria, Private Bag X 20, Hatfield 0028, South Africa

Farai C. Muchadeyi
Agricultural Research Council-Biotechnology Platform, Private Bag X 5, Onderstepoort 0110, South Africa

Gesine Lühken and Julia Küpper
Department of Animal Breeding and Genetics, Justus Liebig University of Gießen, Ludwigstrasse 21a, 35390 Giessen, Germany

Stefan Krebs
Laboratory for Functional Genome Analysis (LAFUGA), Gene Center Munich, LMU Munich, Feodor-Lynen-Strasse 25, 81377 Munich, Germany

Sophie Rothammer and Ivica Medugorac
Chair of Animal Genetics and Husbandry, LMU Munich, Veterinaerstrasse 13, 80539 Munich, Germany

Boro Mioč
Department of Animal Science and Technology, Faculty of Agriculture, University of Zagreb, Svetošimunska cesta 25, 10000 Zagreb, Croatia

Ingolf Russ
Tierzuchtforschung e.V. München, Senator-Gerauer-Strasse 23, 85586 Poing, Germany

Serap Gonen, Mara Battagin, Gregor Gorjanc and John M. Hickey
The Roslin Institute and Royal (Dick) School of Veterinary Studies, The University of Edinburgh, Easter Bush, Midlothian, Scotland, UK

Susan E. Johnston
Institute of Evolutionary Biology, The University of Edinburgh, Charlotte Auerbach Road, Edinburgh EH9 3FL, UK

Mekki Boussaha, Rabia Letaief, Cécile Grohs, Amandine Duchesne, Florence Phocas, Sandrine Floriot, Dominique Rocha and Didier Boichard
GABI, INRA, AgroParisTech, Université Paris-Saclay, 78350 Jouy-en-Josas, France

Pauline Michot, Chris Hozé, Sébastien Fritz and Aurélien Capitan
GABI, INRA, AgroParisTech, Université Paris-Saclay, 78350 Jouy-en-Josas, France
Allice, Maison Nationale des Eleveurs, 75012 Paris, France.

Romain Philippe and Véronique Blanquet
GMA, INRA, Université de Limoges, 87060 Limoges Cedex, France

Diane Esquerré
GenPhySE, INRA, INPT, ENVT, Université de Toulouse, Castanet Tolosan, France

Christophe Klopp
SIGENAE, UR 875, INRA, 31362 Castanet-Tolosan, France

Mogens S. Lund
Department of Molecular Biology and Genetics, Faculty of Science and Technology, Center for Quantitative Genetics and Genomics, Aarhus University, 8830 Tjele, Denmark

Irene van den Berg
Department of Molecular Biology and Genetics, Faculty of Science and Technology, Center for Quantitative Genetics and Genomics, Aarhus University, 8830 Tjele, Denmark
GABI, INRA, AgroParisTech, Université Paris Saclay, 78350 Jouy-en-Josas, France

Didier Boichard
GABI, INRA, AgroParisTech, Université Paris Saclay, 78350 Jouy-en-Josas, France

Jérémie Vandenplas, Jack J. Windig and Mario P. L. Calus
Animal Breeding and Genomics Centre, Wageningen UR Livestock Research, P.O. Box 338, 6700 AH Wageningen, The Netherlands

Ivan Pocrnic, Daniela A. L. Lourenco, Yutaka Masuda and Ignacy Misztal
Department of Animal and Dairy Science, University of Georgia, Athens, GA 30602, USA

Paul M. Hocking
Division of Genetics and Genomics, Roslin Institute and R(D)SVS, University of Edinburgh, Easter Bush, Midlothian EH25 9RG, UK

Imran Zahoor
Division of Genetics and Genomics, Roslin Institute and R(D)SVS, University of Edinburgh, Easter Bush, Midlothian EH25 9RG, UK
Department of Animal Breeding and Genetics, University of Veterinary and Animal Sciences, Lahore 54000, Pakistan

Dirk-Jan de Koning
Division of Genetics and Genomics, Roslin Institute and R(D)SVS, University of Edinburgh, Easter Bush, Midlothian EH25 9RG, UK
Department of Animal Breeding and Genetics, Swedish University of Agricultural Sciences, 750 07 Uppsala, Sweden

Index